Smart Innovation, Systems and Technologies

Volume 28

Series editors

Robert J. Howlett, KES International, Shoreham-by-Sea, UK
e-mail: rjhowlett@kesinternational.org

Lakhmi C. Jain, University of Canberra, Canberra, Australia
e-mail: Lakhmi.jain@unisa.edu.au

For further volumes:
http://www.springer.com/series/8767

About this Series

The Smart Innovation, Systems and Technologies book series encompasses the topics of knowledge, intelligence, innovation and sustainability. The aim of the series is to make available a platform for the publication of books on all aspects of single and multi-disciplinary research on these themes in order to make the latest results available in a readily-accessible form. Volumes on interdisciplinary research combining two or more of these areas is particularly sought.

The series covers systems and paradigms that employ knowledge and intelligence in a broad sense. Its scope is systems having embedded knowledge and intelligence, which may be applied to the solution of world problems in industry, the environment and the community. It also focusses on the knowledge-transfer methodologies and innovation strategies employed to make this happen effectively. The combination of intelligent systems tools and a broad range of applications introduces a need for a synergy of disciplines from science, technology, business and the humanities. The series will include conference proceedings, edited collections, monographs, handbooks, reference books, and other relevant types of book in areas of science and technology where smart systems and technologies can offer innovative solutions.

High quality content is an essential feature for all book proposals accepted for the series. It is expected that editors of all accepted volumes will ensure that contributions are subjected to an appropriate level of reviewing process and adhere to KES quality principles.

Malay Kumar Kundu · Durga Prasad Mohapatra
Amit Konar · Aruna Chakraborty
Editors

Advanced Computing, Networking and Informatics - Volume 2

Wireless Networks and Security Proceedings of the Second International Conference on Advanced Computing, Networking and Informatics (ICACNI-2014)

Springer

Editors
Malay Kumar Kundu
Indian Statistical Institute
Machine Intelligence Unit
Kolkata
India

Durga Prasad Mohapatra
Department of Computer Science and Engineering
National Institute of Technology Rourkela
Rourkela
India

Amit Konar
Department of Electronics and Tele-Communication Engineering
Artificial Intelligence Laboratory
Jadavpur University
Kolkata
India

Aruna Chakraborty
Department of Computer Science and Engineering
St. Thomas' College of Engineering and Technology
West Bengal
India

ISSN 2190-3018 ISSN 2190-3026 (electronic)
ISBN 978-3-319-07349-1 ISBN 978-3-319-07350-7 (eBook)
DOI 10.1007/978-3-319-07350-7
Springer Cham Heidelberg New York Dordrecht London

Library of Congress Control Number: 2014940383

Printed on acid-free paper

Springer is part of Springer Science+Business Media (www.springer.com)

Foreword

The present volume is an outcome, in the form of proceedings, of the 2nd International Conference on Advanced Computing, Networking and Informatics, St. Thomas' College of Engineering and Technology, Kolkata, India, June 24–26, 2014. As the name of the conference implies, the articles included herein cover a wide span of disciplines ranging, say, from pattern recognition, machine learning, image processing, data mining and knowledge discovery, soft computing, distributed computing, cloud computing, parallel and distributed networks, optical communication, wireless sensor networks, routing protocol and architecture to data privacy preserving, cryptology and data security, and internet computing. Each discipline, itself, has its own challenging problems and issues. Some of them are relatively more matured and advanced in theories with several proven application domains, while others fall in recent thrust areas. Interestingly, there are several articles, as expected, on symbiotic integration of more than one discipline, e.g., in designing intelligent networking and computing systems such as forest fire detection using wireless sensor network, minimizing call routing cost with assigned cell in wireless network, network intrusion detection system, determining load balancing strategy in cloud computing, and side lobe reduction and beam-width control, where the significance of pattern recognition, evolutionary strategy and soft computing has been demonstrated. This kind of interdisciplinary research is likely to grow significantly, and has strong promise in solving real life challenging problems.

The proceedings are logically split in two homogeneous volumes, namely, Advanced Computing and Informatics (vol. 1) and Wireless Networks and Security (vol. 2) with 81 and 67 articles respectively. The volumes fairly represent a state-of-the art of the research mostly being carried out in India in these domains, and are valued-additions to the current era of computing and knowledge mining.

The conference committee, editors, and the publisher deserve congratulations for organizing the event (ICACNI-2014) which is very timely, and bringing out the archival volumes nicely as its output.

Kolkata, April 2014

Sankar K. Pal
Distinguished Scientist and former Director
Indian Statistical Institute

Message from the Honorary General Chair

It gives me great pleasure to introduce the *International Conference on Advanced Computing, Networking and Informatics (ICACNI 2014)* which will be held at St. Thomas' College of Engineering and Technology, Kolkata during June 24–26, 2014. ICACNI is just going to cross its second year, and during this small interval of time it has attracted a large audience. The conference received over 650 submissions of which only 148 papers have been accepted for presentation. I am glad to note that ICACNI involved top researchers from 26 different countries as advisory board members, program committee members and reviewers. It also received papers from 10 different countries.

ICACNI offers an interesting forum for researchers of three apparently diverse disciplines: Advanced Computing, Networking and Informatics, and attempts to focus on engineering applications, covering security, cognitive radio, human-computer interfacing among many others that greatly rely on these cross-disciplinary research outcomes. The accepted papers are categorized into two volumes, of which volume 1 includes all papers on advanced computing and informatics, while volume 2 includes accepted papers on wireless network and security. The volumes will be published by Springer-Verlag.

The conference includes plenary lecture, key-note address and four invited sessions by eminent scientists from top Indian and foreign research/academic institutes. The lectures by these eminent scientists will provide an ideal platform for dissemination of knowledge among researchers, students and practitioners. I take this opportunity to thank all the participants, including the keynote, plenary and invited speakers, reviewers, and the members of different committees in making the event a grand success.

Thanks are also due to the various Universities/Institutes for their active support towards this endeavor, and lastly Springer-Verlag for publishing the proceedings under their prestigious *Smart Innovation, Systems and Technologies (SIST) series*.

Wish the participants an enjoyable and productive stay in Kolkata.

Kolkata, April 2014

Dwijesh Dutta Majumder
Honorary General Chair
ICACNI -2014

Preface

The twenty first century has witnessed a paradigm shift in three major disciplines of knowledge: 1) Advanced/Innovative computing ii) Networking and wireless Communications and iii) informatics. While the first two are complete in themselves by their titles, the last one covers several sub-disciplines involving geo-, bio-, medical and cognitive informatics among many others. Apparently, the above three disciplines of knowledge are complementary and mutually exclusive but their convergence is observed in many real world applications, encompassing cyber-security, internet banking, healthcare, sensor networks, cognitive radio, pervasive computing and many others.

The International Conference on *Advanced Computing, Networking and Informatics* (ICACNI) is aimed at examining the convergence of the above three modern disciplines through interactions among three groups of people. The first group comprises leading international researchers, who have established themselves in one of the above three thrust areas. The plenary, the keynote lecture and the invited talks are organized to disseminate the knowledge of these academic experts among young researchers/practitioners of the respective domain. The invited talks are also expected to inspire young researchers to initiate/orient their research in respective fields. The second group of people comprises Ph.D./research students, working in the cross-disciplinary areas, who might be benefited from the first group and at the same time may help creating interest in the cross-disciplinary research areas among the academic community, including young teachers and practitioners. Lastly, the group comprising undergraduate and master students would be able to test the feasibility of their research through feedback of their oral presentations.

ICACNI is just passing its second birthday. Since its inception, it has attracted a wide audience. This year, for example, the program committee of ICACNI received as many as 646 papers. The acceptance rate is intentionally kept very low to ensure a quality publication by Springer. This year, the program committee accepted only 148 papers from these 646 submitted papers. An accepted paper has essentially received very good recommendation by at least two experts in the respective field.

To maintain a high standard of ICACNI, researchers from top international research laboratories/universities have been included in both the advisory committee and the program committee. The presence of these great personalities has helped the conference to develop its publicity during its infancy and promote it quality through an academic exchange among top researchers and scientific communities.

The conference includes one plenary session, one keynote address and four invited speech sessions. It also includes 3 special sessions and 21 general sessions (altogether 24 sessions) with a structure of 4 parallel sessions over 3 days. To maintain good question-answering sessions and highlight new research results arriving from the sessions, we selected subject experts from specialized domains as session chairs for the conference. ICACNI also involved several persons to nicely organize registration, take care of finance, hospitality of the authors/audience and other supports. To have a closer interaction among the people of the organizing committee, all members of the organizing committee have been selected from St. Thomas' College of Engineering and Technology.

The papers that passed the screening process by at least two reviewers, well-formatted and nicely organized have been considered for publication in the Smart Innovations Systems Technology (SIST) series of Springer. The hard copy proceedings include two volumes, where the first volume is named as *Advanced Computing and Informatics* and the second volume is named as *Wireless Networks and Security*. The two volumes together contain 148 papers of around eight pages each (in Springer LNCS format) and thus the proceedings is expected to have an approximate length of 1184 pages.

The editors gratefully acknowledge the contribution of the authors and the entire program committee without whose active support the proceedings could hardly attain the present standards. They would like to thank the keynote speaker, the plenary speaker, the invited speakers and also the invited session chairs, the organizing chair along with the organizing committee and other delegates for extending their support in various forms to ICACNI-2014. The editors express their deep gratitude to the Honorary General Chair, the General Chair, the Advisory Chair and the Advisory board members for their help and support to ICACNI-2014. The editors are obliged to Prof. Lakhmi C. Jain, the academic series editor of the SIST series, Springer and Dr. Thomas Ditzinger, Senior Editor, Springer, Heidelberg for extending their co-operation in publishing the proceeding in the prestigious SIST series of Springer. They also like to mention the hard efforts of Mr. Indranil Dutta of the Machine Intelligence Unit of ISI Kolkata for the editorial support. The editors also acknowledge the technical support they received from the students of ISI, Kolkata Jadavpur University and also the faculty of NIT Rourkela and St. Thomas' College of Engineering and Technology without which the work could not be completed in right time. Lastly, the editors thank Dr. Sailesh Mukhopadhyay, Prof. Gautam Banerjee and Dr. Subir Chowdhury of St. Thomas' College of Engineering and Technology for their support all the way long to make this conference a success.

Kolkata
April 14, 2014

Malay Kumar Kundu
Durga Prasad Mohapatra
Amit Konar
Aruna Chakraborty

Organization

Advisory Chair

Sankar K. Pal	Distinguished Scientist, Indian Statistical Institute, India

Advisory Board Members

Alex P. James	Nazarbayev University, Republic of Kazakhstan
Anil K. Kaushik	Additional Director, DeitY, Govt. of India, India
Atilla Elci	Aksaray University, Turkey
Atulya K. Nagar	Liverpool Hope University, United Kingdom
Brijesh Verma	Central Queensland University, Australia
Debajyoti Mukhopadhyay	Maharashtra Institute of Technology, India
Debatosh Guha	University of Calcutta, India
George A. Tsihrintzis	University of Piraeus, Greece
Hugo Proenca	University of Beira Interior, Portugal
Jocelyn Chanussot	Grenoble Institute of Technology, France
Kenji Suzuki	The University of Chicago, Chicago
Khalid Saeed	AGH University of Science and Technology, Poland
Klaus David	University of Kassel, Germany
Maode Ma	Nanyang Technological University, Singapore
Massimiliano Rak	Second University of Naples, Italy
Massimo Tistarelli	University of Sassari, Italy
Nishchal K. Verma	Indian Institute of Technology, Kanpur, India
Pascal Lorenz	University of Haute Alsace, France
Phalguni Gupta	Indian Institute of Technology Kanpur, India
Prasant Mohapatra	University of California, USA
Prasenjit Mitra	Pennsylvania State University, USA
Raj Jain	Washington University in St. Louis, USA
Rajesh Siddavatam	Kalinga Institute of Industrial Technology, India

Programme Co-chairs

Asit Kumar Das	Bengal Engineering and Science University, Kolkata, India
Ramjeevan Singh Thakur	Maulana Azad National Institute of Technology, India
Umesh A. Deshpande	Sardar Vallabhbhai National Institute of Technology, India

Organizing Chairs

Ashok K. Turuk	National Institute of Technology Rourkela, Rourkela, India
Rabindranath Ghosh	St. Thomas' College of Engineering and Technology, Kolkata, India

Technical Track Chairs

Joydeb Roychowdhury	Central Mechanical Engineering Research Institute, India
Korra Sathyababu	National Institute of Technology Rourkela, India
Manmath Narayan Sahoo	National Institute of Technology Rourkela, India

Honorary Industrial Chair

G.C. Deka	Ministry of Labour & Employment, Government of India

Industrial Track Chairs

Umesh Chandra Pati	National Institute of Technology Rourkela, India
Bibhudutta Sahoo	National Institute of Technology Rourkela, India

Special Session Chairs

Ashish Agarwal	Boston University, USA
Asutosh Kar	International Institute of Information Technology Bhubaneswar, India
Daya K. Lobiyal	Jawaharlal Nehru University, India
Mahesh Chandra	Birla Institute of Technology, Mesra, India
Mita Nasipuri	Jadavpur University, Kolkata, India
Nandini Mukherjee	Chairman, IEEE Computer Society, Kolkata Chapter Jadavpur University, Kolkata, India
Ram Shringar Rao	Ambedkar Institute of Advanced Communication Technologies & Research, India

Web Chair

Indranil Dutta	Indian Statistical Institute, Kolkata, India

Publication Chair

Sambit Bakshi	National Institute of Technology Rourkela, India

Publicity Chair

Mohammad Ayoub Khan	Center for Development of Advanced Computing, India

Organizing Committee

Amit Kr. Siromoni	St. Thomas' College of Engineering and Technology, Kolkata, India
Anindita Ganguly	St. Thomas' College of Engineering and Technology, Kolkata, India
Arindam Chakravorty	St. Thomas' College of Engineering and Technology, Kolkata, India
Dipak Kumar Kole	St. Thomas' College of Engineering and Technology, Kolkata, India
Prasanta Kumar Sen	St. Thomas' College of Engineering and Technology, Kolkata, India
Ramanath Datta	St. Thomas' College of Engineering and Technology, Kolkata, India
Subarna Bhattacharya	St. Thomas' College of Engineering and Technology, Kolkata, India
Supriya Sengupta	St. Thomas' College of Engineering and Technology, Kolkata, India

Program Committee

Vinay. A	Peoples Education Society Institute of Technology, Bangalore, India
Chunyu Ai	University of South Carolina Upstate, Spartanburg, USA
Rashid Ali	Aligarh Muslim University, Aligarh, India
C.M. Ananda	National Aerospace Laboratories, Bangalore, India
Soumen Bag	International Institution of Information Technology, Bhubaneswar, India
Sanghamitra Bandyopadhyay	Indian Statistical Institute, Kolkata, India
Punam Bedi	University of Delhi, Delhi, India
Dinabandhu Bhandari	Heritage Institute of Technology, Kolkata, India

Paritosh Bhattacharya	National Institute of Technology, Agartala, India
Malay Bhattacharyya	University of Kalyani, Kolkata, India
Sambhunath Biswas	Indian Statistical Institute, Kolkata, India
Darko Brodic	University of Belgrade, Bor, Serbia
Sasthi C. Ghosh	Indian Statistical Institute, Kolkata, India
R.C. Hansdah	Indian Institute of Science, Bangalore, India
Nabendu Chaki	University of Calcutta, Kolkata, India
Goutam Chakraborty	Iwate Prefectural University, Takizawa, Japan
Mihir Chakraborty	Indian Statistical Institute, Kolkata, India
Amita Chatterjee	Jadavpur University, Kolkata, India
Debasis Chaudhuri	Indian Institute of Management, Kolkata, India
Lopamudra Chowdhury	Jadavpur University, Kolkata, India
Nabanita Das	Indian Statistical Institute, Kolkata, India
Soumya Kanti Datta	Institute Eurecom, Sophia Antipolis, France
Rajat De	Indian Statistical Institute, Kolkata, India
Utpal Garain	Indian Statistical Institute, Kolkata, India
Ashis Ghosh	Indian Statistical Institute, Kolkata, India
Mukesh Goswami	Dharmsinh Desai University, Gugrat, India
Yogesh H. Dandawate	Vishwakarma Institute of Information Technology, Pune, India
Biju Issac	Teesside University, Middlesbrough, UK
Lakhmi C. Jain	University of South Australia, Adelaide, Australia
R. Janarthanan	T. J. S. Engineering College, Chennai, India
Boleslaw K. Szymanski	Rensselaer Polytechnic Institute, New York, USA
Tienfuan Kerh	National Pingtung University of Science and Technology, Pingtung, Taiwan
Zahoor Khan	Dalhouise University, Halifax, Canada
Dakshina Ranjan Kisku	Asansole Engineering College, Asansole, India
Sotiris Kotsiantis	University of Patras, Hellas, Greece
Dipak Kr. Kole	St. Thomas' College of Engineering and Technology, Kolkata, India
Aswani Kumar Cherukuri	VIT University, Vellore, India
Swapan Kumar Parui	Indian Statistical Institute, Kolkata, India
Prasanta Kumar Pradhan	St. Thomas' College fo Engineering and Technology, Kolkata, India
B. Narendra Kumar Rao	Sree Vidyanikethan Engineering College, Tirupati, India
Flavio Lombardi	Roma Tre University of Rome, Rome, Italy
Subhamoy Maitra	Indian Statistical Institute, Kolkata, India
Pradipta Maji	Indian Statistical Institute, Kolkata, India
Raghvendra Mall	University of Leuven, Leuven, Belgium
Amiya Kr. Mallick	St. Thomas' College fo Engineering and Technology, Kolkata, India
Debaprasad Mandal	Indian Statistical Institute, Kolkata, India
Ujjwal Maulik	Jadavpur University, Kolkata, India
Pabitra Mitra	Indian Institute of Technology, Kharagpur, India

Imon Mukherjee	St. Thomas' College fo Engineering and Technology, Kolkata, India
Nandini Mukherjee	Jadavpur University, Kolkata, India
Jayanta Mukhopadhyay	Indian Institute of Technology, Kharagpur, India
C.A. Murthy	Indian Statistical Institute, Kolkata, India
M. Murugappan	University of Malayesia, Malayesia
Mita Nasipuri	Jadavpur University, Kolkata, India
Rajdeep Niyogi	Indian Institute of Technology, Roorkee, India
Steven Noel	George Manson University, Fairfax, USA
M.C. Padma	PES College of Engineering, Karnataka, India
Rajarshi Pal	Institute for Development and Research in Banking Technology, Hyderabad, India
Umapada Pal	Indian Statistical Institute, Kolkata, India
Anika Pflug	Hochschule Darmstadt - CASED, Darmstadt, Germany
Surya Prakash	Indian Institute of Technology, Indore, India
Ganapatsingh G. Rajput	Rani Channamma University, Karnataka, India
Anca Ralescu	University of Cincinnati, Ohio, USA
Umesh Hodegnatta Rao	Xavier Institute of Management, Bhubaneswar, India
Ajay K. Ray	Bengal Engineering and Science University, Shibpur, India
Tuhina Samanta	Bengal Engineering and Science University, Shibpur, India
Andrey V. Savchenko	National Research University Higher School of Economics, Molscow, Russia
Bimal Bhusan Sen	St. Thomas' College fo Engineering and Technology, Kolkata, India
Indranil Sengupta	Indian Institute of Technology, Kharagpur, India
Patrick Siarry	Universite de Paris, Paris
Nanhay Singh	Ambedkar Institute of Advanced Communication Technologies & Research, Delhi, India
Pradeep Singh	National Institute of Technology, Raipur, India
Vivek Singh	South Asian University, New Delhi, India
Bhabani P. Sinha	Indian Statistical Institute, Kolkata, India
Sundaram Suresh	Nanyang Technological University, Singapore
Jorge Sá Silva	University of Coimbra, Portugal
Vasile Teodor Dadarlat	Technical University of Cluj Napoca, Cluj Napoca, Romania
B. Uma Shankar	Indian Statistical Institute, Kolkata, India
M. Umaparvathi	RVS College of Engineering, Coimbatore, India
Palaniandavar Venkateswaran	Jadavpur University, Kolkata, India
Stefan Weber	Trinity College, Dublin, Ireland

Azadeh Ghandehari	Islamic Azad University, Tehran, Iran
Ch Aswani Kumar	Vellore Institute of Technology, India
Cristinel Ababei	University at Buffalo, USA
Dilip Singh Sisodia	National Institute of Technology Raipur, India
Jamuna Kanta Sing	Jadavpur University, Kolkata, India
Krishnan Nallaperumal	Sundaranar University, India
Manu Pratap Singh	Dr. B.R. Ambedkar University, Agra, India
Narayan C. Debnath	Winona State University, USA
Naveen Kumar	Indira Gandhi National Open University, India
Nidul Sinha	National Institute of Technology Silchar, India
Sanjay Kumar Soni	Delhi Technological University, India
Sanjoy Das	Galgotias University, India
Subir Chowdhury	St. Thomas' College of Engineering and Technology, Kolkata, India
Syed Rizvi	The Pennsylvania State University, USA
Sushil Kumar	Jawaharlal Nehru University, India
Anupam Sukhla	Indian Institute of Information Technology, Gwalior, India

Additional Reviewers

A.M., Chandrashekhar
Acharya, Anal
Agarwal, Shalabh
B.S., Mahanand
Bandyopadhyay, Oishila
Barpanda, Soubhagya Sankar
Basu, Srinka
Battula, Ramesh Babu
Bhattacharjee, Sourodeep
Bhattacharjee, Subarna
Bhattacharya, Indrajit
Bhattacharya, Nilanjana
Bhattacharyya, Saugat
Bhowmik, Deepayan
Biswal, Pradyut
Biswas, Rajib
Bose, Subrata
Chakrabarti, Prasun
Chakraborty, Debashis
Chakraborty, Jayasree
Chandra, Helen
Chandra, Mahesh
Chatterjee, Aditi
Chatterjee, Sujoy
Chowdhury, Archana
Chowdhury, Manish
Dalai, Asish
Darbari, Manuj
Das, Asit Kumar
Das, Debaprasad
Das, Nachiketa
Das, Sudeb
Datta, Biswajita
Datta, Shreyasi
De, Debashis
Dhabal, Supriya
Dhara, Bibhas Chandra
Duvvuru, Rajesh
Gaidhane, Vilas
Ganguly, Anindita
Garg, Akhil
Ghosh Dastidar, Jayati
Ghosh, Arka
Ghosh, Lidia
Ghosh, Madhumala
Ghosh, Partha
Ghosh, Rabindranath
Ghosh, Soumyadeep
Ghoshal, Ranjit
Goswami, Mukesh
Goyal, Lalit
Gupta, Partha Sarathi
Gupta, Savita
Halder, Amiya
Halder, Santanu
Herrera Lara, Roberto
Jaganathan, Ramkumar
Kakarla, Jagadeesh
Kar, Mahesh
Kar, Reshma
Khasnobish, Anwesha
Kole, Dipak Kumar
Kule, Malay
Kumar, Raghvendra
Lanka, Swathi
Maruthi, Padmaja
Mishra, Dheerendra
Mishra, Manu
Misra, Anuranjan
Mohanty, Ram
Maitra, Subhamoy
Mondal, Jaydeb
Mondal, Tapabrata
Mukherjee, Nabanita

Mukhopadhyay, Debajyoti
Mukhopadhyay, Debapriyay
Munir, Kashif
Nasim Hazarika, Saharriyar Zia
Nasipuri, Mita
Neogy, Sarmistha
Pal, Monalisa
Pal, Tamaltaru
Palodhi, Kanik
Panigrahi, Ranjit
Pati, Soumen Kumar
Patil, Hemprasad
Patra, Braja Gopal
Pattanayak, Sandhya
Paul, Amit
Paul, Partha Sarathi
Phadikar, Amit
Phadikar, Santanu
Poddar, Soumyajit
Prakash, Neeraj
Rakshit, Pratyusha
Raman, Rahul
Ray, Sumanta
Roy, Pranab
Roy, Souvik
Roy, Swapnoneel
Rup, Suvendu
Sadhu, Arup Kumar
Saha Ray, Sanchita
Saha, Anuradha
Saha, Anushri
Saha, Dibakar
Saha, Indrajit
Saha, Sriparna
Sahoo, Manmath N.
Sahu, Beeren
Sanyal, Atri
Sardar, Abdur
Sarkar, Apurba
Sarkar, Dhrubasish
Sarkar, Ushnish
Sen, Sonali
Sen, Soumya
Sethi, Geetika
Sharma, Anuj
Tomar, Namrata
Umapathy, Latha
Upadhyay, Anjana
Wankhade, Kapil

Contents

Vehicular, *Ad hoc* and Sensor Networks

Cloud Computing

Efficient Architecture and Computing

Innovative Technologies and Applications

Parallel Processing Concept Based Vehicular Bridge Traffic Problem

Debasis Das[1] and Rajiv Misra[2]

[1] Department of Computer Science and Engineering,
NIIT University, Neemrana, Rajasthan-301705, India
[2] Department of Computer Science and Engineering,
Indian Institute of Technology, Patna
Bihar-800013, India
Debasis.Das@niituniversity.in,
rajivm@iitp.ac.in

Abstract. A Cellular Automata (CA) is a computing model of complex System using simple rules. In this paper the problem space is divided into number of cells and each cell can be constituted of one or several final state. Cells are affected by neighbors with the application of simple rule. Cellular Automata are highly parallel and discrete dynamical systems, whose behavior is completely specified in terms of local relation. In this paper CA is applied to solve a bridge traffic control problem. Vehicular travel which demands on the concurrent operations and parallel activities is used to control bridge traffic based on Cellular Automata technique.

Keywords: Cellular Automata, Simple Rule, Cell, Bridge Traffic Problem.

1 Introduction

Due to the rapid development of our economy and society, more emphasis is required on urban traffic research [8]. Within urban traffic research, the formation and dispersion of traffic congestion is one of the important aspects. Transportation research has the goal to optimize transportation flow of people and goods. As the number of road users constantly increases, and resources provided by current infrastructures are limited, intelligent control of traffic will become a very important issue in the future. Optimal control of traffic lights using sophisticated sensors and intelligent optimization algorithms might therefore be very beneficial. Optimization of traffic light switching increases road capacity and traffic flow, and can prevent traffic congestions. In the recent years there were strong attempts to develop a theoretical framework of traffic science among the physics community. Consequentially, a nearly completed description of highway traffic, e.g., the "Three Phase Traffic" theory, was developed. This describes the different traffic states occurring on highways as well as the transitions among them. Also the concepts for modeling vehicular traffic [8], [20] are well developed. Most of the models introduced in the recent years are formulated using the language of cellular automata [1], [18], [19]. Unfortunately, no comparable framework for the description of traffic states in city networks is present.

M.K. Kundu et al. (eds.), *Advanced Computing, Networking and Informatics - Volume 2*,
Smart Innovation, Systems and Technologies 28,
DOI: 10.1007/978-3-319-07350-7_1, © Springer International Publishing Switzerland 2014

2 Cellular Automata

A cellular automaton [1] is a decentralized computing model providing an excellent platform for performing complex computation with the help of only local information. Researchers, scientists and practitioners from different fields have exploited the CA paradigm of local information, decentralized control and universal computation for modeling different applications [7], [19].CA is an array of sites (cells) where each site is in any one of the permissible states. At each discrete time step (clock cycle) the evolution of a site value depends on some rule (the combinational logic), which is a function of the present state *k* of its neighbors for a k-neighborhood CA. Cellular automata (CA) are a collection of cells such that each adapts one of the finite number of states. Single cells change their states by following a local rule that depends on the environment of the cell. The environment of a cell is usually taken to be a small number of neighboring cells. Fig.1 shows two typical neighborhood [1] options (a) Von Neumann Neighborhood (b) Moore Neighborhood.

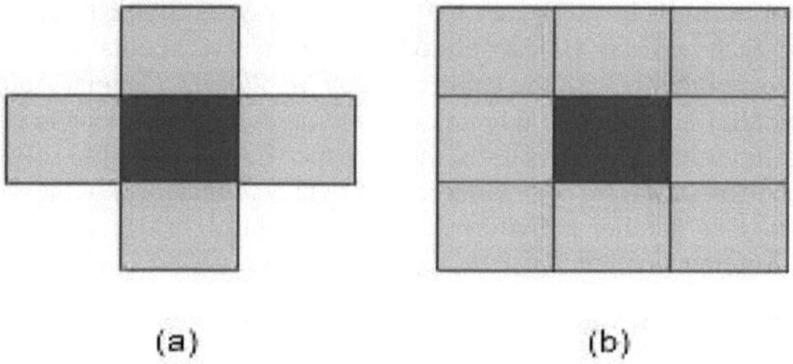

Fig. 1. (a) Von Neumann Neighborhood (b) Moore Neighborhood of the cell is taken to be the cell itself and some or all of the immediately adjacent cells

Typically, a cellular automaton consists of a graph where each node is in a finite state automaton (FSA) or cell. This graph is usually in the form of a two dimensional lattice whose cells evolve according to a global update function applied uniformly over all the cells. As arguments, this update function takes the cell's present state and the states of the cells in its neighborhood as shown in Fig. 2.

3 Related Work

The basic one-dimensional Cellular Automata model for highway traffic flow is based on the rules of the Cellular Automata. The highway is divided into number of cells. Each cell can either be empty or occupied by one car. If a cell is occupied by a car then it is denoted by 1 or otherwise 0.All cars have the same length of cell. In this section we discuss the following CA models [2-10].

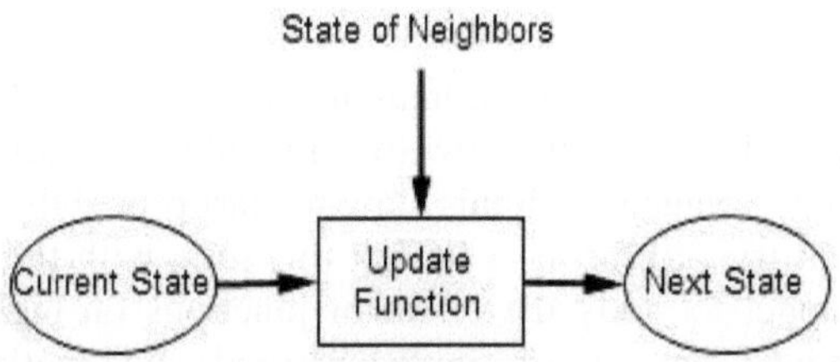

Fig. 2. State Transition Depend on Neighborhood State

3.1 Wolfram's Rule 184 (CA-184) Model

In1983, S. Wolfram's [1], [10] proposed first one dimensional Cellular Automata model with binary state. Here rule 184 is used for traffic flow so, it is called Wolfram's rule 184.For the CA-184 model [2] , we have the following two rules:

(R1) Acceleration and Braking

$v_i(t) \leftarrow \min\{g_{si}(t-1),1\}$

(R2) Vehicle Movement

$x_i(t) \leftarrow x_i(t-1) + v_i(t)$

Rule 1 (R1) is the speed of ith vehicle; this is the current updated configuration of cell. Rule 2 (R2) is for the Vehicle Movement.

3.2 Nagel – Schreckenberg CA Model (NS Model)

In 1992, NS model [11] was framed based on the4 rules of the CA. The rules are:

---

1. if (V < V max) then V = V+1, V is the speed of the car.

If the present speed is smaller than the desired maximum speed, the vehicle is accelerated.

2. if (V > gap) then V = gap, .

If present speed id is greater than the gap in the front, set v=gap.

This rule **avoids** the collision between vehicles.

3. if (V > 0) then V = V-1 with Pbreak.

This rule introduced a random element into the model.

4. X = X + V.

The present position on the road is moved forward by V.

According to these rules the speed and the acceleration/deceleration ratio of vehicles are independent of the speed of the other vehicles at any time.

3.3 BJH Model

In 1996, another model developed by Benjamin, Johnson, and Hui (BJH model) [12] is similar to the NS model, but with 'slow-to-start' rule. The slow-to-start rule allows a stopped vehicle to move again with this slow-to-start probability 1-Ps. If the vehicle doesn't move, then it tries to move again but this time with the probability Ps. The authors used this model to study the effect of junctions on highways, finding that, setting a speed limit near junctions on single lane roads can greatly decrease the queue length to enter the road.

3.4 Fukui-Ishibashi CA Model

In 1996,Fukui and Ishibashi [13] constructed a generalization of prototypical CA-184 CA model. This model has two different categories: 1). Stochastic Fukui--Ishibashi 2). Deterministic Fukui—Ishibashi.

They assume a gradual acceleration of cell per time step and followed the modified rule:

(R1): Acceleration and Braking

$v_i(t) \leftarrow \min\{v_i(t-1) + 1, gs_i(t-1), vmax\}$

These experimental observations have indicated that there is no difference in global system dynamics with respect to either adopting gradual or instantaneous vehicle acceleration.

3.5 The New Time Oriented Cellular Automata (TOCA)

In 1999, The New Time Oriented Cellular Automata (TOCA)[14] model was framed in which the threshold of changing speed is equal to the minimum time headway t. Thus the time headway between two vehicles can never be smaller than the threshold of changing speed.

The TOCA rules can be rewritten as:

---

(1) if $(gap > v \cdot tH)$ and $(v < vmax)$ then $v = v + 1$ with pac

The speed is increased by 1 with the probability pac if the time headway to the vehicle in the front is larger than tH. An average acceleration ratio with the value pac is resulted.

(2) if $(v > gap)$ then $v = gap$

(3) if $(gap < v \cdot tH)$ and $(v > 0)$ then $v = v - 1$ with pdc

The speed is reduced by 1 with the probability pdc if the time headway to the vehicle in the front is smaller than tH. An average deceleration ratio with the value pdc is resulted.

(4) $x = x + v$

3.6 GE Hong Xia, DONG Li-yun, LEI Li, DAI Si-Qiang Model

A modified cellular automata model for traffic flow was proposed by GE Hong Xia, DONG Li-yun , LEI Li and DAI Si-Qiang (2003)[15]. In this model a changeable security gap is introduced. This model is a modified version of the NS model. Discrete random valuesarefrom 0 to Vmax (Vmax = 5). The rules of the extended NS model are set as follows:

---

(1)Acceleration: vn→min(vn + 1, vmax);

(2) Deceleration: vn→min(vn ,$d_n^{(eff)}$) with $d(v_n, d_n^{(eff)}) = d_n + \max(Vanti - gap_{Security}, 0)$, denoting the effective gap, where Vanti = min(d_{n+1}, v_{n+1}) is the anticipated velocity of the leading vehicle in the next time step, $gap_{Security}$ = round (tv_n) is the security gap while considering the the velocity of its following car;

(3) Randomization: vn→max (vn - 1,0) with the probability p .

(4) Update of position: $x_n \rightarrow x_n + v_n$.

The velocity of vehicles is determined by Steps (1) - (3). Finally, the position of the car is shifted in accordance with the calculated velocity in Step (4). r is a parameter determined by the simulation.

3.7 M. Namekawa, F. Ueda, Y. Hioki, Y. Ueda and A. Satoh Model

In 2004, M. Namekawa, F. Ueda, Y. Hioki, Y. Ueda and A. Satoh [16] proposed a model that improves the NS model.

The Simulation time per clock is 0.1 seconds and the speed is 5 km/h. Modified rule of the proposed mode.

---

(1) Acceleration: (v+1)*a<=g and v < vmax →v=v+1
(2) slow down: v*a < g→v =g/a

The distance between the vehicles that was suitable for speed with value to be decided at the minimum distance between the vehicles v*a.

3.8 Clarridge and Salomaa Model

In 2009, Clarridge and Salomaa [17] proposed a Modified version of **BJH model.**

In this model, the cars' velocities are adjusted at each time step according to the following rules. Recall that d is the distance to the next car, v is the velocity of the current car, vnext is the velocity of the next car, pslow is the probability that the slow-to-start rule is applied, and pfault is the probability that the car slows down randomly. We fix vmax =5.

1. Slow-to-Start: As in the BJH rule, if v =0and d> 1 then with probability (1 – p_{slow}), the car accelerates normally (this step is ignored), and with probability p_{slow} the car stays at velocity 0 at this time step (does not move) and accelerates to v =1atthe next time step.

2. Deceleration (when the next car is near): if d<= v and either v<next or v<= 2, then the next car is either very close or going at a faster speed, and we prevent a collision by setting v ← d– 1, but do not slow down more than is necessary. Otherwise, if d<= v, v>= vnext, and v> 2weset v ← min(d–1,v –2) in order to possibly decelerate slightly more, since the car ahead is slower or of the same speed and the velocity of the current car is substantial.

3. Deceleration (when the next car is farther): if v<d<=2v,thenif v>= vnext+4, decelerate by 2 (v ← v–2). Otherwise, if vnext+2 <= v<= vnext+3 then decelerate by 1 (v ← v – 1).

4. Acceleration: if the speed has not been modified yet by one of rules 1-3 andv<vmax and d>v +1, then v ← v +1.

5. Randomization: if v> 0, with probability pfault, velocity decreases by one (v ← v – 1).

6. Motion: the car advances v cells.

This model performs an iteration of cars moving on a road in O(L) time , where L is the length of the road and a parallel implementation based on the constant time.

4 Proposed Model for Vehicular Bridge Traffic Control

It is assumed that the bridge Traffic is caused by cars moving from a dual-lane [21-25] road to a single-lane road, as shown in Fig. 3, where Lane1 and Lane2 are two parallel lanes. These lanes are connected with a bridge lane Bridge at B1. No matter what the time step is, only one car can enter the bridge. After passing through the bridge, at B2 another lanes are Lane3 and Lane4 (as shown in Fig. 3), these two are opposite direction parallels lanes, car can choose Lane 3 and Lane 4 to move to the opposite direction.

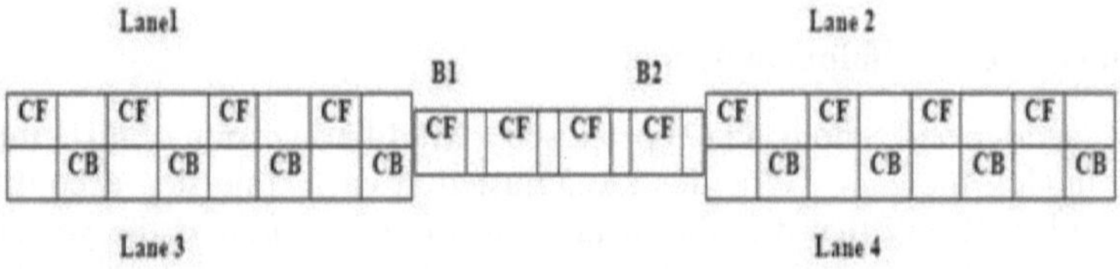

Fig. 3. The model of the Vehicular Bridge Traffic Control System, CF represents cars with forward direction at the instance of time (t) and CB Represent cars with Backward direction at the instance of time (t+1)

Table 1. The Rules that Updated the Next State of the Cellular Automata Cells

Rule	111	110	101	100	011	010	001	000
170	1	0	1	0	1	0	1	0
85	0	1	0	1	0	1	0	1

Table 1 focuses on the situation of the parallel lanes: Lane1, Lane 2, Lane 3, and Lane 4 in all of which no car changes lanes and cars can cross the Bridge only in the forward or backward direction but not in both direction. Cars in each lane simultaneously evolve according to the cellular automata simple rules [1]. Wolfram [1] has investigated cellular automata using empirical observations and simulations. For 2-state 3-neighborhood CA, the evolution of the ith cell can be represented as a function of the present states of (i–1)th, (i)th, and (i+1)th cells as: $x_i(t+1) = f(x_{i-1}(t), x_i(t), x_{i+1}(t))$ *where f*, represents the combinational logic. For a 2-state 3-neighborhood cellular automaton there are 2^3=8 distinct neighborhood configurations or cells are from 0(000) to 7(111) and 2^8=256 distinct mappings from all these neighborhood configurations to the next state, each mapping representing a CA rule ,means 256 CA rules are available, from rule 0 to rule 255 . The next state of the i_{th} cell depends on the present states of its left and right neighbors and on its own present state.

The Bridge Traffic problem[26-28] present in this paper is constructed using cellular automata based on rules 170 and 85.The combinational logic of the rules 170 and 85 [1] for the cellular automata can be expressed as follows:

$$\text{Rule 170: } a_i(t+1) = (a_{(i+1)}(t)) \quad (1)$$

$$\text{Rule 85: } a_i(t+1) = (1 \oplus a_{(i+1)}(t)) \quad (2)$$

i means the cells position and t means time at any instance. The rule specifies the evolution of the cellular automata from the neighborhood configuration to the next state and these are then represented in Table1.

If "1" Represents a car which move the forward or backward direction at any instance of time (t) and "0" represents no car at any instance of time (t) . The rule 170 and rule 85 models for car moving in each lane is shown by Table1.

5 Conclusion

This paper presents a bridge traffic problem based on cellular automata approach. In this paper, we are trying to solve the bridge traffic problem to reduce the number of accident based on cellular automata simple rules(the rules being 170 and 85) with respect to the increasing vehicles. The above knowledge and the proposed model can be taken in planning and controlling the bridge traffic problem in vehicular networks. The processing technique is more effective and efficient for solve the bridge traffic problem in vehicular network.

References

1. Wolfram, S.: A new kind of science Wolfram Media (2002)
2. Adamatzky, A., Lawniczak, R.A.-S.A., Martinez, G.J., Morita, K.: AUTOMATA-2008 Theory and Application of Cellular Automata. Thomas Worsch Editors (2008)
3. Nishinari, K., Schadschneider, A., Chowdhury, D.: Traffic of Ants on a Trail: A Stochastic Modelling and Zero Range Process. In: Sloot, P.M.A., Chopard, B., Hoekstra, A.G. (eds.) ACRI 2004. LNCS, vol. 3305, pp. 192–201. Springer, Heidelberg (2004)
4. Campari, E.G., Levi, G., Maniezzo, V.: Cellular Automata and Roundabout Traffic Simulation. In: Sloot, P.M.A., Chopard, B., Hoekstra, A.G. (eds.) ACRI 2004. LNCS, vol. 3305, pp. 202–210. Springer, Heidelberg (2004)
5. Shoufeng, L., Ximin, L.: Based on Hybrid Genetic Algorithm and Cellular Automata Combined Traffic Signal Control and Route Guidance. In: Chinese Control Conference, pp. 53–57 (2007)
6. Wei, J., Wang, A., Du, N.: Study of Self-Organizing Control of Traffic Signalsin an Urban Network based on Cellular Automata. IEEE Transactions on Vehicular Technology 54(2), 744–748 (2005)
7. Wolfram, S. (ed.): Theory and applications of CA. WorldScientific (1986)
8. Traffic and Granular Flow. Springer (1996,1998,2000,2001)
9. Olariu, S., Weigle, M.C.: Vehicular Networks, From Theory to Practice. CRC Press, Taylor &Francis Group (2009)
10. Wolfram, S. (ed.): Theory and applications of CA. WorldScientific (1986)
11. Nagel, K., Schreckenberg, M.: A cellular automaton model for freeway traffic. J. PhysicsI France 2, 2221–2229 (1992)
12. Benjamin, S.C., Johnson, N.F., Hui, P.M.: Cellular automata models of traffic flow along a highway containing a junction. Journal of Physics A: Mathematical and General 29(12), 3119–3127 (1996)
13. Fukui, M., Ishibasi, Y.: Traffic flow in 1D Cellular Automation Model including cars moving with high speed. Journal of Physics, 1868–1870 (1996)
14. Maerivoet, S., Moor, B.D.: Cellular automata model of road traffic. Physics Reports 419(1), 1–64 (2005)
15. Ge, H.-X., Dong, L.-Y., Lei, L., Dai, S.-Q.: A Modified Cellular Automata Model for traffic flow. Journal of Sanghai Universiy 8(1), 1–3 (2004)
16. Namekawa, M., Ueda, F., Hioki, Y., Ueda, Y., Satoh, A.: General purpose road traffic simulation with cell automation model, pp. 3002–3008. Modelling and Simulation Society of Australia and New Zealand Inc. (2005)
17. Clarridge, A., Salomaa, K.: A Cellular Automaton Model for Car Traffic with a Slow-to-Stop Rule. In: Maneth, S. (ed.) CIAA 2009. LNCS, vol. 5642, pp. 44–53. Springer, Heidelberg (2009)
18. Das, D., Ray, A.: A Parallel Encryption Algorithm for Block Ciphers Based on Reversible ProgrammableCellular Automata. Journal of Computer Science And Engineering 1(1), 82–90 (2010)
19. Das, D.: A Survey on Cellular Automata and Its Applications. In: Krishna, P.V., Babu, M.R., Ariwa, E. (eds.) ObCom 2011, Part I. CCIS, vol. 269, pp. 753–762. Springer, Heidelberg (2012)
20. Das, D., Misra, R.: Parallel Processing Concept Based Road Traffic Model. In: 2nd International Conference on Computer, Communication, Control and Information Technology, pp. 267–271 (2012)

21. Rickert, M., Nagel, K., Schreckenberg, M., Latour, A.: Two lane traffic simulations using cellular automata. Physica A: Statistical Mechanics and its Applications 231(4), 534–550 (1996)
22. Wolf, D.E.: Cellular automata for traffic simulations. Physica A: Statistical Mechanics and its Application 263(1-4), 438–451 (1999)
23. Takayasu, M., Takayasu, H.: 1/F Noise in A Traffic Model. World Scientific (1993)
24. Das, S.: A cellular automata based model for traffic in congested city. In: IEEE International Conference on Systems, Man and Cybernetics, pp. 2397–2402 (2009)
25. Bham, G.H., Benekohal, R.F.: A high fidelity traffic simulation model based on cellular automata and car-following concepts. Transportation Research Part C: Emerging Technologie 12(1), 1–32 (2004)
26. Xiao, S., Kong, L., Liu, M.: A cellular automaton model for a bridge traffic bottleneck. Acta Mechanica Sinic 21(3), 305–309 (2005)
27. Han, Y.S., Ko, S.K.: Analysis of a cellular automaton model for car traffic with a junction. Journal of Theoretical Computer Scienc 450, 54–67 (2012)
28. Esser, J., Schreckenberg, M.: Microscopic Simulation of Urban Traffic Based on Cellular Automata. International Journal of Modern Physics 8(5) (1997)

Analysis of GPS Based Vehicle Trajectory Data for Road Traffic Congestion Learning

Swathi Lanka and Sanjay Kumar Jena

Department of Computer Science and Engineering
National Institute of Technology, Rourkela - 769008, India
swathivanet@gmail.com, skjena@nitrkl.ac.in

Abstract. Successful developments of effective real-time traffic management and information systems demand high quality real time traffic information. In the era of intelligent transportation convergence, traffic monitoring requires traffic sensory technologies. We tabulate various realistic traffic sensors which aim to address the technicalities of both point and mobile sensors and also increase the scope to prefer an optimal sensor for real time traffic data collection. The present analysis extracted data from Mobile Century experiment. The data obtained in the experiment was pre-processed successfully by applying data mining pre-processing techniques such as data transformation, normalization and integration. Finally as a result of the availability of pre-processed Global Position System (GPS) sensors trace data a road map has been generated.

Keywords: Traffic sensor, Traffic flow, GPS probe, Data fusion, Floating car, Fleet management.

1 Introduction

Vehicular Ad-Hoc Network (VANET) is one of the key enabling technologies which can provide the communication between the vehicles which are connected through wireless links [1]. VANET is a component of Intelligent Transportation System (ITS) which can brin0g a noticeable improvement in transportation system towards decreasing congestion and improving safety and traveler convenience. ITS is used to design a smart vehicle. Developing Advanced Driving Assistance Systems (ADAS) aiming to alert drivers about road situation, traffic conditions, and possible traffic congestion with other vehicles has attracted a lot of attention recently [2].

The Advanced Traveler Information System (ATIS) is one of the six components of ITS. ATIS provides solutions for intelligent transportation related applications. It implements emerging computer, communication and information technologies to provide vital information to the users of a system regarding traffic regulation, route and location guidance, hazardous situations and safety advisory and warning messages. ATIS requires a large amount of data for processing, analysis, and storage for effective dissemination of traveler information [3].

M.K. Kundu et al. (eds.), *Advanced Computing, Networking and Informatics - Volume 2*,
Smart Innovation, Systems and Technologies 28,
DOI: 10.1007/978-3-319-07350-7_2, © Springer International Publishing Switzerland 2014

Traffic congestion has a significant negative impact on social and economic activities around many cities in the world. Road traffic monitoring aims to determine traffic conditions of different road links, which is an essential step toward active congestion control. Many tasks, such as trip planning, traffic management, road engineering, and infra-structure planning, can benefit from traffic estimation [4].Traditional approaches for traffic monitoring rely on the use of point traffic sensors, which can mount at a fixed location along the roadway and sense the traffic parameters at the particular location [5], [6]. After traditional approaches, with the increasing growth of mobile technology mobile sensors has got attention, will be placed in a vehicle can collect vehicle related data [6], [7]. Recently in the era of mobile internet services, with the shrinking cost and increased accuracy of GPS, and increasing penetration of mobile phones in the population makes Global Position System (GPS) with Floating Car Data (FCD) as an attractive traffic sensor[8], [9]. Table 1 shows particulars of commercially available traffic sensors.

With the growing prevalence of GPS receivers embedded in vehicles and smart-phones, there have been increasing interests in using their location updates or trajectories for monitoring traffic [10]. Even though GPS is becoming more and more used and affordable, so far only a limited number of cars are equipped with this system, typically fleet management services. Traffic data obtained from private vehicles or trucks is more suitable for estimating traffic under motorways and rural areas [11]. In case of urban traffic, taxi fleets are particularly useful due to their high number and their on-board communication systems already in place. Currently, GPS probe data are widely used as a source of real-time information by many service providers [12].

Existing Conventional traffic congestion detection systems used location based data for congestion detection. However, quantifying congestion is generally carried out using traffic density which is a spatial parameter. Hence spatial data such as travel time helps to detect congestion with a less delay. In our work, we have collected spatio temporal data from mobile based GPS receivers which are attached with each vehicle travelling on the freeway. In this paper, we are particularly interested to collect spatio temporal data and make the raw data set more suitable for efficient congestion learning. The data set is pre-processed in to a human, machine understandable format. The resultant data set can able to improve the effectiveness and the performance of the data mining algorithms and machine learning techniques whenever it applies on the dataset.

The paper is organized as follows: Technicality of various traffic sensors are discussed in Section1. Section 2 designed a three-level structure vehicle activity database format. Mobile Century Data set has discussed in Section 3. This is followed by data pre-processing methods, resultant datasets and realistic road map in Section 4. Section 5 presents conclusions and future work.

Table 1. Technicalities of commercially available traffic sensors

Technology	Sensing parameters	Strengths	Weakness	Suitable applications
Inductive loops [5] (Point sensor)	Vehicle volume, occupancy, time, speed	Conventional standard can obtain accurate occupancy measurements, flexible design can satisfy large variety of applications, adoptable and less sensitive for all weather and lighting conditions	Installation is intrusive to traffic, maintenance and installation cost is more, gives less detection accuracy when large number of vehicles are involved, reinstallation is needed whenever road is repaved.	Traffic flow detection, congestion detection, traffic-density detection
Pneumatic tubes [6] (Point sensor)	Speed, direction of flow, time, volume	Ideal for short term engineering studies, less maintenance and installation cost, portable device can be reused in many locations.	Has limited lane coverage, intrusive to traffic, system damage causes to inaccurate data collection	Vehicle count, traffic flow detection
Video Image Processors [6], (Point sensor)	Road vehicle images, video streams of traffic	Rich array of data collection, can monitor multiple lanes and detection zones with minimum installation and maintenance, insertion and deletion of detection zones is easy	Performance may be affected by weather, vehicle shadows, vehicle projections, occlusions, strong winds, day-night transitions and water, dust on the camera lens. Setup cost is high.	Traffic count, vehicle speed detection, vehicle classification
Acoustic/Ultrasonic Sensors [5] (Mobile sensor)	Occupancy, count, speed	Multiple lane operation is possible, capable to detect high occupancy vehicle with high accuracy, in sensitive to precipitation	Environmental conditions may affect the performance, cold temperature may affect vehicle count accuracy, and occupancy measurement accuracy may be degraded when vehicle travelling with high speed.	Vehicle parking assistance, vehicle detection, pedestrian count
Active/Passive Infrared Sensors [6] (Mobile sensor)	Vehicle position, speed, count,	Can be operated both day and night, multiple lane operation is possible, usage of sophisticated signal processing algorithms gains better accuracy	Sensitive to inclement weather conditions and ambient light, installation and maintenance cost is more	Road obstacle detection, distance measurement

Table 1. (*continued*)

RFID Sensors [7] (Point & Mobile sensor)	Vehicle ID, time	In expensive Less installation and maintenance cost non intrusive to traffic	Only detect equipped vehicles Collect poor array of data Privacy concerns and actors interest is required	Automatic Vehicle Identification, E-Z pass, Electronic Toll Collection
Microwave Radar [6] (Mobile sensor)	Speed, occupancy, presence	A single detector can cover multiple lanes, usage of efficient signal processing techniques increases detection accuracy	Multi path coverage causes redundant vehicle detection, false detection sometimes, unable to detect stopped vehicle	Calculates vehicle speed, vehicle detection
Magnetometer [6] (Point sensor)	Vehicle count, time	Can be used where a point or small-area location of a vehicle is necessary, can be used where loops are not feasible (e.g, bridge decks), insensitive to weather conditions	Installation requires pavement cut, requires multiple units for full lane detection, maintenance cost is more	Vehicle presence detection, vehicle passage detection
GPS with FCD [8],[9] (Mobile sensor)	Longitude, latitude, time, speed	In vehicle sensor simple to install and operate, less maintenance cost, easy penetration due to raped increase of mobile phones, works under all weather and lighting conditions, never suffer with energy consumption problem since GPS will be equipped in a moving vehicle, collects on road real-time information, non intrusive to traffic.	GPS signals may be obstructed by tall buildings and trees, actor interest is required and Signal strength may be degraded under bad weather conditions.	Congestion detection, collusion detection, intersection safety, Road safety

2 Data Base Design and Data Conversion

The initial task of this research paper is to develop a common database format for vehicle activity data, followed by conversion of dataset in to human understandable format. Three-level structure vehicle activity database format is designed and illustrated in Table 2.

The top level of the database lists a program which helps to collect the vehicle activity data. It contains fields such as name of the dataset, the dates of the program, number of vehicles tested, total testing time duration and parameters that are collected(e.g., longitude, latitude, etc). Each entry in this level has a pointer to the second level of the database.

The second level of the database listing data trips for overall program. When a vehicle is travelling on the road, it collects data would correspond to a single trip entry in the second layer of the database. For each trip, various parameters are listed including date, starting and ending times and testing duration per day. Each entry in this level has a pointer to the third level of the database.

The third level represents spatio temporal time series data. The time series data contains the time sequence of position (longitude, latitude), speed and time. Further in this layer of the database, additional parameters may be derived from the existing data for the determination of congestion level of the roadway.

Table 2. Database Topology

Data set title	Date	No of vehicles	No of seconds	Parameters
Mobile Century	2 Days: 8 February 2008, 9 February 2008	100	28,800	longitude, latitude, speed, time
-------------	--------------	--------	-----------	-------------

Run	Date	Starting time	Ending time	Duration(sec)
1	8/2/2008	19:00:00	23:59:58	17,998
2	9/2/2008	00:00:01	02:59:59	10,799

Date	Time	Vehicle ID	Latitude	Longitude	Speed
8/02/2008	19:00:02	1	37.600	-122.064	0.009
9/02/2008	02:59:58	100	37.6002	-122.063	0.013

3 The Data Set

Vehicle trajectories are typically collected from GPS equipped vehicle based mobile phone from Mobile Century experiment [13] took place on February 8th, 2008. It consisted in deploying 100 GPS- equipped Nokia N95 cell phones on a freeway in 100 vehicles during 8 hours (from 8 February at 19:00:00 pm to 9 February at 03:00:00 am). The experiment was conducted on Highway I-880, near Union City, California; between Winton Ave. to the North and Stevenson Blvd. to the South. This 10-mile long section was selected for field experiment. Data has collected on four lane road with a regular time interval of 3 seconds.

4 Data Pre-processing

Several conversion and filtering steps are often necessary for mobile century data. Pre-processing may include (1). Conversion of date and time from Unix Time Zone to local date and time. (2). Conversion of latitude and longitude in to decimal degrees. (3). Constructing new attributes such as vehicle ID. Data fusion is also necessary which can integrate hundred vehicle activity data in to one unified dataset includes all of the data points and time steps from the input data sets.

Pre-processing has done by using data mining pre-processing techniques such as normalization and attribute construction. Normalization technique used unit conversion method. Unit conversion method converted Unix time in to local time and date. Attribute construction must be replacing or adding new attributes inferred by existing attributes. It is necessary to create new attributes that can capture the important information in a data set more effectively than the original ones [14]. In our system vehicle ID is newly constructed attribute. Among hypothesis-driven and data driven methods, data driven method is particularly used for Attribute construction in the present work. The new attributes are then evaluated according to a given attribute quality measure. Table 3 shows pre-processed data for vehicle ID 1 . Finally data from multiple sources have fused and placed in a single data set by using data fusion technique. In order to analyze the road position along with vehicle motion the entire dataset is sorted with respective time has shown in the Table 4. Experimented roadmap has generated with resultant dataset has shown in Fig. 1.

5 Pre-processing Results

Table 3. Pre-processed data for Vehicle 1

V ID	Date & Time	Latitude	Longitude	Speed
1	08-02-08 19:00:02	37.60043	-122.064	0.009
1	08-02-08 19:00:06	37.60043	-122.064	0.01
1	08-02-08 19:00:09	37.60043	-122.064	0.013
1	08-02-08 19:00:12	37.60043	-122.064	0.015
1	08-02-08 19:00:16	37.60043	-122.064	0.016
1	08-02-08 19:00:20	37.60043	-122.064	0.017
1	08-02-08 19:00:24	37.60043	-122.064	0.017
1	08-02-08 19:00:27	37.60043	-122.064	0.015

Table 4. Pre-processed Road based vehicle moment data

V ID	Date & Time	Latitude	Longitude	Speed
1	08-02-08 19:00:08	37.6105	-122.069	5.002
1	08-02-08 19:00:08	37.6220	-122.078	67.776
1	08-02-08 19:00:09	37.6004	-122.064	0.013
1	08-02-08 19:00:09	37.6430	-122.092	52.402
1	08-02-08 19:00:09	37.6141	-122.072	3.143
1	08-02-08 19:00:09	37.6087	-122.068	65.229
1	08-02-08 19:00:09	37.5934	-122.057	68.11
1	08-02-08 19:00:09	37.6005	-122.062	66.612

Fig. 1. Experimented four lane roads on high way I-880, CA. The symbol represents vehicles. Vehicles are positioned on the road by processing on resultant pre-processed Mobile Century data.

6 Conclusion and Future Work

Real-world traffic data is highly susceptible to noise, redundancy and inconsistent data due to their huge size, heterogeneous sources and type of sensory technologies. Low quality traffic data will lead to low quality results processing. Often low quality information leads to in complete control and management. This paper presents a pre-processed Mobile Century data set and a realistic four lane highway I-880 roadmap with positioned vehicles. Roadmap has generated by taking the input as resultant pre-processed dataset. In future work, the resultant data set will be used for road traffic congestion learning for efficient intelligent congestion control under heterogeneous traffic conditions. Future work in our research program will take advantage of this work, and will focus on congestion prediction and detection.

References

1. Hartenstein, H., Kenneth, P.L.: A Tutorial Survey on Vehicular Ad Hoc Networks. IEEE Communication Magazine 46(6), 164–171 (2008)
2. Zhang, J., Fei, Y.W.: Data Driven in Intelligent Transportation System: A Survey. IEEE Transactions on Intelligent Transportation System 12(4) (2011)
3. Kumar, P., Varun, S.: Advanced Traveller Information for Hyderabad City. IEEE Transactions on Intelligent Transportation System 6(1) (2005)
4. Agarwal, V., Venkata, M.: A Cost-Effective Ultrasonic Sensor-Based Driver-Assistance System For Congestion Traffic Conditions. IEEE Transactions on Intelligent Transportation System 10(3) (2009)
5. Ali, S.M., George, B., Vanajakshi, L.: An Efficient Multiple-Loop Sensor Configuration Applicable for Undisciplined Traffic. IEEE Transactions on Intelligent Transportation System 14(3) (2013)
6. Leduc, G.: Road Traffic Data: Collection Methods and Applications. JRC Technical Notes, Joint Research Centre, European Commission (2008)
7. Cheng, W., Cheng, X., Song, M., Chen, B., Zhao, W.W.: On the Design and Deployment of RFID Assisted Navigation Systems for VANET. IEEE Transaction on Parallel and Distributed System (2011)
8. Yong, Z., Zuo, X., Zhang, L., Chen, Z.: Traffic Congestion Detection based on GPS Floating-Car Data. Procedia Engineering 15, 5541–5546 (2011)
9. Vanajakshi, L., Subramanian, S.C.: Travel time prediction under heterogeneous traffic conditions using global positioning system data from buses. IET Intelligent Transportation Systems (2008)
10. Sun, Z., Xuegang, B.: Vehicle Classification Using GPS Data. Transportation Research Record Part C 37, 102–117 (2013)
11. Messelodi, S., Carla, M.M.: Intelligent Extended Floating Car Data Collection. Expert systems with applications 36, 4213–4227 (2009)
12. Zhu, Y., Li, Z.: A compressive sensing approach to urban traffic estimation with probe vehicles. IEEE Transactions on Mobile Computing 12(11) (2013)
13. Herrera, J.C., Daniel, B.W.: Evaluation Of Traffic Data Obtained Via GPS-Enabled Mobile Phones: The Mobile Century Field Experiment. Transportation Research Record Part C 18, 568–583 (2011)
14. Jiawei, H., Micheline, K.: Data mining: Concepts and Techniques, 2nd edn. Morgan Kaufmann Publishers (2006)

Fast Estimation of Coverage Area in a Pervasive Computing Environment

Dibakar Saha, Nabanita Das, and Bhargab B. Bhattacharya

Advanced Computing and Microelectronics Unit
Indian Statistical Institute, Kolkata, India
dibakar.saha10@gmail.com,
ndas,bhargab@isical.ac.in

Abstract. In many applications of pervasive computing and communication, it is often mandatory that a certain service area be fully covered by a given deployment of nodes or access points. Hence, a fast and accurate method of estimating the coverage area is needed. However, in a scenario with a limited computation and communication capability as in self-organized mobile networks, where the nodes are not static, computation-intensive algorithms are not suitable. In this paper, we have presented a simple algorithm for estimating the area covered by a set of nodes randomly deployed over a 2-D region. We assume that the nodes are identical and each of them covers a circular area. For fast estimation of the collective coverage of n such circles, we approximate each real circle by the tightest square that encloses it as well as by the largest square that is inscribed within it, and present an $O(n \log n)$ time algorithm for computation. We study the variation of the estimated area between these two bounds, for random deployment of nodes. In comparison with an accurate digital circle based method, the proposed algorithms estimate the area coverage with only 10% deviation, while reducing the complexity of area computation significantly. Moreover, for an over-deployed network, the estimation provides an almost exact measure of the covered area.

Keywords: Pervasive Computing, Wireless Sensor Networks (WSN), Coverage, Digital Circle, Range.

1 Introduction

In a pervasive computing environment, for tetherless computing and communication, it is often required to place the computing nodes or access points to offer services over a predefined area. In many cases, like vehicular networks, ad-hoc networks, mobile health-care services, surveillance, wireless sensor networks, the nodes are often mobile and have limited power, limited storage and limited computation and communication capabilities. These networks are often self-organized, and can take decision based on their local information only.

A typical wireless sensor network (WSN) consists of spatially distributed autonomous sensor nodes to monitor physical or environmental conditions, such as temperature, sound, pressure etc. Each sensor node has the ability to collect,

M.K. Kundu et al. (eds.), *Advanced Computing, Networking and Informatics - Volume 2*, Smart Innovation, Systems and Technologies 28,
DOI: 10.1007/978-3-319-07350-7_3,

process, and route the sensed data. The streams of sensed data from each node are forwarded cooperatively through several intermediate nodes to finally arrive at the sink node. Sensor networks are used in many applications such as habitat and ecosystem monitoring, weather forecasting, smart health-care technologies, precision agriculture, homeland security and surveillance. For all these applications, the live nodes are required to cover the area to be monitored. Therefore, the classical problems of covering an area with a specific kind of shape such as circle, square, or rectangle, are recently being revisited for modeling and analysis of such networks. In this paper, we address the problem of estimating the area covered by a set of nodes distributed randomly over a 2-D plane. We assume that the nodes are homogeneous, and each of them covers a circular area with a fixed radius. Since estimation of the area covered by an arbitrary set of circles is computation-intensive, it may not be feasible to perform it in real time where the nodes have limited power, storage and computational ability, as in a pervasive computing environment. The area-coverage of the square meshes, hexagonal meshes and honeycomb meshes was studied earlier by Luo *et al.* [1]. Some related theoretical and algorithmic issues concerning rectangle intersection problems were revisited by Six and Wood [2]. Bentley and Wood [3] proposed an optimal algorithm for reporting intersections of n rectangles. In these two papers, the authors proposed an $O(n \log n + k)$ algorithm where k is the number of intersecting pairs. An $O(n\log^2 n)$ algorithm in [4] can be used to construct a generalized Voronoi diagram for a set of n circular discs and to compute the coverage area in terms of circular sectors and quadrangles. A more efficient algorithm with $O(n \log n)$ time complexity and $O(n)$ space complexity for circle intersection/union using a particular generalization of Voronoi diagram called power diagrams was reported later [5]. However, all these algorithms require complex data structures and rigorous computation. In order to reduce computational effort, in literature often it is assumed that the monitoring area is composed of a number of elementary areas or unit square grids [6–9]. In [10], [11], the authors investigated random and coordinated coverage algorithms. Some authors used partitioning techniques to decompose the query region into square grid blocks and studied the coverage of each block by sensor nodes [12–14]. For a more realistic estimation of the covered area, a new $O(n \log n)$ algorithm based on digital geometry is proposed in [15], where a real circle is replaced by a digital circle and discrete domain computation is applied. Though the complexity of the algorithm remains the same as it is in [5], the former uses simple data structures and primitive arithmetic operations only. In this paper, we propose a simpler and faster method of coverage estimation. Given a random distribution of n nodes over a 2-D plane, the circular area covered by each node is approximated within two bounds: (i) by the smallest enclosing square providing an upper approximation, and (ii) by the largest inscribing square providing a lower approximation. For these cases, simpler algorithms have been proposed to estimate the covered area. It is evident that in the first case the algorithm produces an overestimate of the area covered, whereas, by the second approach an underestimate is achieved always. However, it is interesting to study the variation of areas for several random deployment of nodes by varying the number of nodes and the radius. Simulation experiments show that in comparison with an earlier work [15], the proposed algorithms are capable of estimating the coverage area with no more than 10% error. Moreover, for bounded areas, in over-deployed networks, it may provide an almost exact result.

The rest of the paper is organized as follows: Section 2 presents the problem formulation. Section 3 describes the algorithms for finding the intersection points and the area covered by squares. Section 4 shows the simulation results and finally Section 5 concludes the paper.

2 Problem Formulation

Let a set of n nodes $\mathcal{S} = \{s_1, s_2, \ldots, s_n\}$ be deployed randomly over a 2-D region $\mathcal{A}$. Each node covers a circular area with radius r. The problem is to decide whether the region $\mathcal{A}$ is fully covered by the nodes. Note that a computation considering real circles in Euclidean geometry is rather complex. To alleviate this problem, we approximate a circular region as i) the largest square inscribed within it called *inside square*, and ii) as the smallest square enclosing the circle, referred to as *outside square*. With this model, instead of real circular area πr^2 we estimate the covered area in terms of inside squares with area $2r^2$ and outside squares with area $4r^2$ as shown in Fig. 1 and Fig. 2 respectively. Therefore, the problem of measuring the area covered by a set of circles now reduces to the problem of finding the area covered by a set of squares distributed randomly on a 2D plane. To solve this problem, firstly the intersection points among the squares are to be identified. Then the covered area can be represented in terms of some monotone isothetic objects [4], and hence can be computed in linear time only. It is evident that the measured covered area is always underestimated in case-i) and overestimated in case-ii) respectively. However given any random deployment of nodes due to arbitrary overlapping of covered areas it is interesting to study the dependence of the deviations of the estimated areas from the exact covered area with number of nodes and the radius respectively.

The following section describes the details of the area estimation procedure and the respective algorithms.

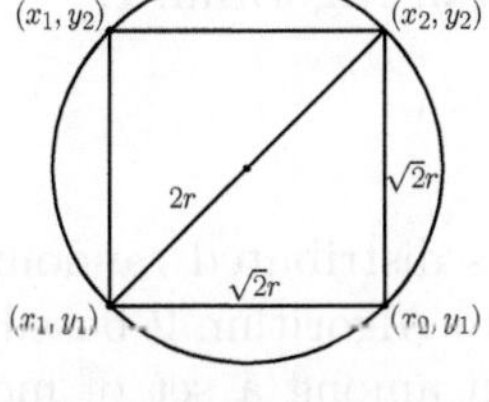

Fig. 1. Maximum square inscribed within a circle

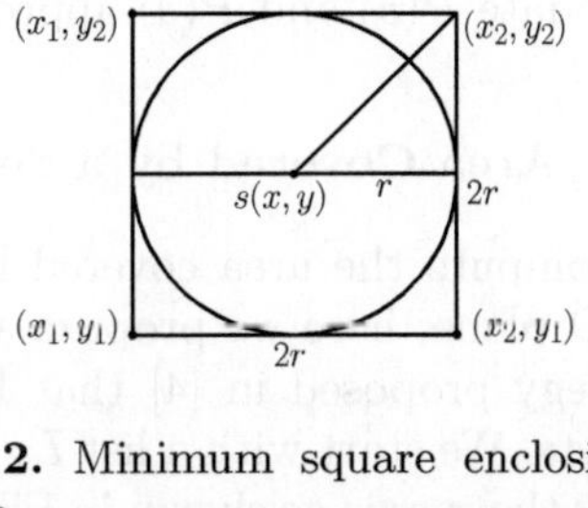

Fig. 2. Minimum square enclosing a circle

3 Area Coverage by Squares

3.1 Intersection of Two Squares

We assume that each square s is defined as a quadruple $s = (x_1, x_2, y_1, y_2)$ where $(x_1, y_1), (x_1, y_2), (x_2, y_1), (x_2, y_2)$ are the bottom-left, top-left, bottom-right and top-right corner points of s respectively, and all co-ordinates are integers.

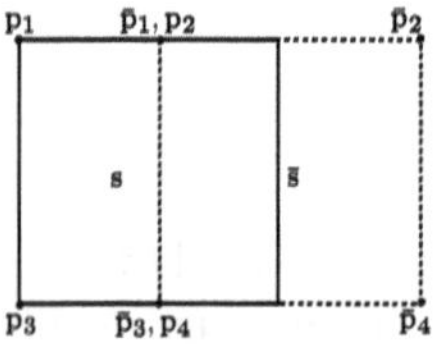

Fig. 3. $x_1 \leq \bar{x}_1$, $y_1 = \bar{y_1}$

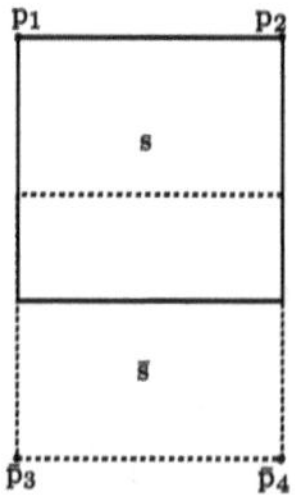

Fig. 4. $x_1 = \bar{x_1}$, $y_1 > \bar{y_1}$

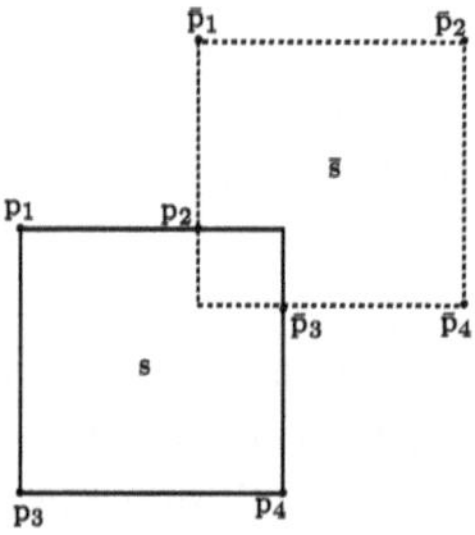

Fig. 5. $x_1 \leq \bar{x}_1$, $y_1 < \bar{y_1}$

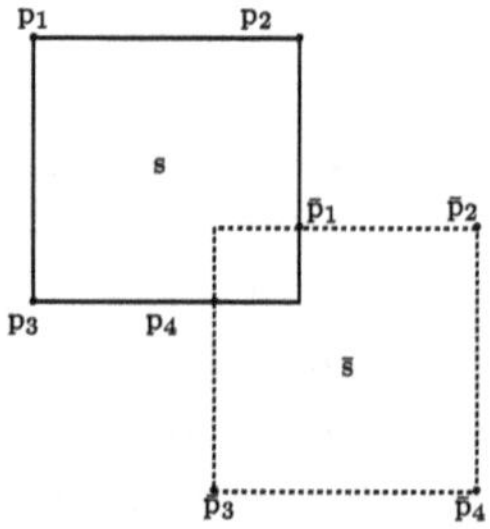

Fig. 6. $x_1 \leq \bar{x}_1$, $y_1 > \bar{y_1}$

Two squares $s = (x_1, x_2, y_1, y_2)$ and $\bar{s} = (\bar{x}_1, \bar{x}_2, \bar{y}_1, \bar{y}_2)$ are said to intersect if and only if the intervals

1) $[\bar{x}_1, \bar{x}_2]$ and $[x_1, x_2]$ overlap, and
2) $[\bar{y}_1, \bar{y}_2]$ and $[y_1, y_2]$ overlap,

where, $[x_1, x_2]$ and $[y_1, y_2]$ define the closed intervals given by the projection of s on the x-axis and the y-axis respectively. Also, each square s maintains a list of four integers, $P(s) : \{p_1, p_2, p_3, p_4\}$. Initially, $p_1 = x_1, p_2 = x_2, p_3 = x_1$ and $p_4 = x_2$ respectively. To find the pair of intersection points between the squares s and $\bar{s}$, without loss of generality, we assume that $x_1 \leq \bar{x}_1$. If the two squares intersect, depending on the relative positions of s and $\bar{s}$ the values of $P(s)$ and $P(\bar{s})$ will change in four different ways as shown in Fig. 3, Fig. 4, Fig. 5, and Fig. 6.

Given two squares s and $\bar{s}$, the procedure to find the intersection points and to update $P(s)$ and $P(\bar{s})$ appropriately is presented in Algorithm 1.

3.2 Area Covered by a Set of Squares

To compute the area covered by a set of n squares distributed randomly over a 2-D plane, here we propose an iterative procedure Algorithm 2 based on the strategy proposed in [4] that finds the intersection among a set of monotone objects. We start with a list L_x of the projections (a_i, b_i) of the squares s_i sorted along the x-axis as shown in Fig. 7. Next we scan L_x and include s_i (if $L_x = a_i$), or delete s_i (if $L_x = b_i$) in L and compute intersections among the newly adjacent

Algorithm 1. Intersection(s,$\bar{s}$, $P(s)$, $P(\bar{s})$)

Input: $s = (x_1, x_2, y_1, y_2)$, $\bar{s} = (\bar{x}_1, \bar{x}_2, \bar{y}_1, \bar{y}_2)$, $P(s) : (p_1, p_2, p_3, p_4)$, $P(\bar{s}) : (\bar{p}_1, \bar{p}_2, \bar{p}_3, \bar{p}_4)$
Output: $P(s), P(\bar{s})$
if $y_1 == \bar{y}_1$ **then**
    if $p_1 \leq \bar{x}_1$ **and** $p_2 \geq \bar{x}_1$ **then** $p_2 = \bar{x}_1$;
    if $p_3 \leq \bar{x}_1$ **and** $p_4 \geq \bar{x}_1$ **then** $p_4 = \bar{x}_1$;
if $x_1 == \bar{x}_1$ **then**
    if $y_1 > \bar{y}_1$ **then**
        $\bar{p}_1 = \bar{p}_2 = p_3 = p_4 = null$;
    else
        $p_1 = p_2 = \bar{p}_3 = \bar{p}_4 = null$;
if $\bar{y}_1 > y_1$ **then**
    if $p_1 \leq \bar{x}_1$ **and** $p_2 \geq \bar{x}_1$ **then** $p_2 = \bar{x}_1$;
    if $\bar{p}_3 \leq x_2$ **and** $\bar{p}_4 \geq x_2$ **then** $\bar{p}_3 = x_2$;
else
    if $p_3 \leq \bar{x}_1$ **and** $p_4 \geq \bar{x}_1$ **then** $p_4 = \bar{x}_1$;
    if $\bar{p}_1 \leq x_2$ **and** $\bar{p}_2 \geq x_2$ **then** $\bar{p}_1 = x_2$;

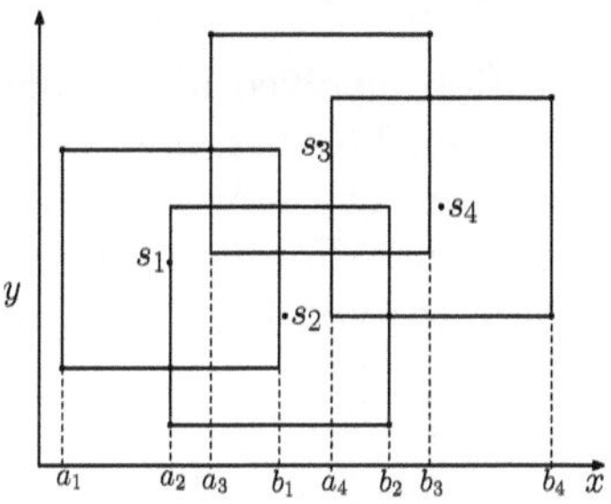

Fig. 7. Projections of squares for four nodes distributed over an area

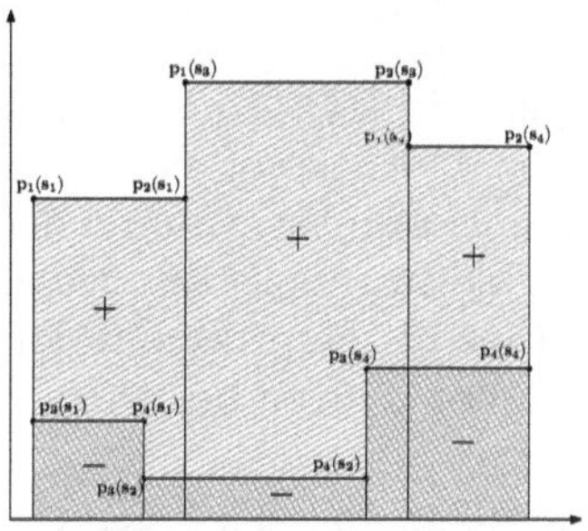

Fig. 8. Area computation by Algorithm 2 for the node distribution in Fig. 7

pairs of squares in L, and update the lists $P(s)$ of relevant squares by Algorithm 1 appropriately. Finally, given $P(s_i)$ for each square s_i, Algorithm 2 computes the area traversing along the closed intervals defined by $P(s_i)$'s appropriately. Fig. 8 shows the area computed by Algorithm 2 for the node distribution shown in Fig. 7.

Complexity Analysis: It is evident from Algorithm 1 that the intersection points between two squares can be computed in constant time. In Algorithm 2, to sort the given set of squares, $O(n \log n)$ time is required, and a linear traversal along the list $P(i)$'s to find the area will require $O(n)$ time. Hence the total complexity of area computation is $O(n \log n)$. Moreover, the procedure requires primitive arithmetic operations such as addition and multiplication of integers, with simple data structures only.

4 Simulation Studies

In order to study the error in the estimated area, we may compute it more accurately assuming that each real circle is approximated by a digital circle on

Algorithm 2. Area computation of a set of squares

Input: Squares $S := \{s_1, s_2, \ldots, s_n\}$
Output: Area total : A_{tot}
Step 1: for *each square* $s_i \in S$ **do**
  Compute a_i and b_i and include in L_x in sorted order along x-axis;
Step 2: for $i = 1$ *to* $2n$ **do**
  if $L_x(i) = a_j$ **then**
    include s_j in L in sorted order along y-axis ;
    By *Algorithm 1*, update $P(s)$ for the newly adjacent pairs of squares in L;
  if $L_x(i) = a_j$ *or* b_j *and* $s_j = L(k)$ *and if both* $L(k-1)$ *and* $L(k+1)$ *exist in* L **then**
    check if intersection points between squares of any pair $(L(k-1), L(k), L(k+1))$ is included within the third one.
  Update the $P(s)$ lists;
  if $L_x(i) = b_j$ **then** delete $L(k)$ from L;
Step 3: for *each square* s_i **do**
  if $p_1(s_i)$ or $p_2(s_i) \neq$ *null* **then** $A_{tot} \leftarrow A_{tot} + (|p_1(s_i) - p_2(s_i)| * (y_2(s_i)))$;
  if $p_3(s_i)$ or $p_4(s_i) \neq$ *null* **then** $A_{tot} \leftarrow A_{tot} - (|p_3(s_i) - p_4(s_i)| * (y_1(s_i)))$;

an integer grid [15]. To avoid the rigorous computation involved in finding the area covered by a set of real circles, the problem is mapped to digital circles [16]. With a large radius (i.e., on a dense grid), a digital circle can represent the real circle closely in respect of covered area as shown in Fig. 9. For completeness, a brief outline of the procedure for computing the area covered by a given set of digital circles is given below.

4.1 Area Covered by Digital Circles

The area covered by a set of n digital circles is basically an isothetic cover, and can be computed in terms of vertical strips, as shown in Fig. 9. To compute the covered area, an iterative procedure is proposed in [15]. In each iteration, for a pair of digital circles, following digital geometry based concepts, the intersection points, as shown in Fig. 10, are computed in constant time. Next, the covered area is represented in terms of a sequence of intersection points defining the boundary of that area. For each circle, a circular list of its intersection points

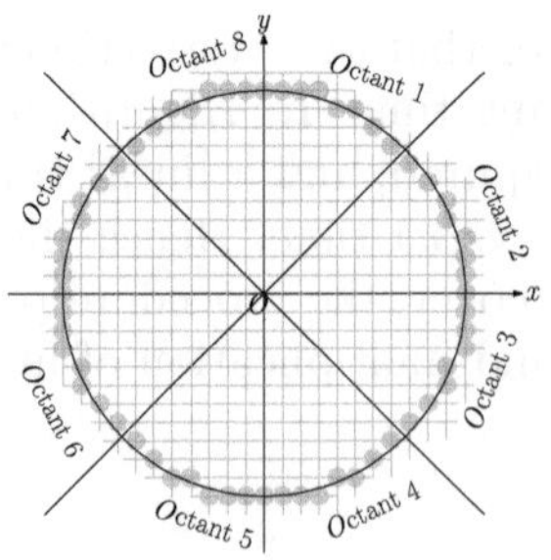

Fig. 9. A digital circle

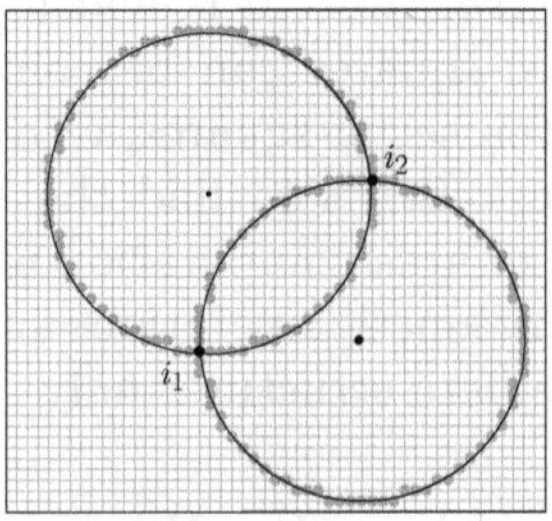

Fig. 10. Intersection between two digital circles

is maintained. The area of a digital circle is computed by traversing the list of intersection points in a cyclic order. In fact, any pair of intersection points i_1 and i_2 defines an arc of a digital circle. The area bounded by the appropriate arcs can be computed by traversing the vertical strips from i_1 to i_2, within the circle. This algorithm also runs in $O(n \log n)$ time; however, the algorithm proposed here using two square approximations has a smaller constant term in the asymptotic complexity, and hence, it needs much less CPU time and memory for practical problems as it involves fewer computations and simpler data structures.

4.2 Results and Discussions

In our simulation study, we assume that n nodes, $5 \leq n \leq 100$, are distributed randomly over a 100×100 grid area. Fig. 11 and Fig. 12 show how the area covered by inside squares, outside squares, digital circles and real circles increases with radius for a single node when the area to be monitored is bounded and unbounded respectively. Fig. 13 and Fig. 14 show how the covered area increases with n for inside, outside squares and digital circles in bounded and unbounded areas respectively. As expected, the outside square grid always overestimates the area and inside squares underestimate the area; the difference is always observed to lie within ± 10 % from that estimated by digital circles. Also for $n \geq 25$, all three shapes estimate the same area in bounded case. It also reveals the fact that the area estimated by outside squares is closer to the real area compared to that achieved by inside squares. Fig. 15 shows the variation of the estimated area with radius. It also shows that the outside square scheme is better compared to the inside square. From the simulation results, it is evident that for unbounded areas, this estimation strategy performs poorly as the radius or as the number of nodes grows. But, for all practical purposes the area to be monitored is a bounded one. Also, it is interesting to observe that for bounded areas, the estimated area can achieve exact results for sufficiently high radius or number of nodes.

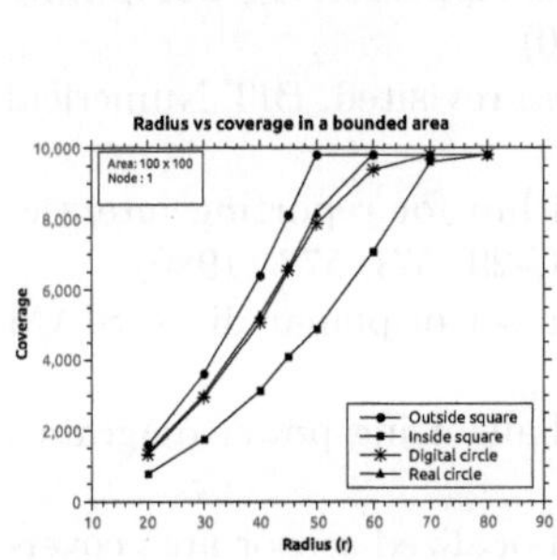

Fig. 11. Radius vs area coverage in a bounded area

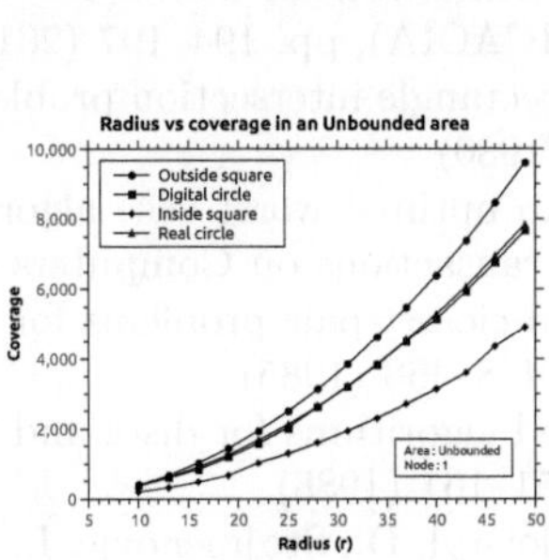

Fig. 12. Radius vs area coverage in an unbounded area

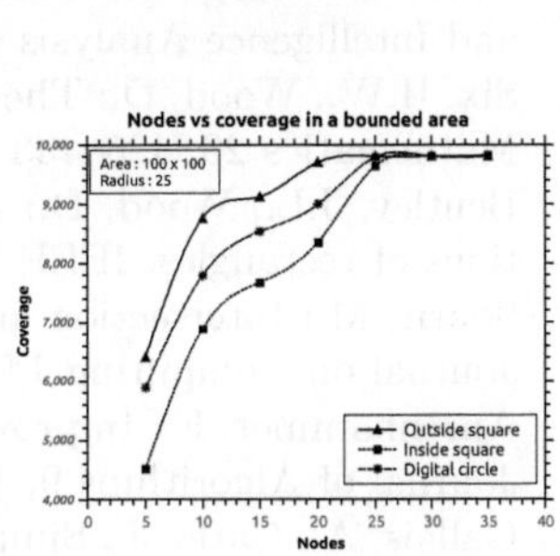

Fig. 13. Nodes vs area coverage in a bounded area

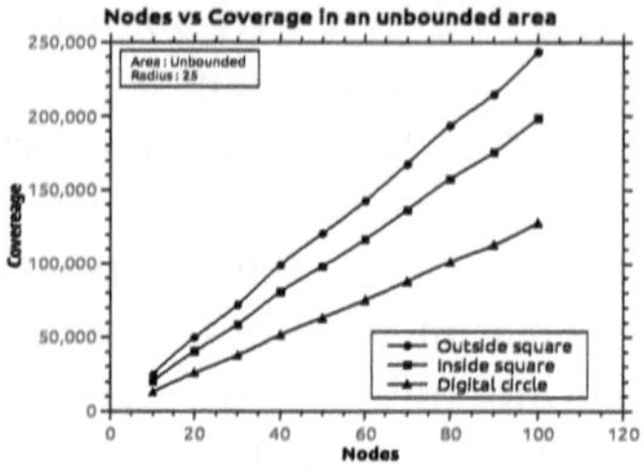

Fig. 14. Nodes vs area coverage in an unbounded area

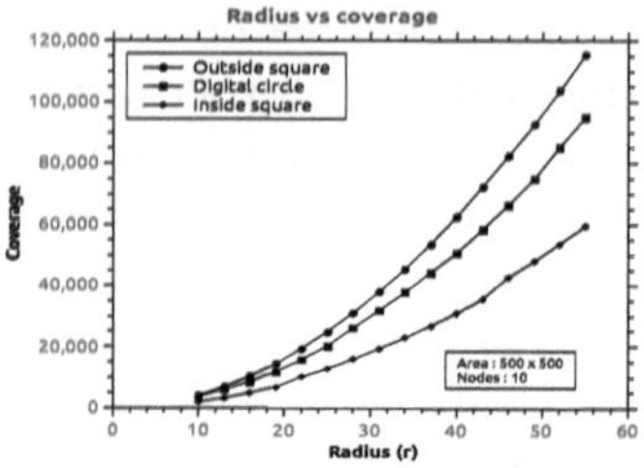

Fig. 15. Radius vs area coverage with multiple nodes

5 Conclusion

In this paper, we have addressed the problem of estimating the area covered by a set of nodes randomly deployed over a 2-D area. It is assumed that each node covers a circular area with a fixed radius r. An upper (lower) approximation of each circular area is provided by the smallest enclosing (largest inscribed) square. Our algorithm provides a fast and simple method of estimating a nearly-accurate area coverage, which has very small computational overhead. Therefore, it will be highly suitable for a low-energy pervasive environment, where dynamic coverage estimation is frequently needed. Experimental results indicate its potential in several promising application areas.

References

1. Luo, C.J., Tang, B., Zhou, M.T., Cao, Z.: Analysis of the wireless sensor networks efficient coverage. In: Proc. International Conference on Apperceiving Computing and Intelligence Analysis (ICACIA), pp. 194–197 (2010)
2. Six, H.W., Wood, D.: The rectangle intersection problem revisited. BIT Numerical Mathematics 20, 426–433 (1980)
3. Bentley, J.L., Wood, D.: An optimal worst case algorithm for reporting intersections of rectangles. IEEE Transactions on Computers C-29, 571–577 (1980)
4. Sharir, M.: Intersection and closest-pair problems for a set of planar discs. SIAM Journal on Computing 14, 448–468 (1985)
5. Aurenhammer, F.: Improved algorithms for discs and balls using power diagrams. Journal of Algorithms 9, 151–161 (1988)
6. Gallais, A., Carle, J., Simplot-ryl, D., Stojmenovic, I.: Localized sensor area coverage with low communication overhead. IEEE Transactions on Mobile Copmuting, 661–672 (2008)
7. Slijepcevic, S., Potkonjak, M.: Power efficient organization of wireless sensor networks. In: Proc. IEEE International Conference on Communications, vol. 2, pp. 472–476 (2001)
8. Pervin, N., Layek, D., Das, N.: Localized algorithm for connected set cover partitioning in wireless sensor networks. In: Proc. 1st International Conference on Parallel Distributed and Grid Computing (PDGC), pp. 229–234 (2010)

9. Huang, C.F., Tseng, Y.C.: The coverage problem in a wireless sensor network. In: Proc. 2nd ACM International Conference on Wireless Sensor Networks and Applications, pp. 115–121. ACM (2003)
10. Hsin, C., Liu, M.: Network coverage using low duty-cycled sensors: Random coordinated sleep algorithms. In: Proc. Third International Symposium on Information Processing in Sensor Networks, pp. 433–442 (2004)
11. Sheu, J.P., Yu, C.H., Tu, S.C.: A distributed protocol for query execution in sensor networks. In: Proc. IEEE Wireless Communications and Networking Conference, vol. 3, pp. 1824–1829 (2005)
12. Saha, D., Das, N.: Distributed area coverage by connected set cover partitioning in wireless sensor networks. In: Proc. First International Workshop on Sustainable Monitoring through Cyber-Physical Systems (SuMo-CPS). ICDCN, Mumbai (2013)
13. Saha, D., Das, N.: A fast fault tolerant partitioning algorithm for wireless sensor networks. In: Proc. Third International Conference on Advances in Computing and Information Technology (ACITY), vol. 3, pp. 227–237 (2013)
14. Ke, W., Liu, B., Tsai, M.: The critical-square-grid coverage problem in wireless sensor networks is NP-complete. Journal of Computer Networks, 2209–2220 (2010)
15. Saha, D., Das, N., Pal, S.: A digital-geometric approach for computing area coverage in wireless sensor networks. In: Natarajan, R. (ed.) ICDCIT 2014. LNCS, vol. 8337, pp. 134–145. Springer, Heidelberg (2014)
16. Bhowmick, P., Bhattacharya, B.B.: Number-theoretic interpretation and construction of a digital circle. Discrete Applied Mathematics 156, 2381–2399 (2008)

A Fault Tolerance Approach for Mobile Agents

Munshi Navid Anjum, Chandreyee Chowdhury, and Sarmistha Neogy

Department of Computer Science and Engineering
Jadavpur University, Kolkata, India
sarmisthaneogy@gmail.com

Abstract. Mobile agent is a program that can migrate autonomously from one environment to another. Many factors affect execution of mobile agents during its life cycle. Errors may occur on the server, or during communication. Error probability further increases with longer path. In mobile agent computing environment any component of the network - node, link, or agent may fail at any time, thus preventing them from continuing their execution. Therefore, fault-tolerance is a vital issue for the deployment of mobile agent systems. Here we propose a scheme to tolerate faults caused by malicious node behavior and link failure using agent cloning. The strategy is shown to prevent the agents from getting lost at irrational nodes (nodes that behave maliciously). The scheme is simulated using IBM Aglet platform and is found to be scalable when the no. of irrational nodes is fixed. Performance improves with more no. of agents.

1 Introduction

A mobile agent is a program that migrates from one environment to another, and is capable of performing appropriately in the new environment [1]. An agent consists of three components: the program which implements it, the execution state of the program and the data [2]. The owner (the node that spawns agents) can decide the route of the mobile agent or the agent itself can decide its next hop destination dynamically depending on context. The migration is similar to remote procedure call (RPC) [3] methods. For instance when a user directs an Internet browser to "visit" a website the browser merely downloads a copy of the site or one version of it in case of dynamic web sites. Similarly, a mobile agent accomplishes a migration attempt through data replication. When a mobile agent decides to migrate, it saves its own state (serialization may be used in case of Java based agents), transports this saved state to the new host, and resumes execution from the saved state. This is called weak migration.

But a mobile agent may also migrate in another way called strong migration [4]. A strong migration occurs when the mobile agent carries out its migration between different hosts while conserving its data, state and code. The platform is the environment of execution. The platform makes it possible to create mobile agents; it offers the necessary elements required by them to perform their tasks such as execution, migration towards other platforms and so on.

M.K. Kundu et al. (eds.), *Advanced Computing, Networking and Informatics - Volume 2*,
Smart Innovation, Systems and Technologies 28,
DOI: 10.1007/978-3-319-07350-7_4, © Springer International Publishing Switzerland 2014

Typical benefits of using mobile agents include [4], [5] reducing the network load, overcoming network latency, executing asynchronously and autonomously, etc.
Since mobile agents migrate from one node to the other collecting and/or spreading meaningful information, loss of a mobile agent results in more data loss as compared to loss of a message. The longer the trail an agent needs to visit, the chance of errors is more. Chances of migrating to a malicious node are even higher. In this paper, a fault tolerance scheme is proposed for the agents in order to prevent them from getting lost while en'route. The scheme uses the concept of cloning. The clone is a copy of a mobile agent, with no critical code and data. Measures are taken so that cloning does not become a significant performance overhead. Security issues are also considered.

The rest of the paper is outlined as follows: Section 2 discusses state of the art. Our work is introduced in Section 3. Implementation of the scheme using Aglet is discussed in Section 4. Finally Section 5 concludes.

2 Related Work

In mobile agent computing environment either the agent may fail due to software failure or the nodes or links in the underlying environment may fail. In either case the agents could not make successful migration. Therefore, fault-tolerance measures should be taken while deploying the mobile agents. Fault tolerance schemes for mobile agents to survive agent server crash failures are complex as execution of agent code at the remote hosts could not be controlled. There are [6], [7] several ideas presented to implement fault-tolerance in mobile agents. We are trying to summarize some of them below:

In witness agent, a failure detection and recovery mechanism is used named witness agent passing [8]. This is done by the virtue of inter agent communication. This communication can be done using two methods, one direct message and other indirect message. Direct messages are passed when (the system assumes that) the agent is at the last visited node. In other cases (when the agent is not present at immediately previous node) indirect messages are used. In this method there is a mailbox at each owner that keeps those unattended messages. This process needs lot of resources and with the increase in the traverse path, more witness agents need to be created, hence consuming even more resources.

CAMA frameworks support fault tolerance at application level. This schema handles faults by introducing three types of operations over the exception namely raise, check and wait [9]. These exceptions are handled through inter agent communication. The advantage of this approach is that the exception handling allows fast operation. It also allows elective error detection and recovery mechanism. Its drawback is that execution of the process can be stalled if any agent raises any exceptions and malfunctions.

Adaptive mobile agents can adapt themselves to the environment. The rules it follows are dependent on the current environment and the working also changes accordingly. Two or more adaptive mobile agents should communicate with each other to acquire the correct role suited for the environment [10]. The roles are also specified about access or restriction to a resource. This control strategy is called Role Based Access Control (RBAC) [11]. The advantage of this technique is that as the mobile agents already resides within the system the communication overhead for inter agent communication is less for that the time required to respond is less for a mobile agent. Increase of routing of adaptive mobile agents, node, link failure or topological changes may produce errors.

Transient errors can be detected before an agent start executing at a node. It is done by comparison of the states of the agent. This technique has the capability to detect and correct more than one error [12]. The time and space overhead are minimal. One issue might be there that if bit errors are not corrected by any duplicates this may block the process.

Unexpected faults may arise in unreliable networks. Then it is not a good approach to create fault tolerance mechanism for every one of them. A unique solution may be created an adaptive mechanism to deal with several types of faults. Chameleon is one of such mechanism for fault tolerance [13]. Flexibility of chameleon gives it its advantage in the case of unreliable networks. The disadvantage of this mechanism is that it becomes blocking if execution at any node fails.

Exactly once protocol guarantees fault less execution in case where the agent needs to be executed only once to yield a correct result, executing more than once may lead to errors. For the execution of each elementary part of an agent a set of resources are needed. The states of these resources are changed for every such part. Resource manager keeps track of these changes [13]. The disadvantage of this technique is that the agents underlying process become atomic; multiple commit or rollback operations increase the complexity.

3 Our Work

In this paper we propose a fault tolerance technique for the mobile agents that can protect the agents from getting lost while en'route. Here we define a mobile agent clone as the copy of the agent which has the same code and data as the original agent at some state. Here it is assumed that the clone is similar to its original copy and hence carries with it critical code and data when it migrates to the next node.

3.1 Problem Description

A node in the network having mobile agent platform spawns mobile agents if it needs to collect (spread) information from (to) N (<= total no. of nodes in the network) hosts in the network. The owner may send an agent with N nodes in its trail or may divide the job to several agents depending on agent performance and network conditions. After a mobile agent visits all the hosts mentioned in its trail and gets required services, it retracts back to the owner. During computation on the hosts, the mobile agent may crash due to hardware/software failures at the hosts or software error at the agent itself. Also the agent may get lost due to link failure or irrational behavior of the intermediate nodes. Any undesirable change of agent code can be detected if the agent's code is digitally signed as in [1].

So, our focus in this work is to protect mobile agents during its life cycle and to minimize the amount of information lost due to failure of an agent.

3.2 The Scheme

In this scheme, we define a mobile agent clone to be a replica of the original. So this clone protects original agent from getting lost in a network during migration and hence protects mobile agents from irrational nodes (hosts). Here a node deploys an agent and assigns its task for which the agent is asked to visit a number of nodes (the owner is interested in) according to some policy. Before migrating to an unknown host, an agent creates its replica. The cloned agent migrates to a number of unknown hosts before a counter (say, SKIP) expires. Here SKIP holds a value that signifies how many nodes (from its trail) the agent will traverse before creating another replica. The value of the counter can be fixed, decided by the underlying application or can be tuned according to agent performance. The variable SKIP is initialized to that value. SKIP is decremented by the cloned agent whenever the cloned agent migrates to a new node, and executes its task successfully. When the counter reaches 0, the cloned agent sends an acknowledgement back to the original agent that was kept at some previously visited host site (owner, for the first time SKIP decrements to 0). That agent upon receiving such acknowledgement kills itself. The agent then creates

```
AgentCode()
  1. Move to a remote host site according to the
     trail given by the owner
     1.1. To migrate to a remote host site a replica
          agent need to be created.
  2. If a reply from its replica is received then
     2.1. This replica kills itself.
     2.2. Return
  3. If time-out occurs then
   //reply from replica is not received in due time
     3.1. The agent retracts back to owner.
     3.2. Reports to the owner about missing reply
          from the previously visited host site.
  4. Decrement SKIP by 1.
  5. Execute its task.
  6. Verify signature of the code.
  7. If SKIP< 0 then
     7.1. Send a reply back to the node where a copy
          is kept/owner
     7.2. Create a replica in the present site.
     7.3. Reset SKIP to the default value.
  8. Otherwise
          Migrate to a new node according to its
          policy
```

another replica and keeps it in the current node and moves on to the next unvisited nodes from its trail. The value of SKIP is reset to the old value. This continues until the trail finishes. If there is no reply from the replica (due to irrational node behavior and agent getting lost) then the agent residing at a previous host site may retract back to owner directly. SKIP is decremented by one at the owner for the next agent that will visit the same trail. Thus more frequently agent cloning happens in order to cope with increasingly hostile/faulty network. Moreover, if an agent retracts back to owner successfully after visiting all the hosts it was asked to, then SKIP is incremented by 1 for the agent that will have the same trail.

In this scheme the replica is kept saved until a reply (that it has executed successfully) from the agent residing at some other node is received. On receipt of a successful reply the replica destroys itself. Let us take an example to describe the situation. Say agent X from node 1 (owner of X) has node nos. 2, 3, 4, 6, 7, 8 in its trail with SKIP=2. So clone X will send a reply to the original X in node 1 after it has reached node 3 whereby SKIP becomes 0. X now keeps itself here and creates a clone, resets SKIP to 2 and resumes its journey.

3.3 Our Scheme to Protect Mobile Agents

This scheme has a fault tolerance strategy. As the replica is kept in one or several nodes, there is a backup even when a link failure occurs or the agent is lost due to irrational nodes. If an agent comes back and reports about a missing reply (according to step 3.2 in AgentCode()), the owner may take necessary steps to inquire about that part of the network and also decrease the default value of SKIP by 1 so that the agents it spawns further may cope with increasingly noisy or hostile behavior of the network. Otherwise it may increase the default value of SKIP by 1 to signify stable environment.

As clones are not created at every node by default, the no. of replies is reduced. Hence the system works reasonably faster.

4 Results

The simulation is carried out in IBM Aglet platform [15]. It runs on Tahiti server. Since it deploys java based mobile agents, it is readily portable to any platform. We have installed Aglet on Linux (Fedora 16) operating system.

Here each node that an agent visits is designated as a port. For example the default port no. 4434 can work as an owner and spawn an agent. Ports can be created by the programmer. These ports are host platforms (nodes) forming a network. Any port can spawn any no. of agents.

Here each node that an agent visits is designated as a port. In our example we have shown an agent spawned by port 4434 visiting a no. of nodes (ports) according to its policy. Fig. 1 shows an example of our implementation on Aglet. The route of an agent is given in Fig. 1. Fig. 1(a) shows that an agent spawned by node 4434 sends an agent to node 9000, and executes its task. In port 9000, no agent copy is kept according to steps 6-7 of AgentCode() and forward to the next node that is port 9001. The counter strucks 0 at port 9001, it is found to be a trusted one as the signature of agent code is not modified and hence a replica is created and stored and the counter is reset as shown in Fig. 1(b). Then the agent visits port 9003 and finally returns to the owner after visiting all the nodes and collecting the relevant information which the owner displays (shown in Fig. 1(c)). This shows rough journey of an agent. It can be influenced by factors like link failure, malicious behavior of both the nodes that an agent is expected to visit and the attacks in-transit etc.

A series of experiments was carried out in Aglet. Some of the results are listed in Fig. 2. Two metric are introduced to measure the performance of the agent based system. One is ratio of successful agents and the other one is ratio of irrational nodes. The first one is defined as follows

$$Ratio\ of\ Successful\ Agents = \frac{No.of\ agent\ ssuccessfully\ returned}{No.of Agents\ Deployed} \quad (1)$$

The ratio of irrational nodes is defined by

$$Ratio\ of\ Irrational\ Nodes = \frac{No.of\ Irrational\ Nodes}{Total\ no\ .of\ Nodes} \quad (2)$$

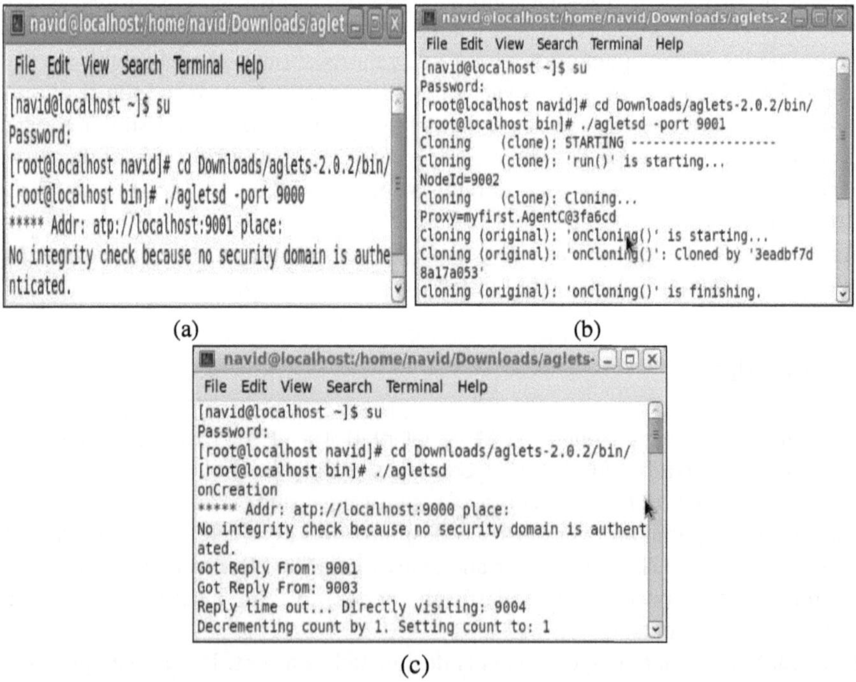

(c)

Fig. 1. (a) Aglet window showing that the host (port 9001) visited by an agent and there is no cloning;(b) Aglet window showing that the host (port 9002) visited by an agent and the agent cloned by itself;(c) Aglet window showing that the agent returns to its owner (port 4434)

In Fig. 2(a) agent success is measured in an increasingly hostile network. Here total no. of nodes in the network is taken to be 20. It can be observed that application performance with cloning (our scheme) is better than without cloning. When all nodes are irrational, it is obvious that no agents can perform its job. But even when 90% of the nodes behave irrationally, that is 18 out of 20 nodes are irrational then also the agents can show some progress with our scheme. However if no fault tolerance measure is applied to the agents, the performance drops to 0 when 60% of the nodes behave irrationally in a network.

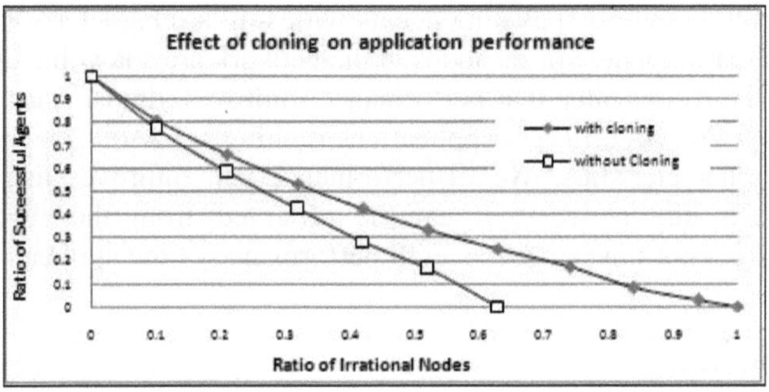

Fig. 2. **(a)** Graph showing that application performance with cloning and without cloning

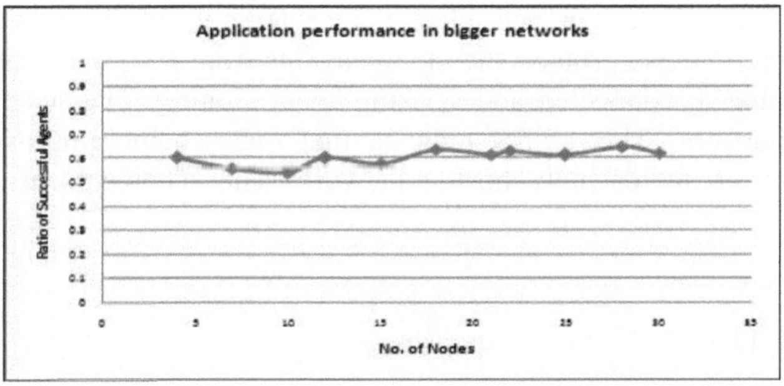

Fig. 3. **(b)** Graph showing that application performance when the no. of nodes are increased

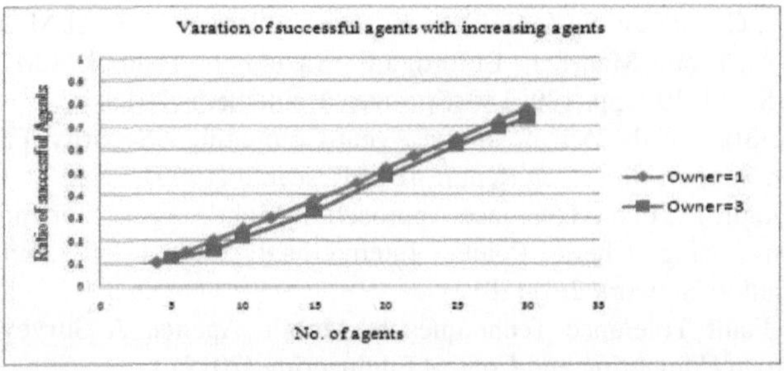

Fig. 4. **(c)** Graph showing application performance when the no. of agents are increased but no. of irrational nodes are fixed

In the next experiment application performance is measured in increasingly larger network. Here ratio of irrational nodes is taken to be 0.21. The graph shown in Fig. 2(b) indicates that application performance changes a little as the network grows in size. Thus the fault tolerance scheme is found to be scalable even when the underlying network is as big as having 30 nodes.

Finally in a network of 20 nodes where no. of irrational nodes is fixed to 5, agent performance is measured with increasing no. of agents. Here it is assumed that the extra agents do not pose significant bandwidth overhead and hence no rational node kills an agent when it is en'route. In Fig. 2(c) the blue line indicates a single agent group that is only a node is spawning agents with similar characteristics. The red one indicates the situation where three nodes in the network are spawning agents for some purpose. The figure indicates that performance improves almost linearly with no. of agents in the system. As different applications running at various nodes spawn agents with differing characteristics, overall performance still improves linearly with total no. of agents in the network. It can be observed from the figure that agent heterogeneity does not much affect overall performance of the applications.

5 Conclusion

In this paper, we propose a fault tolerance scheme for mobile agents working in hostile networks where an agent halts due to the nodes, links, or agent software failure in the network. The fault tolerance scheme presented not only ensures minimum data loss upon failure but also protects the mobile agents from getting lost at the malicious hosts (nodes) and disconnected nodes, using agent cloning. The scheme is tested in IBM Aglet platform. The results indicate that our scheme improves application performance in an increasingly hostile network. The scheme is also found to be scalable.

References

1. Anjum, M.N., Chowdhury, C., Neogy, S.: Securing Network using Mobile Agents. In: Proceedings of 2012 International Conference on Communication, Devices and Intelligent System (CODIS), pp. 266–269 (2012)
2. Chowdhury, C., Neogy, S.: Mobile Agent Security Based on Trust Model in MANET. In: Abraham, A., Lloret Mauri, J., Buford, J.F., Suzuki, J., Thampi, S.M. (eds.) ACC 2011, Part I. CCIS, vol. 190, pp. 129–140. Springer, Heidelberg (2011)
3. Pleisch, S.: State of the Art of Mobile Agent Computing - Security, Fault Tolerance, and Transaction Support. Research Report, IBM Research (1999)
4. Hans, R., Kaur, R.: Fault Tolerance Approach in Mobile Agents for Information Retrieval Applications using Check Points. International Journal of Computer Science & Communication Network 2(3) (2012)
5. Hans, R.: Fault Tolerance Techniques In Mobile Agents: A Survey. In: International Conference on Computing and Control Engineering (2012)

6. Rostami, A., Rashidi, H., Zahraie, M.S.: Fault Tolerance Mobile Agent System Using Witness Agent in 2-Dimensional Mesh Network. International Journal of Computer Science Issue 7(5), 153–158 (2010)
7. Budi, A., Alexei, I., Alexander, R.: On using the CAMA framework for developing open mobile fault tolerant agent systems. In: Proceedings of 2006 International Workshop on Software Engineering for Large-Scale Multi-Agent Systems (2006)
8. Marikkannu, P., Jovin, J.J.A., Purusothaman, T.: Fault-Tolerant Adaptive Mobile Agent System using Dynamic Role based Access Control. International Journal of Computer Applications 20(2) (2011)
9. Ferraiolo, D.F., Kuhn, D.R., Chandramouli, R.: Role-based Access Control. Artech House, Inc. (2007)
10. Kumar, S.G.: Transient Fault Tolerance in Mobile Agent Based Computing. INFOCOMP Journal of Computer Scienc 4(4), 1–11 (2005)
11. Bagchi, S., Whisnant, K., Kalbarczyk, Z., Iyer, R.K.: Chameleon: Adaptive Fault Tolerance Using Reliable, Mobile Agents. In: Proceedings of 16th Symposium on Reliable Distributed Systems (1997)
12. Rothermel, K., Strasser, M.: A fault-Tolerant Protocol for Providing the Exactly-Once Property of Mobile Agents. In: Proceedings ofof 17th IEEE Symposium on Reliable Distributed Systems (1998)
13. Lange, D.B., Oshima, M.: Programming and Developing Java Mobile Agents with Aglets. Addison Wesley (1998)

6. Rostami, A., Rashidi, H., Zahraie, M.S.: Fault Tolerance Mobile Agent System Using Witness Agent in 2-Dimensional Mesh Network. International Journal of Computer Science Issues 7(5), 153–158 (2010)
7. Baig, A., Alagar, V., Alexander, P.: On using the CAMA framework for developing open mobile fault tolerant agent systems. In: Proceedings of 2006 International Workshop on Software Engineering for Large-Scale Multi-Agent Systems (2006)
8. [illegible], F., [illegible], A.A., Pathasarathy, [illegible]: Fault Tolerance Scheme for Mobile Agent [illegible] Dynamic Role based Access Control. International Journal of Computer Applications [illegible] (2011)
9. [illegible], [illegible], S.K., [illegible], P.: [illegible] Mobile Agents [illegible] (2003)
10. [illegible], L.F.G.: [illegible] Fault Tolerance in Mobile Agent Systems. [illegible] Computer Science [illegible] (2001)
11. [illegible], [illegible], Kalbarczyk, Z., Iyer, R.K.: Chameleon: Adaptive Fault Tolerance [illegible] Mobile Agents [illegible] International Symposium on Reliable Distributed Systems (1999)
12. Rothermel, K., Strasser, M.: A Fault-Tolerant Protocol for Providing the Exactly-Once Property of Mobile Agents. In: Proceedings of 17th IEEE Symposium on Reliable Distributed Systems (1998)
13. Lange, D.B., Oshima, M.: Programming and Deploying Java Mobile Agents with Aglets. Addison-Wesley (1998)

Path Discovery for Sinks Mobility in Obstacle Resisting WSNs

Prasenjit Chanak and Indrajit Banerjee

Department of Information Technology
Indian Institute of Engineering Science and Technology (IIEST)
Shibpur, Howrah-711103, India
prasenjit.chanak@gmail.com, ibenerjee@it.becs.ac.in

Abstract. Sink mobility achieves great success in network life time improvement in wireless sensor networks. In mobile sink movement, mobile sink moves through random path or optimized path in obstacle free area. Fixed or constrained path is used in obstacle resisting environment. In fixed or constrained path, sink move through a predefined path, which is designed by the end user. In these strategies, only fixed obstacles are avoided which are previously present in the monitoring area. These strategies are unable to avoid those obstacles which are randomly entered in the monitoring area. In this paper, we propose a quad-tree based online path detection strategy that detects any type of obstacles which enter within the network life time and design a shortest mobile sink movement path avoiding detected obstacles. The proposed scheme divides whole network into different small size region and detects region wise obstacle. On the other hand, proposed scheme constructs region wise shortest path for mobile sink data collection. Simulation results are presented to verify our proposed scheme.

Keywords: Wireless sensor networks, sink mobility; obstacle, quad tree, network life span, energy efficiency.

1 Introduction

Wireless sensor networks provide reliable monitoring from long distance without any human interference. The main requirements of these networks are highly fault tolerant, long life time and low–latency data transfers [1], [2]. The primary goal of this network is to gather relevant data from surrounding and transmits to base station (BS). Deployed sensor nodes are energy constraints, limited computation and low storage capacities. Therefore, energy efficient routing is a very challenging issue in network life time enhancements. Multi-hop communication achieves great success in energy efficient routing strategy design. Nowadays, sink mobility gives better performance in WSN life time enhancements. In mobile sink based data routing techniques, mobile sinks are moved in different paths and collect data from static sensor nodes. WSNs have enabled numerousadvanced monitoring and control application in environmental, biomedical, and numerous other applications.

M.K. Kundu et al. (eds.), *Advanced Computing, Networking and Informatics - Volume 2*, Smart Innovation, Systems and Technologies 28,
DOI: 10.1007/978-3-319-07350-7_5,

In rural application, different types of obstacles are present in the monitoring area [3]. These obstacles are like mountains, buildings etc. Obstacles prevent communication between the nodes. On the basis of movements, obstacles can be classified onto two categories: a) static obstacle and b) moving obstacle. The static obstacles are unable to move any other place i.e. static obstacles are fixed. Once static obstacles have been detected it cannot change position. Therefore, in presence of static obstacle, obstacle moves into a fixed and predefined path easily. On the other hand, moving obstacles are changed its location with respect to random time interval. In mobile sink movement strategy, current obstacle position detection is very important for optimal sink mobility path design. Due to presence of moving obstacle, optimal path construction is a challenging issue in WSNs. Some works have been done on sink movement in obstacle residing. These works only designed a fixed and constant path for sink movements on the basis of static obstacle position. In fixed and constant path movement, sink is unable to avoid moving obstacle and newly introduces obstacle in random time interval in the network monitoring area. In our proposed scheme, according to obstacle position and movement information sink makes its own movement strategy. The optimal sink movement increases network performance.

In mobile sink based data routing scheme, selected number of mobile sinks move in different locations of the network and collect data from static sensor nodes. Therefore, static nodes' energy loss is decreased and increase life time. In sink mobility based data routing scheme, path constriction for mobile sink movement is very important issue in WSN. Luo*et al.* [4] consider a WSN with a mobile base station. The moving base station repeatedly relocates to change the bottleneck nodes closer to the base station. In this technique various types of predetermined strategies are used to search base station movement path and data routing. Somasundara*et al.* [1] approach a cluster based data routing scheme. In this technique sensor nodes are arranged in different cluster. Cluster head collects data from cluster member nodes. Cluster head transmits data to mobile node when it passes by. On the other hand, Bi *et al.*[5] proposed an autonomous moving strategy for mobile sinks in data –gathering sensor networks. In this paper, authors consider a WSN with one mobile sink. Mobile sink moves proactively towards the node that has the highest residual energy in the network. When mobile sink reaches a new location, it broadcasts a message for sensor node data collection. Sensor nodes transmit data to mobile nodes by multi-hop communication.

On the basis of movement strategy, sink mobility can be classified into three categories: a) Random sink mobility, b) Optimized sink mobility, c) Fixed or constrained sink mobility.In random sink mobility strategy, mobile sink randomly moves in arbitrary length and direction paths and collects data from static sensor nodes. In random sink mobility strategy, sink speed is also arbitrary [6], [7], [8]. In optimized sink mobility strategy, mobile sink move through optimal path for data collection. The optimal movement path is designed on the basis of a particular network variable. The sink movement path continuously regulated to ensure optimal network performance [5], [9], [10]. In fixed or constrained sink mobility [11], mobile sink moves through a predefined path. According to obstacle position, end user designs a mobile sink movement path. Mobile sink follows this path and collects data from the static sensor nodes. If any new obstacle entered within the network monitoring area and prevents the sink movement, there have no strategy such to detect the new obstacle and design new sink movement path to avoid the new obstacle.

Based on the above, fixed sink mobility strategy is less effective compared to random and optimized sink mobility strategy. On the other hand, optimized sink mobility is better for sensor nodes' load distribution and network life time enhancement. In outdoor sensor network application, positions of the obstacle and the number of obstacles have been changed randomly. Due to random obstacle position change and number of obstacle variation, sink movement path have been changed randomly. Therefore, fixed sink mobility is not good for obstacle resistant environment.

In this paper, we intend a quad-tree based sink mobility scheme for obstacle resisting environment. The proposed scheme divided whole network into different small size regions. Mobile sinks collect data from static nodes according to region information. On the other hand, proposed scheme detects new and present obstacle position region wise. According to new obstacle position, mobile sink designs a new optimal movement path for data collection. The proposed scheme computes sink movement path in distributed manner. On the other hand, proposed scheme also distributes network load uniformly within the network.

The rest of the paper is organized as follows. Section 2 describes the network model and problem statement. In Section 3, we describe a new sink movement strategy followed by obstacle detection. In Section 4, we evaluate the proposed scheme through simulations and compare it to other sink mobility strategy and finally in Section 5 we conclude our paper.

2 Problem Statement

Assume that *n*numbers of static nodes are deployed in a$M \times M$ [m^2] field and *S* numbers of mobile sinks are deployed random position. Our goal is to identify present obstacle in the $M \times M$ [m^2] field and design optimal sink mobility path avoiding these obstacles. Each node v_i (where $1 \leq i \leq n$) must be covered by one mobile sink S_j (where $1 \leq j \leq n$). Proposed scheme must have following requirements:

1. Mobile sinks movement path must be optimized. Each mobile sink independently makes its movement decisions based only local information.
2. Sensor node detects small size obstacles and movement of the obstacle. According to obstacles' movement and position mobile sink avoid obstacles through optimal way.
3. Obstacle detection and optimized path detection are completely distributed.
4. Optimized sink movement paths constriction process should be efficient in terms of processing complexity and message exchange.
5. Static nodes' load should be distributed uniformly.
6. Mobile sink uniformly cover whole network.

3 Proposed Sinks Movement Scheme

In this section, we describe sink movement scheme in obstacle resisting environment. First, we describe quad-tree based network partition and obstacle detection scheme. Second, we present the sink movement scheme. Finally, we prove that the proposed scheme meets its requirements. Mobile sink follows three types of movement path for data collection within the network.

1. **Foreign trip movement path:** The foreign trip movement paths are used when sink move one level region to another level region.
2. **Internal trip movement path:** The internal trip movement paths are used when any sink move one region to another in same level region.
3. **Local data collection path:** The local data collection paths are used within the region for data collection.

3.1 Quad-Tree Based Network Partition and Obstacle Detection

The proposed scheme split-ups the whole network into δ size region. Each region is supervised by a central node C_i where, $i$is the highest level value. The C_i node of a particular region is connected with a C_{i-1} node of upper region. If δ decreases then node density ρ increases in the network. If ρ is increased, the proposed technique detects minimum size obstacles. The detail process of Quad-tree based logical region partitioning is discussed below.

1. Initially, BS selects a C node within the network. The x, y position of the C node is represented by N.XVAL and N.YVAL where $\mathrm{N.XVAL} = l/2 * level$ and $\mathrm{N.YVAL} = \omega/2 * level$.
2. The C node logically divides the whole network into four regions, NW region, SW region, NE region and SE region. These four regions are directly supervised by the C node. The C node reports about any obstacle information to the BS by single hop communication.
3. The $C_i$node farther selects four C nodes $\{C_{1NW}, C_{1SW}, C_{1NE}, C_{1SE}\}$ in the central position of each region. The C1NW node position in NW region is$\mathrm{N.XVAL_{1NW}} = \frac{\omega}{2*level}$, $\mathrm{N.YVAL_{1NW}} = \mathrm{N.YVAL} + \frac{l}{2*level}$. Similarly C_{1SW}, C_{1NE} and C_{1SE} nodes' positions are selected at the mid position of each region. The $\{C_{1NW}, C_{1SW}, C_{1NE}, C_{1SE}\}$ node farther divide each region into four sub regions NW region, SW region, NE region and SE region.
4. C_{1NW}, C_{1SW}, C_{1NE}, C_{1SE} central nodes are the child node of upper level C node. In similar way, whole network is divided into δ size region. Each small region is supervised by one central node C_i and three member nodes(v_i, v_j, v_k).
5. Each C_i node contents its own region's obstacle information and updates its parent central node $\{C_{i-1}\}$. BS collects obstacle information from the root node C.

Obstacles are detected region wise. Member nodes are detecting obstacles by using their line of communication. The detailed obstacle detection process is discussed below.

1. Initially, any node v_i and two member nodes are participating for obstacle detection process. Each node v_i and member nodes $\{v_j, v_k\} \in V$ are communicated to each other by single-hop communication and make a triangle. Within a region v_i, v_j, v_k are directly communicated to each other by sending the COMM_LIN message and on reply to COMM_LIVE message.
2. If nodes v_i, v_j, v_k complete all communication within a specific time period then nodes v_i, v_j, v_k make decision that no obstacle is present between them.
3. If neighbor nodes are not replying within specific period of time then communicating nodes decide that an obstacle is present between them. Similarly, all nodes send obstacle decision to local C_i node.
4. All static sensor nodes communicate to its two neighbor nodes and make a triangle. If triangle is completed with its neighbor nodes then nodes decide that no obstacle is present between them otherwise an obstacle is found between them and triangle becomes incomplete.
5. When local C_i node gets obstacle information from regional static nodes, it informs to the upper level C_{i-1} node. Recursively every lower level C_i node transmits obstacle information to upper level C_{i-1} node.
6. If obstacle present between them then nodes find out all boundary nodes of the obstacle through shortest cyclic path identification process.

3.2 Optimized Sink Movement

In this section, we describe optimal sink movement strategy. Mobile sink move into different regions and collect data from static sensor nodes. Sink movement path must be optimized in such a way that static nodes' energy loss is minimum. On the other head, mobile sink collect all static nodes' data in every region. When a mobile sink reach into a region, first communicate with local regional central node (C_i). Local central node gives the obstacle information and also gives information regarding the visit of any other mobile sink. Each mobile sink move autonomously, without following a predetermined trajectory. The detailed process of sink movement strategy is discussed below:

1. Sink start movement at random position. When sink start movement, first communicate to nearest central node (C_i) for region information collection. Central node provides obstacle information and visited sink information.
2. If any sink is presently not visiting current region, nearest mobile sink starts to collect data from present region. On the other hand, before starting data collection from static sensor node mobile sink also verify whether any obstacle is present within the region or not.
3. If any obstacle is not present in current region, mobile sink selects a central position within the region for data collection from the static nodes. The central position selects in such a way that the transmission energy loss of the all static sensor nodes within the region is minimum.
4. When mobile sink start to visit a region, C_i set visiting states of that region is "*off*" and the visiting count is "*i*" where $r \leq j \geq 0$ and r is the number of rounds.
5. When mobile sink reach data collection point within the region, first broadcast a advertise message (*ADV_CALL*) for data collection process.

6. When static sensor nodes receive *ADV_CALL* message from mobile sink then transmit a confirmation message (*ADV_CON*) for data transmission.
7. After receiving *ADD_CON* message, mobile sink provide a data transmission time slot to each static sensor node.
8. Each static sensor node transmits data to nearest mobile sink by a given time slot.
9. When mobile sink complete data collection process within the region, mobile sink communicate to C_i node and inform that data collection process of the current region is over and give nearest region information.
10. Local C_i nodes also set visiting states of the current region is *"on"* and increase visiting count is *"j+1"*.
11. After data collection, mobile sink move to other region data collection point and collect data from static sensor nodes in same way.
12. If any region within the local C_i is already visited then $C_i$communicates to C_{i-1} central node.
13. C_{i-1} central node checks another remaining unvisited C_i node and allocates unvisited C_i then requests mobile sink.
14. If C_{i-1} nodes check that all regions have already been visited then communicate with C_{i-2} node. On that condition C_{i-2} find unvisited C_{i-1} similarly this C_{i-1} also find unvisited C_i node and allocate unvisited region.

3.3 Multiple Mobile Sinks Movement Strategy

In multiple mobile sink movement, C_i manages mobile sink in such a way that one region is visited by one mobile sink at a time. On the other hand, upper level C_{i-1} nodes also concurrently manage multiple mobile sink movement strategy. When a mobile sink moves one C_i node to another C_i node then C_i communicates to C_{i-1} node. The C_{i-1} node manages mobile sink in such way that only one mobile sink is monitored by only one lower level C_i nodes. Fig. 1 shows multiple sinks' movement in obstacle free environment. Detailed multiple sinks' movement is given below:

1. In multiple sink movement strategy, under a single C_i maximum four mobile sink collect data from different regions.
2. In Multiple mobile sink movement, when C_i node first time allocates a region then also communicates with C_{i-1}, C_{i-2}.....C nodes.
3. Upper levels C always try to allocate a single sink to a single C_{i+1} node. If number of mobile sink is maximum then Upper level C node allocates maximum four mobile sink in each C_{i+1}.
4. When any mobile sink completes a region then this mobile sink visits another new region which is allocated by the C_i and C_{i-1}, C_{i-2} ...C nodes.

3.4 Sink Movement Strategy in Static Obstacle Resisting Environment

In static obstacle resisting environment, obstacles' positions are fixed i.e. one obstacle has been detected; obstacles are unable to change its position. In static obstacle resisting environment, mobile sink visits each region and overcome obstacle boundary by minimum distance. The detailed process of sink movement strategy in static obstacle resisting environment is discussed below:

1. If any obstacle presents within the region then C_i node informs mobile sink before visiting the region.
2. When any mobile sink starts data collection within the obstacle region, first checks shortest boundary root to overcome the detected obstacle.
3. In obstacle resisting environment, mobile sink selects four data collection points within the region.
4. Four data collection points are selected in such a way that all static nodes transmit data to mobile sink with minimum energy loss.
5. When mobile sink visits four points within the region then all static nodes transmit data to mobile sink which blocked by the any regular or irregular obstacle.
6. When sink visit four data collect points within the region then present obstacles are overcome through shortest path.

Fig. 2 shows multiple sinks' movement in static obstacles resisting environment. In static obstacle resisting environment mobile sink overcome presents obstacle and collect data from each static sensor nodes.

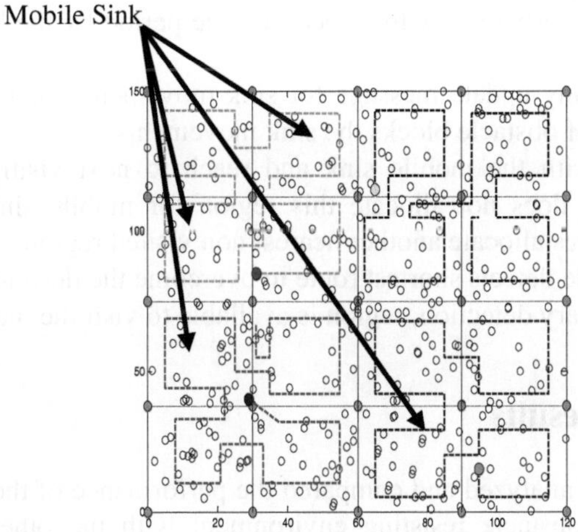

Fig. 1. Multiple mobile sink movement strategy in obstacle free environment

3.5 Sink Movement Strategy in Moving Obstacle Resisting Environment

In moving obstacle resisting environment, obstacle changes its position with respect to time. Therefore, sink also update its movement path with respect to obstacle movement strategy. The detailed process of sink movement strategy in moving obstacle resisting environment is discussed below:

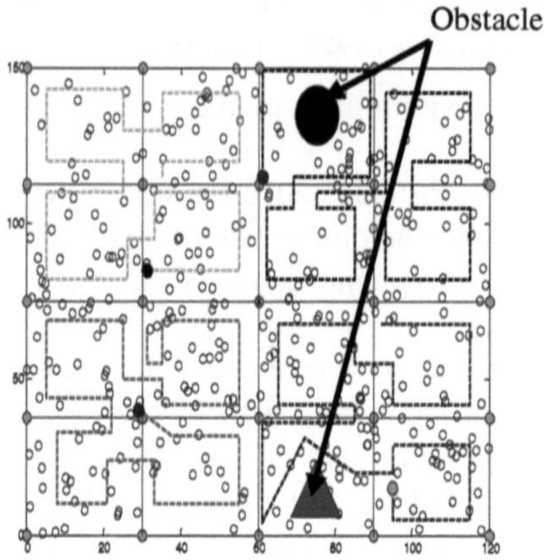

Fig. 2. Multiple mobile sink movement strategy in static obstacle resistant environment

1. If obstacle moves from one region to another region with mobile sink within the same C_i nodes, then mobile sink overcome the obstacle if this obstacle is blocked the sink movement's path.
2. C_i updates obstacle information with a specific time period by the boundary nodes of the obstacle.
3. When C_i nodes get any update message for sink movements then sink change its movement direction if obstacle blocks the sink movements.
4. If obstacle moves with the mobile sink and reach to next visiting region with mobile sink then C_i does not allocate this region for mobile sink visit. In that condition central nodes allocate another nearest non visited region.
5. After that central node detects shortest route to overcome the detected obstacle.
6. After shortest boundary detection, region is available to visit the mobile sink.

4 Simulation Results

In this section, we have analyzed and compared the performance of the proposed sink movement scheme in obstacle resisting environment with the other existing sink movement techniques. In the simulation environments, sensor nodes are randomly placed in a 120 ×120 [m^2] rectangular monitoring area. Deployed mobile sink move with speed of 5 m/s, on the other hand, we also consider that mobile obstacle moves with speed of 3m/s. Energy loss for data sense, receive and transmit is computed as follows:

$$E_t(\beta, d) = \begin{cases} \beta(e_b + e_{fs} d^2) & \text{if } d \le d_0 \\ \beta(e_b + e_{mp} d^4) & \text{if } d > d_0 \end{cases}$$

where, e_{fs} [$J/bit/m^2$] represents the energy dissipated by the amplifier circuit in free space. The multi-path fading channel e_{mp} [$J/bit/m^2$] model is used for the transmitter amplifier. The short distance is defined as $d_0 = \sqrt{e_{fs}/e_{mp}}$ and β is the number of bits transmitted. The energy dissipated by the receiver circuit for receiving β number of bits is represented byE_r (β)

$$E_r(\beta) = \beta e_b$$

We observe the following metrics to evaluate the performance of our proposed sink movement scheme.

- **Total amount of data:** The total amount of data collected by the mobile sink in one round.
- **Total energy consumption:** summation of all sensor nodes' energy loss in one round.
- **Network life time:** Network life time is defined as the number of movement rounds of the mobile sink from the beginning of the data collection phase to the last i.e. nodes' energy exhaustion.

Fig. 3 shows network lifetime competition between proposed scheme, static sink and other sink movement schemes in obstacle free environment. Proposed scheme give better network life time compared to static sink, MASP, 6-positions and distributed schemes. In our proposed scheme, network life time increases due to proper utilization of static nodes energy. On the other hand, static node transmission energy loss is less compared to other techniques.

Fig. 4 shows average energy consumption comparison in obstacle free environment. In our proposed scheme average energy consumption is less compared to other sink mobility scheme. In our proposed scheme, mobile sink collect data from an optimal data collection point. Therefore, each static sensor node transmits data to mobile sink at a minimum energy loss. Similarly, due to minimum distance from static nodes to mobile sink, energy utilization rate is maximum in our proposed scheme.

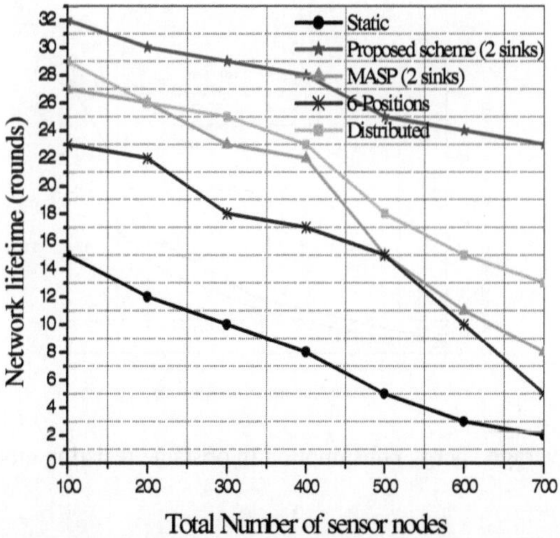

Fig. 3. Network life time in obstacle fee environment

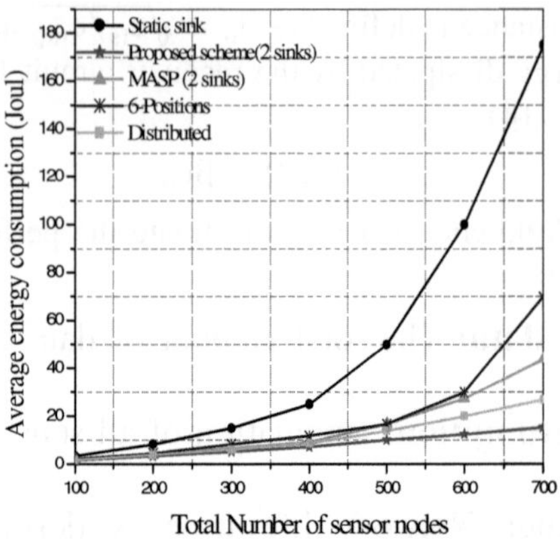

Fig. 4. Average energy consumption in obstacle free environment

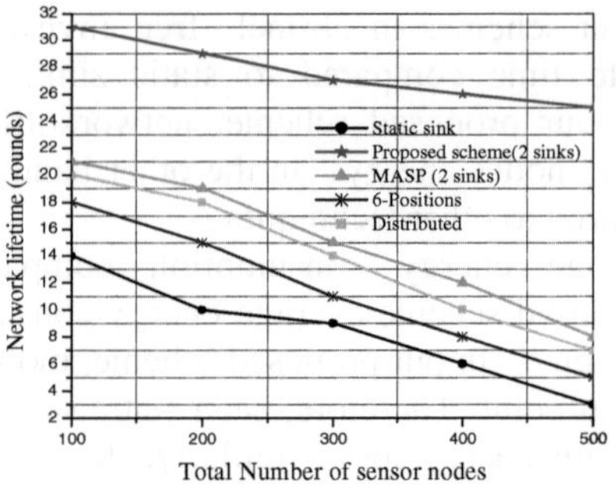

Fig. 5. Network lifetime in obstacle resisting environment

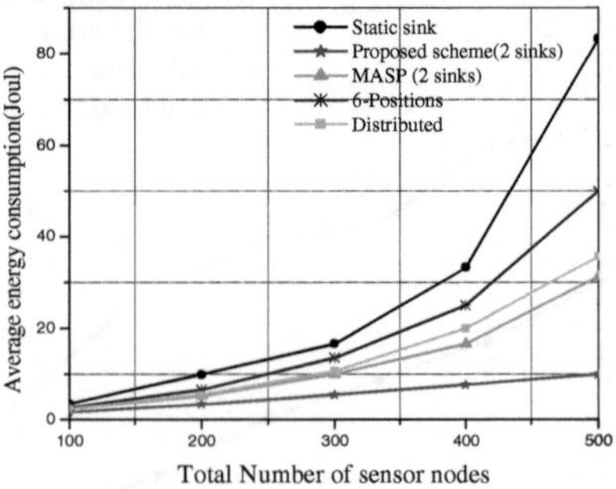

Fig. 6. Average energy consumption in obstacle resisting environment

Fig. 5 illustrates the network lifetime comparison between our proposed scheme and other mobile sink based data collection schemes in obstacle resisting environment. In obstacle resisting environment, our proposed scheme gives better result compared to other mobile sink based data collection schemes. Our proposed scheme detects obstacles region wise and designs sink movement strategy with respect to obstacle information. On the other hand, other sink movement schemes are unable to avoid present obstacles within the monitoring environment.

Fig. 6 shows average energy consumption comparison between proposed scheme and other sink movement scheme in obstacle resisting environment. In our proposed scheme, energy consumption is less compared to other sink movement schemes. In our proposed scheme, message over head is less compared to other sink movement techniques.

5 Conclusions

In this paper we have proposed a quad tree based sink movement scheme in obstacle resisting environment. Proposed scheme divides the whole network into different regions. Each region is supervised by a single central node. Central node contains obstacle information and sinks visit information. According to obstacle position, sink move within the region and collect data from monitoring environment. Our proposed scheme avoids both static and mobile obstacle and collects data from the static nodes. Simulation results show significant improvement in network lifetime and energy consumption compared to other sink movement schemes.

Acknowledgement. This work is partially supported by Major research Project under University GrantsCommission (UGC), Govt. of India (F. No. 42-146/2013(SR)).

References

1. Somasundara, A.A., Kansal, A., Jea, D.D., Estrin, D., Srivastava, M.B.: Controllably mobile infrastructure for low energy embedded networks. IEEE Transaction on Mobile Computin 5(8), 958–973 (2006)
2. Chanak, P., Banerjee, I., Samanta, T., Rahaman, H.: FFMS: Fuzzy Based Fault Management Scheme in Wireless Sensor Networks. In: Mathew, J., Patra, P., Pradhan, D.K., Kuttyamma, A.J. (eds.) ICECCS 2012. CCIS, vol. 305, pp. 30–38. Springer, Heidelberg (2012)
3. Chanak, P., Samanta, T., Banerjee, I.: Obstacle Discovery and Localization Scheme for Wireless Sensor Network. In: Proceedingsof IEEE International Conference on Communication, Devices and Intelligent System (CODIS), pp. 262–265 (2012)
4. Luo, J., Hubaux, J.P.: Joint mobility and routing for lifetime elongation in wireless sensor networks. In: Proceedings of IEEE INFOCOM, pp. 1735–1746 (2005)
5. Bi, Y., Sun, L., Ma, J., Li, N., Khan, I.A., Chen, C.: HUMS: An autonomous moving strategy for mobile sinks in data-gathering sensor networks. In: Proceedings of EURASIP (2007)

6. Chatzigiannakis, I., Kinalis, A., Nikoletseas, S.: Efficient data propagation strategies in wireless sensor networks using a single mobile sink. Computer Communications 31(5), 896–914 (2008)
7. Cheng, Z., Perillo, M., Heinzelman, W.B.: General network lifetime and cost models for evaluating sensor network deployment strategies. IEEE Transactions on Mobile Computin 7(4), 484–497 (2008)
8. Akkaya, K., Younis, M.: Energy-aware routing to a mobile gateway in wireless sensor networks. In: IEEE Global Telecommunications Conference (GLOBECOM), pp. 16–21 (2004)
9. Somasundara, A., Ramamoorthy, A., Srivastava, M.: Mobile element scheduling with dynamic deadlines. IEEE Transactions on Mobile Computing 6(4), 395–410 (2007)
10. Vincze, Z., Vass, D., Vida, R., Vidacs, A., Telcs, A.: Adaptive sink mobility in eventdriven densely deployed wireless sensor networks. International Journal on Ad Hoc and Sensor Wireless Networks 3(2), 255–285 (2007)
11. Tacconi, D., Carreras, I., Miorandi, D., Casile, A., Chiti, F., Fantacci, R.: A system architecture supporting mobile applications in disconnected sensor networks. In: Proceedings of IEEE Globecom, pp. 833–837 (2007)

Alive Nodes Based Improved Low Energy Adaptive Clustering Hierarchy for Wireless Sensor Network

Ankit Thakkar and Ketan Kotecha

Institute of Technology, Nirma University
Ahmadabad, Gujarat, India
{ankit.thakkar,director.it}@nirmauni.ac.in

Abstract. Energy efficiency is one of the important issues in the Wireless Sensor Networks (WSN). In this paper, a decentralized Alive Nodes based Low Energy Adaptive Clustering Hierarchy (AL-LEACH) is presented, that considers number of alive nodes in the network to elect the cluster heads. Alive nodes are used to dynamically compute weights of random numbers. Random number is one of the important parameters to elect cluster heads for the Low Energy Adaptive Clustering Hierarchy (LEACH) protocol. Extensive simulations are carried out to compare our proposed approach AL-LEACH with Low Energy Adaptive Clustering Hierarchy (LEACH), Low energy adaptive clustering hierarchy with Deterministic Cluster-Head Selection (LDCHS) and Advanced LEACH routing protocol for wireless micro sensor networks (ALEACH). Simulation results show that AL-LEACH improves the network life time and number of packets received by Base Station (BS) through balanced energy consumption of the network.

Keywords: Improved LEACH, Alive nodes, Random number, Balanced energy consumption, Cluster head election, Energy efficient routing, Wireless sensor network.

1 Introduction

Energy efficiency is one of the important issues of WSN. To prolong lifetime of the network, long distance transmissions should be avoided. Also, number of transmissions should be reduced, as it affects network lifetime. Cluster based routing protocols provide both these functionalities to enhance network lifetime. Low Energy Adaptive Clustering Hierarchy (LEACH) [1] is one of the prominent cluster based protocols that works in rounds. Each round is divided into two phases: i. Cluster Setup Phase, and ii. Steady State Phase. Cluster Heads (CHs) are elected during Cluster Setup Phase. During this phase, a node assumes a random number between 0 and 1; and the node elects itself as CH, if the generated random number is less than $T(n)$, where $T(n)$ is given by Eq. 1.

$$T(n) = \begin{cases} \frac{p}{1 - p * (r \bmod \frac{1}{p})} & \text{if } n \in G, \\ 0 & \text{otherwise} \end{cases} \quad (1)$$

M.K. Kundu et al. (eds.), *Advanced Computing, Networking and Informatics - Volume 2*, Smart Innovation, Systems and Technologies 28,
DOI: 10.1007/978-3-319-07350-7_6, © Springer International Publishing Switzerland 2014

In Eq. 1, p is the desired percentage of the cluster heads during each round and it is known to the algorithm in advance; G is the set of nodes that had not been cluster head since last $\frac{1}{p}$ rounds, where p is given by Eq. 2.

$$p = \frac{k}{N} \tag{2}$$

Here, N is the total number of nodes and k is the number of cluster heads. Once CHs are elected, they inform their status to other nodes in the network. Non-cluster head nodes associate with one of the CHs, for which minimum communication energy is required. After receiving join messages, CHs prepare TDMA schedule and inform the member nodes. During steady state phase, member nodes transmit data as per the TDMA schedule. CHs perform data aggregation and send it to BS after receiving data from the member nodes.

Low energy adaptive clustering hierarchy with Deterministic Cluster-Head Selection (LDCHS) [2] improves LEACH algorithm by including the remaining energy level available in each node and is given by Eq. 3, where $E_{current}$ and E_{max} denote the current and initial energy of the nodes.

$$T(n) = \begin{cases} \frac{p}{1 - p * (r \bmod \frac{1}{p})} * \frac{E_{current}}{E_{max}} & \text{if } n \in G, \\ 0 & \text{otherwise} \end{cases} \tag{3}$$

Network stuck after certain number of rounds with this threshold value and hence, threshold value $T(n)$ is further modified as shown in Eq. 4.

$$T(n) = \begin{cases} \frac{p}{1-p * (r \bmod \frac{1}{p})} [\frac{E_{current}}{E_{max}} + (r_s \text{ div } \frac{1}{p})(1 - \frac{E_{current}}{E_{max}})] & \text{if } n \in G, \\ 0 & \text{otherwise} \end{cases} \tag{4}$$

where r_s denotes number of consecutive rounds for which the node is not elected as CH. Quadrature-LEACH(Q-LEACH) [3] divides the network area into quadrants and then uses Eq. 1 to elect cluster heads. Coverage based LEACH (CVLEACH) [4] uses Eq. 1 to elect the cluster heads along with the over hearing properties of the node to create non-overlapping cluster regions. Advanced LEACH routing protocol for wireless microsensor networks (ALEACH) [5] improves LEACH protocol by modifying threshold equation $T(n)$, which is given by Eq. 5.

$$T(n) = G_p + CS_p \tag{5}$$

where G_p and CS_p are the general probability and current state probability. General probability is given by Eq. 1 and current state probability is given by Eq. 6.

$$CS_p = \frac{E_{current}}{E_{nmax}} * \frac{k}{N} \tag{6}$$

where k is the desired number of cluster heads, N is the total number of nodes, $E_{current}$ and E_{nmax} denote the current and initial energy of a node. WALEACH [6] improves ALEACH protocol by assigning weight (importance factor) to CS_p

and G_p. WCVALEACH [7] improves WALEACH protocol by assigning weight factor to CS_p and G_p along with over hearing property of the nodes to create non-overlapping cluster regions.

These protocols always use fixed span for the random numbers when nodes participate to become cluster heads. This results into few or no cluster head during a particular round, as the dead nodes are increased in the network. This can be improved by assigning weight (importance factor) to the random numbers. This weight depends on the number of alive nodes in the network and it is computed dynamically for each round. Thus, contribution for this paper can be given as follows:

- Improvement area is identified for LEACH protocol and few of its decedents
- A new method to calculate weighted random numbers is proposed. These weighted random numbers are compared with $T(n)$ to elect CHs, where $T(n)$ is given by Eq. 1

The paper is organized as follows: In Section 2, alive nodes based improved LEACH protocol is proposed; Simulation parameters and result discussion is given in Section 3, and finally concluding remarks are given in Section 4.

2 AL-LEACH: Alive Nodes Based Improved LEACH

A node elects itself as CH, if the random number, *rnd*, is less than threshold value $T(n)$, given by Eq. 1,3–5; where *rnd* is the random number between 0 and 1. The protocols discussed in section 1, have tried to increase the probability of a node to become CH by increasing the threshold value $T(n)$. However, the proposed approach differs from these protocols in the sense that it tries to decrease the importance of the random number *rnd*. This is achieved by assigning weight factor to the generated random number, where weight is proportional to the number of alive nodes in the network for a given round.

The proposed approach is derived from LEACH protocol. Like LEACH, the proposed approach also runs in rounds. During Cluster Head Election phase, each node assumes a random number, *rnd*, between 0 and 1 and it is weighted as per the Eq. 7; where N is the total nodes in the network and *Dead* are the number of dead nodes during that round. A node becomes CH, if RND is less than $T(n)$, where $T(n)$ is given by Eq. 1.

$$RND = rnd * \frac{(N - Dead)}{N} \tag{7}$$

Eq. 7 reduces importance of *rnd*, as the number of dead nodes increase in the network, and thus it increases the probability of the nodes to become CH. The proposed approach works similar to LEACH till the death of first node.

3 Simulation Parameters and Result Discussion

3.1 Simulation Environment

Simulations are carried out in MATLAB and code for LEACH protocol is obtained from csr.bu.edu [8]. The parameters used for simulation are shown in Table 1. Also, first order radio energy model is used as given in [1]. Extensive simulations are carried out by varying the node density and initial energy.

Table 1. Parameters used for simulation

Parameter Name	Value
Node Deployment Area	100m X 100m
Number of Nodes (Excluding BS)	1) 50 2) 100 3) 200
Relative Position of BS	(50,50)
Initial Energy/Node (in Joules)	1) 0.25 2) 0.50 3) 0.75
Simulation Stopping Criteria	5000 Rounds
Transmitter Electronics (ETx_{elec}) Receiver Electronics (ERx_{elec}) ($ETx_{elec} = ERx_{elec} = E_{elec}$)	50 nJ/bit
Energy for Data Aggregation (EDA)	5 nJ/bit/message
Free Space (ϵ_{fs})	10 pJ/bit/m^2
Multi-path Fading (ϵ_{mp})	0.0013 pJ/bit/m^4
Packet Size	4000 bits
Percentage of Cluster Heads	5% [1]
Proposed Approach Compared with	1) LEACH [1] 2) LDCHS [2] 3) ALEACH [5]

3.2 Simulation Metrics

- **Network Lifetime:** Network lifetime is measured using three metrics First Node Dies (FND), Half of the Nodes Alive (HNA) and Last Node Dies (LND). LND refers to the time when 90% of the total nodes die [9].
- **Number of Packets:** It indicates number of packets received by BS from the network. Higher the number of packets received indicate lower die rate of the nodes and consumption of energy [10].
- **Convergence Indicator (CI):** It is given by Eq. 8 [9], where FND, HND and LND refers to time when First Node Dies, Half of the Nodes Die and 90% of the total nodes die. It is used to measure network convergence. Higher the value of CI, better is the balanced energy consumption of the network.

$$CI = \frac{\text{LND - HND}}{\text{HND - FND}} \tag{8}$$

Table 2. Network Lifetime, Convergence Indicator (CI) and Packets received by BS for varying node density and initial energy

Total Nodes	Energy (J/Node)	Protocol	FND (rounds)	HNA (rounds)	LND (rounds)	CI	Packets Received by BS [a]
50	0.25	LEACH	434	604	858	1.494	1643
		LDCHS	434	619	828	1.1297	1617
		ALEACH	453	651	844	0.975	1707
		AL-LEACH	434	621	961	**1.818**	**1865**
	0.5	LEACH	881	1252	1584	0.895	3373
		LDCHS	862	1208	1609	**1.159**	3253
		ALEACH	939	1264	1614	1.077	3359
		AL-LEACH	881	1251	1641	1.054	**3538**
	0.75	LEACH	1321	1848	2258	0.778	4878
		LDCHS	1306	1826	2247	0.810	4878
		ALEACH	1436	1890	2402	**1.128**	4986
		AL-LEACH	1321	1889	2483	1.046	**5092**
100	0.25	LEACH	397	580	645	0.355	2931
		LDCHS	394	589	697	0.554	2950
		ALEACH	419	585	706	**0.729**	2998
		AL-LEACH	397	596	722	0.633	**3259**
	0.5	LEACH	744	1187	1351	0.370	5806
		LDCHS	782	1168	1375	0.536	5879
		ALEACH	849	1190	1330	0.411	5908
		AL-LEACH	744	1181	1425	**0.558**	**6136**
	0.75	LEACH	1114	1760	1972	0.328	8749
		LDCHS	1145	1758	1997	0.390	8789
		ALEACH	1297	1772	2031	**0.545**	8884
		AL-LEACH	1114	1767	2123	**0.545**	**9291**
200	0.25	LEACH	399	591	682	0.474	5989
		LDCHS	409	592	670	0.426	5948
		ALEACH	374	589	684	0.442	5924
		AL-LEACH	399	587	724	**0.729**	**6399**
	0.5	LEACH	799	1171	1332	0.433	11963
		LDCHS	773	1173	1355	0.455	11901
		ALEACH	778	1167	1346	0.460	11866
		AL-LEACH	799	1177	1444	**0.706**	**13001**
	0.75	LEACH	1215	1754	1990	0.438	17757
		LDCHS	1212	1754	2008	0.469	17883
		ALEACH	1124	1767	2009	0.376	17667
		AL-LEACH	1215	1766	2145	**0.688**	**19117**

[a] Measured at the end of the Simulation.

Simulation results are given in Table 2. It can be seen from results that for CI metric, AL-LEACH is preferable as node density increases. The same is evident through Fig. 1-3 Also, it can be concluded from the results that packets received by BS are always more for the AL-LEACH compared to LEACH, LDCHS and ALEACH.

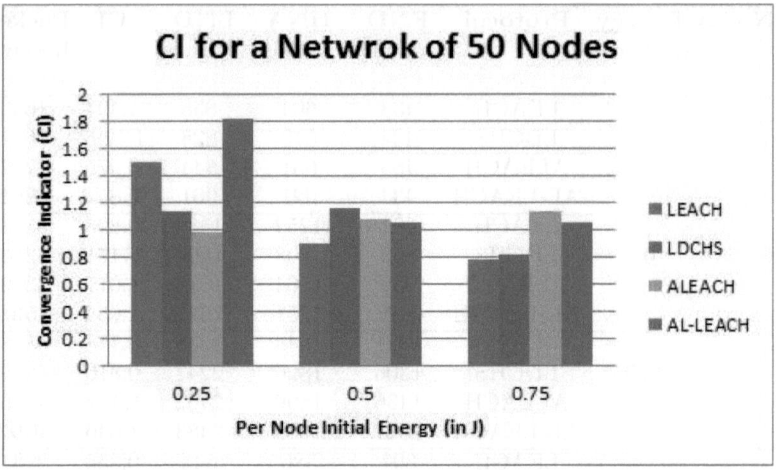

Fig. 1. Convergence Indicator for a network of 50 nodes

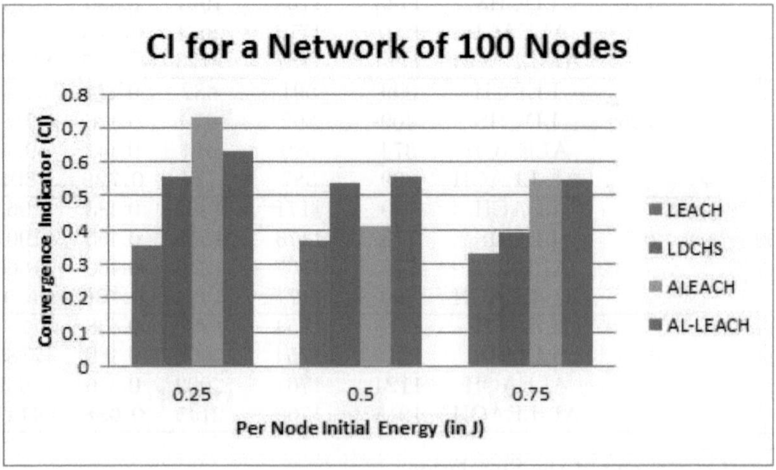

Fig. 2. Convergence Indicator for a network of 100 nodes

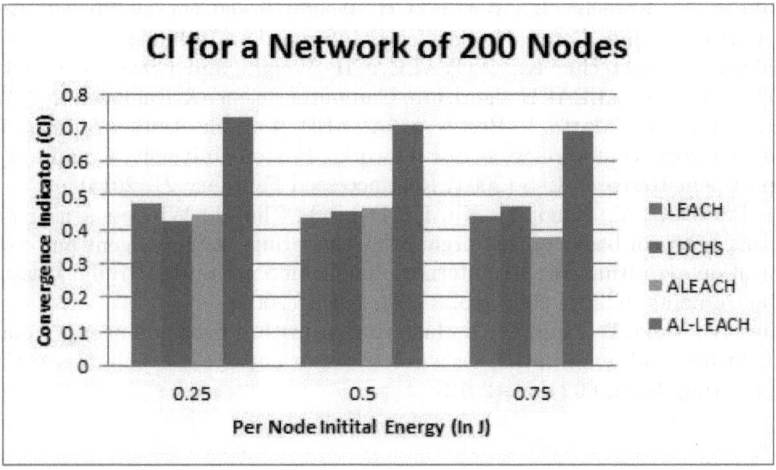

Fig. 3. Convergence Indicator for a network of 200 nodes

4 Conclusion

In this paper, improvement area is identified for LEACH, LDCHS and ALEACH protocols. Also, a solution is proposed in form of AL-LEACH protocol that assigns weights to the random number according to the number of alive nodes in the network. This weighted random number is compared with threshold $T(n)$ to elect cluster head. Simulation results demonstrate that AL-LEACH improves network lifetime. Also, more number of packets are received by BS through balanced energy consumption, which is determined by the Convergence Indicator (CI).

References

1. Heinzelman, W.R., Chandrakasan, A., Balakrishnan, H.: Energy-efficient communication protocol for wireless microsensor networks. In: Proceedings of the 33rd Annual Hawaii International Conference on System Sciences, p. 10. IEEE (2000)
2. Handy, M., Haase, M., Timmermann, D.: Low energy adaptive clustering hierarchy with deterministic cluster-head selection. In: 4th International Workshop on Mobile and Wireless Communications Network, pp. 368–372. IEEE (2002)
3. Manzoor, B., Javaid, N., Rehman, O., Akbar, M., Nadeem, Q., Iqbal, A., Ishfaq, M.: Q-LEACH: A New Routing Protocol for WSNs. arXiv preprint arXiv:1303.5240 (2013)
4. Thakkar, A., Kotecha, K.: CVLEACH: Coverage based energy efficient LEACH Algorithm. International Journal of Computer Science and Network (IJCSN) 1 (2012)
5. Ali, M.S., Dey, T., Biswas, R.: ALEACH: Advanced LEACH routing protocol for wireless microsensor networks. In: International Conference on Electrical and Computer Engineering, ICECE 2008, pp. 909–914. IEEE (2008)

6. Thakkar, A., Kotecha, K.: WALEACH: Weight based energy efficient Advanced LEACH algorithm. Computer Science & Information Technology (CS & IT) 2(4)
7. Thakkar, A., Kotecha, K.: WCVALEACH: Weight and Coverage based energy efficient Advanced LEACH algorithm. Computer Science & Engineering 2(6) (2012)
8. Smaragdakis, G., Matta, I., Bestavros, A.: SEP: A stable election protocol for clustered heterogeneous wireless sensor networks. Boston University Computer Science Department Research Lab (2004) (last accessed February 21, 2014)
9. Qiu, L., Wang, Y., Zhao, Y., Xu, D., Dan, Q., Zhu, J.: Wireless sensor networks routing protocol based on self-organizing clustering and intelligent ant colony optimization algorithm. In: 9th International Conference on Electronic Measurement & Instruments, ICEMI 2009, pp. 3–223. IEEE (2009)
10. Wang, A., Yang, D., Sun, D.: A clustering algorithm based on energy information and cluster heads expectation for wireless sensor networks. Computers & Electrical Engineering 38(3), 662–671 (2012)

Blackhole Attack Defending Trusted On-Demand Routing in Ad-Hoc Network

Swarnali Hazra and S.K. Setua

Computer Science and Engineering, University of Calcutta, India
{swarnali.hazra,sksetua}@gmail.com

Abstract. Ad-hoc networks are vulnerable to blackhole attack. Blackhole attacker drops every incoming legitimate packet to disrupt on-demand routing as well as data delivery in ad-hoc network. The attacker drops received route request packets instead of forwarding them pretending to have valid route to destination. As a result, all data from source will be delivered towards blackhole attacker. In this paper, we have proposed a trusted on-demand routing approach to defend blackhole attacker depending on our trust model with different levels of trust computations. In our approach, blackhole attackers are identified and isolated on context of data forwarding. Simulation and analysis justify our proposal against blackhole attack for on-demand routing in Ad-hoc Network. Simulation results analyses and justifies our trusted proposal against blackhole attack for on-demand routing in Ad-hoc Network.

Keywords: AODV, Blackhole attack, direct trust, indirect trust, trustor, trustee.

1 Introduction

Blackhole attacks are vulnerable to on-demand routing protocols in ad-hoc network [1]. When a source discovers communication route to destination using on-demand routing protocols, source broadcast RREQ (route request) routing packet. Every intermediate node re-sends the received RREQ until destination found. Destination sends back RREP (route reply) routing packet towards source. If an intermediate node has valid communication route to destination, on receiving RREQ, it sends back RREP towards source. Among all received RREP, source always consider highest sequence numbered RREP of shortest hop count for selecting data delivery path. Blackhole attacker drops RREQ instead of re-sending it towards destination and sends back a false RREP of much higher sequence number in comparison with other legitimate RREP. Consequently source becomes fool and considers blackhole forwarded higher sequence numbered RREP for selecting communication route to destination. Consequently, source sends all data packets towards blackhole attacker. On receiving source forwarded data packets via blackhole forwarded RREP traversed path, blackhole attacker drops all data packets. We have considered the popular on-demand routing protocol AODV [7] for this work. We have proposed Context Sensitive Trusted AODV against Blackhole attack (CST-AODV- Blackhole). Trust is the

M.K. Kundu et al. (eds.), *Advanced Computing, Networking and Informatics - Volume 2*, Smart Innovation, Systems and Technologies 28,
DOI: 10.1007/978-3-319-07350-7_7, © Springer International Publishing Switzerland 2014

measure of belief both in positive and negative sense. Positive believe is a measure of expectation fulfillment and negative belief comes from detraction of expectation in between entities. Trust evaluating entity is considered as current trustor (CT), and current trustee (TE) is being evaluated by CT. Our context based trust evaluation approach is the aggregative effect of direct and indirect trusts. Now the inclusion of TE in data delivery path depends on the computed trust value by CT.

Related works are discussed in section 2 and section 3 is focused on Blackhole attack in AODV. In section 4, Trust model is explained. In section 5, CST-AODV-Blackhole is discussed based on trust computation by underlying TOR model. Simulation results of our experiments are presented in section 6. Section 7 concludes the presented work.

2 Related Work

In ad-hoc network several secure routing approaches are proposed against blackhole attack. In [6], authors detect and prevent packet dropping by blackhole attacker with the use of Clusterhead Switch Routing Gateway Protocol. Here the detection and prevention mechanism is proposed on the basis of miss ratio. In [2], a counter algorithm called Receive Reply algorithm, has been proposed to identify blackhole node in ad-hoc network. Here identification process is based on the measure of difference between source sequence number and destination sequence number. In [3], blackhole attacker is identified and isolated based on interrogation of routing table entry about the destination node. Here the security measure is divided into local zone communication and inter zone communication. In [8], authors defined a threshold value with respect to considered constraints and compared destination sequence number with defined threshold value to identify blackhole attack. In [5], a blackhole attacker is blacklisted if the sequence number of forwarded RREP is higher than the defined threshold value. Here, threshold value is defined based on average measure of sequence numbers of all received RREP packets. In [4], suspected nodes are detected on the basis of packet receiving and received packet forwarding aspects. If the suspected nodes are RREP packet senders, then they are identified as blackhole attacker nodes.

3 Blackhole in AODV

Blackhole attacker drops all received legitimate packets. On receiving RREQ, Blackhole attacker drops RREQ instead of re-sending it and falsely sends back a much higher sequence numbered RREP pretending as if it has valid route to destination. AODV rely on the RREP with highest sequence number for selecting data delivery path. Blackhole attacker exploits this mechanism. In Fig. 1, a network part is considered including blackhole attacker. In Fig. 1. S broadcast RREQ to discover a communication route for target D. On receiving this RREQ, D sends back RREP to S via path node3→node1 → S. On the other hand, node2 has the valid route to D, and node2 sends back a RREP to S. In contrast of these two cases, BA sends back false RREP of much higher sequence number on receiving RREQ pretending to have valid route to D and drop that legitimate RREQ instead of re-sending it towards D. S receives RREP from D, from node2 and from BA. RREP from node2 and BA are of same hop count 1. But since S rely on highest sequence numbered RREP, S considers BA forwarded false RREP of much higher sequence number than legitimate nodes. When S sends data packets towards BA, BA drops all data packets.

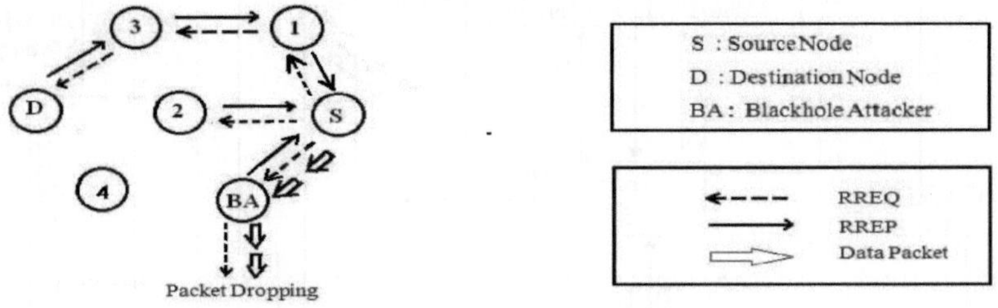

Fig. 1. Example network part including blackhole

4 Trust Model

We have structured Trust Model (Fig.2) with four separate modules like Node Manager, Trust Manager, Decision Manager and On-demand routing protocol Module (AODV for this work). Node Manager Module is responsible for receiving and transmitting routing packets, control packet and are also responsible for transmitting and receiving trust situation based packets respectively towards Trust Manager sub module and from Decision Manager sub module. Trust Manager Module receives trust situation based packets to compute the current trust value about trustee. Trust Manager computes trust value with the help of Trust Engine, Direct Trust Manager and Indirect Trust Manager. Before computing trust value, Event analyzer of trust module analyzes incoming events. Depending on analyzed events, context is analyzed by Context Analyzer of Trust Manager. Decision Manager takes the decision on basis of trust value that computed by Trust Manager. Decision Manager compares the computed trust value with uncertainty point and takes positive or negative decision about trustee. Uncertainty point is application dependent.

5 CST-AODV-Blackhole

If an intermediate node sends RREP pretending that it has valid route for destination, our proposed protocol considers RREP sender node as TE (current trustee). TE sends RREP to its reverse node which is consider as CT (current trustor). CT computes the trust value of TE with different level of trust computations. For indirect evaluation, CT broadcasts recommendation request packet (REQres) for 1-hop recommenders. Against REQres, CT receives recommendations about TE from recommenders.

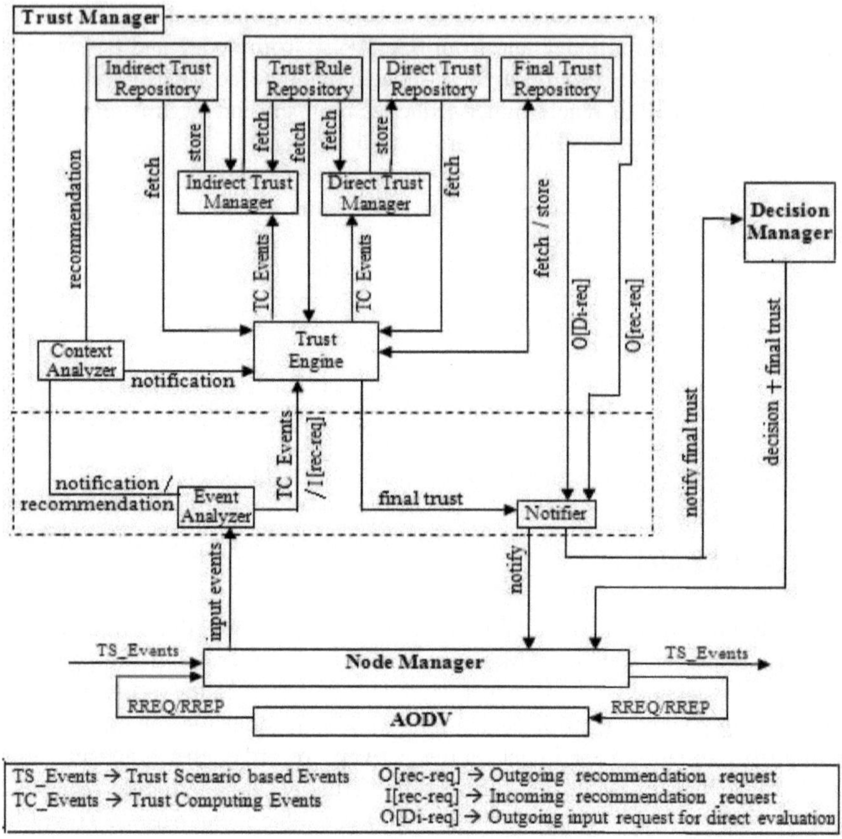

Fig. 2. Trust Model

On the other hand, CT evaluate TE directly on the observation of legitimate packets sending by TE. If TE is blackhole attacker, TE will drop all legitimate packets. To observe that TE is misbehaving or not, CT sends RREP response packet (RREPres) to TE for re-sending it to the next-hop node from TE to target destination (N_H-toD). If TE has valid route to destination, then there exists the N_H-toD. This N_H-toD may be an intermediate node or may be the target destination.

CT evaluates TE using promiscuous mode. Promiscuous mode gives the advantage to CT of listening and interpreting all the packets in its entirety. If TE re-sends CT forwarded RREPres to N_H-toD, CT can listen and interpret that TE re-send RREPres in promiscuous mode. Sequence of packet transfer in our approach is shown in Fig. 3. On the other hand, if TE is blackhole attacker, TE will not re-send RREPres to N_H-toD. As a result, CT cannot listening TE forwarded RREPres in promiscuous mode. We have considered context (C) and misbehavior (M) in our work as follows:

Context (C): This is the event of not listening by CT of RREPres that re-sends by TE (after receiving RREPres from CT) in promiscuous mode within requisite time.

Misbehavior (M): If TE does not re-send RREPres (received from CT) to N_HN-toD, TE's behavior is considered as misbehavior.

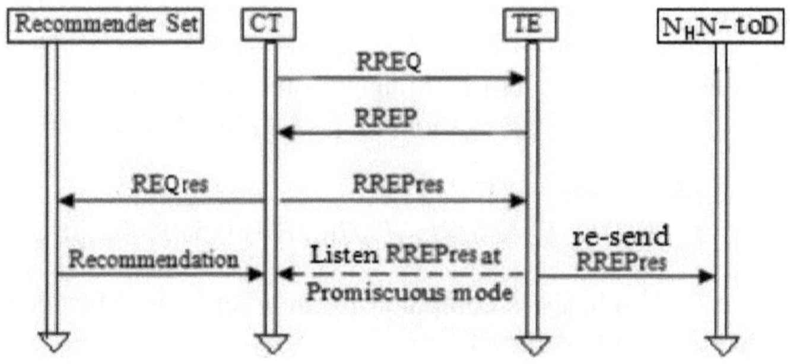

Fig. 3. Packet Transfer sequence in CST-AODV-Blackhole

Symbols are used in CST-AODV-Blackhole as follows.

- $[{}_{CT}T^{C}{}_{TE}]t_{cur}D$: Direct Trust
- $[{}_{CT}T^{C}{}_{TE}]t_{old}N$: Aggregated Old Received Notification
- $[{}_{CT}T^{C}{}_{TE}]t_{cur}R$: Aggregated Current Recommendation
- $[{}_{CT}T^{C}{}_{TE}]t_{cur}I$: Indirect Trust
- $[{}_{CT}T^{C}{}_{TE}]t_{cur}T$: Current Trust
- $[{}_{CT}T^{C}{}_{TE}]t_{old}S$: Aggregated Old Self Evaluated Final Trust
- $[{}_{CT}T^{C}{}_{TE}]t_{old}FT$: Current Final Trust
- $T_{RREPres}$: (time taken for RREPres transmission and reception) + (RREPres travel time) + (MAC and routing layer delays) + (queuing time at receiver node).
- T_P : Processing time taken by TE, depending on different computation.
- T_{Const} : constant times considering network.

For the symbol ${}_{X}T^{C}{}_{Y}$, T denotes trust of trustor X on trustee Y under the context C. Here t_{cur} and t_{old} are the current and old time instants. Different level of trust value varies in between 0.0 and 1.0, highest positive trust for believe case. The trust value 0.5 is considered as uncertainty point for this work.

5.1 Direct Trust

At time of forwarding RPres for TE, CT's Node Manager initializes that time to zero. When TE receives CT forwarded RREPres , TE re-sends that RREPres to TE's next-hop node to target destination (N_HN-toD). After forwarding RREPres, CT waits for listening TE re-send RREPres in promiscuous mode till time ($2T_{RREPres}+T_P+T_{Const}$). If CT can not listen TE re-send RREPres in stipulated time ($2T_{RREPres}+T_P+T_{Const}$), CT understands that TE has dropped CT forwarded RREPres instead of re-sending it to node N_HN-toD. In that case, CT's Direct Trust Manager identifies misbehaviour M on the context C and disbelieves TE. In that case, CT's Direct Trust Manager assign 0.1 as trust value to $[{}_{CT}T^{C}{}_{TE}]t_{cur}D$ for TE. On the other hand, if CT listens TE re-send RREPres in promiscuous mode within stipulated time, CT's Direct Trust Manager can not identify the misbehavior M on the context C and believes TE. In that case, CT's Direct Trust Manager assign 0.9 as trust value to $[{}_{CT}T^{C}{}_{TE}]t_{cur}D$ for TE.

5.2 Indirect Trust

Indirect Trsut Manager of Trust Manager module computes indirect trust ($[{}_{CT}T^{C}{}_{TE}]t_{cur}I$) with weighted aggregation of $[{}_{CT}T^{C}{}_{TE}]t_{old}N$ and $[{}_{CT}T^{C}{}_{TE}]t_{cur}R$ as per equation-(1).

$$[{}_{CT}T^{C}_{TE}]t_{cur}I = 0.5_1 \times [{}_{CT}T^{C}_{TE}]t_{cur}R] + 0.5 \times [{}_{CT}T^{C}_{TE}]t_{old}N\} \quad (1)$$

$[{}_{CT}T^{C}{}_{TE}]t_{cur}R$ and $[{}_{CT}T^{C}{}_{TE}]t_{old}N$ is computed by Indirect Trust Manager according to equation-(2) and equation-(3) respectively.

$$[{}_{CT}T^{C}_{TE}]t_{old}N = \frac{1}{n} \times \{ \sum_{i=1}^{n} [Ti^{C}_{TE}]t_{old}N \times e^{-(t_{old}-t_o)} \} \quad (2)$$

$$[{}_{CT}T^{C}_{TE}]t_{cur}R = \frac{\sum_{i=1}^{n} \{R[{}_{TRi}T^{C}_{TE}]t \times e^{-(t-t_o)}\} \times \{[{}_{CT}T^{C}_{TRi}]t \times e^{-(t-t_0)}\}}{\sum_{i=1}^{n} \{[{}_{CT}T^{C}_{TRi}]t \times e^{-(t-t_0)}\}} \quad (3)$$

In equation-(2), $[Ti^{C}{}_{TE}]t_{old}N$ is the stored notified trusts, where i varies from 1 to n. n is the total number of stored notifications. In equation-(3), $R[{}_{TRi}T^{C}{}_{TE}]t$ is the received recommendation, where TRi is ith recommender and i varies from 1 to n, the total number of recommenders. CT's trust on TRi is $[{}_{CT}T^{C}{}_{TRi}]t$. $e^{-(t-t0)}$ is time decaying parameter, where t_0 is initial time.

5.3 Final Trust

Trust Engine of Trust Manager module computes $[{}_{CT}T^{C}{}_{TE}]t_{cur}T$ with the help of $[{}_{CT}T^{C}{}_{TE}]t_{cur}D$ and $[{}_{CT}T^{C}{}_{TE}]t_{cur}I$ according to equation-(4).

$$[{}_{CT}T^{C}_{TE}]t_{cur}T = 0.8 \times [{}_{CT}T^{C}_{TE}]t_{cur}D + 0.2 \times [{}_{CT}T^{C}_{TE}]t_{cur}I \quad (4)$$

Trust Engine computes final trust $[{}_{CT}T^{C}{}_{TE}]t_{cur}FT$ with the weighted aggregation of $[{}_{CT}T^{C}{}_{TE}]t_{cur}T$ and $[{}_{CT}T^{C}{}_{TE}]t_{old}S$ according to equation-(5).

$$[{}_{CT}T^{C}_{TE}]t_{cur}FT = 0.7 \times [{}_{CT}T^{C}_{TE}]t_{cur}T + 0.3 \times [{}_{CT}T^{C}_{TE}]t_{old}S \quad (5)$$

$[{}_{CT}T^{C}{}_{TE}]t_{old}S$ is computed by Trust Engine according to equation-(6). Here, $[{}_{CT}Ti^{C}{}_{TE}]t_{old}FT$ is i^{th} old self computed final trusts. i varies from 1to n. n is the total number old self computed final trust.

$$[{}_{CT}T^{C}_{TE}]t_{old}S = \frac{1}{n} \times \{ \sum_{i=1}^{n} [{}_{CT}Ti^{C}_{TE}]t_{old}FT \times e^{-(t_{old}-t_o)} \} \quad (6)$$

5.4 Decision and Reaction

Trust Manager Module sends computed final trust to Decision Manager. If $[{}_{CT}T^{C}{}_{TE}]t_{cur}FT$ is greater than 0.5 (uncertainty point), Decision Manager takes belief decision and if $[_{CT}T^{C}{}_{TE}]t_{cur}FT$ is less than uncertainty point, Decision Manager takes disbelief decision. When computed trust value is exactly at uncertainty point, Decision Manager could not justify the belief or disbelief level. In this case, on basis of uncertain decision, Decision Manager takes disbelief decision. CT's Decision Manager sends the taken decision to Node Manager for considering or isolating TE from current route and also send the final trust value as notification to other nodes.

6 Simulation

Simulation result shows that our proposal can efficiently detect blackhole attack in comparison with existing AODV protocol. We have simulated with the ad-hoc network of wireless nodes over 1000m×1000m terrain. For our simulation, we have considered CBR traffic type. From Fig. 4 and Fig. 5, we can see that packet loss rate in our proposal is very low (very nearer to zero percent) where AODV protocol is much higher packet loss rate. Packet loss rate is the function of data packet send by sources and data packets did not received at destination due to packet dropping by blackhole attacker. We have evaluated results after 100 simulation runs.

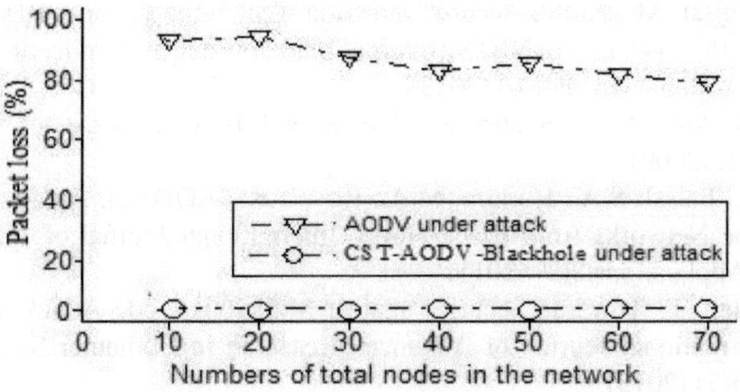

Fig. 4. Packet loss rate vs. total nodes in network

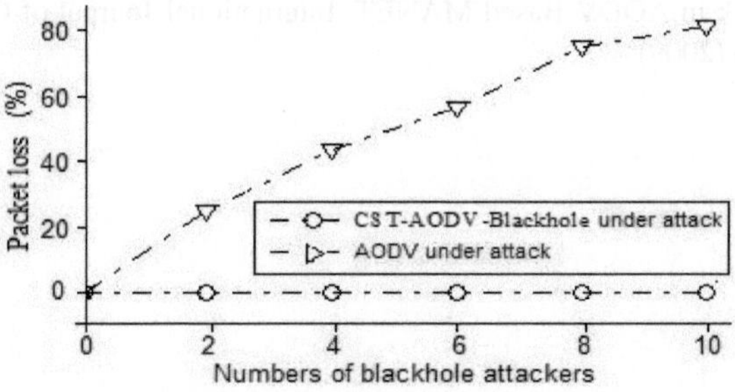

Fig. 5. Packet loss rate vs. numbers of blackhole attackers

7 Conclusion

Our proposal detects blackhole attackers in entire network with high detection rate as shown in the simulation results. Our proposed trust model and trust computation define trust level of relationship between nodes in the network. Depending on trust level, one node believes or disbelieves its trustee. With disbelief of trustor, blackhole attackers are detected and isolated from route. Notification by CT about TE's believe or disbelief level aware the other nodes in the network so that they can consider it during their own computations. In future, we will advance our trust model with respect to other vulnerable attacks in ad-hoc network.

References

1. Menaka, A., Pushpa, M.E.: Effect Of Black Hole Attack On AODV Routing Protocol In MANET. IJCST 4347(2) (2010)
2. Tamilarasan, S.: Securing and Preventing AODV Routing Protocol from Black Hole Attack using Counter Algorithm. International Journal of Engineering Research & Technology 1(5) (2012)
3. Gupta, K., Gujral, M., Nidhi: Secure Detection Technique Against Blackhole Attack For Zone routing Protocol in MANETS. International Journal of Application or Innovation in Engineering & Management 2(6) (2013)
4. Medadian, M., Mebadi, A., Shahri, E.: Combat with Black Hole Attack in AODV Routing Protocol. IEEE (2009)
5. Obaida, M.A., Faisal, S.A., Horaira, M.A., Roy, T.K. ($AODV_R$): An Analytic Approach to Shield Ad-hoc Networks from Black Holes. International Journal of Advanced Computer Science and Applications 2(8) (2011)
6. Kaur, R., Singh, J.: Towards Security against Malicious Node Attack in Mobile Ad Hoc Network. International Journal of Advanced Research in Computer Science and Software Engineering 3(7) (2013)
7. Basagni, S., Conti, M., Silvia Giordano, S.G., Ivan Stojmenovic, I.: Mobile Ad Hoc Networking, ch. 10
8. Payal, W., Raj, N., Swadas, P.B.: DPRAODV: A Dynamic Learning System Against Blackhole Attack in AODV Based MANET. International Journal of Computer Science Issues 2, 809–816 (2009)

Level Wise Initial Energy Assignment in Wireless Sensor Network for Better Network Lifetime

Anindita Ray[1] and Debashis De[2]

[1] Department of Computer Application, B.P. Poddar Institute of Management & Technology Poddar Vihar, 137 V.I.P Road, Kolkata-52, West Bengal, India
[2] Department of Computer Science & Engineering, BF-142, Sector 1, Salt Lake City, West Bengal University of Technology, Kolkata-64, West Bengal, India
{aninrc,dr.debashis.de}@gmail.com

Abstract. One of the basic requirements of Wireless Sensor Network is enhanced network lifetime. Usually Multi-hop communications between sensor nodes and the sink is more energy efficient than single hop communication. However, because the cluster heads (CHs) closer to the sink are loaded with heavy relay traffic, they exhaust much faster than other CHs in the network and ultimately the whole network becomes in operational before time. In this paper we propose a LevelWise Initial Energy Assignment (LWIEA) technique for better network lifetime.For this purpose the whole sensing region is separated in to multiple hierarchical levels and level wise initial energy of sensor nodes have been calculated before real sensor deployment. Applying LWIEA, the resultant heterogeneous sensor network with level wise different initial energy can handle different energy consumption rate in different level. Simulation results show that assigning the initial energies using LWIEA technique effectively improves the network lifetime up to 79% comparedto uniform initial energy assignment strategy with respect to the parameter Full Energy Consumption (FEC) in the network.

Keywords: Initial Energy Assignment, Energy consumption, Network lifetime, WSN.

1 Introduction

Due to the rapid development of wireless communications and integrated circuits, sensor nodes are now smaller and less expensive and for that WSNs are nowadays largely deployed in variety of applications [1] including military surveillance, environmental monitoring, health monitoring and home automation. Clustering is one of the basic approaches for designing energy-efficient, robust and highly scalable distributed sensor networks. Utilizing clusters reduces the communication overhead, thereby decreasing the energy consumption and interference among the sensor nodes. Low Energy Adaptive Clustering Hierarchy (LEACH) [2]is the most primitive cluster-basedroutingprotocol for wireless sensor network.Some of cluster-basedrouting protocols in this group are EECHS [3] , PECRP [4] and EECA-Mhop[5]. One of the

M.K. Kundu et al. (eds.), *Advanced Computing, Networking and Informatics - Volume 2*,
Smart Innovation, Systems and Technologies 28,
DOI: 10.1007/978-3-319-07350-7_8, © Springer International Publishing Switzerland 2014

main reason of poor coverage and lifetime of the network is unbalanced power consumption in WSN. Many algorithms, strategies and mechanisms have been proposed by researchers to improve the network lifetime [5-12]. In [8] it presents a proper Initial energy assignment (IEA) strategy for maximizing network lifetime. According to this strategy sensor node are allocated appropriate initial energy depending on their energy consumption, locations and sensor density.In ACT [9] it presents a cluster-based routing protocol, the aim of which is to reduce the cluster size near the Sink, as cluster heads (CHs) closer to the Sink need to relay more data. In ACT it allows every CH to consume more or less the same amount of energy so that the CHs near the Sink do not exhaust their power so quickly. In this paper we propose a LevelWise Initial Energy Assignment (LWIEA) technique, where the whole sensing region is divided into three equal sized hierarchical levels and level wise initial energy of sensor nodes have been calculated before real deployment. Applying LWIEA, the resultant heterogeneous sensor network can handle different energy consumption rate in different level effectively and improves the network lifetime.

2 Proposed Work

Since the cluster heads (CHs) closer to the Sink are burdened with too many relaying loads, they drain much faster than the other CHs in the network and will become inactive very soon. And since in multi-hop communication the nodes which are closer to sink are responsible for relaying the data packet from the other parts of the sensor network to sink, the data packet will not reach properly to the sink though a significant number of sensor nodes are still alive in the regions farthest away from the sink. For this purpose we propose a LevelWise Initial Energy Assignment (LWIEA) technique where the approximate initial energy of different levels sensor node are calculated first. Then assigning that heterogeneous initial energy in sensor network we have tried to deal with the problem of uneven power dissipation. Applying LWIEA technique it is observed that no part of the resultant heterogeneous sensor network exhaust too early and as a result it effectively improves the network lifetime.

2.1 Assumptions

The following conditions are assumed.

(i) Multi-hop communications have been used to relay data via CHs.
(ii) The whole sensing region is separated with three hierarchical levels of same cluster sizes. The closest Level from Sink is termed as Level1, and then the next one as Level2 and the farthest one is termed as Level3.
(iii) The positions of Sink and sensor nodes are fixed.
(iv) Sensor nodes are uniformly distributed in the region with node density d_n.
(v) The sensing region is of rectangular shape having area 120x90 meter^2.
(vi) Cluster head selection happens based on the residual energy of the sensor nodes.

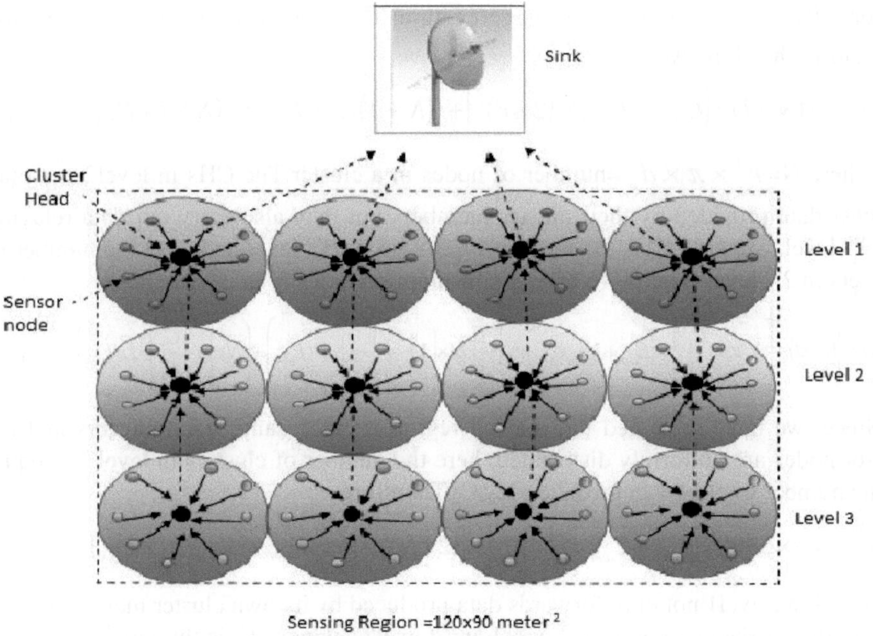

Fig. 1. Multi-hop Sensor network with three equal sized Levels

2.2 Energy Dissipation of Cluster Heads and Cluster Members in Different Levels

In order to calculate the energy dissipation, radio hardware energy dissipation model is used as in [10].According toour proposed approach, each cluster head (CH) in the farthest kth level is concerned only about the data transmitted by the cluster members of its own cluster; its transmission range is $(r_k + r_{k-1})$, where r_k and r_{k-1} are the radius of kth and k-1th level cluster, l is the data packet size and d_n is the node density. Thus here the total energy dissipation of a CH in kth level is

$$ECH_K = (r_k^2 \times \pi \times d_n \times l) \times \left[E_{elec} + E_{amp} \times (r_k + r_{k-1})^2\right] \quad (1)$$

Since the sensing region is separated in three same sized clusters. Here $r_k = r_{k-1} = r$. Energy consumption of CH for transmitting data in level3 is

$$(r^2 \times \pi \times d_n \times l) \times \left(E_{elec} + E_{amp} \times (2 \times r)^2\right) \quad (2)$$

Energy consumption of CHfor receiving andaggregatingdata inlevel3 is

$$(((r^2 \times \pi \times d_n) - 1) \times l \times E_{elec}) + (r^2 \times \pi \times d_n \times l \times E_{DA}) \quad (3)$$

Where E_{DA} is the energy for data aggregation. Therefore Total Energy consumption of a cluster head in level3 is

$$ECH_3 = (N \times l) \times \left[E_{elec} + E_{amp} \times (2 \times r)^2\right] + ((N-1) \times l \times E_{elec}) + (N \times l \times E_{DA}) \quad (4)$$

Where N= $r^2 \times \pi \times d_n$ =number of nodes in a cluster.The CHs in level2 not only process data provided by their cluster members but they also carry out data relaying for 3rd level. Suppose C_3 is the number of clusters in 3rd level and C_2 is the number of clusters in 2nd level. Therefore energy consumption of a CH in 2nd level is

$$ECH_2 = \left\{(N \times l) + \left[N \times l \times \frac{C_3}{C_2}\right]\right\} \times \left[E_{elec} + E_{amp} \times (2 \times r)^2\right] + \left(\left(N \times \left(1 + \frac{C_3}{C_2}\right) - 1\right) \times l \times E_{elec}\right) + \left(N \times \left(1 + \frac{C_3}{C_2}\right) \times l \times E_{DA}\right) \quad (5)$$

Since, we have separated the sensing region in three same sized clusters and the sensor nodes are uniformly distributed, here the number of clusters in level 3 is equal to the number of cluster in level2.So $C_3 = C_2$. Therefore,

$$ECH_2 = (2 \times N \times l) \times \left[E_{elec} + E_{amp} \times (2 \times r)^2\right] + \{(2 \times N) - 1\} \times l \times E_{elec} + (2 \times N \times l \times E_{DA}) \quad (6)$$

Each CH in Level1 not only forwards data produced by its own cluster members but it also performs data relaying for Level2 and Level3.Suppose C_1 is the number of clusters in level1. Therefore energy consumption of a CH in 1st level is

$$ECH_1 = \left\{(N \times l) + \left[N \times l \times \frac{C_2}{C_1}\right] + \left[N \times l \times \frac{C_3}{C_2}\right]\right\} \times \left[E_{elec} + E_{amp} \times (D_b)^2\right] + \left(N \times \left(1 + \frac{C_2}{C_1} + \frac{C_3}{C_2}\right) - 1\right) \times l \times E_{elec} + \left(N \times \left(1 + \frac{C_2}{C_1} + \frac{C_3}{C_2}\right)\right) \times l \times E_{DA} \quad (7)$$

Where D_b =distance of cluster head from Sink.Since C1=C2=C3 we get,

$$ECH_1 = \{3 \times N \times l\} \times \left[E_{elec} + E_{amp} \times (D_b)^2\right] + ((3 \times N) - 1) \times l \times E_{elec} + (3 \times N \times l \times E_{DA}) \quad (8)$$

In each of the level energy consumption of a cluster member will be

$$ECM_K = l \times \left[E_{elec} + E_{amp} \times (r')^2\right] \text{k=1, 2, 3} \quad (9)$$

Where r' is the distance of a cluster member to its nearest cluster head.

2.3 Level Wise Initial Energy Assignment (LWIEA) Technique

As mentioned above cluster heads closer to the Sink are burdened with heavy relay traffic load and the nodes which are farther away relay comparatively less data. Therefore nodes which are closer to the sink should have higher energy level. Our proposed LWIEA technique states that nodes of each of the levels are assigned the initial energy according to their energy consumption per data collection round. For that we have measured the percentage of faster energy consumption occur in the closer levels with respect to the farthest level of the network. According to our assumed network model, if there are k numbers of levels then k^{th} level is the farthest level from the sink node. Here it is assumed that the average distance of a cluster member to its nearest cluster head is $r' = 15$ meter and average distance from the cluster head of closest Level to sink node is $D_b = 50$ meter. Suppose total energy dissipation of a CH in k^{th} level is ECH_K and the energy dissipation of a cluster member of that level is ECM_K. Therefore the total energy dissipation of kth level is

$$ET_K = CH_K \times ECH_K + CM_K \times ECM_K \tag{10}$$

Where CH_K and CM_K are the number of cluster heads and number of cluster members in k^{th} level respectively. And the total energy dissipation of k-1th layer is

$$ET_{K-1} = CH_{K-1} \times ECH_{K-1} + CM_{K-1} \times ECM_{K-1} \tag{11}$$

Therefore (%) of faster energy consumption of level k-1 with respect to k^{th} level is

$$F_{k-1} = ((ET_{k-1} - ET_k)/ET_k) \times 100 \tag{12}$$

To support F_{k-1}% of faster energy consumption occurs in Level k-1 with respect to Level k, the initial Energy of Level k-1 will be

$$E_{k-1} = E_k + F_{k-1}/100 \tag{13}$$

3 Simulation Results and Analysis

For simulation C language and MATLAB 7.7 has been used. Table 1 lists the simulation parameters.

Table 1. Simulation Parameters

Parameter	Value
Total No. of sensor nodes	108
Network size	120X90 m^2
Sink's Location	(60,90)
Cluster Head probability	.11
Initial energy of node	1 Joule
Node Distribution	Randomly distributed
Node density d_n	1 node/100 m^2
Data Packet	400 byte
Radius(r) of each cluster	15 m
E_{elec}	50 nJ/bit
ε_{mp}	10pJ/bit/m^2
No of Levels	3(Three)

The simulation environment is composed of 108 sensor nodes which are distributed in a region of 120mX90m.Since, the sensing region is separated in three same sized clusters and sensor nodes are uniformly distributed, each of the threelevels contain 36 nodes. Here our proposed LWIEA technique is compared with uniform initial energy assignment strategy. In uniform initial energy assignment strategy whereall of the mentioned three levels have same initial energy 1J, thenetwork lifetime of three levels with respect to the number of rounds is obtained as shown in Table 2. Fig.2 shows the amount of remaining energy with respect to the number of rounds in three levels using uniform initial energy assignment strategy.

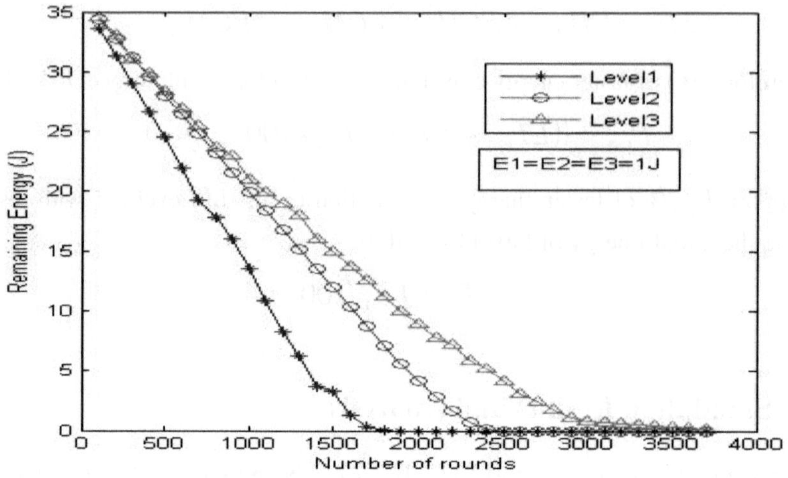

Fig. 2. Remaining Energy with respect to the number of rounds applying Uniform initial energy assignment strategy

Table 2. Network lifetime of three levels with respect to the number of rounds using Uniform initial energy assignment strategy

	Round Half Energy Consumed	Round Full Energy Consumed
Level1 (Initial energy =1J)	780	1900
Level2 (Initial energy =1J)	1050	2460
Level3 (Initial energy =1J)	1300	3680

Applying LWIEA technique, using Eq.(12), it is obtained that the percentage of faster energy consumption in Level2 with respect to Level3 is 53% and the percentage of faster energy consumption in Level1 with respect to Level3 is 100%.So using Eq.(13) the Level2 sensor node's approximate initial energy E_2 is 1.53 J and the Level1 sensor node's approximate initial energy E_1 is 2 J. Using LWIEA algorithm we can get the improvement of network life time of Level1 and Level2 both with respect to half and full energy consumption, which are summarized in Table 3. The percentage of improvement of network life time of level1 and level2 using LWIEA algorithm both with respect to half and full energy consumption is summarized in Table 4. In Fig.3 it shows the amount of remaining energy with respect to the number of roundsin three levels with their modified approximate initial energy.

Table 3. Network lifetime of three levels with respect to the number of rounds applying LWIEA technique

	Round Half Energy Consumed	Round Full Energy Consumed
Level1(Initial energy =2 J)	1350	3400
Level2(Initialenergy =1.53J)	1470	3480
Level3(Initial energy=1J)	1300	3680

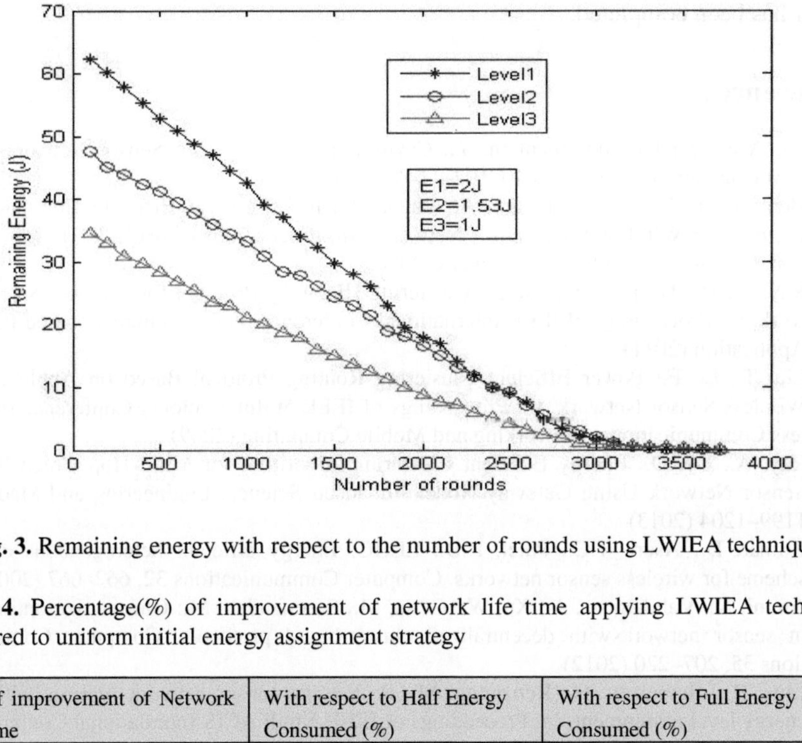

Fig. 3. Remaining energy with respect to the number of rounds using LWIEA technique

Table 4. Percentage(%) of improvement of network life time applying LWIEA technique compared to uniform initial energy assignment strategy

(%) of improvement of Network Lifetime	With respect to Half Energy Consumed (%)	With respect to Full Energy Consumed (%)
Level1	73%	79%
Level2	40%	41.5%

4 Conclusion

In this paper a LevelWise Initial Energy Assignment (LWIEA) technique have been proposed for improved network lifetime.For this purpose the whole sensing region is separated into three hierarchical levels having same size. And level wise approximate initial energy of sensor nodes have been calculated before real sensor deployment.

Applying LWIEA technique, the resultant heterogeneous sensor network with level wise different initial energy can handle different energy consumption rate in different level. Simulation results show that assigning the approximate initial energies using LWIEA technique effectively improves the network lifetime up to 79% compared to uniform initial energy assignment. In future instead of a single sink node multiple sink nodes can also be used for better network lifetime. An efficient sleep wake mechanism can also be used in upcoming days to reduce energy consumption in the sensor nodes so that the overall power consumption of the network can be reduced.

Acknowledgement. Authors are grateful to Department of Science and Technology (DST) for sanctioning a research Project entitled "Dynamic Optimization of Green Mobile Networks: Algorithm, Architecture and Applications" under Fast Track Young Scientist scheme reference no.: SERB/F/5044/2012-2013 under which this paper has been completed.

References

1. Su, A.W., Sankarasubramaniam, Y., Cayirci, E.: A Survey on Sensor Networks. IEEE Communications Magazines 40, 102–114 (2002)
2. Heinzelman, W., Chandrakasan, A., Balakrishnan, H.: Energy-Efficient Communication Protocol for Wireless Microsensor Networks. In: Proceedings of IEEE 33 Hawaii International Conference on System Sciences (2000)
3. Ray, A., De, D.: Energy Efficient Clustering Hierarchy Protocol for Wireless Sensor Network. In: Proceedings of IEEE International Conference on Communication and Industrial Application (2011)
4. Liu, T., Li, F.: Power Efficient Clustering Routing Protocol Based on Applications in Wireless Sensor Network. In: Proceedings of IEEE 5thInternational Conference on Wireless Communications, Networking and Mobile Computing (2009)
5. Ray, A., De, D.: Energy Efficient Clustering Algorithm for Multi-Hop Green Wireless Sensor Network Using Gateway Node. Advanced Science, Engineering and Medicine 5, 1199–1204 (2013)
6. Kumar, D., Aseri, T.C., Patel, R.B.: EEHC: Energy efficient heterogeneous clustered scheme for wireless sensor networks. Computer Communications 32, 662–667 (2009)
7. Amini, N., Vahdatpour, A., Xu, W., Gerla, M., Sarrafzadeh, M.: Cluster size optimization in sensor networkswith decentralized cluster-based protocols. Computer Communications 35, 207–220 (2012)
8. Lijie, R., Zhongwen, G., Renzhong, M.: Prolonging sensor network lifetime with initial energy level assignment. In: Proceedings of IEEE Ninth ACIS International Conference on Software Engineering, Artificial Intelligence, Networking, and Parallel/Distributed Computting, pp. 231–236 (2008)
9. Lai, W.K., Fan, C.S., Lin, L.Y.: Arranging cluster sizes and transmission ranges for wireless sensor networks. Information Sciences 183, 117–131 (2012)
10. Heinzelman, W., Chandrakasan, A., Balakrishnan, H.: An Application-Specific Protocol Architecture for Wireless Microsensor network. IEEE Transactions on Wireless Communication 1, 660–670 (2002)
11. Li, K., Li, J.: Optimal energy allocation in heterogeneous wireless sensor networks for lifetime maximization. Journal of Parallel and Distributed Computing 72, 902–916 (2012)
12. Cheng, X., Du, D.Z., Wang, L., Xu, B.: Relay sensor placement in wireless sensor networks. Wireless Network 14, 347–355 (2008)

Determining Duty Cycle and Beacon Interval with Energy Efficiency and QoS for Low Traffic IEEE 802.15.4/ZigBee Wireless Sensor Networks

Dushyanta Dutta[1], Arindam Karmakar[1], and Dilip Kr. Saikia[2]

[1] Dept. of Computer Science and Engineering, Tezpur University, India
ddutta@tezu.ernet.in, arindam@tezu.ernet.in
[2] Dept. of Computer Science and Engineering, NIT, Meghalaya, India
dks@nitm.ac.in

Abstract. Wireless Sensor Networks (WSNs) are becoming integral part of today's world due to their wide range of applications. WSNs are used in industrial applications such as factory automation and control, environmental monitoring etc. which are of low data rate but demand Quality of Service (QoS) in terms of reliability and timeliness along with energy efficiency. A WSN's lifetime depends on the rate of consumption of energy by the sensors. One way to save energy is the judicious use of active and inactive periods in the duty cycle based Medium Access Control (MAC) protocols. The slotted IEEE 802.15.4 MAC protocol provides for it. The duty cycle mechanism however has the potential to affect the performance of the network in terms of QoS due to possible higher packet collision rate in the active periods with reduction in duty cycle. For such applications an appropriate selection of duty cycle that maintains the required QoS is vital while ensuring longevity of the network. Another important MAC parameter that has a role in the performance of WSNs based on IEEE 802.15.4/ ZigBee is the length of the Beacon Interval (BI). In this paper we present the results of our simulation experiments for determining appropriate Duty Cycle and Beacon Interval that maximizes energy efficiency while ensuring the QoS requirements of the application.

Keywords: IEEE 802.15.4, Quality of Service (QoS), Packet Delivery Ratio (PDR), Energy Efficiency.

1 Introduction

Wireless sensor network (WSN) is a promising technology for today's world due to its wide range of applications such as environmental monitoring, security and surveillance, industrial automation and control. The IEEE 802.15.4 [1] standard has received considerable attention as a major low data rate and low power protocol for wireless sensor networks (WSNs) for applications like process control and factory automation. Use of IEEE 802.15.4 in WSNs is possible due to amalgamation of the IEEE 802.15.4 standard and the ZigBee specifications [2].

M.K. Kundu et al. (eds.), *Advanced Computing, Networking and Informatics - Volume 2*, Smart Innovation, Systems and Technologies 28,
DOI: 10.1007/978-3-319-07350-7_9,

The transceivers in beacon enabled mode of IEEE 802.15.4 MAC are active during a Superframe that is bounded by a Beacon frame generated by the coordinator at regular interval. The interval between two consecutive beacons is the size of the beacon frame which is also called Beacon Interval *(bi)*. The duration of beacon interval is decided by the parameter called *Beacon Order (BO)* and $bi = 15.36 * 2^{BO}$ ms where $0 \leq BO \leq 14$. The size of the active period is also called *Superframe Duration (SD)* and its length is decided by the parameter *Superframe Order (SO)* and $SD = 15.36 * 2^{SO}$ ms, where $0 \leq SO \leq 14$. Thus the *Duty Cycle of(DC)* a network operating in beacon enabled slotted CSMA/CA mode is $2^{(SO-BO)}$.

For a successful deployment of WSNs in industrial environments along with *energy efficiency*, we have to deal with QoS parameters like *successful delivery of data packets (reliability)* and *timeliness*. The defined performance measures represent the prime performance metrics for wireless industrial communication systems which can be referred as *industrial-QoS* [3].

Energy efficiency is extremely important as it determines the sensor network's life. Sensor nodes are typically powered by batteries with a limited amount of energy. Once deployed the node's batteries cannot be replaced or recharged, due to environmental or cost constraints.

Reliability and *timeliness* are very critical issues in industrial environments. Amount of data packets delivered (throughput) to the sink within a predefined deadline (latency), must meet with the requirement of the application. The correct behavior of the sensing system (e.g., the timely and accurate detection of an event) can only be achieved with reliability and timeliness.The IEEE 802.15.4 standard includes a power management mechanism, based on duty cycle, to minimize the power consuming activity of sensor nodes.

In sensor network applications for industrial automation and control, large number of sensor nodes collects data and send these to the servers for processing. Generally there are overlapping areas of coverage by the sensor nodes. Hence, there is in-built redundancy in the network. Further, there is also a redundancy in terms of frequency of sensing of the data. A 100% delivery of packets is therefore not necessary in a WSN. The packet delivery ratio (PDR) tolerance limit will depend upon the redundancy built into the network and the QoS requirement of the application. The choice of duty cycle will have effect on the PDR too.

A MAC parameter that is closely associated with duty cycle in the performance of WSNs is the Beacon Interval. A longer beacon interval implies to longer inactive periods leading to larger accumulation of generated packets to be transmitted during the active period and higher possible contention. A short beacon interval on the other hand leads to more frequent beacons and too short active periods for clearing the accumulated packets. Therefore it is important to choose beacon interval carefully in conjunction with the duty cycle to satisfy the application's requirements.

A WSN application is characterized by its traffic load, required packet delivery ratio (PDR) and a latency limit. The two key MAC parameters that play important role in satisfying the requirements of the application are the *duty cycle (dc)* and the *beacon interval (bi)*. This work focuses on determining the combination of Duty Cycle and Beacon Interval for a WSN an application with given characteristics such that the consumed amount of *energy per byte of data delivered (EPBDD)* is minimized.

Let the objective function $E_{PB}(dc, bi, p, \tau, l)$ denote average energy consumed per byte of data delivered with packet delivery ratio no less than p and latency no more than τ at given traffic load l for duty cycle dc and beacon interval bi. The constrained optimization problem of the network is therefore

$$minimize\ E_{PB}(dc, bi, p, \tau, l) \tag{1}$$

An analytical approach can be used to solve the optimization problem expressed in equation 1. By deriving an analytical model of the sensor network we can calculate the IEEE 802.15.4/ZigBee parameter values that satisfy equation 1. The key inadequacy of this approach is that its validity depends on the accuracy of the underlying analytical model. A model is derived based on some simplifying assumption which may not always be accurate in a real scenario. In this paper we try to solve the above optimization problem by results obtained through simulation in NS2 [4] environment.

The rest of this paper is organized as follows. Section 2 discusses the related work. Section 3 describes the Simulation set-up. Section 4 presents our scheme for selection of duty cycle and beacon interval for low traffic applications and the results of the scheme implemented through simulation. Finally, section 5 draws the conclusions of the work.

2 Related Work

There are several works which discuses the use of low duty cycle as an energy saving mechanism to save energy in IEEE 802.15.4 in WSN. Most of these are studies based on simulations and these highlight the change in the performance of the network in terms of throughput and overall power consumption of the network due to change in duty cycle. One of the earliest works for analysis of the network performance under different duty cycle is[5]. Later [6] studied the impact of BO, SO and traffic load on the performance and energy consumption of the network. Algorithms were developed to adapt duty cycle dynamically based on current network status [7]. The issue of unreliability problem due to inactive period in the time frame is addressed in [8] and later an adaptive and cross-layer framework is proposed in [9] to deal with unreliability problem.

Our study differs from the existing work in that we investigate the possible optimal setting of the two MAC parameters- *duty cycle* and *beacon interval-* to satisfy the QoS requirements in terms of *packet delivery ratio* and *latency* while maximizing the true *energy efficiency*, i.e. minimizing the energy consumed per byte of data successfully delivered *(EPBDD)*.

3 The Simulation Set-up

As stated our scheme to determine the two MAC parameters is based on simulated experiments carried out on NS-2 [4] platform. In all the experiments it is assumed that the IEEE 802.15.4 MAC protocol operating on top of the 2.4 GHz physical layer with a maximum capacity of 250 Kbps. We consider a single hop star topology where all the stations are in radio range of each other and the coordinator is acting as the sink node. The set-up consists of 10 participating nodes placed in a circle of radius 10 m from the sink. The network is operated in the beacon enabled mode and the retransmission mechanism is enabled.

As the duty cycle mechanism is relevant to low traffic rates the data generation rate considered ranged between 0.25 Kbps to 5 Kbps which correspond to 0.1% and 2% of the maximum capacity. The data generation is considered to be Poisson distributed. The size of the packet is considered to be a fixed size of 100 bytes. In the experiments no restriction was put on the queue length. The simulation is conducted in simulation time frame of 1000s and results considered are aggregate value obtained over 100 runs of the simulation.

For computation of the energy consumed a sensor node based on the Chipcon CC2420 radio transceiver is assumed [10].

Table 1. PDR (%) for different traffic loads

dc	1.5625%	3.125%	6.25%	12.5%	1.5625%	3.125%	6.25%	12.5%
BO								
3				97.2496				96.0878
4			95.9049	97.6811			94.7818	97.2056
5		91.6627	98.2135	98.5176		85.7428	98.0233	98.6962
6	83.5067	96.3918	96.7984	97.1859	66.5124	95.5043	96.281	97.1723
7	93.5801	94.899	95.1354	95.0604	90.4898	93.8795	94.6964	94.8805
8	85.7994	85.6503	85.4602	86.269	83.8756	84.4896	84.664	85.1879
9	76.9154	77.2853	77.3321	79.6998	73.3929	73.8127	74.7329	76.925
10	66.2414	67.0596	69.8011	73.5365	62.7681	63.0956	64.9452	69.7861
3				94.4665				87.2728
4			90.1536	97.6122			72.9822	95.7974
5		69.2976	97.05	98.4058		0.868608	92.1261	96.9773
6	1.58228	92.4902	96.0404	96.8061	0.2518	72.9922	92.814	93.5887
7	80.3299	92.0872	92.5257	93.6267	5.6337	83.6533	85.8758	88.2663
8	79.4529	80.4417	81.1682	82.8026	57.435	71.4577	73.219	76.1891
9	67.698	69.33	70.5477	73.9482	58.608	62.0738	64.6191	69.0913
10	57.1974	58.7516	60.9401	66.9251	52.501	54.6787	57.7063	65.9386

4 The Scheme for Determining Duty Cycle and Beacon Interval

The two QoS parameters under consideration in our scheme are Latency and PDR. It is desirable to maximize PDR and minimize latency. However, in a

Table 2. Energy per byte of data delivered (EPBDD) in for different traffic loads

dc	1.5625%	3.125%	6.25%	12.5%	1.5625%	3.125%	6.25%	12.5%
BO								
3				0.132223				0.0750609
4			0.0844868	0.101583			0.0466604	0.0633191
5		0.0483327	0.0572762	0.0754483		0.034045	0.0379117	0.0467251
6	0.0380481	0.0418855	0.0514949	0.0700477	0.0320245	0.0309648	0.0377562	0.0442415
7	0.0370062	0.0416982	0.0505957	0.0699439	0.029384	0.0314588	0.035747	0.0447353
8	0.0480861	0.053921	0.0650524	0.0843591	0.0352067	0.0377241	0.0429165	0.0527567
9	0.0643722	0.0697161	0.0815979	0.102208	0.0434109	0.0465104	0.0518191	0.0623091
10	0.0894155	0.095626	0.104097	0.121894	0.0570514	0.0600595	0.0655194	0.0750757
3				0.0451682				0.0292933
4			0.0335086	0.0376608			0.0279401	0.0265039
5		0.029779	0.0287097	0.0326119		0.0528696	0.0243993	0.0250316
6	0.0539261	0.0262455	0.0280885	0.0320769	0.0558553	0.0248279	0.02385	0.0251337
7	0.0264679	0.026388	0.0284946	0.032601	0.0323686	0.0232268	0.0237338	0.0252225
8	0.0283275	0.0294439	0.032048	0.0367055	0.025619	0.0241211	0.0248731	0.0265703
9	0.032863	0.0336867	0.0366176	0.041916	0.0262408	0.026077	0.0268043	0.0293226
10	0.0408439	0.0413869	0.045693	0.0503829	0.0291434	0.0293019	0.030981	0.0361659

CSMA/CA scenario, desiring higher PDR implies allowing higher latency. Setting a lower latency limit implies lowering the limit of the acceptable PDR. The objective of our study, however is to satisfy the requirement limits for both these two QoS parameters for an application. Along with QoS, minimizing the consumption of energy is required to be done by setting the duty cycle and beacon interval of the MAC layer protocol appropriately. For the purpose we define the following terms:

1. *Reliable Set of Duty Cycles (RSDC)*: Reliable Set of Duty Cycles for a given traffic load is defined to consist of those duty cycles that provide PDR at or above the acceptable limits of the application for some values of beacon interval.
2. *Energy-efficient On-time Reliable Duty Cycle (EORDC)*: Energy-efficient On-time Reliable Duty Cycle is defined for a given traffic load as one within the *RCSDC* for which the *EPBDD* is lowest for some value(s) of beacon interval while the corresponding latency remains within the acceptable limit of the application.
3. *Energy-efficient On-time Reliable Beacon Interval (EORBI)*: Energy Efficient On Time Reliable Beacon Interval is defined for a given traffic load at the *EORDC* as one for which the *EPBDD* is lowest while the PDR remains within the acceptable limits of the application.

The goal is now to determine *EORDC* and *EORBI.* The approach we take is to first determine the *RSCDC* followed by determining of the *EORDC* and the *EORBI.* Prior to these however we need to define what we should consider as the PDR for the network. This is necessary as there is a limit on the acceptable latency. For defining the PDR we observe the following:

1. In a slotted IEEE 802.15.4 network the packets arrive at the destinations during the active period of a cycle. For low duty cycle operations these

deliveries occur in short bursts with latencies that can be expressed in terms of number of beacon intervals. The packets that are delivered in the nth cycle after their transmission from the source will have a latency in the range of $(n-1)bi$ to $((n-1)+dc)bi$, where dc is the duty cycle expressed as a fraction and bi is the beacon interval duration.

2. If we plot throughput in terms of PDR against latency, with the arrival of the packets in bursts the throughput will jump at latency points that are multiples of beacon intervals. The growth will be high initially and then taper down as the PDR approaches a saturation level. We call the latency point where saturation sets in as the *Throughput Saturation Latency (TSL)*. The yield in PDR beyond *TSL* will be marginal in comparison to the increase in latency. Therefore it will be a good tradeoff to consider the PDR at the *TSL* as the PDR of the WSN.
3. The value of *dc* being small it is useful to study *TSL* in terms of number of beacon intervals. We shall therefore express *TSL* in terms of number of beacon intervals.

4.1 Determining Throughput Saturation Latency (TSL)

Accordingly we conduct a set of simulation experiments to determine the *TSL* for different traffic loads at different duty cycles. Considering the low traffic applications and our aim of energy efficiency the ranges of both traffic load and duty cycle are kept low. The traffic load is kept within 2% of the channel capacity, i.e. 5 Kbps, and the duty cycle is kept within 12.5%.

The results are plotted in Fig. 1. It can be observed that for all the cases except one the saturation sets in at $2bi$ and therefore the *TSL* is $2bi$. In all these cases the PDR is >80% at the *TSL*. In the case of duty cycle of 3.125% traffic load of 5 Kbps is very close to the available data rate of the channel and hence the data rate cannot be considered to be low enough for the duty cycle assumed.

The above result can be explained as follows. During the first beacon interval the PDR remains low as a large fraction of the packets get generated within the inactive period and will be deferred to the following active period. Most of these deferred packets get successfully delivered within that active period when the data generation rate is low. As a result the increase in the fraction of packets with latency beyond $2bi$ is therefore marginal. Hence the *TSL* for low traffic rate remains $2bi$ except when the duty cycle is made too low to make the data rate comparable to the bandwidth available.

As these packets get delivered by the end of the active period in the second beacon interval, the maximum latency suffered by these packets will be $(1+dc)bi$. We shall now consider the packets received within this *throughput saturation latency (TSL)* for the PDR of the network. It is important to note that the above result is independent of the energy consumption characteristics of the sensor node.

Based on the above definition of PDR our choice of beacon interval will be restricted to only those for which the allowable latency is higher than $(1+dc)bi$.

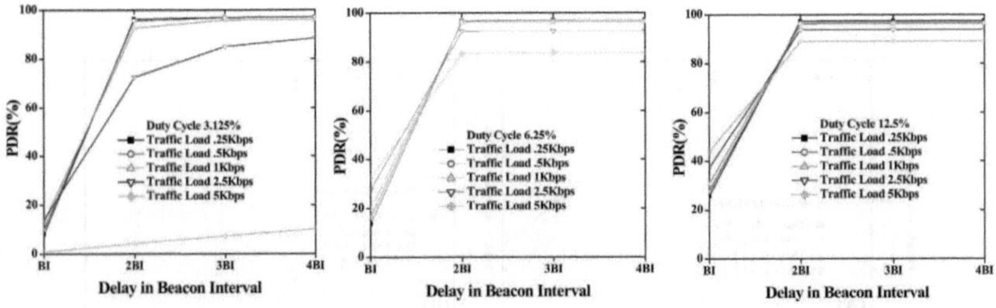

Fig. 1. PDR (%) with in delay of number of beacon intervals

Let us call these the *Candidate Set of BIs (CSBI)*. The *CSBI* shall correspond to those Beacon Orders (BO) for which $(1 + dc) * 15.36 * 2^{BO} < \tau$.

4.2 Determining EORBI and EORDC

For determining the *EORBI* and *EORDC* for an application we carry out a set of experiments with traffic load of the application for the beacon intervals within the *CSBI* of the application. We consider the PDR and the EPB of data delivered against the BOs corresponding to the *CSBI* for different values of duty cycles in the lower range. From the PDR values the *reliable set of candidate duty cycles (RSCD)* becomes apparent as those having PDR above the PDR limit of the application, for some value(s) of BO belong to the set.

In Fig. 2 we plot the PDRs for the WSN specified in the simulation setup described in section 3 for applications with traffic loads of 0.25 Kbps, 0.5 Kbps, 1 Kbps, 2.5 Kbps respectively for the duty cycles of 1.5625%, 3.125%, 6.25% and 12.5% assuming the *CSBI* to correspond to the BO values within the range of 10. These are also tabulated in Table.1 where values with colour black is for traffic load 0.25 Kbps, colour red for 0.5 Kbps, blue for 1 Kbps and violet for 2.5 Kbps.

Let us consider the PDR requirement of the application to be 85%. For an application with traffic load of 0.25 Kbps all the four duty cycles considered here belong to the set *RSCD*. At duty cycle 1.5625% PDR is above 85% for BO values of 7 and 8. At the other three duty cycles PDR is above 85% for BO values less than and equal to 8. For application with 0.5 Kbps traffic load the PDR is above 85% at duty cycles 1.5625%, 3.125%, 6.25% and 12.5% for BO values (7), (5, 6, 7), (4, 5, 6, 7) and (4, 5, 6, 7, 8) respectively. For 1 Kbps traffic load the three duty cycles 3.125%, 6.25% and 12.5% belong to *RSCD* and that happens at BO values (6, 7), (4, 5, 6, 7) and (3, 4, 5, 6, 7) respectively. For traffic load of 2.5 Kbps the duty cycles 6.25% and 12.5% only and that happens at BO values (5, 6, 7) and (3, 4, 5, 6, 7) respectively.

For determining *EORBI* and *EORDC* we now consider the *energy per byte of data delivered (EPBDD)* for the duty cycles in the *RSCD* and at the BOs where PDR is above the required level. For our simulation set-up, the *EPBDD*

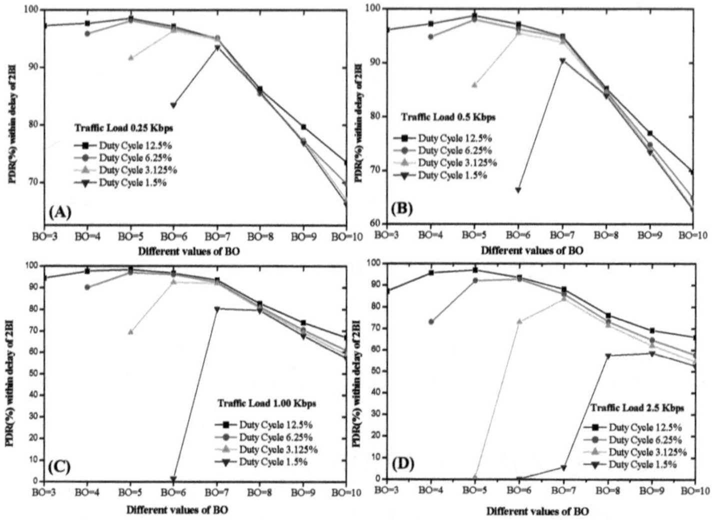

Fig. 2. PDR(%) as a function of traffic load (A) 0.25 Kbps, (B)0.50 Kbps, (C) 1.00 Kbps and (D) 2.50 Kbps

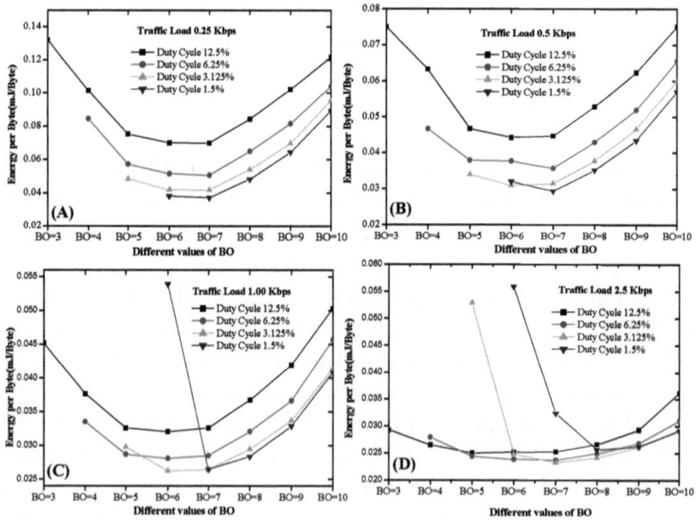

Fig. 3. EPBDD as a function of traffic load (A) 0.25 Kbps, (B)0.50 Kbps, (C) 1.00 Kbps and (D) 2.50 Kbps

for applications with different traffic loads are plotted in Fig. 3. These are also tabulated in Table.2 where values with colour black is for traffic load 0.25 Kbps, colour red for 0.5 Kbps, blue for 1 Kbps and violet for 2.5 Kbps.

For an application with a given traffic load the (BO, DC) pair corresponding to the lowest *EPBDD* gives us the desired *EORBI* and *EORDC* for the application. For the example considered the desired (BO, DC) pairs for applications of traffic loads 0.25 Kbps, 0.5 Kbps, 1 Kbps and 2.5 Kbps are (7, 1.5625%), (7, 1.5625%), (6, 3.125%) and (7, 6.25%) respectively.

5 Conclusion

In this paper we have presented a scheme for determining the duty cycle and beacon interval combination for a beacon enabled IEEE 802.15.4 WSN that minimizes the energy consumption while ensuring that the QoS requirements of the application in terms of latency and packet delivery ratio (PDR) are satisfied. The work focuses on low data rates as is typical of most sensor network applications.

To start with we argued that the best performance in terms of latency and PDR for low data rates can be achieved by limiting the allowable latency to *(1+dc)bi* by restricting the packet retransmissions to within that latency period and the same is validated by simulated experimental results. We then presented a scheme to determine the best possible duty cycle and beacon interval combination to minimize energy consumed per byte of data delivered for an application with given traffic load and having a packet delivery ratio (PDR) requirement within a packet latency limit.

References

1. IEEE TG 15.4 part 15.4: Wireless Medium Access Control (MAC) and Physical Layer (PHY) Specifications for Low-Rate Wireless Personal Area Networks (WPANs). IEEE Std., New York
2. The ZigBee Specification version 2.0, `http://www.zigbee.org/Standards/Downloads.aspx`
3. Zurawski, R.: Embedded Systems Handbook: Networked Embedded Systems, vol. 2. CRC Press (2009)
4. Network Simulator version - 2, `http://www.isi.edu/nsnam/ns/`
5. Lu, G., Krishnamachari, B., Raghavendra, C.: Performance evaluation of the IEEE 802.15. 4 MAC for low-rate low-power wireless networks. In: 2004 IEEE International Conference on Performance, Computing, and Communications, pp. 701–706 (2004)
6. Chen, F., Wang, N., German, R., Dressler, F.: Simulation study of IEEE 802.15. 4 LR-WPAN for industrial applications. Wireless Communications and Mobile Computing 10(5), 609–621 (2010)
7. Gao, B., He, C.: An individual beacon order adaptation algorithm for IEEE 802.15. 4 networks. In: 11th IEEE Singapore International Conference on Communication Systems, pp. 12–16. IEEE (2008)

8. Anastasi, G., Conti, M., Di Francesco, M.: A comprehensive analysis of the MAC unreliability problem in IEEE 802.15. 4 wireless sensor networks. IEEE Transactions on Industrial Informatics 7(1), 52–65 (2011)
9. Di Francesco, M., Anastasi, G., Conti, M., Das, S., Neri, V.: Reliability and Energy-Efficiency inIEEE 802.15. 4/ZigBee Sensor Networks: An Adaptive and Cross-Layer Approach. IEEE Journal on Selected Areas in Communications 29(8), 1508–1524 (2011)
10. Bougard, B., Catthoor, F., Daly, D., Chandrakasan, A., Dehaene, W.: Energy efficiency of the IEEE 802.15. 4 standard in dense wireless microsensor networks: Modeling and improvement perspectives. In: Design, Automation, and Test in Europe, pp. 221–234. Springer (2008)

Tree Based Group Key Agreement – A Survey for Cognitive Radio Mobile Ad Hoc Networks

N. Renugadevi and C. Mala

Department of Computer Science and Engineering
National Institute of Technology, Tiruchirappalli
Tamilnadu, India
{406112002,mala}@nitt.edu

Abstract. Cognitive radio networks solve the spectrum scarcity problem by dynamically utilizing the unused spectrums. To ensure secure and reliable communication, cognitive radio mobile ad hoc networks require more stringent and secure protocols due to their intrinsic nature. Tree based topology for cognitive radio network is widely used as it takes less time for join and leave operations for the users within the channel of the spectrum compared to other topologies. This paper presents a survey of tree based group key agreement schemes applicable to cognitive radio networks.

Keywords: Group Key Agreement, Key tree, Individual and Batch rekeying, Cognitive Radio Mobile Ad Hoc Networks.

1 Introduction

The next generation wireless networks such as *Cognitive Radio Networks (CRN)* can solve the spectrum scarcity problem of the current wireless systems. It employs an intelligent device called *Cognitive Radio (CR)* [1] that is aware of its working environment and capable of detecting the free channels in wireless spectrum. The CR user can dynamically access the spectrums assigned to the licensed user without making any interference to them through the process called *Dynamic Spectrum Access (DSA)* [2]. The CR nodes can employ either *Overlay or Underlay* approach of *Secondary Spectrum Access (SSA)* [3] to share the spectrum bands of licensed users.

Providing security in Cognitive Radio Mobile Ad Hoc Networks (CR-MANETs) is a difficult task due to the extra dynamic functions such as Spectrum sensing, DSA and spectrum handoff performed by CR devices in addition to the normal operations of non-CRN devices. In many operations of CRN, for example, in cooperative spectrum sensing, the CR nodes perform group communication to make the correct sensing decision. In [4], the authors suggested to use a shared 'session or group key' among the CR nodes to achieve the group security. The *Lion Attack* is a cross-layer attack [29] that targets the TCP layer of CRNs through the physical layer in a way that the TCP is not aware of frequent handoffs. This will degrade the overall performance of the TCP layer in CRNs. The countermeasure for the Lion attack has been given in

M.K. Kundu et al. (eds.), *Advanced Computing, Networking and Informatics - Volume 2*, Smart Innovation, Systems and Technologies 28,
DOI: 10.1007/978-3-319-07350-7_10, © Springer International Publishing Switzerland 2014

[30], where the authors have suggested using a common group key to protect the shared control data from eavesdroppers.

As CR-MANETs are sharing the properties of MANETs, *Contributory Group Key Agreement (CGKA)* scheme is more appropriate than centralized and distributed group key management schemes due to lack of Trusted Third Party or Centralized Server. In CGKA, each node should contribute its share and the group key is generated from the secret shares of all members. The CRN nodes are resource restricted devices with less battery power and less computational capability. Hence, the GKA protocol used in the CRNs should be more effective. From the extensive study it is found that, tree-based CGKA methods are more efficient since they reduce the complexity from O (n) to O (log n) while computing the new group key, where n is the group size [9]. Consequently, this paper considers only the existing tree-based CGKA protocols.

The remainder of the paper is organized as follows. Section 2 explains the existing tree-based CGKA protocols. Section 3 presents the comparison table and analyses the performance of the existing protocols and Section 4 concludes the paper.

2 Tree Based CGKA Protocols

As CR-MANETs are dynamic networks without central authority and pre-established infrastructure, the CGKA methods available for wireless networks are applicable to CRNs too. The method of updating the group key, i.e., rekeying is classified as *Individual Rekeying (IR)* and *Batch Rekeying (BR)* based on the rekeying timing. This section describes the tree based CGKA protocols which use Diffie Hellman (DH) [5], Group DH (GDH) [6], Elliptic Curve DH (ECDH) [7] and One Way Function (OWF) [28]. Subsection 2.1 elaborates the IR based CGKA protocols, and the CGKA protocols based on BR are explained in subsection 2.2.

2.1 IR Based CGKA Protocols

In IR based protocols, the group key is updated after each join or leave request and hence it provides both forward and backward secrecy. The following subsections describe the IR protocols which use DH, GDH, ECDH and OWF.

DH and GDH.2 Based Protocols

The following protocols use either a 2 party DH key exchange, where two parties are allowed to derive a shared secret key over an insecure communication channel or GDH.2 [6], an extension of DH that allows more than 2 parties to generate the shared group key.

In [8-9], the first tree based CGKA protocol called TGDH was proposed in which key tree and DHprotocol are combined to generate the group key. The authors compared TGDH with STR [10], and GDH.3 [6] protocol. The TGDH is comparatively efficient in join events, and it is best in leave events. The merge operation of TGDH requires more round and the partition is the most expensive operation in TGDH. The TGDH is more efficient in both computation and communication than GDH. The 'Skinny TRee' (STR) [10] employs the key tree concept used in TGDH.

The computation cost of leave, merge and partition events are higher than in TGDH. The STR is communication efficient and it is more secure against attacks specific to group communication.

The Communication–Computation Efficient Group Key algorithm (CCEGK) [11] was developed for large and dynamic groups that combines two existing protocols such as EGK [12] and TGDH. It outperforms EGK, TGDH and STR protocols.

A novel logical key tree 'PFMH' has been constructed for the protocol called PFMH Tree-based Contributory Group Key Agreement (PACK) [13]. The PFMH is a combined tree structure of Partially Full (PF) key tree and Maximum Height (MH) key tree. The authors introduced a phantom node concept to reduce the cost of single leave event by allowing the existing member to occupy more than one leaf node in the key tree. Unfortunately, the computation and communication cost upon a single member leave is slightly higher than those in TGDH.

A ternary tree based GKA protocol was designed in [14]. It treats every three members as a subgroup and employs GDH.2 [6] to derive a subgroup key. It reduces the number of rounds from $O(\log_2 n)$ to $O(\log_3 n)$. It adopts the merge operation used in CCEGK protocol.

ECDH Based Protocols

The ECDH based protocols perform an efficient point multiplication with smaller key length. The TFAN [15] combines the two protocols such as μTGDH [16] and μSTR [16]. It is efficient in both computation and communication cost. During join event, it requires less computation and communication cost than μTGDH. It needs only a minimum number of serial multiplications than μSTR in leave and refresh operations.

The cluster based GKA [17] scheme for MANETs was introduced to provide reliable communication through redundancy. It uses some main ideas of CCEGK and discusses both balanced and imbalanced algorithm for group operations. It gives the better performance by avoiding "1-affects-n" problem in the group communication.

An authenticated GKA (AECTGDH) [18] considers only the single join and single leave event and it is compared to TGDH operations. It has the same number of rounds and messages as in TGDH, but it is more efficient than TGDH since it uses ECDH instead of DH. It provides implicit key authentication, but it cannot ensure the key confirmation property of GKA.

The DMHBGKA [19] supports secure multicast communication among the users. It performs rotation operations during the join and leave operation to balance the key tree. It requires less number of key operations than GDH. Although it takes more rounds in communication, it has lower message size than GDH and TGDH. The synchronization time and resynchronization time of the group key during join and leave is shorter than DH and GDH since point multiplication needs less time than modular exponentiation.

One Way Function Based Protocols

The following protocol uses a one way function to compute the blinded key from the secret key. A mixing function is used to generate the secret key of a non-leaf node from the blinded keys of its two children. The DISEC [20] developed for secure multicast communication delegates the task of controlling the group and distributing the keys evenly among all group members. It uses binary IDs and key association for each member and this protocol is immune to collusions.

2.2 Batch Rekeying (BR) Based Protocols

In the Interval/Period based BR protocols, join and leave requests are collected during a period of interval called rekey interval and rekeying is performed on a batch of those join and depart requests. It provides a tradeoff between performance and forward, backward secrecy. The DH and OWF based BR protocols are described in the following subsections.

DH Based Protocols

The following protocols change the group key either after a rekey interval or when the join tree is full. Dynamic SubTree (DST) group key agreement (DSGKA) [21], Join-Exit-Tree (JET) [22] and Weighted-Join-Exit-Tree (WJT) [24] use a new logical key tree with Join&Exit tree topology and they use some conditions to activate and deactivate the Join&Exit tree. DST, JET, Join-tree-based CGKA (JDH) [23] and WJT perform rekeying after a fixed number of members have joined in the join tree. In [21-24], the new joining users are first added to the join tree and then they are relocated into the main tree when either the join tree is full or leave operation in the main tree. In DST and JET, the average time cost for join and leave event is O(log(log n)) which is less than O(log n), the cost in TGDH. The protocols DST and JET have less overhead in both computation and communication [21-22]. They give better performance than TGDH in sequential user join event when the group is large. They assume that the users know their self-estimated departure time at the time of joining the group.

JDH protocol also aimed to improve efficiency in rekey operations and it reduced the average join time from O(log(log n)) to O(1) [23]. It adopts the second relocation method employed in DST protocol. It inserts the joining user at the root of main tree which leads to the skewed tree. The time complexity of user leave event is O(log n).

WJT has been developed to present a time efficient CGKA protocol and Exit tree is organized as a weighted tree based on the users' leaving time. It employs a tree balancing algorithm to reduce the height of the main tree so that the average time complexity of join and leave operation has been reduced to O(1) from O(log(log n)) [24]. The location of the users in the key tree is determined based on their staying time. This protocol also leads to a skewed join and exit tree.

Three interval based algorithms such as Rebuild, Batch and Queue-Batch (QB) have been proposed in [25] and it has been stated that the algorithm QB performs well among all. The Queue-merge algorithm of QB inserts a temporary sub tree with joining users into the highest departed position of the key tree. The authors illustrated that the QB outperforms the IR method. The Authenticated Tree-Based Group Diffie–Hellman (A-TGDH) has been proposed to provide key authentication for QB protocol. The Residency-based BR (RBR) [26] combines two protocols such as TGDH and STR to create a new key tree topology using subtree division and residency time classification concepts. It reduces the total rekeying cost by maintaining the location of join and leave member close to the root node of the key tree. In this protocol, the actual join event is postponed until the join tree becomes full and it uses imbalanced skinny main tree. Its rekeying cost is lower than the cost in the existing protocols such as TGDH, PACK and DST.

Table 1. Individual Rekeying Protocols

Sl. No	Protocol	Objective	Metrics used for computation and communication	Advantages	Disadvantages
			DH and GDH.2 based Protocols		
1	TGDH	To design a secure, scalable and fault-tolerant protocol.	Computation: Serial no .of exponentiations, signature generations & verifications. Communication: No. of rounds, Total no. of messages.	All operations need only 3h-3 no. of exponentiations. Efficient in computation and in leave event.	Most expensive partition event.
2	STR	To improve the communication efficiency, to minimize the size of messages and number of rounds.	Computation: No. of sequential exponentiations. Communication No. of rounds, Total number of messages. Robustness.	Communication efficient and higher robustness to network partitions. Fault-tolerant and more suitable for high-delay WAN.	Computationally inefficient.
3	CCEGK	To provide a communication and computation efficient GKA.	Computation: Sequential exponentiation, Signatures and Verifications. Communication: Total rounds, total messages, unicasts and broadcasts messages.	Outperforms TGDH, EGK and STR.	Skewed key tree during join operation.
4	PACK	To achieve performance lower bound in the operations.	Time, No. of exponentiations and rekeying messages in terms of multicast.	O(1) round for single join.	Slightly higher costs than TGDH for single leave event.
5	GKA scheme in [14]	To design a GKA with reduced no. of rounds and messages.	Computation: Sequential exponentiations Communication: No. of rounds Total messages, No of unicast and broadcast messages.	Reduced complexity. Replaces DH and Binary tree with GDH.2 and Ternary tree respectively.	Restrictions in group size, no. of join and leave members.

Table 1. *(continued)*

Sl. No	Protocol	Objective	Metrics used for computation and communication	Advantages	Disadvantages
			ECDH based Protocols		
6	TFAN	To develop both computation and communication efficient GKA for MANETs.	Memory cost: Secret and Public keys. Computation: No. of serial multiplications,Communication: No. of rounds, No. of broadcast and broadcast sizes.	Gives an optimal trade-off between computation and communication. Memory space is lower than µSTR. Most efficient merge and partition event.	Memory space is higher than µTGDH.
7	GKA method in [17]	To present a cluster-based and network topology independent GKA for MANETs.	Computation: Sequential Multiplications Communication: No. of rounds, no. of messages.	Provides redundancy. Gateway nodes use different secret keys in different clusters to provide more security.	Overhead in maintaining more number of keys.
8	AECTGDH	To design an authenticated GKA based on ECDH.	Computation: Execution time. Communication: No. of rounds, No. of messages, Message length.	Provides implicit key authentication. Efficient than TGDH.	Only single join and leave. No key confirmation property.
9	DMHBGKA	To introduce dynamic multicast height balanced GKA for MANETs.	Computation: Key operation, Group key reconstruction, Exponentiations, Node traversal. Communication: No. of rounds, Messages sent per node, Total messages and message size.	Supports securer delivery of multicast data, reduced overhead cost of security management, communication cost and key management functions. Robustness.	Handles only single join and leave events.
			One Way Function based Protocols		
10	DISEC	To design a Scalable Protocol.	No. of keys at each member and in the group. Communication: No. of messages.	Immune to collusions.	No merge and partition operations.

Table 2. Batch Rekeying Protocols

Sl. No	Protocol	Objective	Metrics used for computation and communication	Advantages	Disadvantages
			DH based Protocols		
1	DST/ DSGKA	To develop a time efficient GKA.	Computation: No. of exponentiations. Communication No. of. rounds	Less average time cost.	Requires user's staying time.
2	JET	To achieve better time efficiency.	Computation: Key update time and No. of exponentiations for single user and group. Communication: No. of messages with and without multicast.	Low computation and communication cost.	Requires user's staying time.
3	JDH	To improve the time efficiency in rekey operations.	Computation: Sequential exponentiations. Communication: No. of rounds, unicast and broadcast messages	Reduced average join time from O(log(log n)) to O(1).	Skewed Join tree.
4	WJT	To present time efficient CGKA.	Computation: Averageno. of Sequential exponentiations, averagejoin and leave time.	Reduced average join and leave time from O(log(log n)) to O(1).	Skewed join and exit tree.
5	GKA method in [25]	To develop interval based distributed rekeying alg.	Computation: No. of exponentiations, Communication: No. of. rounds, renewed nodes and broadcast keys.	Reduces cost in a highly dynamic environments.	Height increase during join.
6	RBR	To reduce the total rekeying cost.	Computation: No. of. rounds Communication: No. of messages	Has lowest rekeying cost.	Higher overhead in small group.
			One way Function based Protocols		
7	EDKAS	To minimize the cost of managing the group key.	Computation: No. of mixing functions. Communication:No. of renewed nodes	Immune to collusions.	Height increase during join.
8	DBRM & SDBR	To design collusion resistant algorithm.	Computation: No. of OWF. Communication: No. of renewed nodes	Prevents collusion attack.	Morerenewed node than EDKAS.

One Way Function Based Protocols

EDKAS [27] protocol uses a Distributed One-way Function Tree (DOFT) for generating the group key. It is immune to collusions by maintaining a Responsible Member set for each node to limit the distribution of secret and blinded keys. In this protocol, the leaving node with the lowest ID is selected as the insertion node to insert the temporary key tree. Two algorithms, DBRM and SDBR [28] of a secure distributed BR protocol are used for marking the key tree and re-computing the group key respectively. It avoids a renewed node to be rekeyed more than once. But, the number of renewed nodes is higher than that in EDKAS which increases the communication cost and total rekeying cost also.

3 Comparison of Tree-Based CGKA Protocols

This section compares the various CGKA protocols. Table 1 and Table 2 consolidate the key characteristics of IR and BR protocols respectively. From the performance of both IR and BR based protocols, it is inferred that the BR based protocols outperform the protocols with IR approach. In [25], Lee et al. compared BR with IR and proved that the cost of BR is lower than the IR cost. And also, in dynamic networks with frequent membership changes such as CR-MANETs, IR is not a suitable method. The BR based protocols can be employed in CRNs to achieve the maximum efficiency in rekeying cost and time.

The DH and GDH perform modular exponentiations with expensive computational operations which need high end CPU and memory capabilities. The ECDH performs computationally efficient operations with smaller keys and it is quite suited for smaller and less powerful devices [31]. From the tables, it is concluded that the ECDH based CGKA protocol with batch rekeying technique minimizes both computation and communication complexity. Since the nodes in CR-MANETs are resource restricted battery-powered nodes, the protocols which combine the ECDH and BR concept is more suitable rather than DH and OWF.

4 Conclusion

A survey of group key agreement schemes applicable to cognitive radio networks were presented in this paper. From the survey, it is inferred that ECDH based protocols with batch rekeying perform well by saving the computation cost and bandwidth compared to DH and one way function based protocols. Hence, the tree based protocol with the combination of ECDH and batch rekeying is the best group key agreement scheme for cognitive radio mobile ad hoc networks.

References

1. Mitola, J.: Cognitive radio for flexible mobile multimedia communications. In: IEEE International Workshop on Mobile Multimedia Communications, pp. 3–10 (1999)
2. Akyildiz, I.F., Lee, W.Y., Vuran, M.C., Mohanty, S.: Next generation/dynamic spectrum access/cognitive radio wireless networks: A survey. Computer Networks Journal 50, 2127–2159 (2006)

3. Wyglinski, A.M., Nekovee, M., Hou, T.: Cognitive radio communications and networks: Principles and Practice. ElsevierAcademic Press (2009)
4. Parvin, S., Hussain, F.K., Hussain, O.K., Han, S., Tian, B., Chang, E.: Cognitive radio network security: A survey. Journal of Network and Computer Applications 35, 1691–1708 (2012)
5. Diffie, W., Hellman, M.: New directions in cryptography. IEEE Transactions on Information Theory 22(6), 644–654 (1976)
6. Steiner, M., Tsudik, G., Waidner, M.: Diffie-hellman key distribution extended to group communication. In: 3rd ACM conference on Computer and Communications Security, pp. 31–37. ACM Press (1996)
7. Koblitz, N.: Elliptic Curve Cryptosystems. Mathematics of Computation 48(177), 203–209 (1987)
8. Kim, Y., Perrig, A., Tsudik, G.: Simple and fault-tolerance key agreement for dynamic collaborative groups. In: 7th ACM Conference on Computer and Communications Security, pp. 235–244 (2000)
9. Kim, Y., Perrig, A., Tsudik, G.: Tree-based group key agreement. ACM Transactions on Information and System Security 7(1), 60–96 (2004)
10. Kim, Y., Perrig, A., Tsudik, G.: Communication-efficient group key agreement. In: Dupuy, M., Paradinas, P. (eds.) Trusted Information. IFIP, vol. 65, pp. 229–244. Springer, Boston (2001)
11. Zheng, S., Manz, D., Alves–Foss, J.: A communication-computation efficient group key algorithm for large and dynamic groups. Computer Networks 51(1), 69–93 (2007)
12. Alves-Foss, J.: An efficient secure authenticated group key exchange algorithm for large and dynamic groups. In: 23rd National Information Systems Security Conference, pp. 254–266 (2000)
13. Yu, W., Sun, Y., Liu, K.J.R.: Optimizing the rekeying cost for contributory group key agreement Schemes. IEEE Transactions on Dependable and Secure Computing 4(3), 228–242 (2007)
14. Tripathi, S., Biswas, G.P.: Design of efficient ternary-tree based group key agreement protocol for dynamic groups. In: Communication Systems and Networks and Workshops (2009)
15. Liao, L., Manulis, M.: Tree-based group key agreement framework for mobile ad-hoc networks. In: 20th International Conference on Advanced Information Networking and Applications, vol. 2, pp. 5–9 (2006)
16. Manulis, M.: Contributory Group Key Agreement Protocols, Revisited for Mobile Ad-Hoc Groups. In: IEEE International Conference on Mobile Adhoc and Sensor Systems Conference (2005)
17. Chen, Y., Zhao, M., Zheng, S., Wang, Z.: An Efficient and Secure Group Key Agreement Using in the Group Communication of Mobile Ad-hoc Networks. In: International Conference on Computational Intelligence and Securit, vol. 2, pp. 1136–1142 (2006)
18. Hong, T., Liehuang, Z., Zijian, Z.: A Novel Authenticated Group Key Agreement Protocol based on Elliptic Curve Diffie-Hellman. In: 4th International Conference on Wireless Communications, Networking and Mobile Computing (2008)
19. Lin, H.Y., Chiang, T.C.: Efficient key agreements in dynamic multicast height balanced tree for secure multicast communications in ad hoc networks. EURASIP Journal on Wireless Communications and Networking (2011)
20. Dondeti, L.R., Mukherjee, S.: DISEC: A distributed framework for scalable secure many-to-many communication. In: 5th IEEE Symposium on Computers and Communications, pp. 693–698 (2000)

21. Mao, Y., Sun, Y., Wu, M., Liu, K.J.R.: Dynamic join-exit amortization and scheduling for time efficient group key agreement. In: IEEE INFOCO, vol. 4, pp. 2617–2627 (2004)
22. Mao, Y., Sun, Y., Wu, M., Liu, K.J.R.: JET: Dynamic Join-Exit-Tree Amortization and Scheduling for Contributory Key Management. IEEE/ACM Transactions on Networking 14(5), 1128–1140 (2006)
23. Gu, X., Yang, J., Yu, J., Lan, J.: Join-Tree-based contributory group key management. In: 10th IEEE International Conference on High Performance Computing and Communications, pp. 564–571 (2008)
24. Gu, X., Cao, Z., Yang, J., Lan, J.: Dynamic Contributory Key Management Based On Weighted-Join-Exit-Tree. In: IEEE MILCOM (2008)
25. Lee, P.C., Lui, C.S., Yau, K.Y.: Distributed collaborative key agreement and authentication protocols for dynamic peer groups. IEEE/ACM Transactions on Networking 14(2), 263–276 (2006)
26. Guo, C.J., Huang, Y.M.: Residency-based distributed collaborative key agreement for dynamic peer groups. International Journal of Innovative Computing, Information and Control 8(8), 5523–5542 (2012)
27. Zhang, J., Li, B., Chen, C.X., et al.: EDKAS: A Efficient Distributed Key Agreement Scheme using One Way Function Trees for Dynamic Collaborative Groups. In: IMACS Multiconference on Computational Engineering in Systems Applications, pp. 1215–1222 (2006)
28. Li, B., Yang, Y., Lu, Z., Yuan, B., Long, T.: Secure Distributed Batch Rekeying Algorithm for Dynamic Group. In: IEEE Conference ICCT, pp. 664–667. IEEE Press (2012)
29. El-Hajj, W., Safa, H., Guizani, M.: Survey of Security Issues in Cognitive Radio Networks. Journal of Internet Technology 12(2), 181–198 (2011)
30. León, O., Hernandez-Serrano, J., Soriano, M.: A new crosslayer attack to TCP in cognitive radio networks. In: Second International Workshop on Cross Layer Design (2009)
31. The National Security Agency (NSA)/TheCentral Security Service (CSS), `http://www.nsa.gov/business/programs/elliptic_curve.shtml`

Performance Analysis of Interference Aware Power Control Scheme for TDMA in Wireless Sensor Networks

Ajay Sikandar and Sushil Kumar

School of Computer and Systems Sciences,
Jawaharlal Nehru University
New Delhi-110067
ajay.sikandar@gmail.com, skdohare@yahoo.com

Abstract. Energy saving is an important issue in wireless sensor network. Energy consumed in transmission is higher in the presence of interfering nodes due to more re-transmissions are required for a successful transmission. In this paper, interference aware power control PCTDMA for channel access is proposed. Energy consumption in the presence of interferes is investigated. Mathematical models for channel access and power control are derived. The model is simulated in MATLAB and compared with TDMA. The proposed model consumed low energy as compare to traditional TDMA.

Keywords: MAC, SIR, Wireless sensor networks.

1 Introduction

Wireless sensor networks (WSN) consist of a large number of battery powered sensors. Sensor nodes monitor events of interest in a specified sensing field. Sensors can be deployed either in planned manner or randomly. Planned deployment is used in areas where prior knowledge of the sensing field is available. Sensors are deployed randomly in unknown and hostile sensing field. It is difficult and expansive to replace batteries of sensors when they are deployed randomly. WSN has wide range of potential applications including environment monitoring, habitat monitoring, healthcare, battlefield control, and whether forecasting [1].

The prime objective in wireless sensor network's design is maximizing life time of sensor and network. Since transmission in WSN consumes more energy, it is primary concern while achieving desired network operation. In these situations, the medium access control (MAC) protocol must be energy efficient by reducing the potential energy wasted. When the sensor node receives more than one packet at the same time, collision is occurred. All these collided packets discarded and it requires retransmission. These retransmissions consume additional energy. The second reason of energy consumption is overhearing, meaning that sensor node receives packet that destined for other sensor node. The third reason of energy wastage is due to control packet overhead. Minimizing control packets is one of energy conservation issues. The main concern of MAC is to control channel access, and calculate the channel capacity utilization, network delay and power constraints [2].

M.K. Kundu et al. (eds.), *Advanced Computing, Networking and Informatics - Volume 2*, Smart Innovation, Systems and Technologies 28,
DOI: 10.1007/978-3-319-07350-7_11,

Time division multiple access (TDMA) protocol allows several users to share the same frequency channel by dividing the signal into different time slots. TDMA slots are allocated to sensor nodes that can turn on the radio during the assigned time slots, and turnoff the radio when not transmitting or receiving in the sleep scheduling. It makes sure collision free multiple accesses, that is, a node can transmit or receive in time slots that have been assigned to it. In order to make interference free communication, time slot is assigned to every communication link, thus number of slots equal to the number of communication links of the network [3]. In this protocol, hidden terminal problem can be solved without overhead because it can schedule transmission time of neighboring nodes to occur at different time. In TDMA protocol, several problems occur like time synchronization and hard to be used directly in a scalable network [4]. A unique characteristic of WSN is that transmission by a sensor node will be received by the all sensor nodes with its transmission range. It may be caused signal interference to some sensor nodes that are not intended receiver [5].

Power control based on signal-to-noise ratio (SNR) is a technique to manage co-channel interference in a frequency reuse system. Early studies have focused on signal-to-interference ratio (SIR) balancing that require centralized control for global power adjustment and channel reassignment. In the TDMA, it is important to have energy efficient communication, because high transmit power can increase the signal-to-interference-noise (SINR) at the receiver to enable successful reception on the link, and lower power can diminish interference to other simultaneously utilized link [6].

In this paper we propose interference aware power control scheme for TDMA (PCTDMA). Since the recent technology typically require a large SINR to correctly receive a packet. Our main goal is to reduce interference at receiver node, when one or more sensor nodes are transmitting. Each sensor node has to transmit at high power to meet the SINR requirement, which could increase the interference at return;therefore a mathematical model to minimize the power consumption is presented.

The rest of the paper is organized as follows: In section 2 we present the overview of the related work. In section 3, analytical study of power consumption in the presence of interference is provided. Section 4 presents the simulation result and discussion. Finally we conclude paper in section 5.

2 Related Work

In TDMA,MAC protocols that guarantee collision free communication and fairness. In this paper [7] author suggests TDMA scheme for energy efficiency in order to construct transmission schedule to reduce power consumption at same time minimize end to end transmission time to gateway. TDMA based wake up interval are used for propagating wake up message before data transmission. In this model, TDMA schedule determines the length of listening period, that is as small as possible, and at the same time make sure that wake up packet collision are avoided. In [8], author presented a method for advance energy efficient power optimization, especially for interference limited environment. Authors consider both circuit and transmit power and main aim to make energy efficiency over throughput. In [3]authors address energy efficient sleep scheduling for low data rate WSNs, where the energy consumption of state transition is considered. Author suggests a novelinterference free TDMA sleep scheduling problem called contiguous link scheduling to reduce the frequency of state transition. In [5] authors present efficient centralized and distributed algorithms to obtain a valid link scheduling with theoretically proven performances for more realistic wireless network model. Authors propose two centralized heuristic algorithms: node based scheduling and level based scheduling in [9] Node based scheduling is adapted from classical multi hop scheduling algorithms for general ad-hoc network. Level based scheduling is novel scheduling algorithm for many to one communication in sensor networks. Authors derive upper bounds for these schedules as a function of the total number of packet generated. In [6] author investigates the problem of energy efficiency in TDMA link scheduling with transmission power control using a realistic SINR based interference model. In this scheme work are to be accomplished in four fold. Firstly, express joint scheduling and power control as a novel optimization problem that provides tunable between throughput, energy and latency. Second, author suggest both exponential and polynomial greedy based heuristic algorithm. In third part, authors present the energy latency throughput tradeoff that can be achieved with joint link scheduling and power control. Major energy saving can be obtained significantly sacrificing throughput. Finally minimizing the total energy cost subject to all packets of the links is transmitted within latency bound.

3 Channel Model

In TDMA protocol, the time is divided into slots of equal length. The number of slots in each frame is fixed. It is based on reservation and scheduling. Access point decides which slot is to be used by which sensor node. In this protocol, there is no contention overhead and collision. Protocol has two states: active and sleep. In active state, a sensor node can receive or transmit packets. If no packets are available then node goes to idle mode. In idle mode, node doing nothing but ready to receive or transmit packet. In sleep state, sensor node turnoff its power for a period. The energy consumption of sensor node in sleep state is much less than the energy consumption in active state Fig.1.

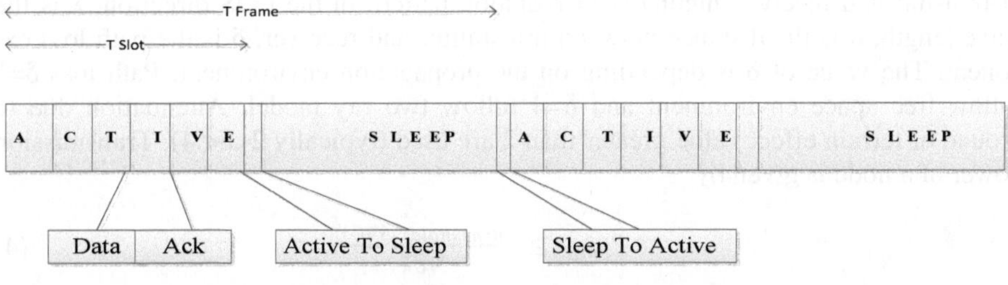

Fig. 1. TDMA based active sleep model

Let there are N number of sensor nodes deployed randomly in the sensing field. Total energy consumption by all sensor nodes EC_{total}can be expressed as

$$EC_{total} = \sum_{i=1}^{N}[P_i^{tx}\left(t_i^{tx} + t_i^{txtosl}\right) + P_i^{rx}\left(t_i^{rx} + t_i^{rxtosl}\right) + t_i^{idle}P_i^{idle} + t_i^{sl}P_i^{sl}] \quad (1)$$

Where, P_i^{tx} and P_i^{rx} are power consumption of i^{th}transmitter and receiver node. t_i^{tx} and t_i^{rx} are time consumption, while transmitting and receiving at node i. $t_i^{tx\ to\ sl}$ and $t_i^{rx\ to\ sl}$ are transition time consumption between transmitting to sleep andreceiving to sleep states respectively. $t_i^{idle}$is time consumption in ideal mode of node i. $P_i^{idle}$is power consumption in ideal mode of node i. $t_i^{sl}$is time consumption in sleep states of node i. $P_i^{sl}$is power consumption in sleep states of node i .Energy saving is accomplished by turning off the transceivers of every sensor in the network during the idle operation. Our main objective is to minimize power consumption. If, the distance between neighboring sensors is less than average distance betweensensors then transmission power can be saved. Power consumption in sleep state is omitted because it is far less than all other states. Therefore the equation (1) can be simplified as

$$EC_{total} = \sum_{i=1}^{N}[P_i^{tx}\left(t_i^{tx} + t_i^{txtosl}\right) + P_i^{rx}\left(t_i^{rx} + t_i^{rxtosl}\right) + n\left(\frac{T_{slot}}{T_F}\right)P_i^{idle} \quad (2)$$

Where n is the number of timeslots in active state, T_F is the total period and T_{slot} is the length of a TDMA slot.

3.1 Power Consumption Model

Let, a transmission is going on from a source to receiver nodes denoted by S and R respectively. The ratio of received power and transmit power using free spacepropagation model [10] is given by

$$\frac{P_R}{P_T} = \left[\frac{\sqrt{G\lambda}}{4\pi d}\right]^{\delta} \quad (3)$$

Where P_R is the received power and P_T is the transmit power,$\sqrt{G}$ is the product of transmit and receive antenna field radiation pattern in the LOS direction, λ is the wave length, d is the distance between transmitter and receiver, δ is the path loss exponent. The value of δ is depending on the propagation environment. Path loss δ=2 follow free space environment and δ=4 follow two ray model. Attenuation due to ground or terrain effect value greater than 2 are used (typically 2<δ< 4). Transmission power of a node is given by

$$P_{Tmin} \geq \frac{P_{Rtheshold} * (4\pi d)^{\delta}}{\sqrt[\delta]{G\lambda}} \quad (4)$$

WhereP_{Tmin} is the minimum transmitting power,$P_{Rtheshold}$ is the minimumrequired received signal. When sender node transmitting data to receiver node.Interference is created by the set of nodes, which is in the transmission range ofreceiver node (cf. Fig. 2).

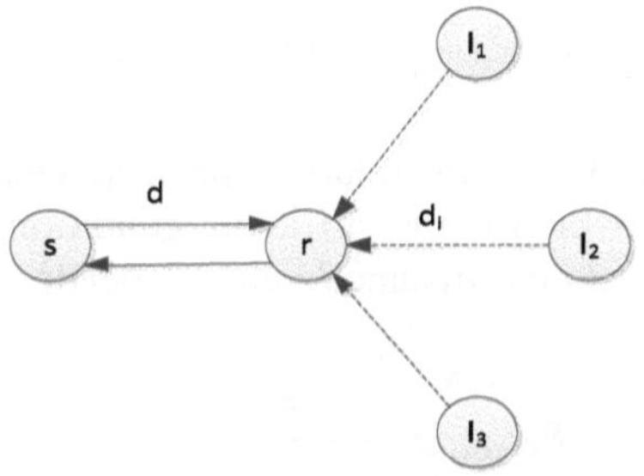

Fig. 2. Interference at Receiver node

Now, signal to interference pulse noise (SINR) ratio is measured as

$$SINR = \frac{P_R}{P_I + N} \tag{5}$$

WhereP_R is the received power signal from transmitting node T. P_I is the received power signal from interfering node I, N is the some noise term. SINR is used as a measure of perceived network interference at receiver node. A transmission issuccessfully received if and only if

$$\frac{P_R}{P_I + N} \geq SINR_{threhold} > 0$$

In order to correctly receive a packet, the receive power level should be greater than the power level of interfering node and noise, so transmitted power of sender node should be greater than or equal to the threshold value. Thus

$$P_{Tmin} \geq \frac{SINR_{Rtheshold} * P_T}{P_R} \tag{6}$$

When node receives a packet, some noise term and interference from other nodes may include at receiver node. The noise power level is much lower than theinterfering signals. Noise can be neglected. We have

$$P_{Tmin} \geq \frac{SIR_{Rtheshold} * P_T}{P_R} \tag{7}$$

Now, calculate SIR value at receiver node, from (1) and (3)

$$SIR = \frac{P_T \left[\frac{\sqrt{G\lambda}}{4\pi d}\right]^{\delta}}{P_{IT} \left[\frac{\sqrt{G\lambda}}{4\pi d_i}\right]^{\delta}} \tag{8}$$

Hered_i is the distance between interfering node and receiving node, P_{IT} is the transmitted power by interfering nodes. All nodes are transmitting same power so power level of transmitting node equal to the transmission power of interfering node. In order to successfully reception of packet, SIR value is greater than equal to certain threshold. After calculating equation (8), we get.

$$SIR = \left(\frac{d_i}{d}\right)^{\delta} \geq SIR_{threhold} \quad (9)$$

This equation shows that Minimum relative distance separation betweensimultaneously transmitting nodes, which is the sufficient condition for a successful reception at receiver node. This relative distance value depends on the $SIR_{threhold}$ and path loss exponent.

$$P_{Tmin} \geq \frac{\left(\frac{d_i}{d}\right)^{\delta} * P_T}{P_R} \quad (10)$$

This equation shows, require power consumption at the receiver node depends on relationship of interference range (distance of the interfering node to receiver node) to the transmission range (distance between the transmitting node and receiving node).

4 Simulation and Result

In this section, outcome of the simulation are carried out to analyze theperformance of channel access model for TDMA in randomly deployed sensornetwork. In the simulation, sensing field of area 500*500 m^2 is assumed. The number of nodes to be deployed is assumed 800. The transition time between the sleep and active state is assumed to 470μ sec. Transmission range of each sensor is set to 15 m. Power consumption in transmitting is assumed to be 17mW. We simulated proposed model in MATLAB.TDMA has been used for comparative analysis.

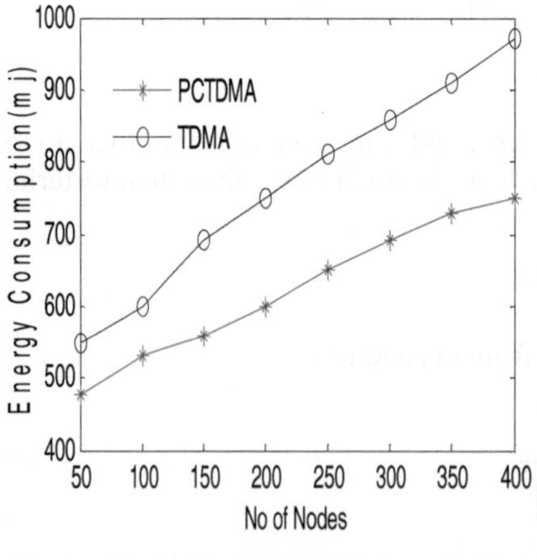

Fig. 3. energy consumption versus number of sensor nodes for PCTDMA and traditional TDMA schemes

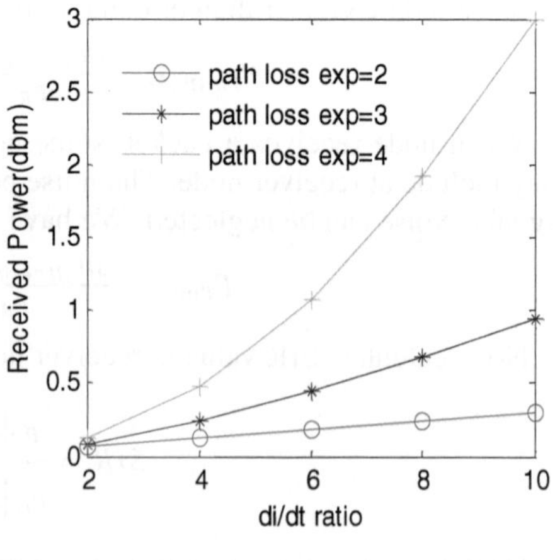

Fig. 4. received power versus ratio of interference range to transmission range for different path loss exponent

Fig. 3 shows that the traditional TDMA wastes more energy due to the idlelistening especially for higher number of sensors. Result shows that PCTDMAconsume less power compare to the traditional TDMA.Fig. 4 shows energyconsumption at receiver node increases with increasing minimum relative distance. It is observed that sensors with lower relative distances should be switched to sleep state to minimize the interference at receiving node that saves the power consumption of the whole networks.

5 Conclusion

In this paper, we have proposed interference aware power control scheme for TDMA (PCTDMA). Mathematical models for channel access and power control are derived. Simulation results show that the traditional TDMA wastes more energy due to the idle listening especially for higher number of sensors. And sensors with lower relative distances should be switched to sleep state to minimize the interference at receiving node that saves the power consumption of the whole networks.

References

1. Akyildiz, I.F., Su, W., Sankaransubramaniam, Y., Cayirci, E.: Wireless sensor network: a survey. Computer Networks 38(4), 393–422 (2002)
2. Demirkol, I., Ersoy, C., Alagoz, F.: MAC Protocols for Wireless Sensor Networks: A Survey. IEEE Communications Magazine 44(4), 115–121 (2006)
3. Ma, J., Lou, W., Wu, Y., Li, X., Chen, G.: Energy Efficient TDMA Sleep Scheduling in Wireless Sensor Network. In: Proceeding IEEE INFOCOM, pp. 630–638 (2009)
4. Wang, W., Wang, H., Peng, D., Sarif, H.: An Energy Efficient Pre-Scheduling For Hybrid CSMA/TDMA in Wireless Sensor Network. In: 10th IEEE Singapore International Conference on Communication Systems (2006)
5. Wang, W., Wang, Y., Yang Li, X., Song, W.Z., Frieder, O.: Efficient Interference-Aware TDMA Link Scheduling for static Wireless Networks. In: Proceeding of ACM Mobile Computting, pp. 262–273 (2006)
6. Lu, G., Krishnamachari, B.: Energy Efficient Joint Scheduling and Power control for wireless sensor network. In: Second Annual IEEE Communications Society Conference on Sensor and Ad Hoc Communications and Networks, pp. 362–372 (2005)
7. Pantazis, N.A., Vergadosb, D.J., Vergados, D.D., Douligeris, C.: Energy efficiency in wireless sensor networks using sleep mode TDMA scheduling. Ad Hoc Network 7(2), 322–343 (2008)
8. Miao, G., Himayat, N., Ye Li, G., Koc, A.T., Talwar, S.: Interference-Aware Energy-Efficient Power Optimization. In: IEEE International Conference on Communications (2009)
9. Ergen, S.C., Varaiya, P.: TDMA Scheduling Algorithms for Wireless Sensor Network. Wireless Networks 16(4), 985–997 (2009)
10. Goldsmith, A.: Wireless Communication. Cambridge University Press (2007)

Cluster Based Controlling of Route Exploring Packets in Ad-Hoc Networks

S. Zeeshan Hussain and Naeem Ahmad

Department of Computer Science
Jamia Millia Islamia University
New Delhi, India
szhussain@jmi.ac.in, naeemahmad.jmi@gmail.com

Abstract. Route Discovery in Ad-hoc Networks is an important process that is used to find the optimal path for data transmission. This route exploring process can be either proactive or reactive depends on the nature of the routing protocol. When a routing protocol initiates a route exploring process, it uses query packets to discover the optimal path. These packets are broadcasted over the network. Since they propagate in the network even after route has been found, these packets become problem of congestion in the network. So, we have proposed a cluster based approach along with Expending Ring Search (ERS) to control these route discovery packets in these networks. In our approach, we divided the whole network into clusters to achieve the scalability of the network and used modified BERS+ to reduce the retransmission of query packets in the network.

Keywords: Route exploring process, Clustering, Multiple class of Congestion, Expending ring search, Unnecessary propagation.

1 Introduction

Self-organizing structure of mobile nodes collectively forms Mobile Ad-hoc Networks (MANETs). These networks allow the mobile nodes to roam freely at their will. This slave less movements causes the change in the structure of the networks. This temporary topology makes routing difficult for the data transmission in MANETs [1]. To achieve adaptability, many routing protocols have been proposed which used different strategies to explore the requested route. All routing protocols used route exploring packets to discover the path for data transmission [2], [3]. These packets cover a large area of the network to find the path and propagate even after route has been discovered. Due to this unnecessary circulation, they pose the problem of congestion. To minimize such type of congestion, many route exploring packets controlling techniques have been proposed [4]. These schemes were based on either selective flooding or bounded flooding. All these schemes used some intermediate nodes unnecessarily. It causes the unnecessary energy consumption of these unnecessary intermediate nodes. Moreover, these schemes also lacked the scalability of the network. Considering

M.K. Kundu et al. (eds.), *Advanced Computing, Networking and Informatics - Volume 2*, Smart Innovation, Systems and Technologies 28,
DOI: 10.1007/978-3-319-07350-7_12, © Springer International Publishing Switzerland 2014

these issues, we have proposed a scheme which used network scalability aspect of clustering and packets controlling property of ERS technique. Though, it is not a first contribution in cluster based routing algorithm. Various other cluster based routing schemes have been proposed [5]. But Cluster based implementation of ERS has not yet been done to make the comparison with other schemes quantitatively. In section 3, we have shown that this cluster based implementation is far better than simple Blocking-ERS+ scheme to control the unnecessary circulation with least use of intermediate nodes. BERS+ [4] is a standard technique among the ERS based schemes which are analyzed in [6]. But Cluster based implementation of BERS+ used less number of intermediate nodes than simple BERS+. It helps nonparticipating intermediate nodes to save their energy. On the other hand, we used Distributed Weighted Clustering Algorithm (DWCA) technique [7] which is based on the combined weight metric of mobile nodes. It is described in detail in Section 4.1.

Rest of the papers is summarized as follows: Section 2 described the previous work related to congestion control techniques. Section 3 demonstrated the mathematical comparison with other algorithm. Section 4 represented the design mechanism of clustering, routing packets, and tables. In Section 5, we discussed the methodology used in route exploring process, route maintenance scheme and also about packets controlling technique. In Section 6, we concluded our work.

2 Related Work

Every person wants to connect his call at once and also communicate smoothly without any interference. It is only possible when all time one of the communication lines is free for transmission without congestion. Multiple class of congestion [8] exists in networks. It can be either due to heavy data transmission or route discovery process. To make collision and congestion free networks, several routing strategies pertaining to the selective and limited flooding have been proposed. The objective of these schemes was to transmit the route exploring packet to all nodes with minimum conveying nodes, so that the route exploring expenses can be curtailed. These schemes are broadly classified into reliable and unreliable schemes [5]. In early stage of this field, reliable schemes were proposed to lessen the broadcast expenses of route exploring process. Flooding and all Self-pruning schemes belong to this category. FRESH [9], DREAM [10], probabilistic scheme [11], Query Localization [12], Location Aided routing [13], HoWL [14], MPR [15], WRS [16], all are self-pruning schemes so called selective flooding schemes. In MPR [15], only neighbor nodes those belong to the multi-point relay set retransmited the packet and other discarded it. This scheme was somewhat better than flooding. Like MPR [15], WRS [16], and [11], were also selective flooding based schemes. But these schemes required a large storage space at each node of the network and also prone to unnecessary circulation of packets. In [9], FRESH scheme was proposed in which route was found using on anchor nodes. Due to repeated use of the common intermediate nodes between anchor nodes, it consumed too much energy of nodes. It was also a time taken practice that wasted its time to search the anchor node.

On the other hand, various bounded broadcasting schemes have been proposed to reduce the retransmission of the query packet. Limited Broadcast Algorithm (LBA) [17], Limited Hop Broadcast Algorithm (LHBA) [18], TTL sequence based ERS [19], Blocking-ERS [20], Blocking-ERS+ [4], and Blocking-ERS* [21] belong to this category. All these schemes (except [19]) were based on chasing strategy. They allowed the route exploring packets within the limited area of the network. LBA [17] was channel capacity based scheme in which chase packet had higher priority than request packet. It caused the larger end to end delay. This flaw of LBA [17] was overcome in LHBA [18]. Now, it was destination node initiated. Due to limited propagation of chase packets, this scheme controls query packets of one part of the network. To overcome flaw of LHBA [18], a TTL sequence-based ERS [19] algorithm was proposed. This scheme did not use any chase packet to stop the query packets. In this scheme, source node flooded the query packets with specified TTL value. Only those intermediate nodes participate that fall within this searching ring of TTL value. If intermediate nodes failed to find path, source node again broadcasted the query packets with increased TTL value. Likewise, at each failure, it expanded its searching area as ripple across the water. Like FRESH [9], it also consumed too much power of intermediate nodes. Blocking-ERS [20] and BERS* [21] were extended version of [19]. They saved too much energy of the mobile nodes and also reduced the end to end delay. They introduced the slight delay in the propagation of the route exploring packets rather than broadcasting the packet repeatedly. The main drawback of both was that they were not adaptive with mobility of the destination node because of the limited journey of request packets. This shortcoming was improved in [4]. In [4], Blocking ERS+ was proposed that enabled the route request packets to travel beyond the searching ring. Route request packets were broadcasted by the intermediate nodes without waiting. It introduced time delay beyond the maximum hop count if route founder did not lie within the searching ring. This scheme has many benefits over aforementioned techniques such as minimum chasing time, minimum end to end delay, and maximum reception ratio. To make more efficient BERS+ scheme, we made control packet destination initiated rather than source initiated. It helps to control the packet with least time. We also used clustering to reduce the retransmission of the route exploring packets. The efficiency of our scheme can be measured mathematically. It is shown in next section.

3 Mathematical Modeling of Comparative Advantage

We suppose that network is made of N nodes which are grouped into C distributed clusters. For simplicity, we took cluster as a circle. These clusters are arranged like arrangement of seeds on the carom board at the start of the game see Fig. 2(a). Centered cluster contains 6 adjacent clusters. At each ring, numbers of clusters are twice the number of clusters in the previous ring. Calculation of each cluster searching ring is done based on this network configuration shown in Table 1. Thus total number of nodes in the network i.e.

Table 1. Calculation of nonparticipating nodes with clusters

R_c	C_u	**Participating nodes**
1	1+6	$\sum_{i=1}^{7}(k_i - l_i)$
2	1+6+12	$\sum_{i=1}^{19}(k_i - l_i)$
3	1+6+12+24	$\sum_{i=1}^{43}(k_i - l_i)$
:	:	
n	1+6.(2^n - 1)	$\sum_{i=1}^{C_u}(k_i - l_i)$

$$N = \sum_{i=1}^{C} k_i$$

Let k_i be the number of nodes in the ith cluster. Of them, l_i nodes are non-participating nodes of the route discovery in i^{th} cluster where $l_i \leq k_i$. Then participating nodes will be $k_i - l_i$ in i^{th} cluster. R_c is a cluster searching ring. C_u represents number of used clusters in a particular cluster searching ring.

R_s represents a simple searching ring and i^{th} simple ring contains n_i nodes. Let source and destination are n-hop away. It creates n searching rings in the route discovery. Thus Total number of participating nodes in n^{th} simple searching ring are $\sum_{i=1}^{n} n_i$. Let n simple rings are contained by m cluster rings. So, m^{th} cluster ring contains $C_u = 1 + 6.(2^m - 1)$ clusters. Thus total number of nodes within the mth cluster searching ring is $\sum_{i=1}^{C_u} k_i$.

Total number of participating nodes in m^{th} cluster ring are $\sum_{i=1}^{C_u}(k_i - l_i)$ which will always be less than the number of participating nodes in n^{th} simple searching ring i.e. $\sum_{i=1}^{n} n_i$. Since Blocking ERS+ used every node in the route discovery of the searching ring, no nonparticipating intermediate node will be in the searching ring. Thus $\sum_{i=1}^{n} l_i = 0$ for Blocking ERS+. This difference can be seen in the Venn diagram of Fig. 1. It shows that Blocking ERS+ is more energy consuming than our approach which is cluster based BERS+. From Fig. 1, it is also clear that the cluster based BERS+ used very less number of intermediate nodes during route discovery. It also saved more energy of the intermediate nodes than BERS+.

4 Design Mechanism

In our approach, fundamental prerequisites for route exploring process are clustering of the network, routing packets, and routing tables. These three are important outfits to make routing easy. In this section, we discussed the design mechanism of these outfits.

4.1 Clustering Used

We used DWCA [7] which is an energy efficient clustering and free from ripple effect of clusters. It has a high cluster stability and constant convergent time

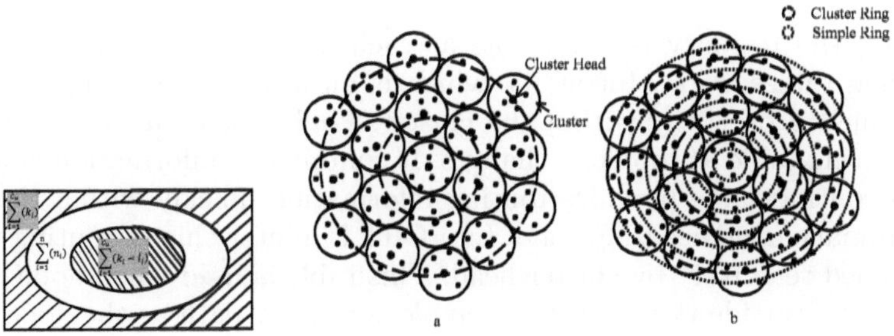

(a) (b)

Fig. 1. (a) Venn diagram of number of nodes, (b) BERS+ with clustering, and without clustering

of $O(1)$ [5]. In DWCA, we divide the whole network into distributed clusters. Cluster head is chosen provisionally with maximum battery power and strong connectivity, and low mobility. This cluster formation process is very similar to [7]. Initially, each node computes its weight using attributes parameters like degree of the node, residual time of battery of a node, and mobility. When each node has done this computation, each node starts competition with its neighbors to be the cluster head. This competition is done up to m-hop away neighbor nodes. Maximum weighted node among them is chosen as cluster head and remaining nodes become ordinary nodes. To continue the clustering, uncovered weighted node starts competition with its neighbor to be a cluster head of the next cluster. This clustering process is continued until each node belongs to exactly one cluster. After clustering process, each cluster head aggregates the cluster information that is maintained proactively.

4.2 Packets Used in Route Exploration

Packet formation is an essential course of route exploring process. Different types of packets are designed for different purposes.

Our scheme of route exploring is very similar to BERS+. But design mechanism of packet is different. We used two in one strategy of LHBA [18] to design the route reply and controlling packet. One packet is used for both purposes. In our approach, we used 5 types of packets in route exploring process. Type 0 is for internal RREQ packet which is used for proactive process of intra cluster routing. Type 1 is for external RREQ packet which is used for reactive process of inter cluster routing. Since reactive approach of route exploring gives the broadcast storm problem [22], it needs attention to control the packet from unnecessary circulation in the network. So, type 2 is for RRCL which is a combo of two packets. This packet works as a RREP packet as well as CTRL packet. Type 3 is for notification packet, and type 4 is for route error packet.

4.3 Routing Tables

To balance the memory requirements at each node in the networks, we used two routing tables. First, IntraCRT which is an intra cluster routing table. It is maintained proactively within the cluster. Each master node updates intra cluster routing table as topology changes. It contains all information of the cluster. This routing table provides routing information quickly within the cluster as any ordinary node needs. Second, InterCRT is a inter cluster routing table. It is maintained reactively by cluster head. This table is used when node needs to send the data outside the clusters. It employs routing path as the node needs.

5 Methodology

In this section, we will discuss the working model of our routing protocol. Our routing scheme is the hybrid approach of modified BERS+ and Cluster based routing. We used packet controlling feature of the BERS+ and network scalability aspect of cluster based routing. We used 5 types of packets in the route exploring process which work on the flag type. This process contains two tasks: First, route discovery and second, route maintenance. These are discussed in next two subsections.

5.1 Route Discovery Scheme

Primary task of the routing protocol is route exploring process. In our approach, this is done at two levels: First, Route discovery within the Cluster that has pure proactive approach, and second, Route discovery between Clusters that has pure reactive nature.

Route Discovery within a Cluster. is pure proactive level. Each ordinary node of the cluster has routing information within the intra routing cluster table. This routing information is updated periodically by the sentinel node of the cluster. Since each node knows about other ordinary nodes, every node can send data without interacting with the sentinel node. It is as usual as previous cluster based routing strategies used [7]. It also does not require any controlling scheme for routing packets.

Route Discovery between Clusters. is a pure reactive level. At this level, we used BERS+ [4] with destination initiated controlling feature of LHBA [18]. Whenever source node wishes to send the data outside of the cluster, it sends a type 1 external request packet to the cluster head. Upon receiving request packet, Cluster head immediately starts the route exploring process. First, it checks its InterCRT table to initialize the hop count field in the type 1 packet and forwards it to the adjacent cluster heads via gateway nodes. Adjacent cluster heads receive the packet and search the desirable path in their interCRT table. If it fails to find the path, it cashes the path travelled by the type 1 packet in

```
if Type 2 packet received Or Type 1 packet is Duplicate then
|  Discard the type 1 packet;
end
else if Present node is Destination cluster head then
|  Create type 2 reply packet;
|  Broadcast it;
end
else
|  if hopcount ≤ MaxCount then
|  |  Broadcast the type 1 packet;
|  end
|  else
|  |  while waiting for time t do
|  |  |  if  Discard Type 1 packet;
|  |  |  then Type 2 packet received
|  |  |   |  B
|  |  |  end
|  |  |  roadcast the type 1 packet;
|  |  end
|  end
end
```

Algorithm 1. Processing of Type 1 Packet

its interCRT table and continues the searching process. At each attempt failed, intermediate cluster heads expand searching area as ripple across the water (see Fig. 2(a)). If hop count field exceeds, intermediate cluster heads introduced the delay in the processing of type 1 packet. This delay helps to catch the type 1 packets from unnecessary propagation. This process is continued until the route is found. Receiving type 1 packet, Cluster head takes action defined in Algorithm 1.If any intermediate cluster head finds the route, it sends out a type 2 reply packet back to the source cluster head following the reverse path. Source cluster head receives the type 2 packet from the destination cluster head that has destination node in its IntraCRT table. It caches all routing information in the InterCRT table before sending the packet to the source node. Source node manipulates the path to reduce the length of the route because of cluster head as it receives the reply packet. This manipulation is done based on the routing information stored in the IntraCRT. It transmits data following the given path with slight manipulation.

Controlling Technique. Each intermediate cluster head expands the searching ring as its attempt fails. When TTL field of type1 packet reaches to zero, next intermediate cluster heads introduce the slight delay in the processing of type 1 packet. This delay helps type 2 packet to catch them. We used this type 2 reply packet for dual purpose. First, it informs the sender about the path. Second, it controls the unnecessary circulation of the type 1 packet in the network. Upon receiving the type 2 packet, each intermediate cluster head will take action defined in Algorithm 2.

```
if Type 2 packet is Duplicate then
|   Discard it;
end
else if Present node is Source cluster head then
|   Send Reply Packet to Source Node;
|   Update MaxCount field in IntraCRT;
|   Broadcast type 2 packet;
end
else
|   if Present cluster head in the header of the packet then
|   |   Update MaxCount field in IntraCRT;
|   |   Broadcast type 2 packet;
|   end
|   else if Type 1 packet received And Broadcasted it then
|   |   Broadcast type 2 packet;
|   end
|   else
|   |   Discard it;
|   end
end
```

Algorithm 2. Processing of Type 2 Packet

5.2 Route Maintenance

Route can be no longer valid due to the mobility of nodes. So we require a route repair strategy. Route maintenance within the cluster is very simple because of periodic messaging. Route is recovered through proactive information sent by cluster head. On the other hand, this strategy is very cumbersome between clusters due to reactive approach. In this strategy, we used bypass scheme to recover the path and to continue the data sending process. Whenever link breakage happens during the transmission, the node that finds the link breakage immediately uses a bypass scheme of [12] to find the alternate path. This node also sends back type 3 notification packet to the source node about the new path. As source node receive type 3 packet, it drops the previous route. Now it uses new alternate path for data transmission. If bypass scheme fails, the node sends back a type 4 error packet to the source node to inform about link breakage. If source still has some data for transmission, it again starts the route exploring process. The major advantage of our route repair scheme is clustering. Every ordinary node knows about other ordinary nodes within the cluster. So, alternate route is always available and recovers it too early as compared to the ordinary schemes.

6 Conclusion

MANETs are temporary networks in which topology changes due to random movement of mobile users. But nature of the human is that they like to live in a clique. This behavior helps to cluster the MANETs. As we know Clustering has many advantages for MANETs such as saving battery, low routing overhead, reduces the size of the routing table, and improves throughput and scalability. In MANETs, congestion problem due to unnecessary propagation of packets has become hazardous problem. So, we proposed a routing framework for MANETs based on distributed clustering using modified BERS+. In this scheme, we used clustering to achieve the scalability and slight improved version of BERS+ to control the unnecessary circulation of the packets. We used type 2 packet for dual purpose instead of using separate reply, and chase packets. It makes the controlling of the packets destination initiated so that the controlling of packets can be fast. From mathematical interpretation, we observed that Clustering based implementation of modified BERS+ used less number of intermediate nodes than simple BERS+. It also helps nonparticipating intermediate nodes to save their energy.

References

1. Yousefi, S., Mousavi, M., Fathy, M.: Vehicular ad hoc networks (vanets): challenges and perspectives. In: 2006 6th International Conference on ITS Telecommunications, pp. 761–766 (2006)
2. Perkins, C., Bhagwat, P.: Highly dynamic destination-sequenced distance-vector routing (dsdv) for mobile computers. In: ACM SIGCOMM Computer Communication Review, vol. 24, pp. 234–244. ACM (1994)
3. Perkins, C., Royer, E.: Ad-hoc on-demand distance vector routing. In: Second IEEE Workshop on Mobile Computing Systems and Applications, pp. 90–100. IEEE (1999)
4. Al-Rodhaan, M., Mackenzie, L., Ould-Khaoua, M.: Improvement to blocking expanding ring search for manets. Department of Computing Science. University of Glasgow, Glasgow (2008)
5. Abbasi, A., Younis, M.: A survey on clustering algorithms for wireless sensor networks. Computer Communications 30(14), 2826–2841 (2007)
6. Barjini, H., Othman, M., Ibrahim, H., Udzir, N.: Shortcoming, problems and analytical comparison for flooding-based search techniques in unstructured p2p networks. Peer-to-Peer Networking and Applications 5(1), 1–13 (2012)
7. Choi, W., Woo, M.: A distributed weighted clustering algorithm for mobile ad hoc networks. In: International Conference on Internet and Web Applications and Services/Advanced International Conference on Telecommunications, p. 73. IEEE (2006)
8. Karenos, K., Kalogeraki, V., Krishnamurthy, S.: Cluster-based congestion control for supporting multiple classes of traffic in sensor networks. PhD thesis, University of California, Riverside (2005)
9. Dubois-Ferriere, H., Grossglauser, M., Vetterli, M.: Age matters: efficient route discovery in mobile ad hoc networks using encounter ages. In: Proceedings of the 4th ACM International Symposium on Mobile Ad hoc Networking & Computing, pp. 257–266. ACM (2003)
10. Basagni, S., Chlamtac, I., Syrotiuk, V., Woodward, B.: A distance routing effect algorithm for mobility (dream). In: 4th Annual ACM/IEEE International Conference on Mobile Computing and Networking, pp. 76–84. ACM (1998)
11. Preetha, K., Unnikrishnan, A., Jacob, K.: A probabilistic approach to reduce the route establishment overhead in aodv algorithm for manet. arXiv preprint arXiv:1204.1820 (2012)

12. Castaneda, R., Das, S., Marina, M.: Query localization techniques for on-demand routing protocols in ad hoc networks. Wireless Networks 8(2-3), 137–151 (2002)
13. Ko, Y.B., Vaidya, N.: Location-aided routing (lar) in mobile ad hoc networks. Wireless Networks 6(4), 307–321 (2000)
14. Minematsu, M., Saito, M., Hiroto, A., Tokuda, H.: Efficient route discovery scheme in ad hoc networks using routing history. IEICE Transactions on Communications 88(3), 1017–1025 (2005)
15. Qayyum, A., Viennot, L., Laouiti, A.: Multipoint relaying for flooding broadcast messages in mobile wireless networks. In: Proceedings of the 35th Annual Hawaii International Conference on System Sciences, pp. 3866–3875. IEEE (2002)
16. Aitha, N., Srinadas, R.: A strategy to reduce the control packet load of aodv using weighted rough set model for manet. The International Arab Journal of Information Technology (2009)
17. Gargano, L., Hammar, M.: Limiting flooding expenses in on-demand source-initiated protocols for mobile wireless networks. In: 18th International Parallel and Distributed Processing Symposium, p. 220 (2004)
18. Zhang, H., Jiang, Z.P.: On reducing broadcast expenses in ad hoc route discovery. In: 25th IEEE International Conference on Distributed Computing Systems Workshops, pp. 946–952. IEEE (2005)
19. Chang, N., Liu, M.: Revisiting the ttl-based controlled flooding search: Optimality and randomization. In: 10th Annual International Conference on Mobile Computing and Networking, pp. 85–99. ACM (2004)
20. Park, I., Kim, J., Pu, I.: Blocking expanding ring search algorithm for efficient energy consumption in mobile ad hoc networks. In: Third Annual Conference on Wireless On-demand Network Systems and Services, pp. 191–195 (2006)
21. Pu, I., Shen, Y.: Enhanced blocking expanding ring search in mobile ad hoc networks. In: 3rd International Conference on New Technologies, Mobility and Security, pp. 1–5. IEEE (2009)
22. Tonguz, O., Wisitpongphan, N., Parikh, J., Bai, F., Mudalige, P., Sadekar, V.: On the broadcast storm problem in ad hoc wireless networks. In: 3rd International Conference on Broadband Communications, Networks and Systems, pp. 1–11 (2006)

An Efficient Route Repairing Technique of Aodv Protocol in Manet

Ashish Patnaik[1], Kamlesh Rana[1], Ram Shringar Rao[2], Nanhe Singh[3], and Uma Shankar Pandey[3]

[1] Galgotia's College of Engineering and Technology, Greater Noida, India
[2] Ambedkar Institute of Advanced Communication Technologies & Research, Delhi, India
[3] School of Open Learning, University of Delhi, India
{ashishpatnaik89,ranakamles, nsingh1973,uspandey1}@gmail.com, rsrao08@yahoo.in

Abstract. A mobile Ad Hoc Network (MANET) is an autonomous system of mobile nodes connected by a wireless link. A MANET does not have any access point or does not depend on any central administrator. In a large scale Ad Hoc Networks, there is a continuous node mobility may cause radio links to be broken frequently. This further leads to increase in packet dropping rate, end to end delay, reduction in the packet delivery rate and finally complete efficiency of network degraded. Therefore to overcome from these consequences we propose Ad hoc On Demand Route Repair Technique (AODVRRT) which introduces a new mechanism so that link failure can be predicted and accordingly perform a rapid local path repair. So by repairing the path this protocol not reduces the end-to-end latency and packet loss rate, but it also increases the efficiency of the network. Finally, simulation is done to measure performance.

Keywords: MANET, Routing Protocols, AODV, OSLR, DSDV, AODVRRT, Path Loss, End-to End Delay, Packet Delivery Ratio.

1 Introduction

A Mobile ad hoc network (MANET) is a system comprises of many number mobile nodes in which data packets are forwarded using the nodes in a network so that it can communicate outside the direct range of wireless transmission. Since MANET is infrastructure less so it does not require any centralized administration such as access points or any fixed base stations, and it can be set up quickly and inexpensively. Thus MANET allows a group of user to communicate with each other over a slow wireless connection. Since it has a decentralize network all the nodes in a network will carry the activity of topology discovery and message delivery by themselves. MANET is a collection of mobile node which is self-configured and connected by wireless links to forms a differently arbiter topology. Thus in this all the nodes can act as a router which further incorporates in moving the data packets and these routers are free to move within the network with the adaptable change of topology.

M.K. Kundu et al. (eds.), *Advanced Computing, Networking and Informatics - Volume 2*, Smart Innovation, Systems and Technologies 28,
DOI: 10.1007/978-3-319-07350-7_13, © Springer International Publishing Switzerland 2014

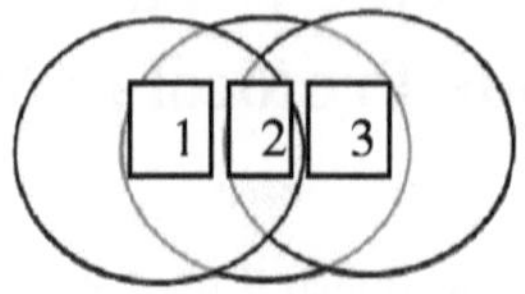

Fig. 1. A simple Mobile ad-hoc network with three nodes

In Fig.1, node 1 and node 3 must discover the route through node 2 for communication. The nominal range of each node's radio transceiver are indicated by the circles. Node 1 and node3 are not within the transmission range directly of each other, since node 1's circle does not cover node 3's .So node 2 will be involved if node 1 want to send a data to node 2 for forwarding the data packets from 1 to 3. Here node 2 acts as a intermediate node. So node 1 could not send data packet to node 3 if it will not involve node 2.Thus MANETs also have some important characteristics such as Dynamic topologies, Bandwidth-constrained & variable capacity links, Frequent routing updates etc. Its major applications are Disaster relief operations, Military or police exercises, Adhoc Mettings etc.

2 Related Work

Jain *et al.*[1] proposes a new local repair scheme so that the deficiency of existing local repair can be removed. This improved local repair scheme concerns about the end to end delay in transmission and overhead requirement. During local repair, Ant algorithm is used by the repairing node for finding new route for next to next node in the link. This reduced size of F-ANT and B-ANT will give significant overhead reduction.

Naidu and Chawla [2] gives the idea about the existing Local Repair Trial method in AODV which is extended in order to achieve broadcasting and minimizing the flooding. This protocol first creates the mobile nodes so that the broadcasting can be done easily and if any link failure is detected local repair technique can be applied. In this Diameter Perimeter Model is used to increase the number of intermediate nodes. Finally this paper presents a solution to minimize route overheads of AODVLRT. This also analyses the enhanced AODVLRT with the existing local repair technique. This technique consists of three modules. First the perimeter routing is used for broadcasting. Secondly, local repair method is used to minimize the flooding and lastly, there is an increase in the number of intermediate nodes from source to destination.

Youn *et al.*[3] proposes a new local repair scheme using promiscuous mode is used. This scheme is mainly composed of two parts: adaptive promiscuous mode and quick local repair scheme. Adaptive promiscuous mode is to repeat the switching processes between promiscuous mode and non promiscuous mode to overcome energy limit caused by using promiscuous mode in overall time and quick local repair scheme is to fast perform the local re-route discovery process with the information of the active connection in the local area acquired by promiscuous mode.

In Subburam *et al.*[4], the authors have proposed an on-demand delay and bandwidth based quality of service (QoS) routing protocol (AODV-D) to ensure that delay does not exceed a maximum value and the minimum available bandwidth is required to send the packets. Moreover, their proposed routing protocol will follow the concept of unicast-type two hop local route repair protocol to recover the lost links efficiently while increasing network reliability, increasing utilization, minimizing the number of control messages and shortening the repair delay.

Rao *et al.*[5] have proposed an algorithm to controlled congestion by applying efficient local route repair method. When the link failure is detected at intermediate node, the hop count of destination is compared with source hop count and if destination is closer to breakage link than source, local repair is done. When local repairing in process the packet should stored in buffered queue and as alternative route has found the packet then transmitted to destination.

3 Routing Protocols

Routing is the process of sending data from a source to a destination in a network. During the transfer of the information at least one intermediate node in the network is encountered. Generally this process involves two concepts: firstly determination of optimal routing paths and transfer of the data packets through a network. This process of transferring data packets through an internetwork is called as packet switching and the determination of route could be very complex task. There are several metrics which are used by the routing protocol as a standard measurement to calculate the best path for data packets routing to the destination such as the number of hops, which are used by the routing algorithm to find the optimal path for the data packet to the destination. In the path determination process, routing algorithms try find out and maintain routing tables, which contain the complete route information for the data packets. The information of routing is varied from one routing mechanism to another. The IP-address prefix and the next hop are the entries which were filled in the routing table .In a routing table, Destination/next hop associations tells the router that a particular destination can be reached by sending the packet to a router representing the next hop on the way to final destination and set of destination for a valid routing entry can be specified by IP-address prefix. Static routing and dynamic routing are the main two classification of the routing. Static routing refers to the routing strategy being stated manually or statically, in the router. All the routing table usually written by a networks administrator are maintained by the static routing.

4 AODV Routing Protocol

The Ad hoc On-Demand Distance Vector (AODV) routing protocol allows dynamic, self-initiating and multi-hop routing between participating nodes which are willing to establish and maintain an ad hoc network. AODV allows mobile nodes to establish a route quickly for new destinations, and maintenance of the routes to destinations is not required by the nodes. Whenever there is any link breakages and changes in network topology then it is managed by the AODV in a timely manner. Whenever there is any link breakdown, AODV allows the affected set of nodes to be notified so that they can be able to invalidate the routes using the lost link. One distinguishing feature of AODV is to use of a destination sequence number for each route entry. The destination sequence number is generated by the destination so that it can be included with any route information it sends to requesting nodes. Using destination sequence numbers ensures freedom from loops

Route Discovery: In AODV Routing protocol first of all route discovery process is initiated by the source node to reach the destination node and it search those routes which it does not have in its cache. In this procedure ROUTE REQUEST packet is broadcasted, so that these packets were flooded across the network. These route request packets contain the source node address, destination node address, path identifier, path record and sequence of hop count towards reaching the destination. Thus sequence numbers are used to prevent the duplications. When there is a valid path to reach destination then the ROUTE REPLY packet is send either by the intermediate node that contain path to reach destination or either by the destination node.

Route Maintenance: One of the main features of the AODV routing is the path maintenance process. Thus in this each node monitor the path operation and informs the sender about the error in the routing. If there is any route breakdown due to link failure then the ROUTE ERROR message is send by the detector node to the source node which then removes the all path in its cache and proceed with the new route discovery process.

5 Problem Statement

Since in a large scale Ad Hoc Networks, there is a continuous terminal mobility which may cause radio links to be broken frequently due to dynamic topology and asymmetric links.

i) **Dynamic Topology:** Since Ad hoc networks does not contain the constant topology; so due to this the mobile node might move in a network and their medium characteristics might get change. In ad-hoc networks, these changes in topology are reflected in the routing tables and some routing algorithms had to be adapted. For example in a fixed network infrastructure routing table updating takes place for every 50sec. This updating frequency might be low for ad-hoc networks.

ii) **Asymmetric links:** Generally most of the wired networks have symmetric links which are always fixed. But this is not a case with ad-hoc networks as the nodes are mobile and changing their position constantly within network.

So, this leads to increase in end to end delay, packet dropping rate, throughput and can reduce the packet delivery rate.

6 Proposed Work

Generally routing protocols are constrained on the condition of Shortest Path .This constraint ignores the path stability. Therefore from the point of stability, these routing protocols overlook the stability of route in a network and the shortest path maybe the constraint of the network. To overcome this problem, we present the AODVRRM.

6.1 Ad-Hoc on Demand Distance Vector Route Repair Technique (AODVRRT)

Ad-hoc On Demand distance vector Route Repair Technique(AODVRRT) increases the performance of the network whenever active route fails. This concerns with the mechanism of route repair based on the stability estimation method. Thus in this for finding route stability initially, every node begins to estimate the radio links stabilities of its neighbours and each node periodically broadcasts Hello message (HELLO) for keeping the track of link stabilities between a node and its neighbours. In this protocol, when a Hello messages is received by the node, it first checks the distance between the itself and the neighbouring node because it is aware of distance, it evaluates the stability of radio link to the broadcasting neighbour. This information for estimating multi-hop route-stabilities is recorded in follow-up manner. In path discovery process, RREQ is broadcasted by the source node that has a new link stability field. The intermediate node rebroadcast only the RREQ with the value having maximum in route stability among received RREQs.

6.2 Route Repair Technique

Consider a data is flowing from source to destination. Fig. 2. shows the three possible message exchange routes are 1-2-4-5-8, 1-3-4-5-8 and 1-6-7-8. The AODV algorithm adopts to find only the path with shortest route and it fails to repair the route, but our proposed AODVRRM works in the following way:

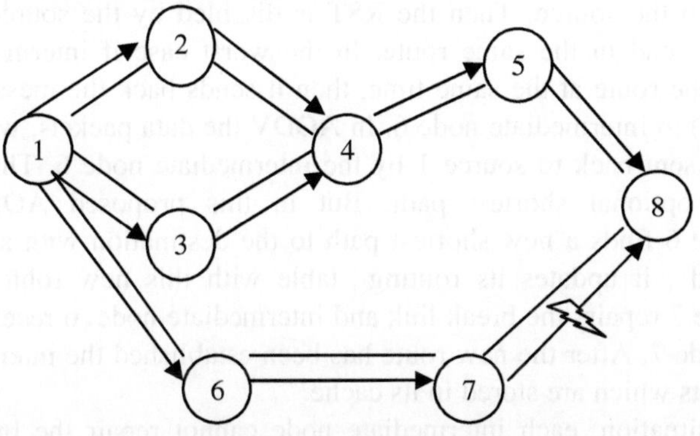

Fig. 2. AODVRR messaging

1. First of all RREQ is send by the Source node to destination node and waits for RREP.
2. RREP is returned by the Destination node if it have a valid RREQ; and at the return process, all nodes in this path check its route stability in the RREQ.
3. Several shortest paths were selected by Source node and if there are multiple shortest paths or only single shortest path, source node will also select a next shortest path from the remaining paths.
4. The stability of each route is calculated by the Source node on different selected paths.
5. Based on the following proportion, source node dynamically calculates route stability on the different paths based on the following conditions:

Therefore in these conditions for find the stability of the path each node maintains two tables NPL (Neighbour Power List) and PDT (Power Difference Table)

- Node Power List contains the last received signal strength for packets originating from each neighbor .Whenever a packet is received and any Hello interval in the network is occurred then this table is updated.
- Power Difference Table contains the rate at which power is changing between each pair of neighbors. PDT describes whether the link signal strength is increasing or decreasing between each pair of neighbors. This table is also update whenever a packet is received.

In Fig. 2, the shortest and the maximum stability path is 1-6-7-8. If data is flowing (1-6-7-8),the intermediate node 7 detects the link break and the message of route error (RERR) will not be send to the source node 1. Instead, node 7 sends a Route Repairing (RR) message back to the previous intermediate node 6. After sending Route repair to 6, RREQs is broadcasted to repair the break route by the intermediate node 7. The intermediate node 6 now sends information to the source node 1 to stop the data flow in the route through Route Stop (RST) message and it stores the data packets in its cache, and then the timer started and waits for the node 7 to give the acknowledgement. In the mean time if the link failure is repaired by the node 7 then it broadcast the Route Repair Ok (RROK) message to the intermediate node 6. Then the data packets which are stored in its cache are sent and the Route Repair Ok (RROK) message is sent to the source. Then the RST is disabled by the source node 1 and starts the data to send in the same route. In the worst case if intermediate node 7 unable to repair the route at the same time, then it sends back the message of Route Repair Fail (RRF) to intermediate node 6. In AODV the data packets, which is stored in the cache, are sent back to source 1 by the intermediate node 6. Then the source finds the other optional shortest path. But in this proposed AODVRRM the intermediate node 6 finds a new shortest path to the destination with and maximum stability path and , it updates its routing table with this new route, even if the intermediate node 7 repairs the break link and intermediate node 6 received a RROK message from node 7. After the new route has been established the intermediate node 6 sends the packets which are stored in its cache.

In the worst situation, each intermediate node cannot repair the break link and cannot find a new route to the destination. Then, the source node will receive a Route repair message. In this case, AODVRRM source node selects the already found the shortest path with maximum stability route broadcasts a data packets to the destination. When all these scenarios fail then the AODV again starts new route discovery process.

7 Simulation Result

The simulation results show overall performance of the protocol within the network. Through these performance measures the efficiency of the protocol can be judged. Thus this section evaluates the performance of AODVRRM composed with AODV by computer simulation using NS-2. Our detailed simulation setup is shown below in Table 1. In Fig. 3 we conclude that there is an occurrence of packet loss with the different arrival rates. The packet loss rate of AODVRRM and AODV is approximately same for a small arrival rates. The loss of rate of packets is increasing when the data arrival rate is more than the 20 packets per second. So through this result we conclude that the when there is a large data arrival rate then the packet loss rate of AODVRRM is lower than the AODV.

Table 1. Simulation Setup

Parameter	Values
Topology	700 ×700
Simulation Time	100 seconds
Pause Time	10 ms
Number of Nodes	1,6,15,20,30,32
Transmission Range	300 m
Traffic Size	CBR
Packet Size	100 bytes
Packet Rate	10 packet/s
Maximum Speed	20 m/s
Routing Protocol	AODV
X Dimension of Topography	700
Y Dimension of Topography	700

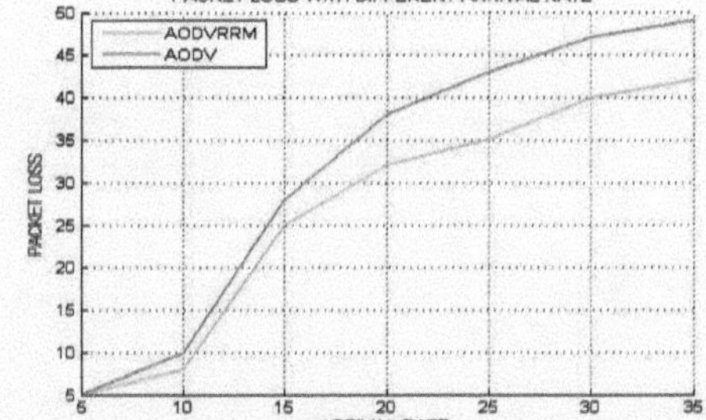

Fig. 3. Packet loss vs arrival rates

Through Fig. 4, we conclude that there is an average end-to-end delay with the arrival rates. When the arrival rate is less i.e. 30 then end to end delay is almost same, but when the arrival rate reaches 35 packets per second then the average end to end delay increases drastically. Thus this has been shown that AODVRRM has better performance than the AODV. In Fig. 5, we varied the packet delivery ratio with the number of nodes. In this we see that as the number of nodes are increasing the packet delivery ratio is also increasing and hence we see that the performance of the AODVRRM is more efficient then the performance of the AODV.

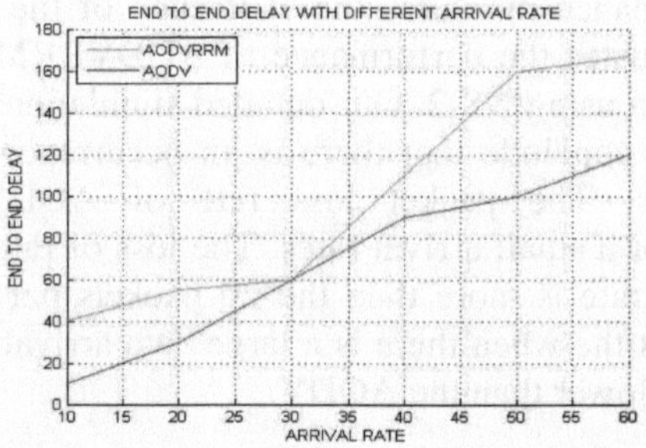

Fig. 4. End-to-End delay vs arrival rates

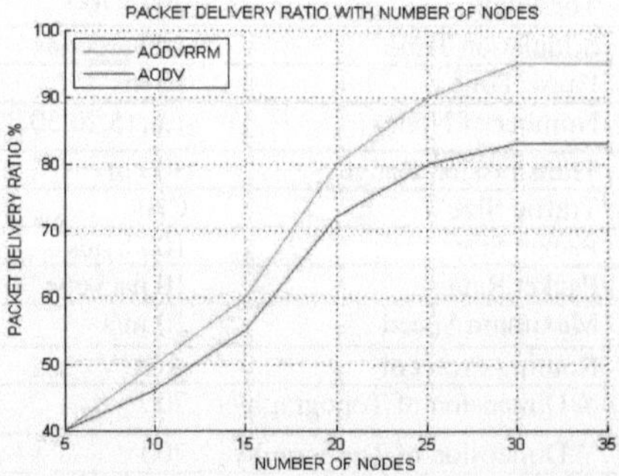

Fig. 5. Packets delivery ratio vs node number

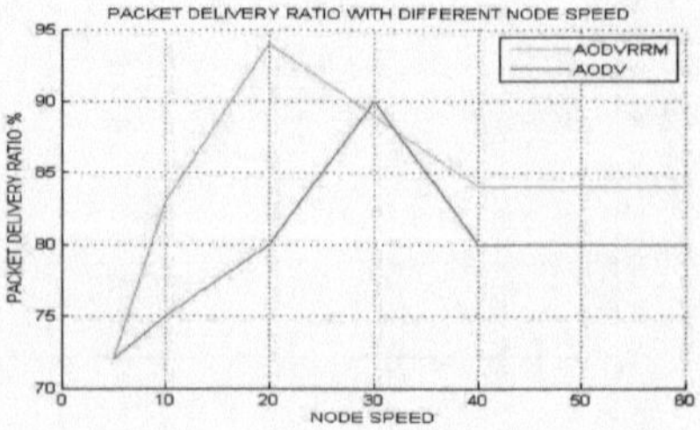

Fig. 6. Packets delivery ratio vs nodes speed

In Fig. 6. we varied the packet delivery ratio with the nodes speed. Thus in this we conclude that the all nodes are moving in a efficient and specified speed. Thus we see AODVRRM is quite better.

8 Conclusion

After the deep research on Ad Hoc On demand routing protocol (AODV), this thesis proposes a new routing algorithm AODVRRM, in which adding stability field to the Route Request message avoids selecting the routes which are analogous and unstable during establishing a new route discovery; and adding the mechanism of route repair to the Route request message instead of starting a new routing discovery as far as possible. The simulation of this protocols has been carried out using Ns-2 simulator and found out that packet loss rate, throughput and the end-to-end latency is improved and network resources are utilised efficiently.

References

1. Jain, J., Gupta, R.: On Demand Local Link Repair Algorithm For AODV Protocol. International Journal of Computer Application 35(5), 20–25 (2011)
2. Naidu, P.P., Chawla, M.: Extended Ad Hoc on Demand Distance Vector Local Repair Trail for MANET. International Journal of Wireless & Mobile Networks (IJWMN) 4(2), 235–250 (2012)
3. Youn, J.-S.: Quick Local Repair Scheme using Adaptive Promiscuous Mode in Mobile Ad Hoc Networks. Journal of Networks 1(1), 1–11 (2006)
4. Subburam, S., Khader, P.S.A.: Efficient Two Hop Local Route Repair Mechanism Using Qos-Aware Routing for Mobile Ad Hoc Networks. Indian Journal of Science and Technolog 5(11), 3651–3659 (2012)
5. Rao, K.S., Shrivastava, L.: Efficient Local Route Repair Method in AODV to Reduce Congestion in MANET. Corona Journal of Science and Technolog 1(1), 35–38 (2012)

Random Backoff Sleep Protocol for Energy Efficient Coverage in Wireless Sensor Networks

Avinash More and Vijay Raisinghani

School of Technology, Management and Engineering
MPSTME, NMIMS University,
Mumbai - 400056, India
avinash.more@nmims.edu, rvijay@ieee.org

Abstract. A Wireless Sensor Network is a collection of many small sensor nodes. Every sensor node has a sensing range and a communication range. Coverage of a sensor node means the sensing region within which an event can be observed or detected. Most protocol designs for energy efficient coverage optimization maintain an adequate working node density. However, they ignore the residual energy level of the nodes. In this paper, we propose Random Backoff Sleep Protocol(RBSP) which ensures that the probability of neighbor nodes becoming active is inversely related to the residual energy level of the current active node. This will help in increasing the network lifetime by balancing energy consumption among the nodes. RBSP uses dynamic sleeping window, for the neighbor nodes, based on the amount of residual energy at an *active* node. Simulation results show that our scheme achieves more power saving and longer lifetime compared to Probing Environment and Adaptive Sleeping protocol(PEAS).

Keywords: Coverage, Random Backoff, Sleeping window, Wireless Sensor Networks.

1 Introduction

A typical Wireless Sensor Network (WSN)[1],[2] is an adhoc network composed of small sensor nodes which cooperatively monitor some physical environment. Each sensor node has a sensing range or sensing coverage range[3],[4],[5] which is the region or area that a node can observe or monitor. Sensing coverage for a WSN could be interpreted as the collective coverage of all the sensors in the WSN. Sensing coverage ensures proper monitoring and radio coverage ensures proper data transmission within the WSN. Sensing coverage and radio coverage both are important for ensuring that the coverage of the region is adequate and the sensors are able to transmit data to the sink. It is important to minimize the number of active nodes, while still achieving maximum possible sensing and radio coverage. The aim here is to ensure that sufficient number of nodes are available for the longest possible time while ensuring proper functioning of the WSN.

M.K. Kundu et al. (eds.), *Advanced Computing, Networking and Informatics - Volume 2*, Smart Innovation, Systems and Technologies 28,
DOI: 10.1007/978-3-319-07350-7_14,

Sensor nodes have limited energy, usually supplied by a battery. In view of the limited battery life, it is essential to make these nodes energy efficient. Energy saving is important for applications that need to operate for a longer time on battery power. Most of the existing work[6],[7],[8],[9],[10],[11], for coverage optimization, obtained by node scheduling, does not consider the residual energy of the nodes. For example, in Probing Environment and Adaptive Sleeping(PEAS)[6], a sleeping node occasionally enters probing mode and broadcasts messages(probes) within its local probing range and checks whether an active (working) node exists within its probing range. The probing node enters the active state only when it receives no replies from its working neighbors, else it goes back to sleep mode. The probing node calculates a random sleeping time before the next round of probing, based on the reply message received from the active node.

The aim of PEAS is to maximize network coverage and connectivity by waking up minimum number of nodes. The authors show that the network lifetime increases linearly with the number of nodes. In PEAS, the wakeup rate is randomized and spread over time based on an *exponential function.* This causes unnecessary waking up of nodes, due to which energy consumption increases and network lifetime decreases. PEAS is useful for a network where the node density is high. If the node density is not high enough then some of the probing nodes may enter the active state which would lead to a reduction in the network and node lifetime. To avoid these shortcomings we propose *Random Backoff Sleep Protocol* (RBSP).

Random Backoff Sleep Protocol(RBSP) is a probe based protocol which utilizes the information about residual energy level in the active node. This is in contrast to PEAS, which ignores this information. Further, RBSP does not use any exponential function for the wakeup time. PEAS uses an *exponential function* to compute the random backoff time[6]. This exponential function causes the intervals between successive wakeups of the sleeping nodes to increase. A *sensing void* (uncovered area) could get created if an active node dies, and the sleeping node has not woken up in time. RBSP protocol employs a novel backoff algorithm for calculation of sleeping time period. The proposed protocol uniformly chooses a random value of sleeping window based on residual energy of the active node. Using this mechanism, when an active node has high residual energy, the probability of a neighbor node turning *on* is low. Similarly, when an active node has low residual energy, the probability of a neighbor node turning *on* is high. This will help in balancing the energy consumption among the nodes. Due to this, we expect the network lifetime to increase substantially.

The rest of paper is organized as follows: in Section 2, we review some coverage optimization protocols used in wireless sensor networks. In Section 3, we present the details of our protocol – RBSP. Section 4 contains performance evaluation using simulations. Finally, we present our concluding remarks in Section 5.

2 Related Work

Many research efforts have been made to exploit the inherent coverage redundancy to extend the lifetime of wireless sensor networks. Ye *et al.* [6] present Probing Environment and Adaptive Sleeping(PEAS) which is a distributed protocol, based on probing to extend network lifetime by turning on minimum number of active nodes. PEAS is a location independent protocol. Gui *et al.* [7] propose Probing Environment and Collaborating Adaptive Sleeping(PECAS) which is an extension to PEAS [6]. PECAS does not allow active nodes to operate continuously until energy depletion. Occurrence of sensing void is reduced in PECAS because a active node schedules itself to enter into sleep mode after some specified time.

Yun-Sheng *et al.* [8] propose Controlled Layer Deployment(CLD) which uses deterministic node deployment and is based on PEAS. CLD [8] helps to achieve a longer network lifetime as compared to PEAS[6].

Xing *et al.*[9] present Coverage Configuration Protocol(CCP) which is a decentralized protocol. CCP requires lesser number of active nodes. CCP is a location dependent protocol. Zhang *et al.*[10] introduce Enhanced Configuration Control Protocol(ECCP) which provides a mechanism to avoid sensing voids in the network. However, it requires more number of active sensor nodes as compared to CCP. Honghai *et al.* [11] present an Optimal Geographical Density Control algorithm that determines the minimum number of active nodes for full coverage. When OGDC is compared with PEAS, it requires 50% lesser active nodes for full coverage.

Chen *et al.* [12] present *Span* which is a distributed, randomized algorithm where nodes make local decisions whether to sleep, or become active as a *coordinator*. Network lifetime increases due to *Span*. Kijun *et al.* [13] propose MAC protocol which is based on a backoff algorithm for wireless sensor networks which used dynamic contention period based on residual energy at each node. In case of all the above protocols, the residual energy of the active node is not considered for determining the sleep schedules. In case of reference [13] the residual energy is considered for medium access and not for planning the coverage. In the section below we discuss RBSP's random backoff sleep cycle, state transition diagram and finally details of the working of the protocol.

3 Random Backoff Sleep Protocol

We propose Random Backoff Sleep Protocol(RBSP) for node scheduling. The wakeup rate of RBSP is based on residual energy of an *active* node. At each active node a *sleeping window* is dynamically computed based on the amount of residual energy of the active node. The probability of neighbor nodes becoming active is inversely related to the residual energy level of the current active node. Neighbor nodes use the sleeping window information from the active node to determine its sleep time.

Fig. 1 gives a simple example for illustration. We have considered three cases for RBSP, in case-I we assume that, for time interval T0 to T1, node A is active and its residual energy is only 10%. Hence the sleeping window of active node A is very small due to which, wakeup rate of neighboring sleep nodes is also very high. In case-II for the time interval T to T', node B is active and its residual energy is 40%. As a result, the sleeping window of active node B is slightly larger as compared to that of node A. This causes the wakeup rate of neighboring sleep nodes to be moderate. Therefore, the probability of sleeping nodes turning *on* is also moderate. Similarly, for case-III, the time interval t to t', node C is active, and its residual energy is -90%. This causes the sleeping window of active node C to be very large. Therefore, the wakeup rate of neighboring sleeping nodes is very low. Due to this the probability of sleeping nodes turning *on* is very less.

While in PEAS (case-IV), node A is active at T0, a sleeping node B wakes up at T1 and a sleeping node C wakes up at T2. In PEAS wakeup rate of sleeping nodes is not based on residual energy of active node. PEAS uses an *exponential function* to compute the sleep interval[6]. Due to this, initially the sleeping nodes would wakeup frequently and later at a slower pace. This could create a sensing void if the active node dies when the sleep intervals are wide. Also the frequent wakeups could cause energy loss.

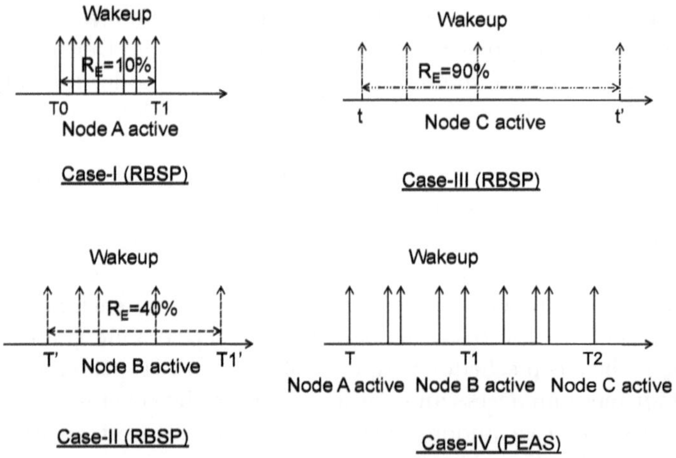

Fig. 1. RBSP and PEAS wakeup cycle

3.1 State Transition of RBSP

Each node in RBSP has three operating states which are similar to PEAS [6]: *SLEEP*, *FLOAT* and *ACTIVE.* The state transition diagram for all three modes is shown in Fig. 2. In the SLEEP state, a node turns its radio off to conserve energy. Each node in FLOATING state broadcasts HELLO message within its sensing range *Rs*, where *Rs* is the maximum sensing range within which an event can be observed or detected. The ACTIVE node continuously senses the physical environment and communicates with other sensor nodes. Each node in RBSP has three operating states which are similar to PEAS [6]: *SLEEP*, *FLOAT* and *ACTIVE*. The state transition diagram for all three modes is shown in Fig. 2. In the SLEEP state, a node turns its radio off to conserve energy. Each node in FLOATING state broadcasts HELLO message within its sensing range *Rs*, where *Rs* is the maximum sensing range within which an event can be observed or detected. The ACTIVE node continuously senses the physical environment and communicates with other sensor nodes.

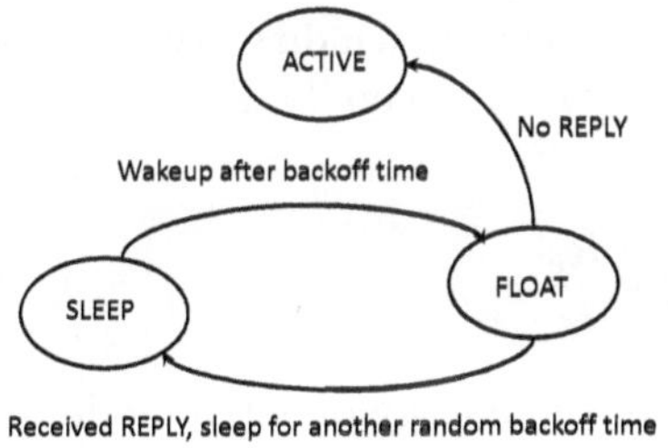

Fig. 2. State Transition Diagram of RBSP

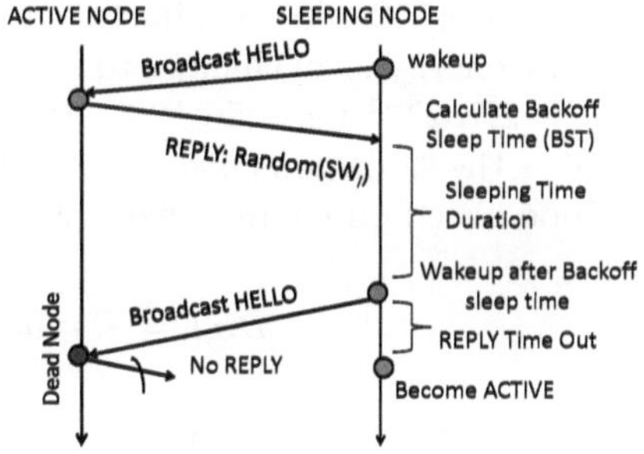

Fig. 3. Flow diagram of RBSP

The flow diagram of RBSP is shown in Fig. 3. Nodes are initially in sleeping state. Each node sleeps for a random backoff time based on *sleeping window* of active node. After the node wakes up, it enters into a floating state. The Floating node broadcasts HELLO message within its sensing range *Rs*. Any active node(s) within that sensing range responds with a REPLY message, which includes a unique random number from the *sleeping window* based on its residual energy. If the floating node hears a REPLY, it goes back to sleep mode for another random period of time, generated according to equations 1 and 2. If floating node does not hear any REPLY, it enters into active state. The floating node computes the *Reply Time (RT)* based on the time interval from sending the HELLO packet to the receipt of the REPLY message. The floating node maintains a timer with the value *Reply Time Out (RO) = 2 * RT*. If a reply is not received within the reply time out period, then the floating node enters into active state. Thus using RBSP, each sleeping node determines whether any active node is present within its sensing range or not. Any node once enters into active state, it remains active until it consumes all of its energy. RBSP's working mechanism and the computations at the nodes is explained below.

3.2 Working Mechanism of Random Backoff Sleep Protocol

In our protocol, each node has 10 energy levels depending on its residual energy. The energy level i and the sleeping window SW_i, corresponding to the energy level of a node, are shown in the equation below.

$$i = ceil(\frac{b\%}{10})$$
$$SW_i = 2^i to 2^{i-1} \tag{1}$$

where, b is the battery level of node in percentage. Each node initially starts from energy level $i = 10$ where its sleeping window is $SW_{i=10} = 2^{10} to 2^{10-1}$, i.e. (1024-512). When active node consumes more than 10% of its initial energy, its energy level changes to $i = 9$ and its sleeping window size decreases to $SW_{i=9} = 2^9 to 2^{9-1}$, i.e. (512-256). Similarly, if the active node consumes 20% of its initial energy, its energy level changes to $i = 8$ and its sleeping window size decreases to $SW_{i=8} = 2^8 to 2^{8-1}$ i.e. (256-128). In this way, the sleeping window size becomes smaller as the node consumes more power. The Backoff Sleep Time (BST) used by a node based on energy level i is given by

$$BST = Random(SW_i) * \frac{R_E}{I_E} * \eta \tag{2}$$

where, R_E is the residual energy and I_E is the initial energy of active node. η is a tunable parameter having unit of time and depends on the application of the sensor network. In the next section, we evaluate the performance of RBSP and compare it with PEAS.

4 Performance Evaluation

We have implemented RBSP and PEAS in ns-2[14]. The energy model in this protocol is similar to PEAS[6], where Sleep:Idle:Tx:Rx as 0.03mW:12mW:60mW: 12mW. We assume that the maximum sensing range is 5 meters and is equal to the transmission range. The initial energy of each node is set at 1 Joule. We run the simulation for 150 sec. The packet size of HELLO and REPLY messages are 20 bytes each. We have deployed 100 sensor nodes over $50 \times 50m^2$ network field. We vary node density fraction from 0.02 to 0.1 in order to calculate number of active nodes, where the node density fraction is the ratio of number of deployed nodes to the total area of the network field. Nodes are randomly deployed in the field and remain stationary after deployment.

Fig. 4 shows the number of active nodes with respect to time. Number of active nodes in case of RBSP is comparable to PEAS.

Fig. 5 shows the number of active nodes with varying fraction of node density. The RBSP and PEAS maintain adequate active nodes in order to monitor the intended network field. As we increase the node density fraction, active nodes vary in linear proportion to the number of deployed nodes. Again the number of active nodes in case of RBSP is comparable to PEAS.

Fig. 6 shows the average energy consumption of the network with respect to time. The average energy consumption is the ratio of the total energy consumption to the total number of nodes in the network. We can see that the average energy consumption of RBSP is less as compared to that of PEAS. The energy

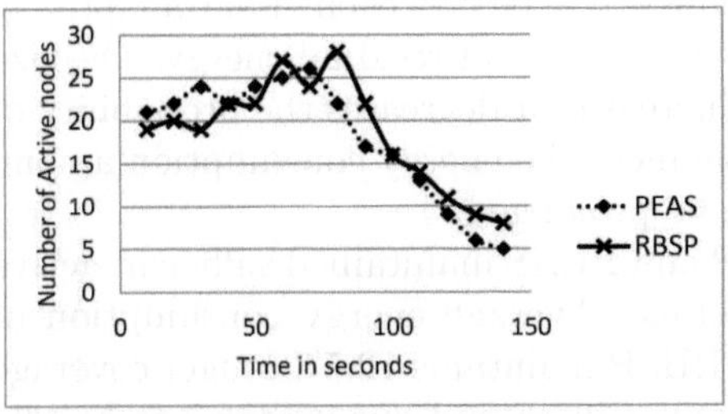

Fig. 4. Number of active nodes

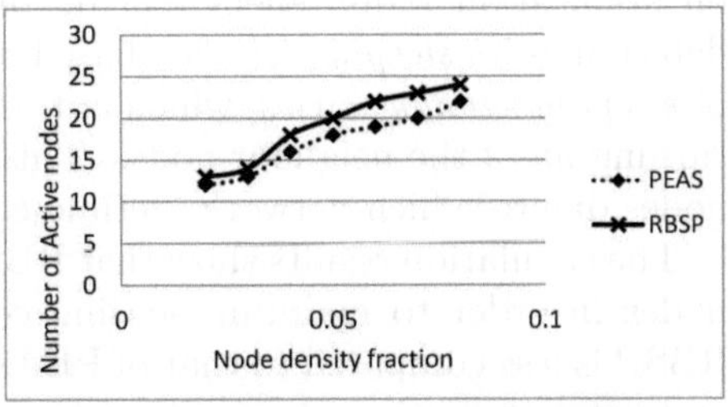

Fig. 5. Number of active nodes

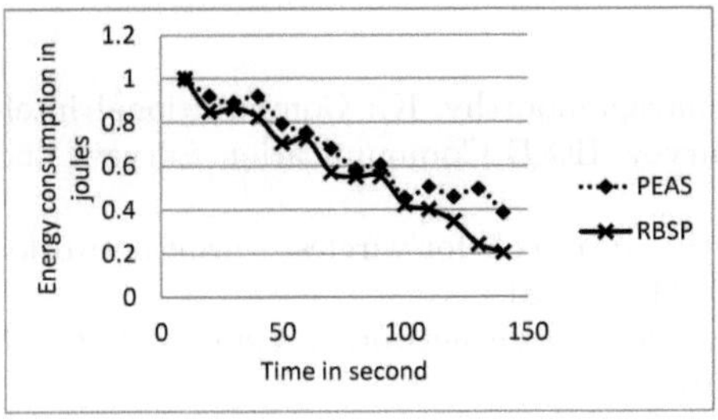

Fig. 6. Average energy consumption

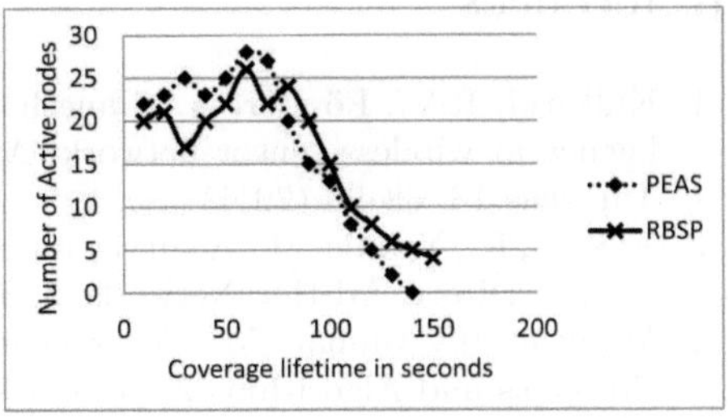

Fig. 7. Coverage lifetime

consumption of RBSP is less, due to changeable sleeping window determined on the basis of residual energy of active nodes. If the node has more residual energy, it requires fewer wakeups of sleeping nodes, due to which average energy consumption is less and network lifetime is more.

In Fig. 7, we can see the coverage lifetime for RBSP and PEAS. We assume that the presence of at least one active node in the network is sufficient to maintain minimum coverage in the region. For the case of a well-planned deployment, it is worth noting that the ratio of the entire sensing area to the maximum sensing area per node is about $\frac{50*50}{\pi*(5)^2} \approx 31$, which implies that at least 31 nodes are required to cover the entire area. At the time instant of 150 seconds, PEAS does not have any active node to monitor the field but in case of RBSP, four nodes are in active state to monitor the area. Hence, RBSP maintains adequate number of nodes active for a longer period of time, approximately 12.5% longer than that of PEAS. Therefore, RBSP has 12.5% more coverage lifetime as compared to PEAS.

5 Conclusion

We have proposed a Random Backoff Sleep Protocol(RBSP) which is a location free protocol that depends on the residual energy of ACTIVE nodes. Moreover, by simply varying the tunable parameter η, different sleeping time intervals based on application requirement can be provided. In RBSP each node adaptively determines its *sleeping window* based on the amount of residual energy. The size of *sleeping window* varies, which in turn increases or decreases the probability of turning *on* of the neighbor nodes. This balances the energy consumption among nodes due to which network lifetime can be prolonged.

The simulation results show that RBSP and PEAS maintained sufficient active nodes in order to maintain sensing coverage. Average energy consumption of RBSP is less compared to that of PEAS. RBSP maintains 12.5% longer coverage and network lifetime. In our future work we will extend our protocol to handle node failure probability which could creates sensing void.

References

1. Kulkarni, R.V., Förster, A., Ganesh Venayagamoorthy, K.: Computational intelligence in wireless sensor network: A survey. IEEE Communication Surveys and Tutorials 13, 68–96 (2011)
2. Akkaya, K., Younis, M.: A survey on routing protocols for wireless sensor networks. Science Direct, Ad Hoc Networks 3, 325–349 (2005)
3. Mulligan, R., Ammari, M.: Coverage in wireless sensor networks: A survey. Network Protocols and Algorithms 2, 27–53 (2010)
4. Akyildiz, I., Su, W., Sankarasubramaniam, Y., Cayirici, E.: Wireless sensor networks: A survey. Computer Networks 38, 393–422 (2002)
5. Ghosh, A., Das, S.: Coverage and connectivity issues in wireless sensor networks: A survey. Science Direct, Pervasive and Mobile Computing 4, 303–334 (2008)
6. Fan Ye, F., Zhong, G., Cheng, J., Lu, S., Zhang, L.: Peas: A robust energy conserving protocol for long-lived sensor networks. In: 23rd International Conference on Distributed Computing Systems (DCS 2002), pp. 28–37 (2002)
7. Gui, C., Mohapatra, P.: Power conservation and quality of surveillance in traget tracking sensor networks. In: Proceedings of 10th Annual International Conference on Mobile Computing and Networking (ACM MobiCom 2004), Pennsylvania, USA, pp. 129–143 (2004)
8. Yen, Y., Hong, S., Hang, R., Chao, H.: An energy efficient and coverage guaranteed wireless sensor network. In: Proceedings of IEEE Wireless Communications and Networking Conference, WCNC, pp. 2923–2930 (2007)
9. Xing, G., Wang, X., Zhang, Y., Lu, C., Pless, R., Gill, C.: Integrated coverage and connectvity configuration for energy conservation in sensor networks. ACM Transactions on Sensor Networks 1, 36–72 (2005)
10. Zhang, S., Yuhen Liu, Y., Pu, J., Zeng, X., Xiong, Z.: An enhanced coverage control protocol for wireless sensor networks. In: 42nd Hawaii International Conference on System Sciences, HICSS 2009, pp. 1–7 (2009)
11. Zhang, H., Hou, J.: Maintaining sensing coverage and connectivity in large sensor networks. Ad Hoc and Sensor Wireless Networks 1, 14–28 (2005)
12. Li, W., Bao, J., Shen, W., Chen, B., Jamieson, K., Balakrishan, H., Morris, R.: Span: An energy-efficient coordination algorithm for topology maintenence in ad hoc wireless networks. Wireless Networks 8, 481–494 (2002)

13. Cho, C., Pak, J., Kim, J.-n., Lee, I., Han, K.-J.: A random backoff algorithm for wireless sensor networks. In: Koucheryavy, Y., Harju, J., Iversen, V.B. (eds.) NEW2AN 2006. LNCS, vol. 4003, pp. 108–117. Springer, Heidelberg (2006)
14. ns-2: Network Simulator (2013), http://www.www.isi.edu/nsnam/ns (accessed December 24, 2013)

Distributed Diagnosis of Permanent and Intermittent Faults in Wireless Sensor Networks

Manmath Narayan Sahoo and Pabitra Mohan Khilar

Department of Computer Science & Engineering
National Institute of Technology Rourkela, Odisha - 769 008, India
{sahoom,pmkhilar}@nitrkl.ac.in

Abstract. Faults are inevitable in Wireless Sensors Networks (WSNs) because of physical defects caused due to environmental hazards, imperfection or hardware and/or software related glitches. If faults are not detected and handled properly the consequences may be inexorable in case of safety critical applications. This paper presents a distributed fault diagnosis algorithm to handle both permanent and intermittent faults in WSNs. The proposed diagnosis algorithm is based on the comparison of test results and residual energy estimations by neighboring sensor nodes. The intermittent faults are handled by iterating the comparisons for r rounds. The basic *time-out* mechanism is adopted to handle permanently faulty sensor nodes.

Keywords: wireless sensor networks, fault diagnosis, intermittent faults.

1 Introduction

A WSN is a distributed, self configurable, ubiquitous and infrastructure less[1] network, without any centralized administrations. It is often composed of many tiny, low-cost, battery-powered sensor nodes. Each node is aided with sensing, data processing, and communicating capabilities. The application of WSNs have tremendously grown up over last few decades. Environmental monitoring, transportation, crisis management, and military surveillance applications are name to few. A sensor node may have faults and measurement errors due to physical defect, imperfection or hardware and/or software related glitches. The harsh operational environment further aggravates the problem. In order to provide the quality of service (QoS), it is highly required to detect faulty sensors and let all fault-free sensors to receive these faulty events. This makes the network still operational in presence of faults, of course with degraded performance. The *distributed fault diagnosis* is intended to draw a consensus among the fault-free sensors about the status of all faulty sensors in the system. It acts as a basis for designing dependable systems by isolating the faulty sensors from the network. This paper considers the problem of distributed fault diagnosis in WSNs.

Fault diagnosis has been a focused area of research since last few decades and was first explored by Preparata *et al.* in [1] for a wired network with point

[1] Without any fixed infrastructures such as access points or base stations.

M.K. Kundu et al. (eds.), *Advanced Computing, Networking and Informatics - Volume 2*, Smart Innovation, Systems and Technologies 28,
DOI: 10.1007/978-3-319-07350-7_15, © Springer International Publishing Switzerland 2014

to point communication links. Since then, many variants of this model have been proposed. Comparison based model; the most favorable fault diagnosis mechanism has been the key discussion in [2, 3], where the decisions about the fault status of nodes are based on the comparison outcomes of the results of the same task executed by different nodes. The distributed fault diagnosis protocols for Mobile ad hoc Networks (MANETs) are extensively investigated in [4–6]. However, due to the harsh operational environments, sensor nodes fail more frequently than the nodes in other platforms. This makes the task of fault diagnosis more challenging.

Jaikaeo *et al.* have proposed a centralized fault diagnosis algorithm in [7] addressing the response implosion problem in sensor network diagnosis, thus reducing the traffic at central manager. Lee *et al.* in [8] have discussed another centralized fault management scheme that uses a central manager provided with a global view of the network to reliably execute predefined corrective and preventive management maintenance. Nevertheless, the scheme suffers with certain limitations. It is non-scalable and cannot be advantageous for larger networks; central manager is the bottleneck due to high traffic. MANNA: a management architecture for fault detection in event driven WSNs is presented in [9]. This scheme puts an external manager having the global knowledge of the network to detect the faulty events. However, it suffers from the disadvantages of a centralized approach. According to Ding *et al.* in [10], *Neighbor coordination* is another interesting approach to detect faulty nodes in sensor networks. Based on this approach, a sensor is assumed to be faulty if it deviates significantly from the median of readings of neighboring sensors. In the fault detection scheme presented by Chessa *et al.* in [11], a fault-free initiator starts the diagnosis process by accumulating information from its neighbors and the process continues until all the faulty nodes are identified. However, authors have considered no fault types other than crash fault. In [12] Chen *et al.*, have discussed a comparison based distributed diagnosis protocol for WSNs. This scheme is developed on the basis of the comparison results of own sensed data and neighbor's data. However, the scheme suffers from high communication complexity and hence not energy efficient. Authors in [13], have presented a probabilistic approach to diagnose faulty sensors in intermittent fault environment. Nevertheless, the scheme seems to be complex in terms of diagnosis time, message exchanges and more importantly energy consumption. For faulty sensor identification considering transient faults, a comparison based method that uses time redundancy have been discussed in [14] by Lee and Choi. Some more fault management schemes are briefed in the survey [15].

In this paper we present an efficient Fault Diagnosis Algorithm (FDA) for static topology WSNs, in presence of permanent and intermittent faults. The rest of the paper is organized as follows. Section 2, describes the network and fault model for WSNs. The proposed FDA is presented in Section 3. In Section 4, we discuss the simulation results for the algorithm, concluding in Section 5.

2 Network and Fault Model

We consider a WSN, consisting of n sensor nodes. The sensor nodes are assumed to be homogeneous and stationary. A permanently faulty node does not change its state until it is repaired and/or replaced. In contrast, an intermittently faulty node fluctuates between fault-free and being faulty, irregularly. The proposed FDA eyes on the detection of nodes with following fault types:

- permanent or intermittent faults in sensors
- permanent fault in communication unit

The sensor nodes with permanently faulty communication units are to be excluded from the network. However, the nodes with malfunctioning sensors still remain associated with the network since they have the ability to relay data packets among the nodes.

The undirected graph $C = (S, L^t)$, where S is the set of sensor nodes and L^t denotes the set of logical links between sensors at any given time t, represents the *communication graph* or *topology* of sensor network at time t. Sensor nodes S_i and S_j are said to be *adjacent* or *1-hop neighbors*, if they are in the transmission range of each other. $N^t_{S_i}$ denotes the set of nodes adjacent to S_i at time t, called the neighborhood set of S_i.

A *test graph*, $T = (S', L'^t)$ can be constructed from the communication graph by excluding the nodes with permanently faulty communication units and the links associated with those nodes. So $S' \subseteq S$, $L'^t \subseteq L^t$, and T is a sub-graph of C. Each link, $l^t_{(S_i,S_j)} \in L'^t$ is labelled by a binary value $c^t_{(S_i,S_j)}$. Without loss of generality we consider the test graph and the communication graph to be the same. We consider that the maximum number of faulty neighbors for any node $S_i \in S$ is $(\lceil |N^t_{S_i}|/2\rceil - 1)$. The links of the communication system are assumed to be error free.

3 Proposed Fault Diagnosis Algorithm

The proposed diagnosis algorithm is based on the comparison of sensor measurements by neighboring sensor nodes. Let $x^t_{S_i}$ denotes the sensor measurement of node S_i at a given time t. By considering the spatial correlation in sensor networks, the measurement difference of two fault-free neighboring sensors is presumed to be very small. However, if at least one of them is faulty then the difference is significant. Hence, if $l^t_{(S_i,S_j)} \in L^t$ then

$$\mid x^t_{S_i} - x^t_{S_j} \mid \begin{cases} \leq \delta_1, & \text{both } S_i \text{ and } S_j \text{ are fault-free} \\ > \delta_1, & \text{either or both of } S_i \text{ and } S_j \text{ is/are faulty.} \end{cases} \tag{1}$$

To aid the diagnosis process, the residual energy estimations by neighboring sensor nodes are also compared. Let $E^t_{(S_i,S_j)}$ be the estimation of node S_j about the residual energy of node S_i and $E^t_{S_i}$ be the own observed residual energy of S_i, at time t. Hence, if $S_i \in N^t_{S_j}$ then

$$| E^t_{S_i} - E^t_{(S_i,S_j)}| \begin{cases} \leq \delta_2, & \text{both } S_i \text{ and } S_j \text{ are fault-free} \\ > \delta_2, & \text{either or both of } S_i \text{ and } S_j \text{ is/are faulty.} \end{cases} \tag{2}$$

In Equations (1) and (2), δ_1 and δ_2 are two predefined thresholds. These thresholds may vary depending on the application. Now for each $l^t_{(S_i,S_j)} \in L^t$, $c^t_{(S_i,S_j)}$ can be defined as follows

$$c^t_{(S_i,S_j)} = \begin{cases} 0, & | x^t_{S_i} - x^t_{S_j}| \leq \delta_1 \text{ and } | E^t_{S_i} - E^t_{(S_i,S_j)}| \leq \delta_2 \\ 1, & \text{Otherwise.} \end{cases} \tag{3}$$

In Equation (3), $c^t_{(S_i,S_j)} = 0$, signifies both S_i and S_j are fault-free. But if at least one of S_i and S_j is faulty, then $c^t_{(S_i,S_j)} = 1$. Each sensor node, $S_i \in S$ maintains a boolean status register $StatR_{S_i}[]$ of size n, keeping the fault status of all the nodes in the network. Initially all the neighbor nodes are assumed to be fault free (0) and the status of all non neibhoring nodes are unknown (-1).

In each round, up to total of r rounds, each sensor node $S_j \in S$ sends its own observed sensor reading and expected residual energy of $S_i \in N^t_{S_j}$ to S_i i.e. it sends a message $M = (x^t_{S_j}, E^t_{(S_i,S_j)})$ to S_i. Upon receiving the message M from its neighbor S_j, node S_i performs the threshold test defined in Equation (3) and increments $StatR_{S_i}[j]$ by 1, if at least one of the test conditions fails. At the end of r rounds each sensor finds a partial diagnosis about the neighbors. Of course at this point the sensor node S_i does not have the fault status of non neighboring nodes. In order to reach a general consensus, all nodes in the network exchange their status registers. There may be a situation, when an intermittently faulty sensor node S_j sends to S_i, sensor measurement and expected residual energy of S_i, both correctly, in all r rounds; in which case S_i misdiagnoses S_j as fault-free. To overcome this situation we follow a majority voting as defined in Equation (4). We consider the maximum number of neighbors to which S_j may send such correct values in all r rounds is $\lceil n^+_{S_j}/2\rceil - 1$, where $n^+_{S_j}$ represents the number of fault-free neighbors of S_j.

$$StatR_{S_i}[j] = \begin{cases} 0, \; if \left(\sum\limits_{\substack{(S_k \in N^t_{S_j}) \\ (StatR_{S_i}[k] \leq 0) \\ (StatR_{S_k}[j]=0)}} 1 \right) \geq \left\lceil \frac{1}{2} \left(\sum\limits_{\substack{(S_k \in N^t_{S_j}) \\ (StatR_{S_i}[k] \leq 0)}} 1 \right) \right\rceil \\ 1, \; if \left(\sum\limits_{\substack{(S_k \in N^t_{S_j}) \\ (StatR_{S_i}[k] \leq 0) \\ (StatR_{S_k}[j]=0)}} 1 \right) < \left\lceil \frac{1}{2} \left(\sum\limits_{\substack{(S_k \in N^t_{S_j}) \\ (StatR_{S_i}[k] \leq 0)}} 1 \right) \right\rceil \; and \; StatR_{S_i}[j] = 0. \end{cases} \tag{4}$$

Algorithm 1. Proposed fault diagnosis algorithm

Data: $C = (S, L^t)$: The communication graph.
// The test graph and communication graph are considered to be same.
r: Maximum number of rounds.
Result: $StatR_{S_i}[]$ for each node $S_i \in S$
Initialization: NR=0; FFCount=0; NFNbrCnt=0; FFNbrStatSum=0; IFNbrCnt=0;
for *each* $S_i \in S$ *and* $S_j \in S$ **do**
  if $S_i == S_j$ *or* $l_{(S_i,S_j)} \in L^t$ **then**
    $StatR_{S_i}[j] = 0;$
  else
    $StatR_{S_i}[j] = -1;$
  end
end
repeat

  for *each* $S_j \in S$ *and* $S_i \in N^t_{S_j}$ **do**
    S_j sends a message $M = (x^t_{S_j}, E(S_i, S_j)^t)$;
  end
  if *a node* S_i *receives a message* M *from* $S_j \in N^t_{S_i}$ **then**
    if $|x^t_{S_i} - x^t_{S_j}| > \delta_1$ *or* $|E^t_{(S_i,S_j)} - E^t_{S_i}| > \delta_2$ **then**
      $StatRS_i[j] += 1;$
    end
  end
until *(++NR $\neq$ r)* ;
for *each* $S_i \in S$ **do**
  S_i broadcasts its status register $StatR_{S_i}[]$ to other nodes in the network;
end
for *each* $S_i \in S$; $S_j \in S$ *and* $S_i \neq S_j$ **do**
  for *each* $S_k \in N^t_{S_j}$ **do**
    if $StaR_{S_i}[k] == 0$ *and* $StatR_{S_k}[j] == 0$ **then**
      FFCount++;
    end
    if $StatR_{S_i}[k] \leq 0$ **then**
      NFNbrCnt++;
    end
  end
  if *FFCount* $\geq \lceil$*NFNbrCnt++/2*$\rceil$ **then**
    $StatR_{S_i}[j] = 0;$
  else if $StatR_{S_i}[j] = 0$ **then**
    $StatR_{S_i}[j] = 1;$
  end
end
for *each* $S_i \in S$; $S_j \in S$ *and* $S_i \neq S_j$ **do**
  for *each* $S_k \in N^t_{S_j}$ *and* $StaR_{S_i}[k] == 0$ **do**
    FFNbrStatSum+=$StatR_{S_k}[j]$;
    IFNbrCnt++;
  end
  if *FFNbrStatSum=* $(r \times$ *IFNbrCnt*$)$ **then**
    $StatR_{S_i}[j] = 2;$ // $StatR_{S_i}[j] = 2$ indicates S_j is permanently faulty.
  else if *FFNbrStatSum>0* **then**
    $StatR_{S_i}[j] = 1;$ // $StatR_{S_i}[j] = 1$ indicates S_j is intermittently faulty.
  end
end

In Equation (4), 0 and 1 indicates S_j to be fault-free and intermittently faulty respectively. There may be the case, when an intermittently faulty sensor node S_j sends sensor measurement and expected residual energy of S_i, either or both incorrectly, to the node S_i in all r rounds; in which case S_i misdiagnoses S_j as permanently faulty. To handle this situation and to determine the actual fault type, we follow Equation (5).

$$StatR_{S_i}[j] = \begin{cases} 1, & if \;\; 0 < \left(\sum_{\substack{(S_k \in N^t_{S_j}) \\ (StatR_{S_i}[k]=0)}} StatR_{S_k}[j] \right) < r \left(\sum_{\substack{(S_k \in N^t_{S_j}), \\ (StatR_{S_i}[k]=0)}} 1 \right) \\ 2, & if \;\; \left(\sum_{\substack{(S_k \in N^t_{S_j}) \\ (StatR_{S_i}[k]=0)}} StatR_{S_k}[j] \right) = r \left(\sum_{\substack{(S_k \in N^t_{S_j}), \\ (StatR_{S_i}[k]=0)}} 1 \right) \end{cases} \quad (5)$$

The values 1 or 2 of $StatR_{S_i}[j]$ in Equation(5) signifies S_j to be intermittently faulty or permanently faulty respectively. The proposed FDA is more precisely described in Algorithm 1.

4 Simulation Analysis

To support the feasibility of the proposed FDA, simulations are performed using the OMNET++ simulator. The results are compared with that of the detection algorithm discussed by *Lee* and *Choi* in [14]. Based on the faulty behaviour, the proposed FDA classifies the sensor nodes into three different classes: permanent fault class, intermittent fault class, and fault-free class. Two performance measures are used for evaluation, (i) Classification Accuracy (CA): The ratio of the number of nodes classified in to a particular class to the total number of nodes of that class, and (ii) False Alarm Rate (FAR): The ratio of the sum of the number of faulty nodes classified as fault-free and the number of fault-free nodes classified as faulty to the total number of nodes in the network.

A simulation scenario is created for a sensor network with 1000 nodes randomly deployed over 1000 × 1000 m^2 area.Each sensor node is equipped with AA battery with default initial energy 18720 Joule. With proper adjustment of the transmission range (common for all nodes), the desired value of average node degree (d) can be obtained. In the simulation, the sensor nodes are randomly chosen to have permanently faulty sensors with probabilities 0.02, 0.04, 0.06, 0.08, 0.10 and 0.12 respectively. We also consider that p_{if} is 150% of p_{pf}, in each case. Here, p_{if} denotes the probability of a node being intermittently faulty, and p_{pf} represents the same for a node being permanently faulty. The values of δ_1 and δ_2 are considered to be 4 and 2 respectively. In order to evaluate *Lee* and *Choi's* algorithm, we consider the same simulation scenario with $\theta_1 = \lceil d/2 \rceil$ and $\theta_2 = 2$ as the values of thresholds used in their algorithm. The FDA is run for $r(= 10)$ rounds to handle intermittent faults. The obtained simulation results

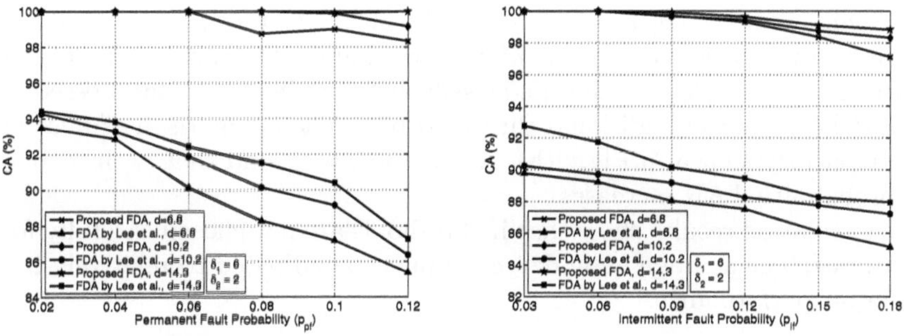

(a) (b)

Fig. 1. Comparison of CA for (a) permanently faulty, and (b) intermittently faulty nodes

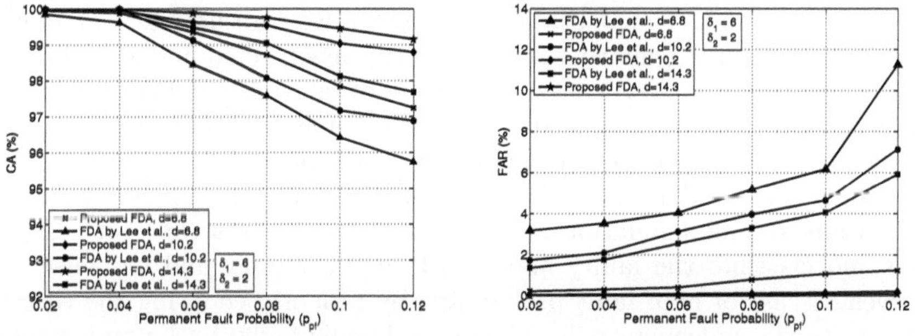

(a) (b)

Fig. 2. Comparison of (a) CA for fault-free nodes, and (b) False Alarm Rate

for CA and FAR for different values of d are compared as depicted in Fig. 1, and Fig. 2.

It can clearly be observed that the CA decreases with lower node degrees; since, in case of sparse networks the fault-free sensor nodes may not always form a connected graph for fault diagnosis purpose. In such scenarios, all neighbors of a particular node may be faulty at the same time, leading to misdiagnosis of the node. Such scenarios arise with more counts for low d and high fault probability, in which case the performance even degrades.

Fig. 1(a) depicts the comparison of classification accuracy for permanently faulty nodes with d values 6.8, 10.2, and 14.3. In some rounds, if a permanently faulty node produces a sensor measurement that does not differ from the sensor measurements of its fault-free neighbors by a minimum threshold δ_1, then it is not classified as permanently faulty. The additional threshold test on residual energy in the proposed FDA handles such cases and improves the performance.

An intermittently faulty node that generates incorrect sensor measurements in less than or equal to θ_2 rounds are not classified as intermittently faulty in the fault detection algorithm by Lee *et al.* For low value of δ_1, fault-free nodes may be miss-diagnosed as faulty. Such miss-classification scenarios are suppressed in the proposed FDA by the additional threshold test. The comparison of false alarm rates are clearly shown in Fig. 2(b). As obvious, we found that with increase in fault probability, FAR increases.

The simulation results show that if thresholds are not chosen carefully for the applications. The average node degree, d must be adjusted to relatively high to have better performance.

5 Conclusions

In this paper we propose a distributed fault diagnosis algorithm for WSNs, in order to handle sensor nodes having permanently fault sensor or intermittently faulty processing unit. The algorithm is based on two threshold tests: (i) on sensor measurements of neighboring nodes, and (ii) on expected and actual residual energy of the sensor nodes. Two special cases of intermittent faults are considered: One, where an intermittently faulty node sends both sensor measurement and expected residual energy of neighboring nodes correctly to some of its neighbors in all r rounds; Another, where at least one of these values are incorrect in all r rounds. The simulation experimental results vows that the algorithm detects and classifies the faulty nodes with high accuracy and low false alarm rate, even in case of high fault probability, by properly choosing the threshold values. In future, endeavour will be made to handle faults in dynamic topology environment.

References

1. Preparata, F., Metze, G., Chien, R.: On the connection assignment problem of diagnosable systems. IEEE Transactions on Electronic Computers EC-16, 848–854 (1967)
2. Malek, M.: A comparison connection assignment for diagnosis of multiprocessor systems. In: Proceedings of the 7th Annual Symposium on Computer Architecture, ISCA 1980, pp. 31–36. ACM (1980)
3. Chw, K., Hakimi, S.: Schemes for fault-tolerant computing: A comparison of modularly redundant and t-diagnosable systems. Information and Control 49, 212–238 (1981)
4. Chessa, S., Santi, P.: Comparison-based system-level fault diagnosis in ad hoc networks. In: Proceedings of the 20th IEEE Symposium on Reliable Distributed Systems, pp. 257–266 (2001)
5. Elhadef, M., Boukerche, A., Elkadiki, H.: Diagnosing mobile ad-hoc networks: two distributed comparison-based self-diagnosis protocols. In: Proceedings of the 4th ACM International Workshop on Mobility Management and Wireless Access, MobiWac 2006, pp. 18–27. ACM (2006)

6. Elhadef, M., Boukerche, A., Elkadiki, H.: A distributed fault identification protocol for wireless and mobile ad hoc networks. Journal of Parallel and Distributed Computing 68, 321–335 (2008)
7. Jaikaeo, C., Srisathapornphat, C., Shen, C.: Diagnosis of sensor networks. In: IEEE International Conference on Communications, ICC 2001, vol. 5, pp. 1627–1632 (2001)
8. Lee, W., Datta, A., Cardell-oliver, R.: Winms: Wireless sensor network-management system, an adaptive policy-based management for wireless sensor networks. Technical report, The University of Western Australia (2006)
9. Ruiz, L., Siqueira, I., Oliveira, L., Wong, H., Nogueira, J., Loureiro, A.: Fault management in event-driven wireless sensor networks. In: Proceedings of the 7th ACM International Symposium on Modeling, Analysis and Simulation of Wireless and Mobile Systems, MSWiM 2004, pp. 149–156. ACM (2004)
10. Ding, M., Chen, D., Xing, K., Cheng, X.: Localized fault-tolerant event boundary detection in sensor networks. In: Proceedings of the 24th Annual Joint Conference of the IEEE Computer and Communications Societies (INFOCOM 2005), vol. 2, pp. 902–913 (2005)
11. Chessa, S., Santi, P.: Crash faults identification in wireless sensor networks. Computer Communications 25, 1273–1282 (2002)
12. Chen, J., Kher, S., Somani, A.: Distributed fault detection of wireless sensor networks. In: Proceedings of the 2006 Workshop on Dependability Issues in Wireless Ad Hoc Networks and Sensor Networks, DIWANS 2006, pp. 65–72. ACM (2006)
13. Khilar, P., Mahapatra, S.: Intermittent fault diagnosis in wireless sensor networks. In: 10th International Conference on Information Technology (ICIT 2007), pp. 145–147 (2007)
14. Lee, M., Choi, Y.H.: Fault detection of wireless sensor networks. Computer Communications 31, 3469–3475 (2008)
15. Yu, M., Mokhtar, H., Merabti, M.: Fault management in wireless sensor networks. IEEE Wireless Communications 14, 13–19 (2007)

Efficient Coverage Protocol for Homogeneous Wireless Sensor Networks

Ramachandra Polireddi and Chinara Suchismita

National Institute of Technology Rourkela
Odisha - 769008, India
polireddyram@gmail.com, suchismita@nitrkl.ac.in

Abstract. In this paper, a localization technique is used to reduce the number of active sensors at an instant of time in a region for wireless sensor networks(WSNs). The active nodes in a region is calculated baed on the K-hop Independent Dominating Set(K-IDS). The proposed algorithm is evaluated using Castalia Simulator.The simulation results are analysed using network parameters namely,energy consumption,number of active nodes in a cluster and packet delivery ratio. The simulation results indicates that the proposed algorithm outperforms LEACH and MOCA protocls.

1 Introduction

WSN is a collection of hundreds of nodes connected to each other through short range of wireless links. WSN consists of two types of nodes sensor nodes and Base stations. Sensor nodes senses the environment and forward the data to the base station[1]. These nodes have mobility, organise themselves into a network according to the location and environment. Such nodes act autonomously, but cooperatively to form a logical network. But a WSN comprises of a potentially large set of nodes that may be spread over a wide range of geographical area, indoor and outdoor. The processing power, memory and battery power are important factors. WSNs enable numerous sensing and monitoring services in areas of vital importance such as efficient industry production, safety and security at home as well as in traffic and environmental monitoring.WSNs active when an event detected. Early research studies in WSNs targeted military applications, especially for battlefield monitoring [2]. Nowadays, WSNs are now being deployed in civilian areas, mining areas, building control (i.e., monitoring the A.c's automatic power on /off),locking the car doors and being used for habitat observation [3],health monitoring [2], object tracking [4], etc.

Actually sesnors having different sensing range and transmission range [5]. The diagram of sensor is shown below.

Sensors are deployed in harsh and unattend environment. Sensors are equipped with battery,replacing the battery after the deployment is difficult. The energy efficiency is an important feature to enhance the network life time.

In this paper, we develop a Efficient coverage protocol(ECP). It is a clustering based protocol that maximizes the network life time.The key features of

M.K. Kundu et al. (eds.), *Advanced Computing, Networking and Informatics - Volume 2*, Smart Innovation, Systems and Technologies 28,
DOI: 10.1007/978-3-319-07350-7_16, © Springer International Publishing Switzerland 2014

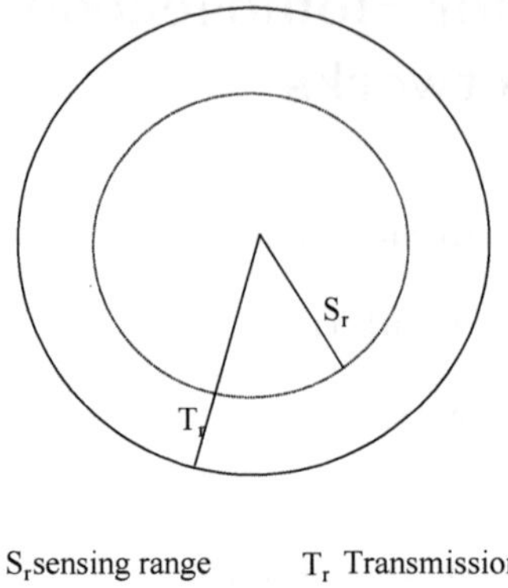

Fig. 1. Sensor

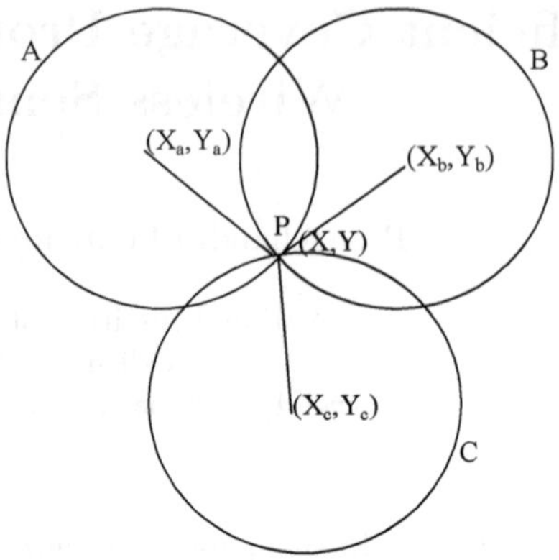

Fig. 2. Trilateration

ECP are: network coverage, network connectivity, maximum energy cluster head and active nodes. This protocol implementation uses the same radio model as in LEACH which is first order radio model[6]. The remainder of the paper is organized as follows. Section 2 describes the related work, Section 3 designing ECP protocol and Section 4 presents the simulation results.Finally, the paper is concluded in Section 5.

2 Related Work

Clustering is the best technique for reducing the energy consumption, improving the network lifetime and provides the scalability to the sensor network [7]. In hierarchical network model, sensors are logically partitioned into clusters. Each cluster has the cluster head(CH). The CH performs the data aggregation and fusion on the data which is coming from the cluster members.Clustering process is two level process. The CH's present in the higher level and cluster members are present in the lower level. CH's transmit the data to the BS directly or through intermediatery nodes, i.e., Gateway nodes.

LEACH [6] is the first protocol based on clustering.In this CH is elected randomly by the use of probability. In this all nodes have the equal probability to become a cluster head.The cluster heads are elected for each round randomly.The CH performs the major roles like TDMA scheduling to cluster members, data aggregation and send to the BS.

In LEACH-C [8], clustering protocol CH's are selected based on residual energy of sensors.The higher residual energy node becomes CH.Where as in LEACH, the lower energy sensor is also a CH. In this protocol, MST (minimum spanning tree) used for data forwarding to the BS. In both LEACH and LEACH-C consider the energy distribution, communication overhead and reduces the number of transmissions. However, these algorithms don't considers the coverage of sensors. The network life time can be enhanced by reducing the overlapping sensing ranges of the sensors.

In MOCA protocol [9], it uses the timers for synchronization among sensors. In this sensors are organises into overlapping clusters. There is no central control for election of CH but this performed in distributed manner. It uses the parameters like cluster radius, node transmission range and cluster head probability. In this author, consider only non overlapping CH's and didn't consider the minimum number of non overlapping cluster members.

Ruay-Shiung [10] proposed a clustering protocol gives uniform clusters based on hop-count and residual energy of sensors.The maximum residual energy and hop-count sensor node becomes CH. This algorithm, do nt consider the overlapping coverage of sensors.

3 Designing ECP

In this section, we are assuming different types of sensors are deployed uniformly in the region $M \times M$ region. We assume that 5% of sensors are deployed with GPS and the remaining sensors without GPS. Now the trilateration method used for finding the positions of sensors. Trilateration method [11] uses the relative position mechanism.

Cluster Formation:

In this phase sensors sends the hello message to their k-hop neighbours. The hello message contains location information, attribute type and residual energy. Due to this each and every sensor gets the neighbours information regarding residual energy and number of neighbours. Then selecting the node which has the highest number of neighbours and residual energy as a CH. The CH sends the advertisement message to all k-hop neighbours. Then each and every sensor in the cluster knows their CH. Cluster formation is shown in Fig. 3.

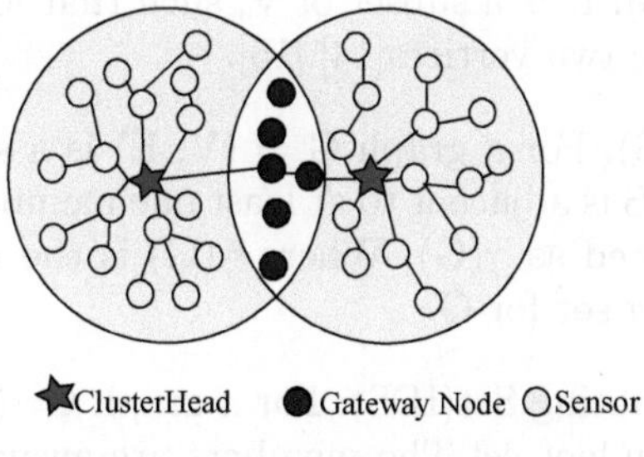

Fig. 3. Cluster formation

3.1 Determination of Active Nodes and Non Active Nodes

Our aim is to minimize the overlapped sensing range among sensors and maintaining the coverage over the sensors deployment region. Sensors deployed uniformly in the sensing area.So many sensors sensing range is overlapped by one another. Due to this so much of redundant data is forwarded to the CH and CH's

Algorithm 1. cluster_formation

SetupPhase :
sends the hello message to neighbours
Select's the node which has more number of neighbour's and residual energy level
if *Two* or more nodes have the equal number of neighbour's and residual energy level **then**
  Then choose any node as CH (cluster head)
else
  The numbers of neighbour's are equal and different residual energy levels
  Then choose highest residual energy node as CH(cluster head)
  The CH node broadcast the message to the neighbour's
  CH cerate's a TDMA schedule for each node in the cluster
  According to their TDMA slot transfer the data and communicates with cluster head
end if

Algorithm 2. *Data* Aggregation at Cluster Head

*CH*aggregates the data coming from the cluster members
Average is used for the DataAggregation.

forwarded the aggregated data to the BS through the intermediatery sensors and CH's. Determination of number of active nodes has been proven to be NP-Hard problem [12]. Sensor network consists of n number of nodes and each node have the transmission range. Each node act as a vertices(V) and transmission range act as edge(E) in graph terminology [13]. In this, each node connected to other nodes via a transmission range, i.e., edge. Edges represent the communication between the nodes. Some useful basic definitions are given below.

- Independent Set(IS): It is a subset of V, such that for every two vertices no edge connecting the two vertices[14],[15].
- Dominating Set(DS): For a graph G = (V, E) is a subset S of V such that every vertex not in S is adjacent to at least one member of S. The domination number is represented as $\gamma(G)$. Where $\gamma(G)$ is the number of vertices in a smallest dominating set for G.
- Independent Dominating Set(IDS): For a graph G=(V,E),IDS is both dominating and independent set.The members are away from each other by at least one hop distance.It is represented by i(G). Determination of IDS is NP-Hard[12].
- K-IDS(Independent Dominating Set): It is DS such that any two members are away from each other by at most k-hop distance.
- 1-Independent Dominating Set(1-IDS): It is DS such that any two members are away from each other by 1-hop distance.

The realtion ship between IS,DS and IDS is γ(G) $\leq$ i(G) $\leq$ α(G).

Calculation of IS, DS, and IDS for the Fig. 4

IS :1,3,5,7,9
DS :3,6,1
IDS :1,3,5,7,9
1-IDS :2,4,6,8,12

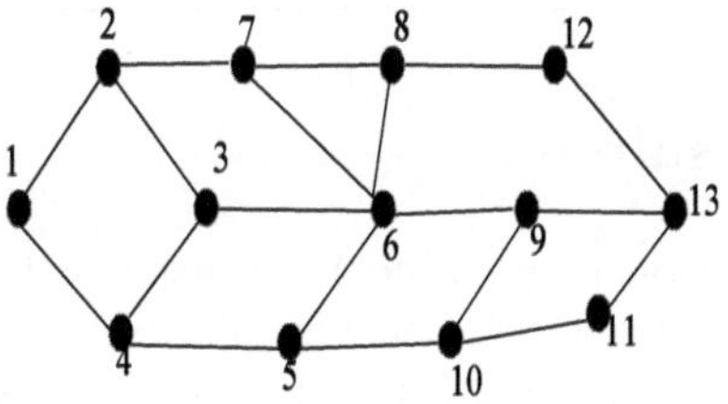

Fig. 4. Graph representation of sensors in a cluster

Algorithm 3. Active nodes determination

$BroadCast(msg(I, Hello, Neighbour(i))$
while $True$ **do**
  $Received\_msg(\forall j, Hello, i)$
  $Received\_msg(\forall j, Active, i)$
  $Received\_msg(\forall j, Sleep, i)$
  if $I \neq Active$ AND $Received\_msg(j, Active, i)$ **then**
    $sleep(i) \leftarrow True$
    $forward\_msg(i, Sleep, Neighbour(i))$
  end if
  if $I \neq Active$ AND $received\_msg(j, Sleep, i)$ **then**
    $Active(i) \leftarrow True$
    $Forward\_msg(i, Active, Neighbours(i))$
  end if
  if $I \equiv Active$ AND $Residual\_energy \leq threshold$ **then**
    $Sleep(i) \leftarrow True$
    $BroadCast\_msg(I, sleep, Neighbours(i))$
    Goto Step 6 to 7
  end if
end while

Table 1. Simulation parameters

Parameter	Value
Network area	$50 \times 50m^2$
Number of sensors	100
E_{elec}	5 nJ/bit
ε_{fs}	10pJ/bit/m^2
ε_{mp}	0.0013 pJ/bit/m^4
E_0	2J
E_{DA}	5nJ/bit/message
message size	2000bits
d_0	30m

4 Simulation Results

The parameters as mentioned in Table 1 are used in the simulation. The network area is $50 \times 50m^2$ with 100 sensors. The sensor transmitting radius is 10m.The numbers of clusters are 4.Each cluster consists of 13 cluster members approximately.For this we are giving some mathematical model. Let us consider network area is $L \times Lm^2$. The 'n' number of sensors are deployed in a sensing region.Each sensor transmission range is T_r.The area of sensing field is L^2. Each cluster range/area isΠT_r^2. Therefore the number of cluster members present in each cluster is $n \times \Pi \times T_r^2 \div L^2$.

4.1 Energy Consumption Model

$$E_{TX}\left(L,d\right) = \begin{cases} L * E_{Elec} + L * \varepsilon_{fs} * d^2 & d < d_0 \\ L * E_{Elec} + L * \varepsilon_{mp} * d^4 & d > d_0 \end{cases} \tag{1}$$

where E_{TX} is energy consumed by the transmitter, L is length of message, E_{elec} is consumed by electronics in sensor, ε_{fs} is free space channel model, d is distance between transmitter and receiver, $2 and 4$ are path loss exponents ε_{mp} is multipath channel model, d_0 is threshold distance.

$$E_{RX}(L,d) = L * E_{Elec} \tag{2}$$

where E_{RX} is energy consumed by the receiver, L is the length of message, E_{Elec} is energy consumed by the electronics in sensor. Initially the energy of each sensor is E_0. Then total energy is $E_{Total} = NE_0$.

After the deployment of sensors are logically partitioned into clusters.The clustering process is done in rounds based on the residual energy of the sensors.Each cluster consists of active and non-active sensors. The energy consumption is calculated for active sensors in each round as follows.

$$E_{round} = L(2{*}K{*}n_a{*}E_{Elec}{+}K{*}n_a{*}E_{DA}{+}K{*}\varepsilon_{mp}{*}d_{toBS}^4{+}K{*}n_a{*}\varepsilon_{fs}{*}d_{toBS}^2) \tag{3}$$

where n_a is active nodes, K is the number of clusters, l is the number of bits in message and E_{DA} is energy consumption during data aggregation and the distance from CH to base station(d^2_{toBS}).

The energy consumed by the CH in a cluster is calculated as follows

$$E_{CH} = L * E_{Elec} * (n_a - 1) + L * E_{DA} * n_a + L * E_{Elec} + L * \varepsilon_{mp} * d^4_{toBS} \quad (4)$$

Fig. 7 shows the simulation results of Number of active nodes in a cluster for the different rounds upto 8 rounds. In this we compare proposed ECP protocol with the existing protocols like LEACH, MOCA. The results shows that LEACH have the more number of active nodes. Due to this more energy consumed and number of live nodes decreases with the time. Therefore the network life time decreases, where as ECP have the less number of active nodes and also cover the whole sensing region with minimum number of active nodes and the remaining nodes in non active state. Therefore the energy consumption reduced and enhance the network life time.

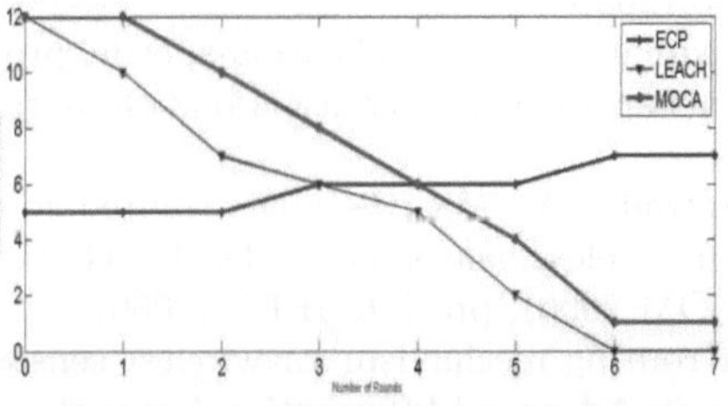

Fig. 5. Comparison of algorithms based on active nodes

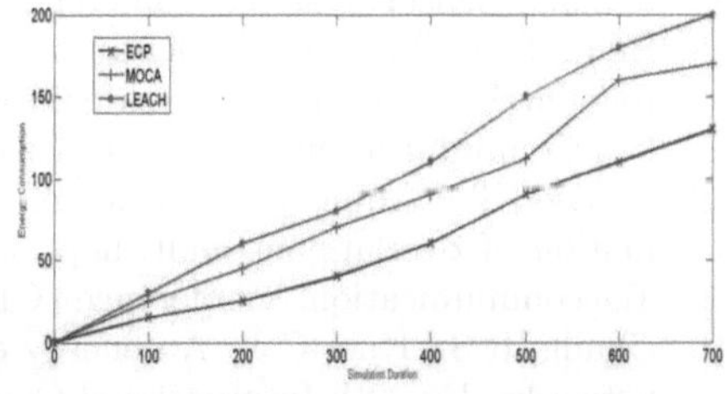

Fig. 6. Energy consumption comparison of algorithms

Fig. 8 shows the energy consumption of different protocols like LEACH, MOCA, and ECP for the different simulation duration periods. The figure itself describes the ECP consumes less energy when compared to other protocols like LEACH and MOCA.

5 Conclusion

The efficient coverage mechanism reduces the energy consumption of network and enhance the network life time. In this paper, we proposed a Energy efficient Coverage Protocol. ECP reduces the overlapping sensing ranges of sensors and also maintains the network connectivity among sensors. The active nodes determined by the use of K-IDS mechanism. The simulation results shows that ECP protocol outperforms LEACH and MOCA in enhancing the Network life time and reducing the overlapping sensing ranges.

References

1. Al-Karaki, J., Kamal, A.: Routing techniques in wireless sensor networks: a survey. IEEE Wireless Communications 11(6), 1–23 (2004)
2. Akyildiz, I., Su, W., Sankarasubramaniam, Y., Cayirci, E.: Wireless sensor networks: a survey. Computer Networks 38(4), 1–29 (2002)
3. Mainwaring, A., Culler, D., Polastre, J., Szewczyk, R., Anderson, J.: Wireless sensor networks for habitat monitoring. In: Proceedings of the 1st ACM International Workshop on Wireless Sensor Networks and Applications, pp. 1–10 (2002)
4. Salatas, V.: Object tracking using wireless sensor networks. PhD thesis, Monterey, California. Naval Postgraduate School (2005)
5. Wu, J., Yang, S.: Energy-efficient node scheduling models in sensor networks with adjustable ranges. International Journal of Foundations of Computer Science 16(1), 1–14 (2005)
6. Heinzelman, W., Chandrakasan, A., Balakrishnan, H.: Energy-efficient communication protocol for wireless microsensor networks. In: 33rd Annual Hawaii International Conference on System Sciences, pp. 1–10 (2000)
7. Li, C., Ye, M., Chen, G., Wu, J.: An energy-efficient unequal clustering mechanism for wireless sensor networks. In: 2005 IEEE International Conference on Mobile Ad hoc and Sensor Systems Conference, pp. 1–5 (2005)
8. Heinzelman, W., Chandrakasan, A., Balakrishnan, H.: An application-specific protocol architecture for wireless microsensor networks. IEEE Transactions on Wireless Communications 1(4), 1–11 (2002)
9. Youssef, A., Younis, M., Youssef, M., Agrawala, A.: Wsn16-5: Distributed formation of overlapping multi-hop clusters in wireless sensor networks. In: Global Telecommunications Conference (GLOBECOM 2006), pp. 1–6. IEEE (2006)
10. Chang, R.S., Kuo, C.J.: An energy efficient routing mechanism for wireless sensor networks. In: 20th International Conference on Advanced Information Networking and Applications, vol. 2, pp. 1–5. IEEE (2006)
11. Mao, G., Fidan, B., Anderson, B.: Wireless sensor network localization techniques. Computer Networks 51(10), 1–24 (2007)
12. Efrat, A., Har-Peled, S., Mitchell, J.: Approximation algorithms for two optimal location problems in sensor networks. In: 2005 2nd International Conference on Broadband Networks, pp. 1–9. IEEE (2005)
13. Shin, I., Shen, Y., Thai, M.: On approximation of dominating tree in wireless sensor networks. Optimization Letters 4(3), 393–403 (2010)
14. Akkaya, K., Senel, F., McLaughlan, B.: Clustering of wireless sensor and actor networks based on sensor distribution and connectivity. Journal of Parallel and Distributed Computing 69(6), 573–587 (2009)
15. Gary, M., Johnson, D.: Computers and intractability: A guide to the theory of np-completeness (1979)

A Novel ICI Self-cancellation Scheme for Improving the Performance of OFDM System

Ravi Prakash Yadav[1], Ritesh Kumar Mishra[2], and Shipra Swati[3]

[1] Department of Electronics and Communication Engineering
National Institute of Technology, Patna, India
[2] Department of Electronics and Communication Engineering
National Institute of Technology, Patna, India
[3] Department of Computer Science and Engineering
PICT, Pune, India
{er.ravipdeo,shipra.er}@gmail.com,
ritesh@nitp.ac.in,

Abstract. Orthogonal frequency division multiplexing (OFDM) is a promising technique for fourth generation (4G) broadband wireless communication systems. However its performance is degrade due to Doppler frequency drift or frequency drift between transmitter and receiver oscillator which causes frequency offset. This leads to a loss in the orthogonality between sub-carriers and results in inter-carrier-interference (ICI). In this paper we proposed a novel ICI self-cancellation scheme for ICI mitigation in OFDM system and compared it with the standard OFDM system and conventional self-cancellation scheme in terms of carrier-to-interference ratio (CIR) and bit-error rate. This scheme works in two very simple steps. At the transmitter side, one data symbol is modulated onto four sub-carriers with appropriate weighting coefficients. At the receiver side by linearly combining the received signals on these sub-carriers leads to a sufficient reduction in ICI. The simulation result shows that the proposed scheme outperforms the existing method.

Keywords: Inter-carrier-interference (ICI), OFDM, ICI self-cancellation, carrier-to-interference ratio (CIR).

1 Introduction

OFDM is widely known as the promising communication technique in the broadband wireless communication systems.Currently, OFDM is being used in many wireless communication systems, such as Digital Video Broadcasting (DVB) systems, HIPERLAN2 (High Performance Local Area Network), Worldwide Interoperability for Microwave Access (Wi-Max). In OFDM, a high data rate channel is divided in to many low data rate sub-channels and each sub-channel is modulated in different sub-carriers. Due to this each channel experience a flat-fading and equalization at the receiver is lesscomplex. So it provides high spectral efficiency and robustness to multipath interference. In OFDM sub-channels are orthogonal, but due to frequency offset

M.K. Kundu et al. (eds.), *Advanced Computing, Networking and Informatics - Volume 2*,
Smart Innovation, Systems and Technologies 28,
DOI: 10.1007/978-3-319-07350-7_17,

orthogonality is lost which causes inter-carrier-interference (ICI) and it degrades the performance of OFDM system [1-2].

Various schemes have been investigated to mitigate ICI in OFDM system, such as frequency domain equalization [3], time domain windowing [4], pulse-shaping [5], self-cancellation [6-10], frequency offset estimation and correction technique [11-12] and so on. Among the schemes the ICI self-cancellation is a simple way for ICI reduction. Several self-cancellation schemes have been developed such as data-conversion [6], real constant weighted data-conversion [7], plural weighted data-conversion [8], data-conjugate [9], weighted- conjugate transformation [10].

In this paper we present a theoretical expression for CIR for the proposed scheme and its performance is compared with conventional self-cancellation scheme and standard OFDM system in terms of carrier-to-interference ratio (CIR) and bit-error rate (BER).

2 OFDM System Description and ICI Analysis

Fig.1 illustrates the block diagram of the baseband, discrete time FFT-based OFDM system. Firstly a stream of input serial bit is converted in to parallel by S/P, each parallel bit then mapped in to symbols using MPSK modulation, then the symbols are modulated by IFFT on N-parallel sub-carriers and transmitted after adding cyclic prefix and converted in to serial data. The addition of cyclic prefix is used to cancel inter-symbol-interference (ISI). At the receiver side, the cyclic prefix is removed from received data after S/P, then perform FFT, de-mapped in to bits and back to serial data using P/S.

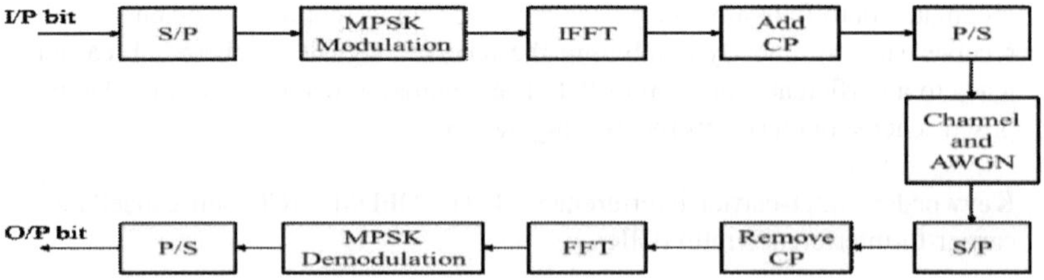

Fig. 1. Block diagram of baseband OFDM System

In OFDM system the time domain transmitted signal could be expressed as:

$$x(n) = \frac{1}{N}\sum_{k=0}^{N-1} X(k)\, e^{j2\pi kn/N} \quad (1)$$

where $x(n)$denotes the nth sample of OFDM transmitted signal, $X(k)$ denotes the modulated symbol for the kth sub-carrier ($k = 0,1,\dots,N-1$) and N is the total number of sub-carriers.

The received signal in time domain is given by:

$$y(n) = x(n)e^{j2\pi n\varepsilon/N} + \omega(n) \tag{2}$$

where ε denotes the normalized frequency offset and is given by $\Delta f.NT_s$. $\Delta f$is the Doppler frequency shift and T_s is sub-carrier symbol period, $\omega(n)$ is additive white Gaussian noise (AWGN) introduced in the channel.

The effect of this frequency offset on the received symbol stream can be understood by considering the received symbol $Y(k)$ on the k^{th}sub-carrier.

$$Y(k) = X(k)S(0) + \sum_{\substack{l=0 \\ l\neq k}}^{N-1} X(l)S(l-k) + W(k) \tag{3}$$

where $W(k)$ is the FFT of $\omega(n)$,the first term in right hand side of eqn.3 is desired signal and second term is interference signal. $S(l-k)$ are the complex coefficients for the ICI components in the received signal. The ICI components are the interfering signals transmitted on sub-carrier other than the kth sub-carrier.

The $S(l-k)$ can be expressed as:

$$S(l-k) = \frac{\sin[\pi(l+\varepsilon-k)]}{N\sin[\pi(l+\varepsilon-k)/N]}\exp\left[j\pi\left(1-\frac{1}{N}\right)(l+\varepsilon-k)\right]. \tag{4}$$

The carrier-to-interference ratio (CIR) is the ratio of the signal power to the power in the interference components. It serves as a good indication of signal quality. The derivation assumes that the standard transmitted data has zero mean and the symbols transmitted on the different sub-carriers are statistically independent. The desired signal is transmitted on sub-carrier "0" is considered, then the CIR of standard OFDM system is simplified as:

$$\text{CIR} = \frac{|S(k)|^2}{\sum_{\substack{l=0 \\ l\neq k}}^{N-1}|S(l-k)|^2} = \frac{|S(0)|^2}{\sum_{l=1}^{N-1}|S(l)|^2}\,. \tag{5}$$

3 ICI Self-cancellation Scheme

The difference between ICI coefficient of two consecutive sub-carriers is very small [6]. This forms the basis of ICI self-cancellation scheme. Fig. 2 shows the block diagram of typical ICI self-can0cellation scheme. The main idea of this scheme is that one data symbol is modulated onto two consecutive sub-carriers with predefined weighting coefficients. If the data symbol 'a' is modulated onto the first sub-carrier then '-a' is modulated onto the second sub-carrier. Hence, the ICI signals generated between the two sub-carriers almost gets "self-cancelled", hence the name self-cancellation.Assume that transmitted symbols are constrained such that:

$$X'(k) = X(k)\,,\ X'(k+1) = -X(k), \quad (k = 0,2,\ldots..N-2)$$

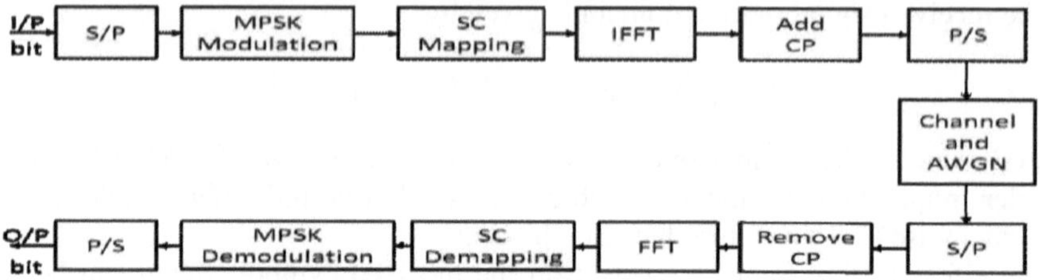

Fig. 2. Block diagram of ICI self-cancellation system

Using Eq. 4, this assignment of transmitted symbols allowsthe received signal on subcarriers k and $k+1$ to be written as:

$$Y'(k) = \sum_{l=0,l=even}^{N-2} X(l)[S(l-k) - S(l+1-k)] + W(k) \quad (6)$$

$$Y'(k+1) = \sum_{l=0,l=even}^{N-2} X(l)[S(l-k-1) - S(l-k)] + W(k+1) \quad (7)$$

For the majority of $(l-k)$values, the difference between $S(l-k)$ and $S(l+1-k)$ is very small. Then the ICI signals generated by the sub-carrier l will be cancelled out significantly by the ICI generated by sub-carrier $l+1$. This is called ICI cancelling modulation.

The ICI can be further reduced by using ICI cancelling demodulation. The demodulation is suggested to work in such a way that each signal at the $(k+1)^{th}$ sub-carrier (now k denotes even number) is multiplied by "-1" and then summed with the one at the kth sub-carrier. Then the resultant data sequence is used for making symbol decision. It can be represented as:

$$Y''(k) = \frac{1}{2}[Y'(k) - Y'(k+1)]$$

$$= \frac{1}{2}\{X(k)[2S(0) - S(1) - S(-1)] + \sum_{\substack{l=0,l=even \\ l\neq k}}^{N-2} X(l)[2S(l-k) - S(l-k-1) - S(l-k+1)] + W(k) - W(k+1)\} \quad (8)$$

According to definition of CIR, the CIR of ICI self-cancellation scheme can be represented as:

$$\text{CIR} = \frac{|2S(0)-S(1)-S(-1)|^2}{\sum_{l=2,l=even}^{N-2} |2S(l)-S(l+1)-S(l-1)|^2} \quad (9)$$

4 Proposed ICI Self-cancellation Scheme

We know that the difference between the ICI coefficient S(l-k) and S(l+1-k) is very small. So in proposed scheme onedata symbol is modulated onto four sub-carriers, that is, the data symbol 'a' is modul0ated onto the first and third sub-carriers, '-a' is modulated onto the second sub-carrier and '$ae^{-j\pi/2}$' is modulated onto the fourth sub-carrier. It means that the data modulated within $(k+1)^{th}$ sub-carrier is negative of the modulated data within k^{th} sub-carrier and data modulated within $(k+3)^{th}$ sub-carrier is rotated phase $-\pi/2$ of the modulated data within $(k+2)^{th}$ sub-carrier. Assume that transmitted symbols in proposed scheme are constrained such that:

$$X'(k) = X(k),\ X'(k+1) = -X(k),\ X'(k+2) = X(k),\ \ X'(k+3) = e^{-j\pi/2}X(k),\quad (k = 0,4,8,\ldots\ldots\ldots, N-4)$$

The received signal on sub-carrier $k$becomes:

$$Y'(k) = \sum_{l=0,4,8,..}^{N-4} X(l)[S(l-k) - S(l+1-k) + S(l+2-k) + e^{-j\pi/2}S(l+3-k)] + W(k) \tag{10}$$

The received signal on sub-carrier$k+1$ becomes:

$$Y'(k+1) = \sum_{l=0,4,8,..}^{N-4} X(l)[S(l-k-1) - S(l-k) + S(l+1-k) + e^{-j\pi/2}S(l+2-k)] + W(k+1) \tag{11}$$

The received signal on sub-carrier $k+2$ becomes:

$$Y'(k+2) = \sum_{l=0,4,8,..}^{N-4} X(l)\big[S(l-k-2) - S(l-k-1) + S(l-k) + e^{-j\pi/2}S(l+1-k)] + W(k+2) \tag{12}$$

The received signal on sub-carrier $k+3$ becomes:

$$Y'(k+3) = \sum_{l=0,4,8,..}^{N-4} X(l)[S(l-k-3) - S(l-k-2) + S(l-k-1) + e^{-j\pi/2}S(l-k)] + W(k+3) \tag{13}$$

This is the ICI cancelling modulation of the proposed scheme. To further reduce the ICI, the proposed ICI cancelling demodulation works as follows:

- Each signal at the sub-carrier $(k+1)$ is multiplied by "-1" and added to the one at the sub-carrier k, then the resultant data sequence is represented by:

$$Y_1(k) = Y'(k) - Y'(k+1)$$

$$= \sum_{l=0,4,8,..}^{N-4} X(l)\big[2S(l-k) - 2S(l+1-k) + S(l+2-k) - S(l-k-1) + e^{-j\pi/2}\{S(l+3-k) - S(l+2-k)\}] + W_1(k) \tag{14}$$

- We rotate the phase ($-\pi/2$) of the signal on $(k+3)$ sub-carrier and subtract it from the signal on $(k+2)$ sub-carrier, then the resultant data sequence is given by:

$$Y_2(k) = Y'(k+2) - e^{-j\pi/2}Y'(k+3)$$

$$= \sum_{l=0,4,8,..}^{N-4} X(l)\big[2S(l-k) - S(l-k-1) + S(l-k-2) + e^{-j\pi/2}\{S(l+1-k) - S(l-k-1) + S(l-k-2) - S(l-k-3)\}] + W_2(k) \tag{15}$$

- Then the decision variable can be represented by:
- $Y'''(k) = \frac{1}{4}[Y_1(k) + Y_2(k)]$

$$= \frac{1}{4}\sum_{l=0,4,8,..}^{N-4}[X(l)[4S(l-k) - 2S(l+1-k) + S(l+2-k) - 2S(l-k-1) + S(l-k-2) + e^{-j\pi/2}\{S(l+1-k) - S(l+2-k) + S(l+3-k) - S(l-k-3) + S(l-k-2) - S(l-k-1)\}]] + W'''(k)$$

$$= \frac{1}{4}[X(k)\Big[4S(0) - 2S(1) + S(2) - 2S(-1) + S(-2) + e^{-\frac{j\pi}{2}}\{S(1) - S(2) + S(3) - S(-3) + S(-2) - S(-1)\}\Big] + \sum_{\substack{l=0,4,8,..\\ l\neq k}}^{N-4} X(l)[4S(l-k) - 2S(l+1-k) + S(l+2-k) - 2S(l-k-1) + S(l-k-2) + e^{-\frac{j\pi}{2}}\{S(l+1-k) - S(l+2-k) + S(l+3-k) - S(l-k-3) + S(l-k-2) - S(l-k-1)\}]] + W'''(k) \tag{16}$$

The theoretical CIR of proposed scheme can be derived as:

$$\mathrm{CIR} = \frac{\left|\begin{matrix}4S(0)-2S(1)+S(2)-2S(-1)+S(-2)+\\ e^{-j\pi/2}[S(1)-S(2)+S(3)-S(-3)+S(-2)-S(-1)]\end{matrix}\right|^2}{\sum_{l=4,8,..}^{N-4}\left|\begin{matrix}4S(l)-2S(l+1)+S(l+2)-2S(l-1)+S(l-2)+\\ e^{-j\pi/2}[S(l+1)-S(l+2)+S(l+3)-S(l-3)\\ +S(l-2)-S(l-1)]\end{matrix}\right|^2} \tag{17}$$

5 Simulation Results

The proposed scheme is compared with conventional ICI self-cancellation scheme in terms of CIR and BER. The simulation parameters for the proposed scheme are shown in Table 1.

Table 1. Simulation Parameters

Parameter	Specification
FFT size	256
Sub-carriers	256
Modulation	QPSK
Frequency offset	0.15, 0.18, 0.2
OFDM symbols for one loop	5000

From Fig.3, it is clear that proposed scheme provides more than 13-dB CIR improvement over standard OFDM system and close to self-cancellation scheme for normalized frequency offset $0 < \varepsilon < 0.2$.

Fig. 4-6 show the BERcomparison of standard OFDM system, self-cancellation and proposed scheme for QPSK with different value of frequency offset. Figs. show that even at larger frequency offset the proposed scheme provides better BER performance as compared to standard OFDM system and self-cancellation scheme.

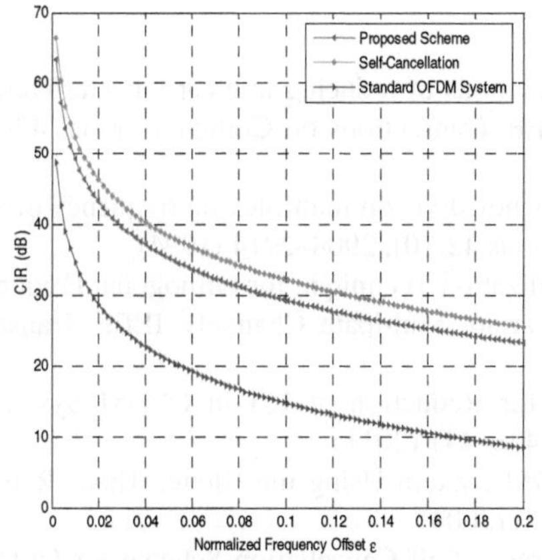

Fig. 3. CIR comparison

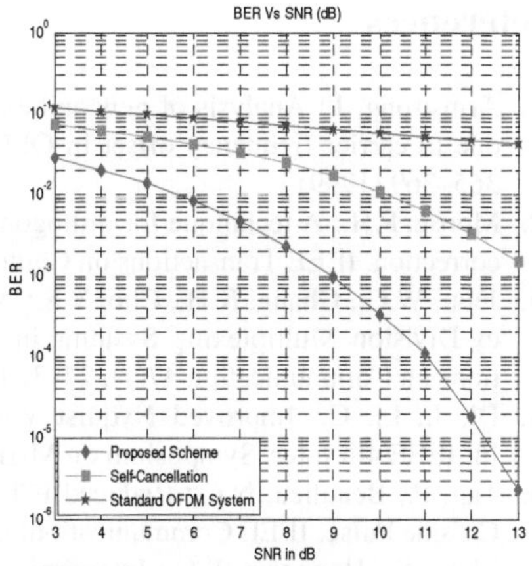

Fig. 4. BER versus SNR for $\varepsilon = 0.15$

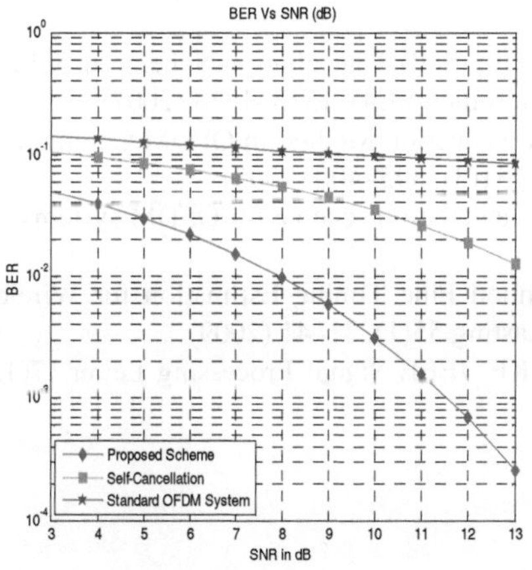

Fig. 5. BER versus SNR for ε =0.18

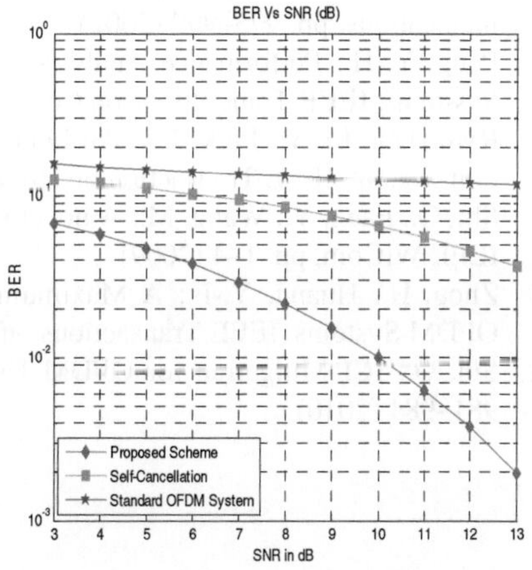

Fig. 6. BER versus SNR for ε =0.2

6 Conclusion

In this paper, a novel ICI self-cancellation is proposed to mitigate the effect of ICI caused by frequency offset in OFDM system. The proposed scheme provides significant CIR improvement over standard OFDM system that with no ICI Self-Cancellation scheme, which has been studied theoretically and by simulations. Also proposed scheme gives better BER performance as compared to standard OFDM system and conventional ICI self-cancellation scheme. Although the bandwidth efficiency of proposed scheme is reduced, due to four sub-carriers is used for transmitted data, itcan be solved by using larger signal alphabet size.

References

1. Armstrong, J.: Analysis of new and existing methods of reducing intercarrier interference due to carrier frequency offset in OFDM. IEEE Transactions on Communications 47(3), 365–369 (1999)
2. Moose, P.H.: A technique for orthogonal frequency division multiplexing frequency offset correction. IEEE Transactions on Communications 42(10), 2908–2914 (1994)
3. Jeon, W.G., Chang, K.H., Cho, Y.S.: An Equalization Technique for Orthogonal Frequency-Division Multiplexing Systems in Time-Variant Multipath Channels. IEEE Transactions on Communication 47(1), 27–32 (1999)
4. Di, J., Li, C.: Improved Nyquist Windows for Reduction of ICI in OFDM Systems. In: IEEE 4th Inter. Symposium on MAPE, pp. 438–441 (2011)
5. Tan, P., Beaulieu, N.C.: Reduced ICI in OFDM System Using the 'Better Than' Raised Co-sine Pulse. IEEE CommunicationLetters 8(3) (2004)
6. Zhao, Y., Haggman, S.G.: Intercarrier Interference Self-Cancellation Scheme for OFDM Mobile Communication Systems. IEEE Transactions on Communications 49(7) (2001)
7. Fu, Y., Ko, C.C.: A new ICI self-cancellation scheme for OFDM systems based on a gener-alized signal mapper. In: Proceedings of the 5th Wireless Personal Multimedia Communications, pp. 995–999 (2002)
8. Peng, Y.-H.: Performance Analysis of a New ICI-Self- Cancellation Scheme in OFDM Sys-tems. IEEE Transactions on Consumer Electronics 53(4), 1333–1338 (2007)
9. Ryu, H.G., Li, Y., Park, J.S.: An Improved ICI Reduction Method in OFDM Communication System. IEEE Transaction on Broadcasting 51(3), 395–400 (2005)
10. Shi, Q., Fang, Y., Wang, M.: A novel ICI self-cancellation scheme for OFDM systems. In: IEEE WiCom, pp. 1–4 (2009)
11. Zhou, H., Huang, Y.-F.: A Maximum Likelihood Fine Timing Estimation for Wireless OFDM Systems. IEEE Transactions on Broadcasting 55(1), 31–41 (2009)
12. Shi, Q.: ICI Mitigation for OFDM Using PEKF. IEEE Signal Processing Letter 17(12), 981–984 (2010)

Ultra High Bit-rate Fiber Optic Hybrid (WDM and TDM) Communication System Design and Simulation

Utpal Das[1], P.K. Sahu[2], and Bijayananda Patnaik[3]

[1] Department of Electronics and Telecommunication,
Silicon Institute of Technology, Bhubaneswar, India
[2] School of Electrical Sciences, Indian Institute of Technology, Bhubaneswar, India
[3] Department of Electronics and Telecommunication,
International Institute of Information Technology, Bhubaneswar, India
utpal.das@silicon.ac.in

Abstract. An ultra-high bit-rate fiber optic hybrid system is proposed in this work. The proposed system utilizes the advantages of both wavelength division and time division multiplexing techniques. The performance has been investigated for various advanced data formats such as carrier-suppressed return-to-zero (CSRZ), duo-binary return-to-zero (DRZ) and modified duo-binary return-to-zero (MDRZ). MDRZ format is proved to be the best among all. The proposed system performance has been investigated at various bit rate of 10Gbps, 40Gbps and 100Gbps, for over the selected transmission lengths of 1650 Km, 400 Km, 100 Km respectively.

Keywords: Hybrid fiber optic system, WDM, TDM, MDRZ, DRZ, CSRZ.

1 Introduction

Bandwidth demanding applications, such as high-definition television (HDTV), telemedicine, broadband Internet service, E-commerce, and other activities are major source behind the explosive growth of internet. Therefore, high capacity networks are needed in order to satisfy the tremendous growth in bandwidth requirements. Radio-over-Fiber (RoF) systems with high data rates are used for wireless broadband communication applications. The main advantages of RoF is low attenuation, large bandwidth, immunity to radio frequency interference, reduced power consumption, dynamic resource allocation, multi-operator support and multi-service operation, etc. The demand for network bandwidth is largely due to growth in traffic such as video on demand, video conferencing, High-Definition Television(HDTV),e-learning, etc., [1]. Several promising solutions have been devised over the last decade such as signal multiplexing in wavelength, frequency, time and in code has been investigated for utilizing the higher band width of optical systems optimally. Amongst many of these multiplexing methods time division multiplexing (TDM) and wavelength division multiplexing (WDM) are the most promising multiplexing systems in optical communication. In Time-division multiplexing(TDM)'N' independent information

Smart Innovation, Systems and Technologies 28,
DOI: 10.1007/978-3-319-07350-7_18, © Springer International Publishing Switzerland 2014

streams each running at a data rate of R b/s are combined into a single information streams operating at a higher rate of N x R b/s and offers an enormous transmission capability. The two key limitations in very high bit rate time division multiplexing system involves pulse spreading caused by fiber dispersion and crosstalk in the de-multiplexer [2]. On the other-hand, wavelength-division multiplexing (WDM) supports several non-overlapping wavelengths bands with each wavelength supporting a single communication channel operating at different frequencies. This is used for long distances communication over a single fiber to serve more number of users but the performance is severely limited by the non-linearity i.e. stimulated Raman scattering, four wave mixing, cross phase modulation [1], [3-5].So to achieve a common goal in terms of high bandwidth capacity of WDM and bandwidth efficiency of TDM, there is an effort to design a hybrid network that would have positive characteristics of both WDM and TDM [6], [9].

Modulation plays a vital role on system performance. Return-to-zero (RZ) and non-return-to-zero (NRZ) are the two widely used modulation formats in conventional standard fiber transmission network [7]. But NRZ and RZ formats are not suitable for long distance hybrid systems (WDM+TDM), as they are highly susceptible to nonlinear effects. An alternatives to RZ and NRZ, several advanced modulation formats like carrier-suppressed return to-zero(CSRZ), Duo-binary return to-zero(DRZ)and Modified duo-binary return to-zero(MDRZ) modulations have been used [10]. For longer transmission distance CSRZ format is superior over NRZ and RZ with respect to signal degradation due to Kerr nonlinearities and chromatic dispersion [1] ,[5].Both DRZ and MDRZ signals suppress all discrete frequency tones that appear in the conventional RZ signal spectrum. MDRZ has the least timing jitter and amplitude distortion [6].

The proposed system is a 4-channels hybrid system designed employing WDM and TDM technique, with 200 GHz channel spacing. In this paper CSRZ, DRZ and MDRZ modulation formats are analyzed on the basis of Q-value and eye opening penalty for different transmission distances.

2 Transmitters of Modulation Formats

2.1 Carrier Suppressed Return-to-Zero (CS-RZ) Format

CSRZ (67% RZ) is a special form of RZ where the carrier is suppressed. The major features of CSRZ format is that it reduces the nonlinear impairments in a channel and improves the spectral efficiency in high bit rate systems. In CSRZ signal has 'π' phase shift between adjacent bits. This phase alternation, in the optical domain, produces no DC component; thus, there is no carrier component for CSRZ. Phase alternating between adjacent bit slots reduces the fundamental frequency components to half of the data rate. CSRZ has better tolerance to chromatic dispersion due to its lower optical power, allowing for more channels to be multiplexed in the channel. In addition, the carrier suppression also minimizes the effect of four wave mixing (FWM) in WDM systems [8]. Fig. 1(a) shows the block diagram of the 40Gbps CSRZ transmitter. The generation of a CSRZ optical signal requires two MZ modulators as shown in this figure. The first MZ modulator encodes the NRZ data. Then the generated NRZ optical signal is modulated by the second MZ modulator that is driven by a clock at the half bit-rate (20 GHz). That introduces a π phase shift between any two adjacent bits and the spectrum gets modified such that the central peak at the carrier frequency is suppressed as shown in Fig. 1(b).

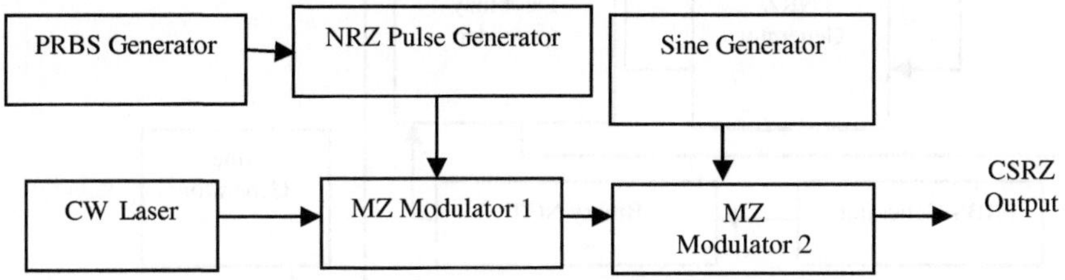

Fig. 1. (a) Block diagram of CSRZ Transmitter

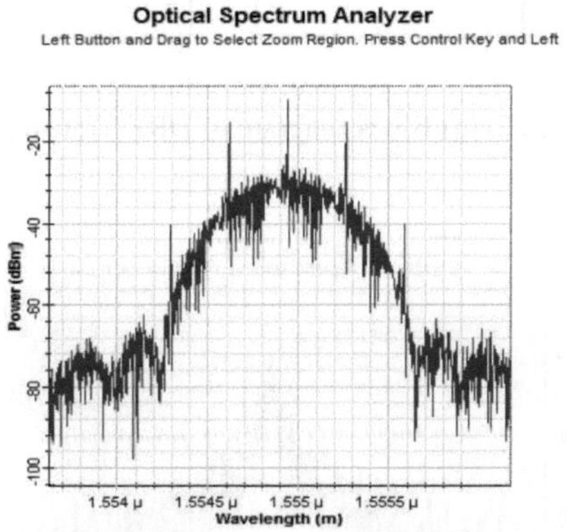

Fig. 1. (b) Frequency Spectrum of 40Gbps CSRZ signal

2.2 Duo-binary Return to Zero (DRZ) Format

Optical duo-binary (DRZ) modulation is a scheme to transmit R bits/s using less than R/2 Hz of bandwidth. In DRZ signaling, a phase change occurs whenever there is odd numbers of 0-bits between two successive 1-bits.The two main advantages of this modulation format are increased tolerance to the effects of chromatic dispersion (CD) and improved narrowband optical filtering. Fig.2(a) shows the schematic of the 40Gbps duo-binary transmitter. The duo-binary was generated by first creating an NRZ duo-binary signal using a duo-binary pre-coder, NRZ generator and a duo-binary pulse generator. The generator drives the first MZM, and then cascades this modulator with a second modulator that is driven by a sinusoidal electrical signal with the frequency of 40GHz, Phase=-90^0.The duo-binary pre-coder used here is composed of an exclusive-or gate with a delayed feedback path. DRZ formats are very attractive, because their optical modulation bandwidth can be compressed to the data bit rate B, that is, the half-bandwidth of the NRZ format2B [11] as shown in Fig. 2(b).

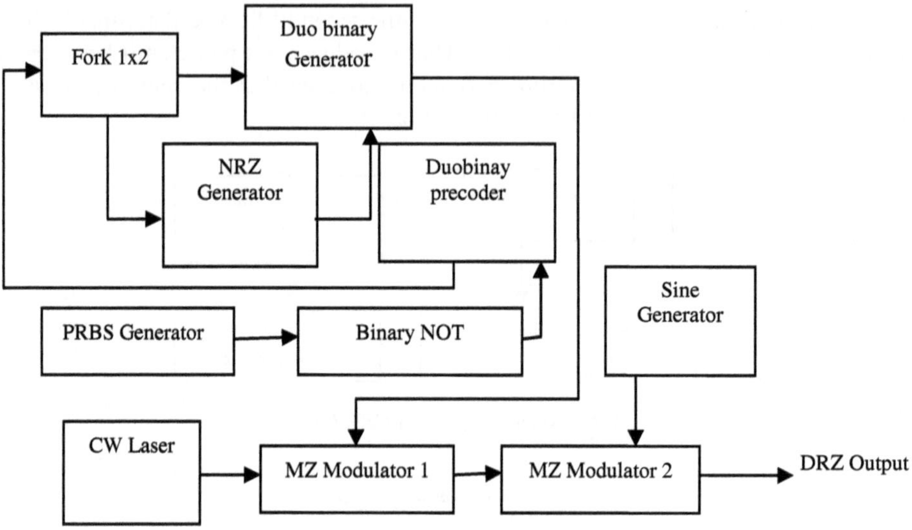

Fig. 2. (a) Block diagram of DRZ Transmitter

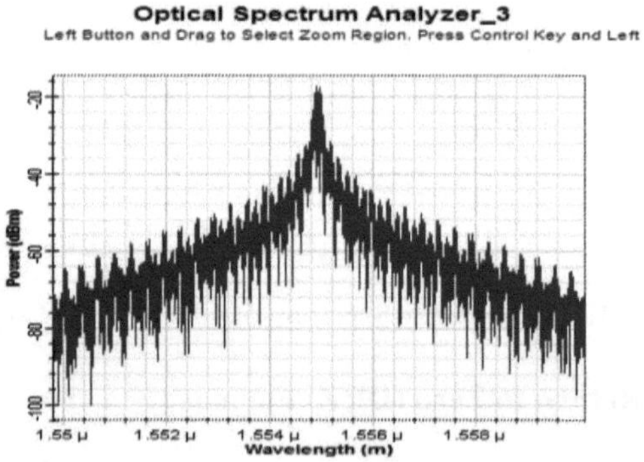

Fig. 2. (b) Frequency Spectrum of 40Gbps DRZ signal

2.3 Modified Duo-binary RZ (MDRZ) Format

Modified duo-binary (MDB) modulation format is characterized by phase inversion in the pulses triggered by the presence of a logical "one" in the previous bit slot. Duo-binary spectrum does not have any DC content. MDRZ has opposite phase in adjacent "1"s and due to this fact self-phase modulation, cross-phase modulation and intra-channel four-wave mixing in WDM transmission systems can bereduced. Fig.3 (a) shows the schematic of the 40Gbps modified duo-binary transmitter also called carrier- suppressed duo-binary format. The generation of MDRZ signal is almost identical to the DRZ signal, except that the delay and add circuit is replaced by a delay and subtract circuit. In the modified duo-binary signal the phase is alternated between 0 and π for the bits'1'. The phase of all the "zero" bits are kept constant and a180^0 phase variation between all the consecutive "ones" is introduced [11]. The optical signal spectrum of Fig.3 (b) shows that the carrier of the duo-binary signal has been suppressed.

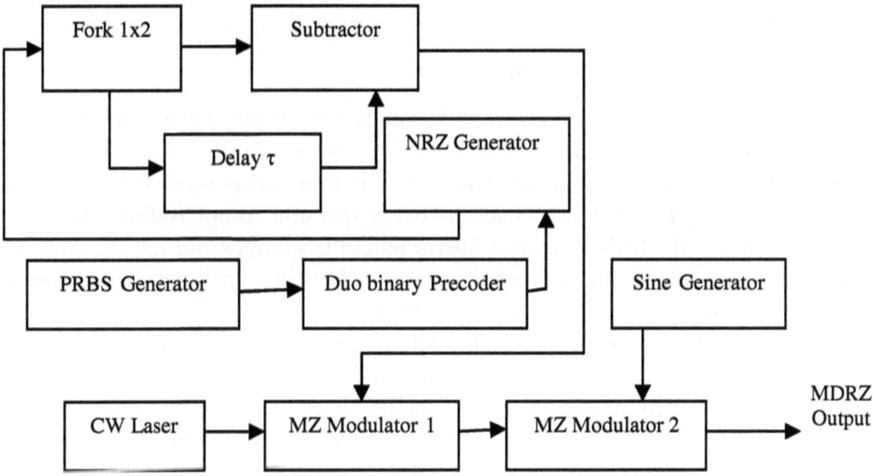

Fig. 3. (a) Block diagram of MDRZ Transmitter

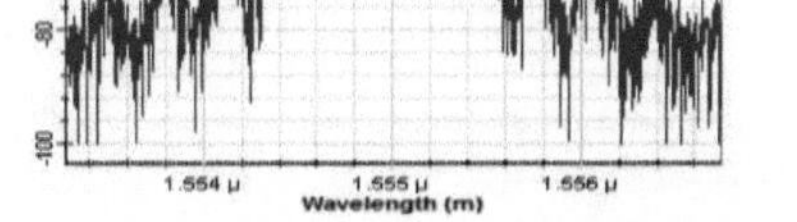

Fig. 3. (b) Frequency spectrum of 40Gbps MDRZ signal

3 Simulation

The proposed system consists of a transmitter, fiber Spools and an optical receiver as shown in Fig. 4.Transmitter consists of 4 number of hybrid (WDM+TDM) channels. The multiplexed channels are operating at 10Gbps, 40Gbps and 100Gbps respectively with the central frequency of the first channel is 192.8THz. Optical multiplexer supports different modulation formats e.g., carrier-suppressed return- to- zero (CSRZ), duo-binary return- to- zero (DRZ) and modified duo-binary return- to- zero (MDRZ). Standard single mode fiber (SSMF) is used to achieve long distance communication in this setup. The simulation parameters used is given in Table 1 considering bit rates of 10Gbps,40Gbps and 100Gbps respectively. The transmission link is designed suitably, so that the first-order dispersion is compensated exactly (D =0), that is [5],

$$D_{SMF}L_{SMF} = D_{DCF}L_{DCF} \tag{1}$$

D stands for first-order dispersion and L stands for length of the respective fiber. Fiber parameters used in the system model is as given in Table 2, which is commercially available and suitable. The EDFA is used for compensating the linear loss and the noise figure of which is set to 6dB.Dispersion compensating fiber (DCF) of 10 Km is used to the SSMF fiber of 50Km length to compensate for the dispersion and the nonlinearities. The model is simulated using symmetrical dispersion compensation technique over N spans of SSMF of 50Km each [12].

In the receiver the signal is passed through an optical de-multiplexer, detected by PIN detector, filtered, applied to power meter and BER analyzer. The Responsivity and dark current of PIN diode considered is 1A/W & 10nA respectively and a 4th order filter is used. 3R regenerator is used to generate an electrical signal connected directly to the BER analyzer which is used to visualize the graphs and results such as eye diagram, eye opening, BER, Q value of the setup.

Table 1. Simulation parameters

Sequence length	128
Samples/bit	64
Channel spacing	200GHz
Capacity	4-channel X PGbps, P=10, 40 and 100
Distance	50 Km X N spans N=2,8,33
Input Power	3dBm

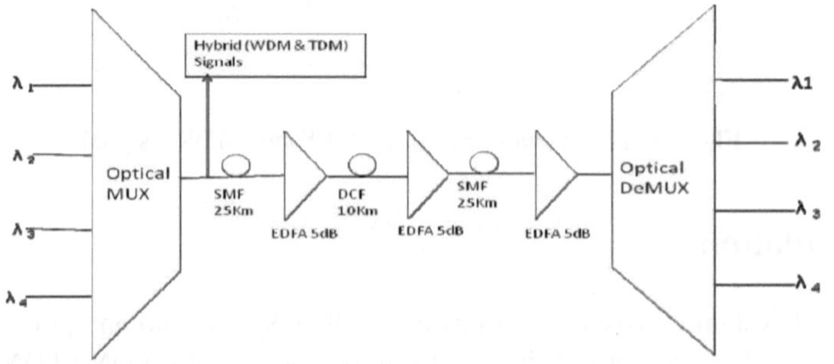

Fig. 4. Schematic of the hybrid (WDM+TDM) setup using symmetric compensation scheme

Table 2. Fiber Parameters

Fiber	Atten. (dB/km)	Disp. (ps/km-nm)	Disp. slope (ps/km-nm^2)	Effective core area(μm^2)
SMF	0.2	17	0.075	70
DCF	0.5	-85	-0.3	22

4 Results and Discussion

The proposed ultra-high bit-rate fiber optic hybrid system performance has been analyzed for bit rates of 10Gbps, 40Gbps and 100Gbps respectively using three different modulation formats. Fig. 5(a),(b) indicates Q-factor vs different modulation formats for a link distance of 1650 Km for bit rate of 4X10Gbps. In terms of Q value the MDRZ and DRZ out-performs the CSRZ formats with zero time delay and half bit rate delay. DRZ and MDRZ are less affected by linear crosstalk and inter-channel FWM (IFWM) induced timing jitter; as both these formats suppress all discrete frequency tones that appear in conventional RZ signal spectrum, whereas Fig.5(c) shows the transmission performance of 4X10Gbps hybrid system using three modulation formats for varying lengths. The transmission distances exceeding 1400km the Q value for CSRZ drops below the minimum specified value of 6.8dB due to cross talk and IFWM. DRZ format can be used for faithful transmission up to a distance of 1600 Km, whereas MDRZ format seems to be most robust against the fiber nonlinearities and optical Kerr's effect thus it can be used over a transmission distance up to 1650km. The results are also supported by the eye openings for Channel 1 as shown in Fig. 6(a), (b), and (c) for MDRZ, DRZ and CSRZ format respectively. The worst eye opening is shown in Fig. 6(c) by CSRZ format for the distance of 1650 Km. This is due to inter channel XPM and FWM, which results in spectral broadening caused by the phase variation. The best eye opening is shown in Fig. 6(a) for MDRZ format as it is more robust against linear crosstalk and IFWM effects. We have also investigated the transmission performance of 4X40Gbps and 4X100Gbps bit rate for all the three modulation formats with frequency range of 192.8–193.4THz using 200GHz frequency spacing between the adjacent channels.

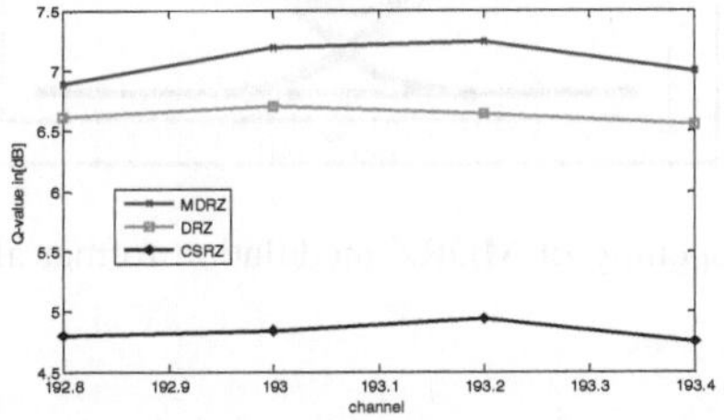

Fig. 5. (a) Q-factor vs different modulation formats over the transmission length of 1650 Km with zero time delay using bit rate of 4X10Gbps

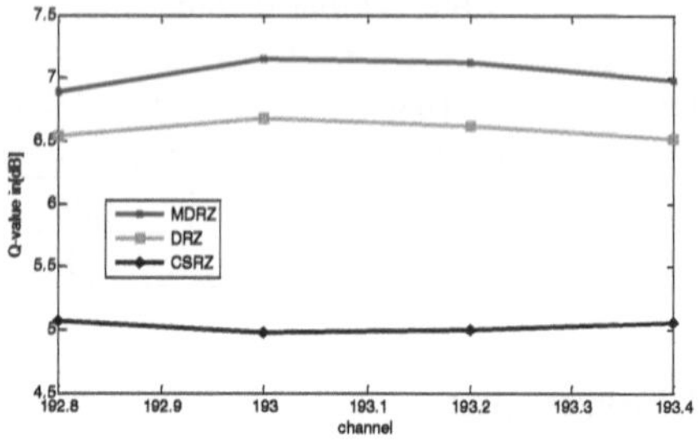

Fig. 5. (b) Q-factor vs different modulation formats over the transmission length of 1650 Km with half bit-rate delay using bit rate of 4X10Gbps

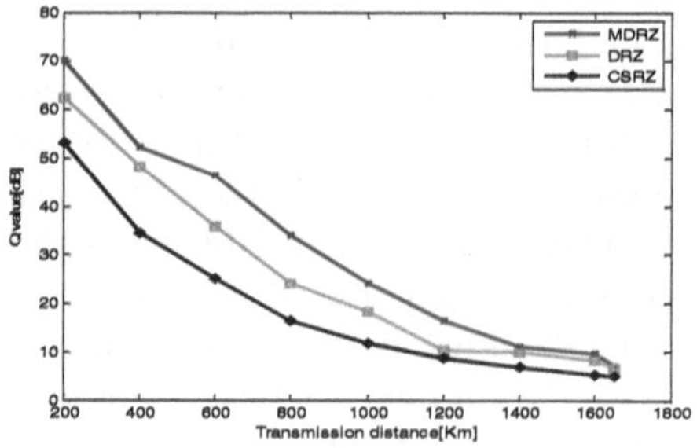

Fig. 5. (c) Q-factor vs Fiber length over 1650 Km for bit rate of 4X10Gbps

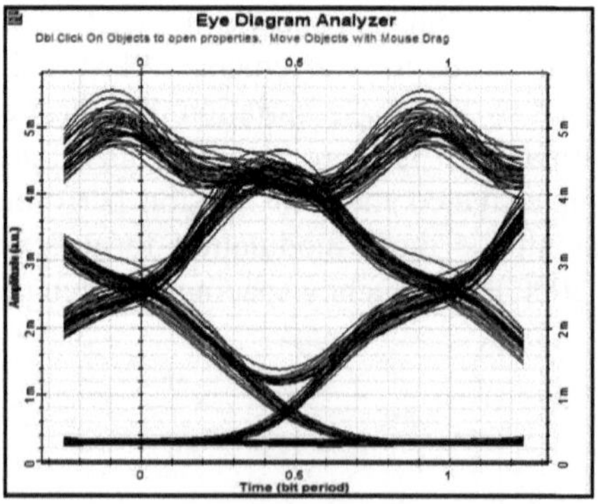

Fig. 6. (a) Showing eye opening of MDRZ modulation format after link distance of 1650Km for channel 1

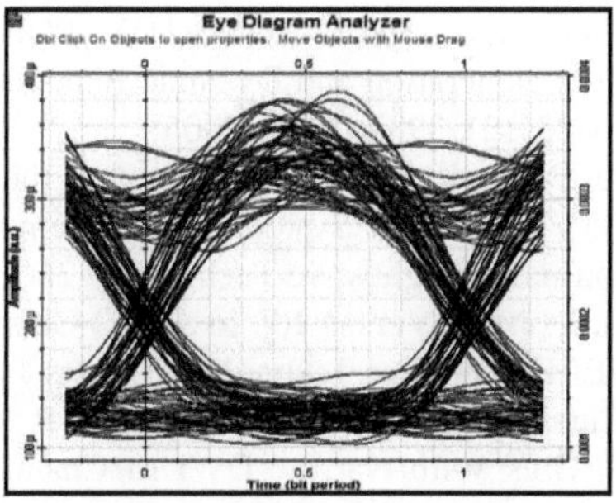

Fig. 6. (b) Showing eye opening of DRZ modulation format after link distance of 1650Km for channel 1

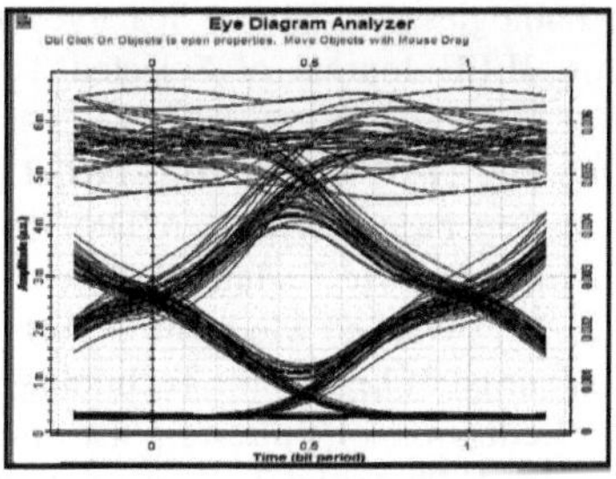

Fig. 6. (c) Showing eye opening of CSRZ modulation format after link distance of 1650Km for channel 1

5 Conclusion

A 4-channel hybrid system is proposed with bit rates of 10Gbps, 40Gbps and 100Gbps, for 200 GHz channel spacing. Employing MDRZ modulation format a link distance of 1650 km is obtained for the proposed set-up. Superior performance has been observed for MDRZ modulation format comparing to other formats as it suppresses all the discrete frequency tones. Whereas CSRZ only suppresses the optical carrier tone and creates side- band tones spaced at odd multiples of B/2 on both sides of carrier frequency.

References

1. Memon, M.I., Mezosi, G., Li, B., Lu, D., Wang, Z., Sorel, M., Yu, S.: Generation and Modulation of Tunable mm-Wave Optical Signals Using Semiconductor Ring Laser. IEEE Photonics Technology Letters 21(11), 733–735 (2009)
2. Murakami, M., Matsuda, T., Maeda, H., Imai, T.: Long-Haul WDM Transmission Using Higher Order Fiber Dispersion Management. IEEE Journal of Lightwave Technology 18(9), 898–900 (2000)

3. Bock, C., Prat, J., Walker, S.D.: Hybrid WDM/TDM PON Using the AWG FSR and Featuring Centralized Light Generation and Dynamic Bandwidth Allocation. IEEE Journal of Lightwave Technology 23(12), 3981–3988 (2005)
4. Oh, J.M., Koo, S.G., Lee, D., Park, S.-J.: Enhancement of the Performance of a Reflective SOA-Based Hybrid WDM/TDM PON System With a Remotely Pumped Erbium-Doped Fiber Amplifier. IEEE Journal of Lightwave Technology 26(1), 144–149 (2008)
5. Hayee, M.I., Willner, A.E.: NRZ Versus RZ in 10–40-Gb/s Dispersion-Managed WDM Transmission Systems. IEEE Photonics Technology Letters 11(8), 991–993 (1999)
6. Bosco, G., Carena, A., Curri, V., Gaudino, R., Poggiolini, P.: On the Use of NRZ, RZ, and CSRZ Modulation at 40 Gb/s With Narrow DWDM Channel Spacing. IEEE Journal of Lightwave Technology 20(9), 1694–1704 (2002)
7. Hodžic´, P., Konrad, B., Petermann, K.: Alternative Modulation Formats in N X 40 Gb/s WDM Standard Fiber RZ-Transmission Systems. IEEE Journal of Lightwave Technology 20(4), 598–607 (2002)
8. Kani, J.-I.: Enabling Technologies for Future Scalable and Flexible WDM-PON and WDM/TDM-PON Systems. IEEE Journal of Selected Topics in Quantum Electronics 16(5), 1290–1297 (2010)
9. An, F.T., Kim, K.S., Gutierrez, D., Yam, S., Hu, E(S.-T.), Shrikhande, K., Kazovsky, L.G.: SUCCESS: A Next-Generation Hybrid WDM/TDM Optical Access Network Architecture. IEEE Journal of Lightwave Technology 22(11), 2557–2569 (2004)
10. Winzer, P.J., Essiambre, R.-J.: Advanced Modulation Formats for High-Capacity Optical Transport Networks. IEEE Journal of Lightwave Technology 24(12), 4711–4728 (2006)
11. Cheng, K.S., Conradi, J.: Reduction of pulse-to-pulse interaction using alternative RZ formats in 40-Gb/s systems. IEEE Photonics Technology Letters 14(1), 98–100 (2002)

P2P-VoD Streaming: Design Issues and User Experience Challenges

Debjani Ghosh, Payas Rajan, and Mayank Pandey

Motilal Nehru National Institute of Technology Allahabad, Allahabad, India
{debjani.gautam,payas.rajan}@gmail.com,
mayankpandey@mnnit.ac.in

Abstract. Peer-to-Peer based Video-on-Demand (P2P-VoD) applications are becoming very popular for scalable video distribution in both academic and commercial environments. Asynchronous arrival of peers who wish to watch videos from beginning, as well as willing to avail interactive services like jump, rewind, fast-forward etc. in these systems makes it challenging to design and deploy them. A lot of research has already been done on the architectural design issues of P2P-VoD systems. In this paper, we present a survey on approaches which address some existing design issues like alleviating the server stress due to asynchrony, building efficient P2P-VoD applications with interactive services and handling peer churn as well as the time-varying nature of network. In addition to these issues, we also discuss the challenges in deployment of P2P-based VoD systems with additional user experience features like on-demand watching of segmented scenes, on-the-fly creation of playlists etc. over best-effort Internet.

1 Introduction

In recent times, we have seen a significant rise in the number of smart phones, tablets, smart TVs and high speed Internet connections which provide almost everything on the consumer's screen. With the rapid advancement of Internet technologies, applications are no more limited to browsing and e-mail, and are extended to real time multimedia streaming. We have seen a rapid surge in demand for the real time multimedia streaming applications both in entertainment and academic world. Real time multimedia streaming applications over Internet Protocol are classified into two forms: live and on-demand. Video conferencing, live broadcasting of important events are some popular applications of live streaming, whereas VoD is more suitable for some other applications like on-demand distance learning, and pull-based content distribution such as on-demand watching of media using Internet Protocol television. It provides the user additional flexibility in terms of independence of watching time, location and device, and thus a lot of research has been dedicated in this area. For streaming, three components are involved: source of media, network and the client. Source of media may be centralised or distributed. In a centralised architecture, clients directly

M.K. Kundu et al. (eds.), *Advanced Computing, Networking and Informatics - Volume 2*, Smart Innovation, Systems and Technologies 28,
DOI: 10.1007/978-3-319-07350-7_19, © Springer International Publishing Switzerland 2014

connect to a single source server. It is successful till the server has enough capacity to handle the incoming connections. But the situation worsens when the demand for media increases beyond the server's capacity. These situations can take place because of a sudden increase in demand of a particular stream, and due to creation of such hotspots, the entire system can cease to function. Thus, the Quality of Service (QoS) like scalability, robustness and fault tolerance suffers in a server-client architecture.

Distributed architectures like Content Delivery Network (CDN), proxy servers and P2P technologies address the issues of the server-client architecture. Architecture of CDN and proxy servers is similar: the content delivery nodes in CDN, and proxy servers in a proxy servers network are arranged at the edges of the network and the source media server pushes the contents to each of these nodes, which in turn serve it to their nearby clients. The major difference between these two architecures is their buffering capacity. Each CDN node carries the full content of streaming, whereas buffering capacity of proxy servers is less, thus limiting the scalability. Both the architectures are very costly to implement, but the CDN tends to be costlier of the two, though it provides better QoS metrics like lower startup delay and interactive service latency etc. P2P architecture utilizes the upload bandwidth and storage capacity available with the clients in order to disseminate the contents. It has a better fault tolerance, requires no dedicated infrastructure, and has a lower deployment cost. Due to these benefits, streaming using this architecture has attracted significant research in this area. However, streaming in a distributed environment requires media synchronisation, efficient handling of peer churn and strict bounds on desired QoS parameters to provide a good Quality of Experience (QoE). Moreover, while designing P2P-based VoD systems, asynchronous arrival of each peer as well as provision of interactive services like forward, rewind, pause, seek etc. makes the design complex. In this paper, we present a survey on certain existing approaches of addressing some design issues of P2P-VoD systems such as: reducing server load, provision of interactive services and handling peer churn as well as the best-effort nature of network. In addition to interactive services, VoD systems today are aiming to provide some advanced User Experience (UX) features. There is a multitude of such UX features that can be provided. In this paper, we discuss some such features like: on-demand provision of segmented scene of on going video stream, automatic on-the-fly creation of playlists, video tagging etc. Also, we discuss some challenges in providing these UX features. The organization of the paper is as follows: Section 2 presents the survey on designing issues and existing approaches to solve these issues. Section 3 points out some challenges in designing the P2P-VoD system with UX features. We conclude the paper in Section 4.

2 P2P-VoD Streaming

All P2P systems consist of various components, which are responsible for different functions like content discovery, distribution, delivery and distributed coordination among the peers. However, the overlay structure varies among different systems. For building real time streaming applications using a P2P architecture, [1] discussed two types of overlay structures. These are termed as tree-based and mesh-based overlays. P2P tree-based streaming is conceptually similar to Application Level Multicast (ALM) [2], in which the participating peers hold the incoming stream in their buffers and the buffered stream is then served to their child peers. In P2P mesh-based streaming systems, there is no fixed hierarchy among the peers, and each peer can connect to multiple peers and can share the data with them.

Since VoD streaming applications provide interactive services like forward, rewind, fast-forward etc., fast content discovery is essential to maintain acceptable QoE. For content discovery, it is essential to match the buffer contents of the supplying peers and the demand of the requesting peer. Since the buffer contents can change rapidly due to different interactive actions on the stream, all the dynamic relationships need to be tracked in an efficient way. Due to this design complexity, designing VoD streaming systems becomes much more challenging. Some popular P2P-VoD streaming systems which use tree and mesh overlays are listed in Table 1.

Table 1. P2P-VoD System

System	Topology
Joost	Mesh
PPLive	Mesh
PPStream	Mesh
SopCast	Mesh
Tvants	Mesh
P2Cast	Tree
BiToS	Mesh

Table 2. Source Search Cost

Approach	Search Cost
DHT based	O(log N)
Dynamic Skip List	O(log N)
Ring Based	O(log (T/w))
Instant Leap	O(log 1)
Tree-based	O(log S)
vEB-tree-based	O(log log S)
Overlapping relation	O(log N)

The authors of [3],[4] compared tree and mesh based overlays for P2P VoD streaming. In [3], it is pointed out that mesh overlay has some disadvantages such as presence of high playback delay, need for developing data scheduling algorithms for retrieving data from peers, and traffic overhead due to exchange of messages. On the other hand, [4] proved through experiments that mesh-based systems appear to be a better choice than tree-based systems, especially for larger overlays and higher streaming rates. According to them, mesh-based systems provide a consistently higher application throughput when the number of overlay nodes or the streaming rates increase, and they perform better under churn and large flash crowds. Both the topologies have their trade-offs, so different topologies may be needed to achieve desired functional as well as non-functional requirements.

In [5], several key design issues are discussed including file segmentation strategy, replication strategy, content discovery and management, piece or chunk selection policy, transmission strategy and authentication. These are essential design requirements for building streaming applications with a better QoS. In the following sub sections, we present a survey on some other design issues and their existing solutions:

2.1 Reducing Server Stress due to Asynchronous Arrival

The most important design issue in a VoD system is handling the asynchronous arrival of peers who wish to watch the video from beginning. This limits the scalability of the system. This problem can be addressed by deploying the VoD system using CDN, proxy servers, P2P systems etc. P2P architecture provides independence from need of dedicated infrastructure and hence has a lower cost. But, many authors [6–8],[9],[10] believe that this problem cannot be solved by P2P technology alone, and the combination of CDN or proxy servers can make the systems more efficient. [6] proposed a "Hybrid Content Distribution Network" which is a combination of CDN and pure P2P network aiming to improve system scalability as well as downloading time and service capacity. In a similar manner, several researches have attempted to reduce server workload by proposing systems that utilize both proxy servers and P2P technology. Some notable examples can be found in [7–10].

Though the combination of these technologies can enhance the performance of VoD systems, yet we can not ignore the organizations or industries who want to build their systems using P2P architecture. Many researches [11, 12, 16–19] discussed some mechanisms through which the scalability problem caused due to pure P2P architecture can be addressed. [11, 12] have proposed caching mechanisms that adjust the buffer space in order to improve scalability. In addition to these mechanisms, many have worked on stream reuse techniques such as "batching" [13], "chaining" [14], and "patching" [15]. In *batching*, the requests issued for same video within the same interval of time are first enqueued and the group of requests is served afterwards by a single multicast stream wheras in *chaining*, clients who arrive early transmit video streams to the new clients, forming chains across the network. In *patching*, extra short streams namely the patches are added to the active multicast streams so that the late arriving clients can watch the beginning part of the stream.

However these techniques have their own advantages as well as disadvantages, for example conventional patching requires large client cache space. So many researches [16],[20] have proposed improvements in these stream reuse techniques. In [16], conventional patching scheme is improved by introducing a novel patching scheme called "Client-Assisted Patching", where patching eliminates the service latency caused by batching. In [20], conventional batching is modified by providing a mechanism called "Distributed Peer-to-Peer Batching" (DPPB), which disperses the duty of multicast delivery from the central server to a number of high-end peers, which have a higher storage and bandwidth capacity than normal peers. In addition to the improvements in stream reuse techniques, [21] utilizes the combination of batching, chaining and patching to deliver popular videos with low start up latency while using the smallest number of server channels.

Some researches [17–19] have discussed caching strategy and prefetching strategy for reducing the server workload and improving the scalability of VoD system on P2P network. The authors of [18] have used these strategies to force a peer to prefetch a segment on joining for caching purpose regardless of whether that segment is currently demanded by that peer or not. Also, the peer has the option of replacing the segment that is currently in its cache with the last segment that it finished streaming. In [19], the caching strategy is discussed and two fundamental problems are addressed: how a host finds video pieces which are scattered in the whole system, the part of the video a host should cache and what existing data should be discarded to make necessary space. The implementation of caching strategy also necessitates cache update. In [22], a cache replacement algorithm called "Peer Cache Adaptation" (PECAN) is proposed, where each peer adjusts its cache capacity adaptively in order to meet the server's upload bandwidth constraint.

2.2 Interactive Services in VoD

Implementing interactive services in P2P-VoD streaming systems is a major challenge, as the interactive actions like fast-forward, jump etc. can strongly increase the burden on whole system. From the survey, we observed that certain P2P-VoD systems like VMesh [23], BulletMedia [24], Dynamic Skip List(DSL) [25], DP^n2P^m [26], P2PVR [27], P-Chaining [28], DirectStream [29] and BitTorrent-based VoD system [30] provide fully interactive VoD services. Many researches [23–25, 31–35] have discussed different approaches for building an interactive P2P-VoD system with the goal of minimizing the source searching cost. The authors of [25] have used a dynamic skip list approach to construct an interactive P2P-VoD system. The proposed system allows the peers to connect sequentially according to their playback progress at the lower layer of the skip list. Also, a peer is allowed to randomly connect to a few non-adjacent peers in the upper layers. In [31], an AVL tree is adopted for building an interactive VoD system that can achieve a search efficiency sublinear to the number of peers. Moreover, the authors of [32] have adopted the "vEB(van Emde Boas) tree-based architecture" in order to reduce the source search cost.

In [33], the authors have utilized a ring-assisted overlay management scheme, where each peer maintains a set of concentric rings with different radii and places neighbors on the rings based on how similar their cached contents are. The authors of [34] have given a method named "InstantLeap", where peers are grouped according to their playback locality, so that each peer strategically connects to a number of peers with similar playback progress, as well as to some other peers watching different parts of the video. The authors of [35] have utilized content based overlay named "Overlapping Relation Network" (ORN), where the overlapping relationships between buffer fragments of peers are used to build the interconnections among the peers in order to enhance the discoverability. The authors of [23, 24] have utilized a Distributed Hash Table (DHT) in order to discover supplier peer. Table 2 gives a comparison among various approaches for building interactive P2P-VoD systems, based on the source search cost. The symbols N, T, w and S in Table 2 denote the number of nodes, video size, buffer size and number of segments respectively.

In addition to source searching in interactive P2P-VoD systems, reducing the response time due to interactive functions is also necessary. The authors of [36] have addressed this issue by a scheme named "Interleaved Video Frame Distribution" (IVFD). In this scheme, video stream data is divided into M interleaved groups, which can be stored in M parent peer groups, and each child peer then selects one parent peer from each group. Since the client is connected to M parent peers in order to obtain the intermittent video data, no parent peer reseeking process is necessary and thus the response time is reduced.

2.3 Dynamicity of Peers and Best-Effort Internet

Deploying P2P-VoD systems over a best-effort packet switched network like the Internet is challenging as these networks were originally designed for data traffic and are more suitable for applications such as e-mail, file transfer, web surfing etc. In addition to these challenges, peer churn needs to be handled in VoD applications. Many researches [37–39] have addressed these challenges by using network coding techniques such as Scalable Video Coding(SVC), Multiple Description Coding(MDC) etc. SVC provides post-encoding adaptation of the transmitted bit stream, whereas MDC is based on redundant coding approach and is mainly applied over the error-prone networks [38]. In MDC, the source streams are independently encoded with multiple descriptions, which are then sent over multiple paths to the receiver and the receiver then independently decodes and plays them.

To handle the time varying nature of the network and the peers,the authors of [37] have used SVC to maximize the overall playback quality. The authors of [37] have discussed that SVC supports layered description of streams, and they utilize this property by encoding the video stream into base layer and several enhancement layers, which progressively improve the reconstructed video quality. With the help of these layered streams, a new peer can successfully initialize video playback by starting to play the base layer, thus reducing the start-up delay of playback. Also, the occurrences of "frame freezing" due to temporal network congestion or insufficient peer bandwidths can be minimized by dynamically adding or dropping layers.

While streaming, the transmission of a bit stream depends on the capacity of three main components i.e. sender, network and receiver; and the bit rate depends on weakest among the three. Due to best-effort nature of Internet, it is important to generate unbalanced descriptions of different resolution of the same stream, where the low-resolution description can be used to improve the playback quality of the high-resolution description. [38] achieved unbalanced descriptions by a multiple description scalable video coder, where the coder provides a clustering algorithm that can solve the redundancy-rate allocation optimization problem. While generating multiple descriptions of a stream, the authors of [40] pointed out that some key points must be remembered: increasing the multiple description coding substreams reduces the frame loss rate, increasing the group of picture's length increases the frame loss rate, increasing the good state transition probabilities and decreasing the bad state transition probabilities results in lower frame loss rate.

In addition to the above issues addressed by network coding techniques, the authors of [39] also used these techniques to resolve the throughput maximization problem of P2P-VoD systems. The authors described that the random selection of the encoding coefficients in an appropriate field can eliminate the redundant packets with very high probability and therefore a peer with a higher throughput can reconstruct the video at a higher quality. The authors of [41] have proposed 'Coded VoD', which also uses network coding techniques with the aim of mitigating the asymmetric interests among peers in the P2P-VoD systems.

To resolve the bursty packet loss problem caused by departure of peers, the authors of [42, 43] have adopted an error protection technique known as "Forward Error Correction" (FEC). In [42], multisource structure and a distributed FEC scheme is combined in such a manner that each peer connects to multiple parents according to the prespecified FEC packets. However, the authors of [43] believe that FEC scheme is not beneficial for best-effort Internet as successive lost packets decrease the recovery performance of FEC protection. They addressed this problem by dispersing continuous lost packets to different FEC blocks by using Concurrent Multipath Transmission (CMT), combining FEC with path interleaving. Moreover, the authors of [44] have discussed that the error protection schemes overburden the network resources. They adopted an error recovery mechanism known as "Automatic-Repeat-Request" (ARQ) and proposed a meta-heuristic combinatorial optimization technique called "Ant Colony Optimization" (ACO) for solving the packet loss problem. Recently, the authors of [45] have proposed a XOR-based frame loss protection scheme to mitigate the effects of peer churn and network packet loss.

It has been observed from the survey that many existing mesh-based P2P-VoD system use network coding techniques. Some examples are rstream [46] which uses fountain code technique [47], and the systems proposed in [48, 49]. Recently, the authors of [50] have presented a scalable video streaming system based on end-system multicast, where the scalable network coding technique with push-based content distribution is extended to perform prioritized streaming with error and congestion control.

3 Challenges

Although P2P architecture has some distinct advantages, designing applications using this architecture presents significant legal, commercial and technical challenges, such as: membership control, protecting data stored in the system or downloaded by peer, incentives and managing the systems. A detailed description of these challenges can be found in [51]. The authors of [5] have discussed some key challenges encountered by P2P streaming systems like enhancing QoS including smoothness of display, lowering the frequency of visual distortions and start-up delay, handling dynamicity of peers and network, and developing security mechanisms to authenticate content in order to make the system resistant to pollution attacks.

User Experience(UX) has been defined in a multitude of ways in different researches and by different organisations. The ISO 9241-210 standard [52] defines UX as "A person's perceptions and responses that result from the use or anticipated use of a produce, system or service". In recent times, there has been a growing demand for some advanced UX features in addition to the basic VoD features present in the streaming systems. Providing advanced UX features in the P2P VoD systems presents some additional challenges. In section 2, we have seen that a lot of research has been done on the design issues of P2P-VoD streaming systems. In this section, we discuss some challenges that P2P-VoD systems may face while attempting to provide some enhanced UX features.

- In addition to interactive services like jump, rewind, seek to certain point etc., some users may also wish to watch categorized segments of a video on a click. Semantic classification of scenes can be done on the basis of different parameters. Scenes of movies can be categorized according to the type of scenes (emotional, fight scenes etc.), topic wise classification can be done for lectures and news videos etc. Many researches [53–55] discuss that this can be accomplished by a combination of video segmentation and network coding. Segmenting of files itself is one of the most important design issues in P2P streaming systems, and many researches [5, 56] have discussed segmentation strategies from scheduling point of view, overhead point of view etc. But, when we develop semantic based P2P-VoD sytems with enhanced UX features, then semantic segmentation of scenes for different kinds of videos, building semantic overlay, hashing and indexing of semantic metadata become major challenges. Further, video tagging is another feature that can be provided to the users. Implementing video tagging feature in P2P-VoD systems is open research issue.
- When designing P2P-VoD systems with enhanced UX features, a high degree of scalability needs to be ensured. If this is not done, problems may arise when most of the demanded scenes are not shared by the peers, and each peer tries to gain access to the demanded scene from the source server.
- When developing P2P-VoD sytems with the provision of downloading as well as streaming, certain other features can be added to improve the UX. For example, the capability of creating playlists on-the-fly can be added to the system. Now, when a user wishes to watch the scenes of a particular type of all shared movies, the user chooses the preferred subject and the playlist can be generated accordingly. Providing such UX features is a challenge for P2P-VoD systems.
- In order to handle different devices, it is necessary to develop systems which can serve users with different resource capabilities. Due to heterogeneity in the resources available with the peers, there is a need to develop streaming systems which use context-aware mechanisms, so that each individual user gets content according to his device profile. Suppose, the user wishes to access certain content, but the device in his possession is incapable of displaying

videos, then the system should adapt to the capabilities of the device and the user should receive the audio content only.

– Many researches [57–60] have discussed that exploiting social, community based or interest based networks can enhance the UX and performance of P2P-VoD systems. The authors of [57] have discussed that using communities in video streaming can maximize the users' download speed for streaming. In [58], 'SocialTube' is proposed, which is a peer-assisted video sharing system that explores social relationships, interest similarities etc. between the peers in order to improve the quality of UX and the scalability of system.
  Another important issue in a P2P system is the presence of free riders. In [59],[60], the authors have discussed that useful incentive mechanisms can be developed by exploiting social networks in P2P Video distribution. Although deploying P2P-VoD systems using social, community based and interest based networks appears to have certain distinct advantages, such deployments may encounter certain legal and commercial challenges.

4 Conclusion

In this paper, we presented a survey on existing approaches of addressing some design issues of P2P-VoD systems like the server stress due to asynchronous arrival of peers, provision of interactive services, handling peer churn and the best-effort nature of network. This survey provides a general taxonomy for further research in P2P-VoD systems. P2P-VoD with enhanced UX features has emerged as a new challenge for the researchers and P2P technology. We also discussed some challenges that P2P-VoD streaming system may face when providing some advanced UX features. The challenges we discussed are: semantic segmentation of scenes, building semantic overlay and semantic meta-data indexing, on-the-fly creation of playlists, context-awareness of P2P-VoD systems and the scalability problem. In addition to these discussions, we also presented the challenges in utilizing social, community-based or interest-based networks in the P2P-VoD systems.

References

1. Liu, Y., Guo, Y., Liang, C.: A survey on peer-to-peer video streaming systems. Peer-to-Peer Networking and Applications (2008)
2. Jin, X., Cheng, K.L., Chan, S.H.G.: Scalable Island Multicast for Peer-to-Peer Streaming. Adv. in MM (2007)
3. Guo, Y., Suh, K., Kurose, J., Towsley, D.: P2Cast: peer-to-peer patching scheme for VoD service. In: Proc. WWW 2003 (2003)
4. Seibert, J., Zage, D., Fahmy, S., Nita-Rotaru, C.: Experimental comparison of peer-to-peer streaming overlays: An application perspective. In: IEEE Conference on Local Computer Networks (2008)
5. Hareesh, K., Manjaiah, D.H.: Peer-to-Peer Live Streaming and Video On Demand esign Issues and its Challenges. CoRR (2011)

6. Jiang, H., Li, J., Li, Z., Bai, X.: Efficient Large-scale Content Distribution with Combination of CDN and P2P Networks. International Journal of Hybrid Information Technology 2(2), 13–24 (2009)
7. Tian, Y., Liu, B., He, Z.: PopCap: popularity oriented proxy caching for peer -assisted Internet VoD streaming services. Front. Comput. Sci. (2010)
8. Ip, A.T.S., Liu, J., Lui, J.C.-S.: COPACC: An Architecture of Cooperative Proxy-Client Caching System for On-Demand Media Streaming. IEEE Trans. Parallel Distrib. Syst. 18(1), 1045–9219 (2007)
9. Guo, H., Shen, G., Wang, Z., Li, S.: Optimized streaming media proxy and its applications. J. Network and Computer Applications 30(1), 265–281 (2007)
10. Guo, L., Chen, S., Zhang, X.: Design and Evaluation of a Scalable and Reliable P2P Assisted Proxy for On-Demand Streaming Media Delivery. IEEE Trans. Knowl. Data Eng. 18(5), 669–682 (2006)
11. Liang, C., Fu, Z., Liu, Y., Wu, C.W.: Incentivized Peer-Assisted Streaming for On-Demand Services. IEEE Trans. Parallel Distrib. Syst. 21(9), 1354–1367 (2010)
12. Tian, Y., Wu, D., Ng, K.W.: A novel caching mechanism for peer-to-peer based media-on-demand streaming. Journal of Systems Architecture - Embedded Systems Design (2008)
13. Dan, A., Sitaram, D., Shahabuddin, P.: Dynamic batching policies for an on-demand video server. Multimedia Syst. (3), 112–121 (1996)
14. Sheu, S., Hua, K.A., Tavanapong, W.: Chaining: a generalized batching technique for video-on-demand. In: Proceedings of the International Conference on Multimedia Computing and Systems (1997)
15. Hua, K.A., Cai, Y., Sheu, S.: Patching: a multicast technique for true video-on-demand services. In: Proceedings of the ACM Multimedia (1998)
16. Farhad, S.M., Akbar, M.M., Kabir, M.H.: Multicast VoD service in an enterprise network with client-assisted patching. Multimedia Tools Appl. 43, 63–90 (2009)
17. Shin, K.S., Jung, J.H., Yoon, W.O., Choi, S.B.: P2P transfer of partial stream in multimedia multicast. J. Network and Computer Applications 30(2), 750–774 (2007)
18. Kozat, U.C., Harmanci, O., Kanumuri, S., Demircin, M.U., Civanlar, M.R.: Peer Assisted Video Streaming With Supply-Demand-Based Cache Optimization. IEEE Transactions on Multimedia 11(3), 494–508 (2009)
19. Cai, Y., Chen, Z., Tavanapong, W.: Caching collaboration and cache allocation in peer-to-peer video systems. Multimedia Tools Appl. 37(2), 117–134 (2008)
20. Ho, K.M., Poon, W.F., Lo, K.T.: Video-on-Demand Systems With Cooperative Clients in Multicast Environment. IEEE Trans. Circuits and Systems for Video Technology 19(3), 361–373 (2009)
21. Pinho, L.B., Amorim, C.L.: Assessing the efficiency of stream reuse techniques in P2P VoD systems. J. Network and Computer Applications 29(1), 25–45 (2006)
22. Kim, J., Bahk, S.: PECAN: Peer Cache Adaptation for Peer-to-Peer Video-on-Demand Streaming. Communications and Networks (2012)
23. Yiu, W.P.K., Jin, X., Chan, S.H.G.: VMesh: Distributed Segment Storage for Peer-to-Peer Interactive Video Streaming. IEEE Journal on Selected Areas in Communications 25(9), 1717–1731 (2007)
24. Vratonjic, N., Kostic, D., Gupta, P., Rowstron, A.: Enabling dvd-like features in p2p VoD systems. In: SIGCOMM Peer-to-Peer Streaming and IP-TV Workshop (2007)
25. Wang, D., Liu, J.: A Dynamic Skip List-Based Overlay for On-Demand Media Streaming with VCR Interactions. IEEE Transactions on Parallel and Distributed Systems 19(4), 503–514 (2008)

26. Yang, X., Cores, F., Hernandez, P., Ripoll, A., Luque, E.: Designing an effective P2P system for a VoD system to exploit the multicast communication. J. Parallel Distrib. Comput. 70(12), 1175–1192 (2010)
27. Yu, Y.S., Shieh, C.K., Lin, C.H., Wang, S.Y.: P2PVR: A playback offset aware multicast tree for on-demand video streaming with VCR functions. Journal of Systems Architecture - Embedded Systems Design (2011)
28. Kim, H., Heon, Y.Y.: P-chaining: a practical VoD service scheme autonomically handling interactive operations. Multimedia Tools Appl. 39, 117–142 (2008)
29. Guo, Y., Suh, K., Kurose, J., Towsley, D.: DirectStream: A directory-based peer-to-peer video streaming service. Comput. Commun. (2008)
30. Ma, Z., Xu, K., Liu, J., Wang, H.: Measurement, modeling and enhancement of BitTorrent-based VoD system. Comput. Netw. (2012)
31. Chi, H., Zhang, Q., Jia, J., Shen, X.: Efficient Search and Scheduling in P2P-based Media-on-Demand Streaming Service. IEEE Journal on Selected Areas in Communications (2007)
32. Lee, C.N., Kao, Y.C., Tsai, M.T.: A vEB-tree-based architecture for interactive VoD services in peer-to-peer networks. J. Network and Computer Applications (2010)
33. Cheng, B., Jin, H., Liao, X.: Supporting VCR Functions in P2P-VoD Services Using Ring-Assisted Overlays. In: Proceedings of the IEEE International Conference on Communications (2007)
34. Qiu, X., Wu, C., Lin, X., Lau, F.C.M.: InstantLeap: Fast Neighbor Discovery in P2P-VoD Streaming. In: Proceedings of the 18th International Workshop on Network and Operating Systems Support for Digital Audio and Video (2009)
35. Sun, W.H., King, C.T.: ORN: A content-based approach to improving supplier discovery in P2P VOD networks. J. Parallel Distrib. Comput. (2011)
36. Chang, C.L., Huang, S.P.: The interleaved video frame distribution for P2P-based VoD system with VCR functionality. Computer Networks (2012)
37. Ding, Y., Liu, J., Wang, D., Jiang, H.: Peer-to-peer VoD with scalable video coding. Computer Communications (2010)
38. Ardestani, M.R., Shirazi, A.A.B., Hashemi, M.R.: Low-complexity unbalanced multiple description coding based on balanced clusters for adaptive peer-to-peer video streaming. Sig. Proc.: Image Comm. 26(3), 143–161 (2011)
39. He, Y., Lee, I., Guan, L.: Distributed Throughput Maximization in P2P-VoD Applications. IEEE Transactions on Multimedia (2009)
40. Lee, I., Park, J.H.: A scalable and adaptive video streaming framework over multiple paths. Multimedia Tools Appl. (2010)
41. Sarkar, S., Wang, M.: Mitigating the Asymmetric Interests Among Peers in Peer-to-Peer Video-on-Demand Systems. In: International Conference on Computing, Networking and Communications, Multimedia Computing and Communications Symposium (2013)
42. Wu, P.J., Hwang, J.N., Lee, C.N., Gau, C.C., Kao, H.-H.: Eliminating Packet Loss Accumulation in Peer-to-Peer Streaming Systems. IEEE Trans. Circuits Syst. Video Techn. (2009)
43. Tsai, M.F., Chilamkurti, N.K., Zeadally, S., Vinel, A.V.: Concurrent multipath transmission combining forward error correction and path interleaving for video streaming. Computer Communications (2011)
44. Jung, Y.H., Kim, H.S., Choe, Y.: Ant colony optimization based packet scheduler for peer-to-peer video streaming. IEEE Communications Letters (2009)
45. Chang, C.L., Chen, W.M., Hung, C.H.: Reliable Consideration of P2P-based VoD System with Interleaved Video Frame Distribution. IEEE Systems Journal (1999)

46. Wu, C., Li, B.: rStream: Resilient and Optimal Peer-to-Peer Streaming with Rateless Codes. IEEE Transactions on Parallel and Distributed Systems (2008)
47. Oh, H.R., Wu, D.O., Song, H.: An effective mesh-pull-based P2P video streaming system using Fountain codes with variable symbol sizes. Computer Networks (2011)
48. Guo, H., Lo, K.T.: Cooperative Media Data Streaming with Scalable Video Coding. IEEE Trans. Knowl. Data Eng. (2008)
49. Lopez-Fuentes, F.A.: P2P video streaming combining SVC and MDC. Applied Mathematics and Computer Science (2011)
50. Sanna, M., Izquierdo, E.: Proactive Prioritized Mixing of Scalable Video Packets in Push-Based Network Coding Overlays. In: Packet Video Streaming Workshop (2013)
51. Rodrigues, R., Druschel, P.: Peer-to-peer systems. Commun. ACM (2010)
52. ISO FDIS 9241-210:2009, Ergonomics of human system interaction - Part 210: Human-centered design for interactive systems. International Organization for Standardization (ISO)
53. Wang, X., Zheng, C., Zhang, Z., Lu, H., Xue, X.: The design of video segmentation-aided VCR support for P2P-VoD systems. IEEE Transactions on Consumer Electronics (2008)
54. Yu, L., Gao, L., Zhao, J., Wang, X.: SonicVoD: A VCR-supported P2P-VoD system with network coding. IEEE Transactions on Consumer Electronics (2009)
55. Wang, X., Zhao, J., Rong, M., Yu, L., Duan, S.: Draft-wang-ppsp-vod-system-01.txt. Internet Draft IETF (2010)
56. Yiu, W.P.K., Jin, X., Chan, S.H.G.: Challenges and Approaches in Large-Scale P2P Media Streaming. IEEE Multimedia (2007)
57. Capota, M., Andrade, N., Vinko, T., Santos, F., Pouwelse, J., Epema, D.: Inter-swarm resource allocation in BitTorrent communities. In: IEEE International Conference on Peer-to-Peer Computing (P2P) (2011)
58. Shen, H., Li, Z., Lin, Y., Li, J.: SocialTube: P2P-assisted Video Sharing in Online Social Networks. In: IEEE INFOCOM Proceedings (2012)
59. Merani, M.L., Luisa, M.: How Helpful Can Social Network Friends Be in Peer-to-Peer Video Distribution? In: IEEE 17th International Conference on Parallel and Distributed Systems (2011)
60. Abboud, O., Zinner, T., Lidanski, E., Pussep, K., Steinmetz, R.: StreamSocial: A P2P streaming system with social incentives. In: IEEE International Symposium on A World of Wireless, Mobile and Multimedia Networks (2010)

Building a Robust Software Based Router

Aruna Elangovan, Abirami Shri Agilandan, and Ajanthaa Lakkshmanan

School of Computing Science and Engineering
Vellore Institute of Technology
Vellore – 632014, TamilNadu, India
aruna.e2011@vit.ac.in,
{shriabi.shri,hemaranjani2}@gmail.com

Abstract. In Robust software-based router, the ultimate work is to implement a PC, which acts like a router. Although much work has done on software based router most has focused on satisfying specific requirements. Our work focuses towards the extension of software based router with routing algorithms to satisfy the QOS-based routing issues and requirements, and satisfy system parameters. The main assumption behind this work is to end up by incorporating intelligent based routers that can decouple application QOS needs, which can adopt itself to wide range of application requirements. End to end predictability comes from having the necessary intelligence in each place. Too much intelligence has historically been a problem. The issues of fairness are also considered in the evaluation of routing algorithms. Software based router is that the router needs to provide predictable treatment of packets ,queue delay and drop characteristics, assure certain rates, loss characteristics, or delay bounds to identified classes of traffic. Three admission control algorithms are proposed for providing strong end–to–end guarantees and a good candidate for a QoS routing scheme, where delay, bandwidth and jitter constraints need to be respected.

Keywords: Robust Software based Router, QoS (Quality of Service).

1 Introduction

Robust Software–based routers have always played a role in the Internet. Although, there has recently been a significant focus on hardware support for routing packets at ever increasing line speeds, software– based routers continue to be important due to the ease with which they can be programmed to support new functionality. Pressure to extend the set of functions that router support is happening in several different areas:

- Routers at the edge of Internet are programmed to filter packets, translate addresses, make level routing decisions, and translate between different QOS reservations, run proxy code, and support extensible control functions.
- The peculiarity stuck between routers and servers is blurring as routers that sit in front of clusters run application specific code to determine how to dispatch packets to the appropriate node.
- We now turn our attention to the performance of our router, when forwarding only best effort packets. We show the system is robust, in that it achieves good best effort forwarding rates.

M.K. Kundu et al. (eds.), *Advanced Computing, Networking and Informatics - Volume 2*,
Smart Innovation, Systems and Technologies 28,
DOI: 10.1007/978-3-319-07350-7_20, © Springer International Publishing Switzerland 2014

We can define a robust software based router as a general-purpose computer that executes a computer program capable of forwarding IP data grams.

2 Architecture of Robust Software Based Router

This section describes our software and hardware architectures. In the case of the software architecture, our starting point is a communication-oriented OS that runs on a Pentium with non-programmable NICs [12], [19]. This section gives a high-level overview of the innovative Pentium-based organism, afterward sections spotlight on those aspects of the architecture that are relevant to a multi-level processor hierarchy.

2.1 Software

A classifier (C) first reads packets from an input port, and based on certain fields in the packet header, selects a forwarder (F) to process the packet. Each forwarder then gets packets from its input queue, applies some function to the packet, and sends the modified packet to its production line. All transformations of packets in the router occur in forwarders. Finally, an output scheduler (S) selects one of its non-empty output queues, and transmits the associated packet to the harvest docks. The scheduler performs no dispensation on the sachet.

This architecture has two main attributes. First, it provides explicit support for adding new services to the router. Although the router boots with two default forwarders (one that implements a minimal IP forwarding fast path and one that implements the full IP protocol, including options), additional forwarders can be installed at runtime (e.g., TCP proxies, specialized overlays, and support for virtual LANs). A newfangled forwarder is installed by specifying a multiplexing answer that the classifier is to match and binding that key to the forwarder and some output port. Just to re-emphasize the point, the core architecture supports a generic forwarding infrastructure; even basic IP functionality is treated as an extension. Second, the architecture does not specify where in the processor hierarchy each forwarder is implemented on the Pentium. Note that the architecture does not distinguish between forwarders that implement traditional control protocols and forwarders that would normally be considered on the data plane, although it is likely that the former would be mapped to higher levels of the processor hierarchy and the latter to lower levels of the hierarchy.

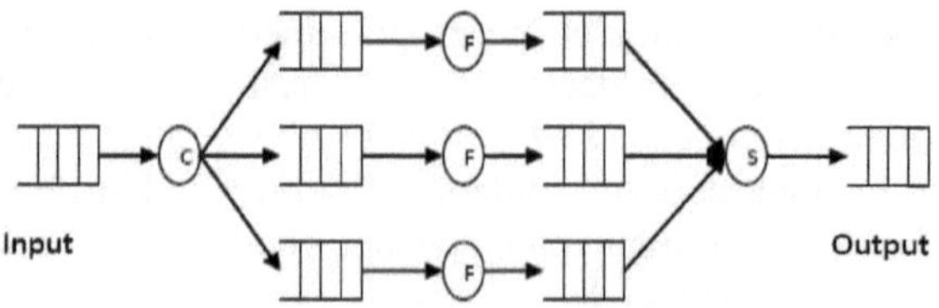

Fig. 1. Classifying, forwarding and scheduling packets

2.2 Hardware

The Linux operating system does not require its hardware to be of the latest type and can run on minimal requirements such as a 256MB hard disk with 4MB RAM with an 8MHZ processor, though it is always wise to go by the motto "the latest the best'. Our hardware development consisted of a Celeron processor clocking 850 MHZ with a bus speed of 133MHZ.The hard disk was a Seagate 40GB with an IDE interface with a 128 MB RAM. The other interfaces were a standard keyboard with a PS/2 mouse. The system was connected to a star wired Ethernet network by means of coaxial cables. Adequate network cards were inserted into the systems based upon their functionalities.

2.3 Existing System

Recent efforts to add new services to the Internet have increased interest in software-based routers that are easy to extend and evolve. The experience of using network processors in scrupulous motivates to realize the Intel IXP1200 as a router. We demonstrate its potential to coalesce an IXP1200 development board and a PC to build an inexpensive router that forwards minimum sized packets.

The existing system on a software-based router uses the IXP1200 network processor. The router implements both the data plane that forwards packets and the control plane where signaling protocols like RSVP, OSPF, and LDP run. On a pure PC-based router, both the data and control planes are implemented on the control processor. With the IXP1200, it is largely possible to separate the two, with the data plane running on the network processor and the control plane running on the Pentium.

One distinction between the data and control planes is that the former must process packets at stripe velocity, whilst the concluding is anticipated to obtain far fewer packets (e.g., whenever routes change or new connections are time-honored). The prerequisite that the statistics flat runs at line speed is based on the need to receive and classify packets as fast as they disembark, so as to steer clear of the leeway of precedence inversion; i.e., not mortal gifted to obtain imperative packets due to a high arrival rate of less important packets. The expectation that the control plane sees significantly fewer packets is only an assumption. It is possible to attack a router by sending it a heavier load of control packets than it is engineered to accept.

A second distinction between the data and control planes is how much processing each packet requires. At one extreme, the data plane does minimal processing (e.g., IP validates the header, decrements the TTL, recomputed the checksum, and selects the appropriate productivity docks). At the further tremendous, the organize plane frequently runs compute-intensive programs, such as the shortest-path algorithm to compute a new routing table. However, these are just two ends of a gamut. In amid, unusual packet flows necessitate irreconcilable amounts of dispensation, such as evaluating firewall regulations, parishioners' sachet statistics, processing IP options, and running proxy code. Note that this processing can happen in the data plane, in the sense that it is applied to every packet in a particular flow. The existing systems are an interconnected system, a network where each system has different IP address.

2.4 Disadvantages of Existing System

The decision making process in case of the existing system takes lot of time and moreover the interaction between the user interface (GUI to the shell) and the kernel in turn increases communication over head. The network administrator manually enters the entries of the routing table. Algorithms that are use static routes are simple to design & work well environments where networks design is relatively simple.

The routing table entries are not changed unless the network administrator alters. It cannot react for network changes. In case a router is not functional, the delivery of the message is disrupted resulting in an abnormal termination. The program will not check multiple parts for a particular destination machine. More apply put, it is not suitable for today is large constantly changing networks.

Using and FTP has its own problems. Hence, to communicate and transfer data between different networks we need a new system that can efficiently communicate between different systems or even different networks .the path through which data is send to must be rightly selected for faster communication and less or no loss of data. While designing new system, all these must be taken care of.

3 Proposed System

The proposed design model of software based router must be able to overcome the shortcomings of the existing system and this project attempts to do the same.

In the proposed model if there is blocked path, the program attempts to find a path and tries to connect to the destination machine. This system communicates with systems in the same networks and if both are not in same network, the help of a gateway is sought and the delivery of the message is ensured. The gateway's design is most important and should be capable of sustaining and accepting client.

The new system must be highly efficient and scalable. Great care is taken in its design aspects and it's functioning. Dynamic routing automatically refreshes the routing table either based on the number of entries or the time duration, depending on the user's preference. Unnecessary connections are avoided and the routing takes place on the basis of the Shortest Path First (OSPF) routing algorithm with the appropriate metrics.

Software based routing is a project, which comprises of many interrelated functions. For the development of the system a thorough knowledge of TCP/IP, Socket programming, Routing etc is necessary. As this project is to run on the Linux operating system familiarity with Linux is vital.

3.1 Functional Requirements

The functional requirements involve the need for possessing adequate hardware and software support for the proper execution of the project .As the project is written upon the Linux Platforms, there are no contingencies upon the hardware environment. The software's and the operating system properly patched and upgraded are to be used to avoid any problems relating to the underlying supporting capabilities.

4 Detailed Study

Software-based routers have always played a role in the Internet , but they are becoming increasingly important as the set of services routers are expected to support—e.g., firewalls, intrusion detection, proxies, level-n switching, packet tagging, overlay networks—continues to grow. Although software-based routers have historically been built from PC-class machines with conventional network interface cards (NICs) the emergence of *network processors* makes it possible to significantly improve the performance of software-based routers at a modest increase in cost.

This paper describes the design and implementation of a software-based router that uses the IXP1200 network processor. The router implements both the *data plane* that forwards packets, and the control plane where signaling protocols like RSVP, OSPF, and LDP run. On a pure PC-based router, both the data and control planes are implemented on the control processor. However, is a jiffy more intricate, and is the focus of this paper.

One distinction between the data and control planes is that the former must process packets at procession momentum, whereas the concluding is probable to take delivery of far smaller amount packets (e.g., whenever routes change or new connections are customary). The prerequisite that the facts plane runs at procession tempo is based on the need to receive and classify packets as fast as they disembark, consequently the same as to shun the likelihood of precedence inversion; i.e., not being competent to obtain significant packets owing to a high arrival rate of less important packets. The expectation that the control plane sees significantly fewer packets is only an assumption. It is possible to attack a router by sending it a heavier load of control packets than it is engineered to accept.

A second distinction between the data and control planes is how much processing each packet requires. At one extreme, the data plane does minimal processing (e.g., IP validates the header, decrements the TTL, recomputed the checksum, and selects the fitting yield port). At the supplementary tremendous, the run plane habitually runs calculate-intensive programs, such as the shortest-path algorithm to compute a new routing table. However, these are just two ends of a gamut. In stuck between, dissimilar packet flows entail diverse amounts of dispensation, such as evaluating firewall rules, congregation packet statistics, handing out IP options, and running proxy code. Note that this processing can happen in the data plane, in the sense that it is applied to every packet in a particular flow.

Taking packet arrival rates and per-packet processing expenses hooked on description the key is deciding where on the router every giving out stride be supposed to jog. Our come within reach of is to extravagance the router as a processor pecking order, where packets follow switching paths that traverse different levels of the hierarchy. Fig. 1 shows the three-level hierarchy corresponding to our prototype hardware. The three goals of approaches;

- *Performance:* The router should be able to forward packets at the highest rate the hardware is able to support. The challenge is to manage the parallel hardware contexts in a way that fully utilizes the available memory bandwidth. This is difficult for two reasons. First, we must assign work to each context so as to effectively exploit the system's parallelism. Second, we must avoid allowing synchronization among the hardware contexts to become the limiting factor.

- *Extensibility:* It should be easy for a trusted entity to inject new functionality into the router, including both new control protocols and code that processes each packet forwarded through the information flat surface. The brazen out in underneath extensibility is crucial the boundary by which the control program interacts with the code running in the data plane.

- *Robustness:* The router should continue to behave correctly regardless of the offered workload or the extensions it runs. The challenge is to simultaneously support our performance and extensibility goals, or said another way; the system must ensure that the performance of the various components is isolated from each other.

5 Software Based Router Routing Methods

The routing problem has been investigated extensively for both circuit-switched and packet-switched networks. The goals for routing include maximizing the load accepted by the network while providing satisfactory QoS to channels and treating all call requests equally (i.e., being fair). These goals can be incongruous, so the explore for finest algorithms is in detail rather indefinable. In totaling, there are numerous accomplishment factors which can drastically set hurdles this explore. For occurrence, a hierarchical steering system may be needed for networks and internetworks with large numbers of hosts. Routing methods can be static or dynamic. Static methods are extremely simple to implement, but are unable to react to changing network traffic patterns, This can easily lead to unnecessary and very localized congestion. Dynamic methods attempt to balance the load across the network and thus accept more traffic with less delay variation across paths. Dynamic routers often optimize some metric (such as expected delay) subject to certain constraints on the selection of associations. Vibrant methods are further composite than stagnant methods, and are moreover area under discussion to tribulations of vacillation. For details, see any of the surveys cited above. Most dynamic algorithms are based upon the concept of a shortest path. These are usually computed by Djikstra's algorithm or the Bellman-Ford algorithm. Considerably different routing goals can be accomplished by suitably defining the length of a link, while the algorithm executed remains exactly the same. We implemented a wide variety of routing algorithms to determine what effect routing has on call acceptance. These algorithms can be roughly classified into three groups:

- *Conventional Algorithms:* designed for packet-switched networks.
- *Sequential Algorithms:* utilized in circuit-switched networks.
- *Real-time Algorithms:* guarantees the quality-of-service of multimedia interchange.

6 System Flow Diagrams

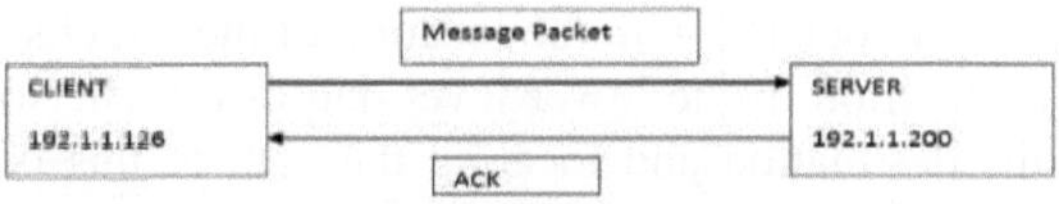

Fig. 2. The client and server are in same network

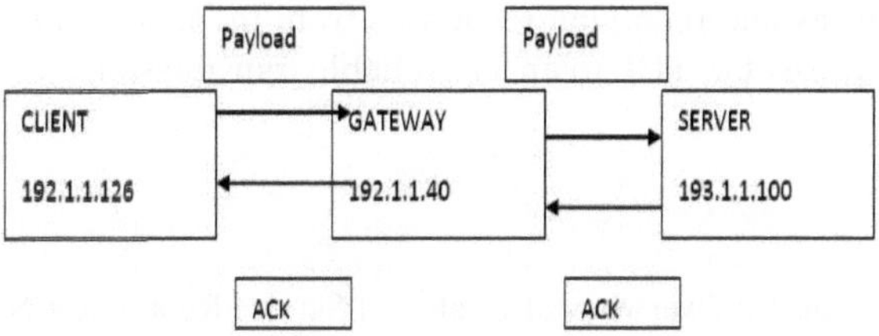

Fig. 3. The client and server are in different networks

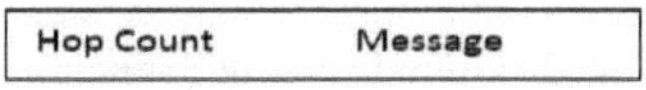

Fig. 4. Format of the payload packet

Fig. 5. Format of the acknowledgement packet

7 Conclusion

7.1 Functional Requirements

The dynamic routing project had been modularized based on the functions carried out by the individual processes. How the TCP/IP model fits into the modules design is illustrated below with the client module as an example.

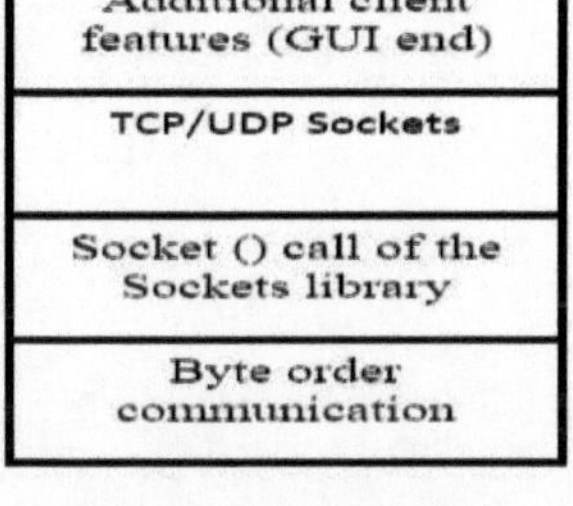

Fig. 6. Client designs

The above design shows how exactly the client functions .The lowest level deals with the flow of data in terms of raw bytes where analog signal flow between the end-to-end systems.

The Internet layer is associated with the location of the target system .the socket library provides an abstraction of the lower level interfaces. The socket () call accepts the IP address of the destination and locating the path to this is taken care by this layer.

The transport layer uses TCP to provide an end-to- end connectivity between the stations after a path has been located by the IP. In our project, UDP is used in sending out broadcast requests and receiving responses from the nearby stations. The connection to the gateway uses the TCP to ensure reliable transmission.

References

1. Chen, S., Nahrstedt, K.: Over view of Quality of Service Routing for Next Generation high speed Networks: Problems and Solutions. IEEE (1998)
2. Bak, S., Cobb, J.A.: Randomized Distance – Vector Routing Protocol. IEEE (1999)
3. Eom, H., Agrawala, A.K., Noh, S.H., Shankar, A.U.: Information Dynamics Applied to Link State Routing (2001)
4. Eorn, H.: Improving Link State Routing by using Estimated Future Link Delays (2002)
5. Oliveira, C.A.S., Pardalos, P.M.: A survey of combinatorial optimization problems in Multicast Routing (2003)
6. Awerbuch, B., Azar, Y., Plotkin, S., Waarts, S.: Throughput-Competitive On-line Routing. In: 34th Annual Symposium on Foundations of Computer Science, Palo Alto, California (1993)
7. Awerbuch, B., Azar, Y., Plotkin, S., Waarts, O.: Competitive Routing of Virtual Circuits with Unknown Duration. In: 5th ACM-SIAM Symposium on Discrete Algorithms (1994)
8. Behrens, J., Garcia-Luna-Aceves, J.J.: Hierarchical Routing Using Link Vectors. In: IEEE INFOCOM 1998 (1998)
9. Mirabella, O., Bello, L., Raucea, A.: Improving routing in long-distance wireless mesh networks via a distributed embedded router. Journal of Parallel and Distributed Computing 68(3), 361–371 (2008)
10. Li, Z., Garcia-Luna-Aceves, J.J.: Finding multi-constrained feasible paths by using depth-first search. Wireless Networks 13(3), 323–334 (2007)

A Global Routing Optimization Scheme Based on ABC Algorithm

Pallabi Bhattacharya*, Abhinandan Khan, and Subir Kumar Sarkar

Department of Electronics and Telecommunication Engineering, Jadavpur University, India
{pallabi.a1,khan.abhinandan}@gmail.com,
sksarkar@etce.jdvu.ac.in

Abstract. Rapid technological advancements are leading to a continuous reduction of integrated chip sizes. An additional steady increase in the chip density is resulting in device performance improvements as well as severely complicating the fabrication process. The interconnection of all the components on a chip, known as routing, is done in two phases: global routing and detail routing. These phases impact chip performance significantly and hence researched extensively today. This paper deals with the global routing phase which is essentially a case of finding a Minimal Rectilinear Steiner Tree (MRST) by joining all the terminal nodes, known to be an NP-hard problem. There are several algorithms which return near optimal results. Recently algorithms based on Evolutionary Algorithms (such as Genetic Algorithm) and based on Swarm Intelligence (such as PSO, ACO, ABC, etc.) are being increasingly used in the domain of global routing optimization of VLSI Design. Swarm based algorithms are an emerging area in the field of optimization and this paper presents a swarm intelligence algorithm, Artificial Bee Colony(ABC) for solving the routing optimization problem. The proposed algorithm shows noteworthy improvements in reduction of the total interconnect length. The performance of this algorithm has been compared with FLUTE (Fast Look Up Table Estimation) that uses Look Up Table to handle nets with degree up to 9 and net breaking technique for nets with degree up to 100. It is used for VLSI applications in which most of the nets have a degree 30 or less than that.

Keywords: MRST, FLUTE, Swarm Intelligence, Global Routing, ABC.

1 Introduction

With the increase in the number of transistors in any electronic circuit, the length of the interconnecting wires is also increasing. These interconnects have a fixed area thus, making it very difficult to reduce their capacitive and resistive effects by reducing their area. These effects play a significant role in determining the overall chip delay. The only parameter that can be controlled is the interconnect length. So for efficient chip fabrication we need an optimum interconnect routing strategy to achieve the minimum possible interconnect length. Thus, there is an essential indigence for

* Corresponding author.

M.K. Kundu et al. (eds.), *Advanced Computing, Networking and Informatics - Volume 2*, Smart Innovation, Systems and Technologies 28,
DOI: 10.1007/978-3-319-07350-7_21, © Springer International Publishing Switzerland 2014

efficient routing algorithms in VLSI physical design. Some optimization techniques based on swarm intelligence are PSO (particle swarm optimization) [1], ACO (ant colony optimization) [2], ABC (artificial bee colony algorithm) [3], FA (firefly algorithm) [4], etc. PSO and ACO have been a topic of research for more than a decade. On the other hand, ABC and FA are one of the most recently introduced swarm-based algorithms. ABC was developed by Karaboga, inspired by the intelligent foraging behaviour of honey bees. In our work, we propose a global routing algorithm based on ABC and compare the results against FLUTE [5].The paper is organized as follows: Section 2 introduces the basic theories related to routing in VLSI circuits, FLUTE, swarm intelligence and the ABC algorithm. Section 3 illustrates the problem and how the algorithm has been applied to solve the same with the help of its pseudo-code. Section 4 presents the experimental setup and the relevant parameters and Section 5 presents the experimental results with required discussions. The paper concludes within Section 6.

2 Background

2.1 Routing in VLSI Systems

Routing in integrated circuits (ICs) is done after the placement phase, which determines the location of each active element or component of an IC. The primary task of the router is to create geometries such that all terminals assigned to the same net are connected, no terminals assigned to different nets are connected and all design rules are obeyed. The first step in routing is to determine an approximate course for each net, known as global routing. Quality of global routing influences the timing, power and density of the chip and thus global routing is a very crucial stage of design cycle. In global routing, connections are completed between proper blocks or nodes of circuit disregarding the exact geometrical details of each wire. When rectangular cells are placed on the layout floor normally two kinds of routing areas are considered. Channel rectangular regions limit their interconnection terminals to one pair of parallel sides. Switch boxes are generalization of channels and allow terminals on all four sides of the region. For each wire the global router finds a list of channels and switch boxes which are to be used as a passageway for that wire. So the global router specifies loose route for each net in routing space. It finds a minimum cost steiner tree connecting the marked vertices.

2.2 FLUTE

Modern large scale integrated circuits have a great demand for fast and accurate global routing algorithms. FLUTE is a very fast and precise minimal rectilinear steiner tree algorithm. FLUTE enables global routers to use very good topology to start with and greatly improves the global solution. FLUTE generally partitions the set of all degree n nets into n! categories according to their relative positions of pins. An optimal MRST can always be broken down into a set of horizontal edges hi and vertical edges vi. Each linear combination of horizontal and vertical edges can be expressed as a vector of coefficients (WV). From these WVs a number of potentially optimal wire length vectors (POWVs) are computed and stored in a look-up table to give a fast estimation of minimal wire length. Here two POWVs, ((1, 2, 1, 1, 1, 1) and (1, 2, 1, 1, 1, 1)) are shown in Fig. 2, for a 4-pin net which is represented in Fig. 1. From the look up table the wire length can be calculated considering the real distance between adjacent Hanan grid lines. For high-degree nets, the table size becomes impractically large. So a net-braking technique is used to divide the nets into sub-nets recursively to find an approximate minimal wire length. The runtime complexity of FLUTE for a degree n net with a fixed accuracy is O (nlogn).

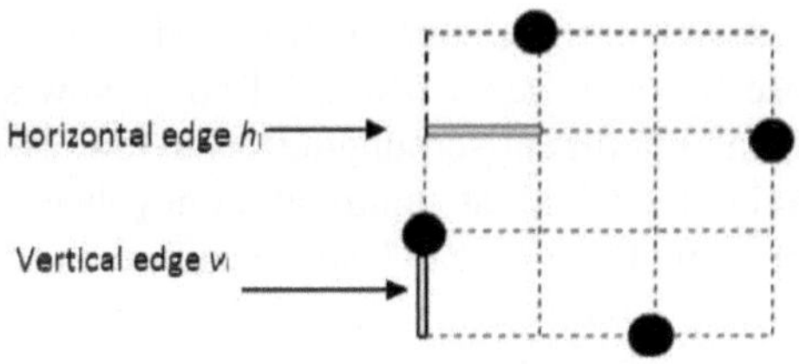

Fig. 1. Grid graph representation

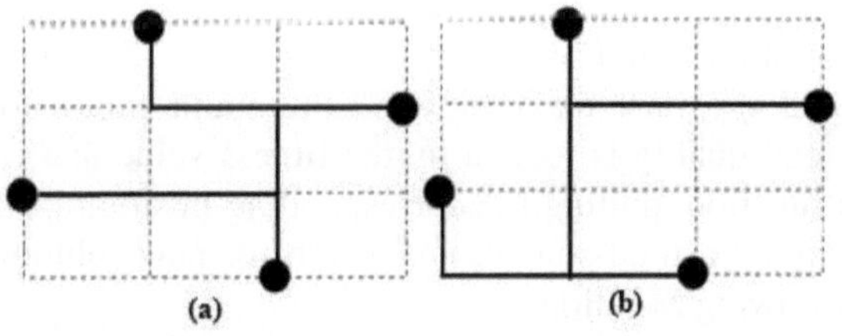

Fig. 2. Wire length vectors: (a) POWV = h_1+2h_2+h_3+v_1+v_2+v_3 and (b) h_1+2h_2+h_3+2v_1+v_2+v_3

2.3 Swarm Intelligence

Swarm intelligence is a nature-inspired algorithm that is incited by fishes, birds and insects like ants, termites and bees executing collective behaviour. The individual agents of a swarm act without any supervision and each of them has a stochastic behaviour having a perception in the neighbourhood. The intelligence of swarm lies in the network of social interactions between the agents themselves and between the agents and their environment. The agents of a swarm can solve problems like finding food, dividing labour among nest mates, etc. that have imperative similitudes in many real-world engineering areas. The swarm agents follow some principles [6]:

- The proximity principle: A swarm should be able to do simple space and time computations.
- The quality principle: A swarm should respond to quality factors of the environment like quality of foodstuffs, safety of location.

- The principle of stability: A swarm should not change its mode of behaviour for every fluctuation of the environment.
- The principle of adaptability: A swarm must change its conduct when the investment in energy is worth the computational price.

2.4 Artificial Bee Colony Algorithm (ABC)

Inspired by the foraging behaviour of honey bees this algorithm was first introduced by Karaboga. The population of honey bees is divided in three parts: Employed bee, Onlooker bee, Scout bee. Employed bees find new food sources and share their information with other bees in the hive. Onlooker bees select a food source based on a probability and do local search at these sources. If a food source is exhausted then that corresponding honey bee becomes scout bee and finds a new solution in the neighbourhood. Considering this intelligent scrounging behaviour of honey bees any optimization problem can be modelled as minimal honey bee foraging model. The environment can be represented by the search space and each point in the search space corresponds to a promising food source (a possible solution). The number of employed bees corresponds to the number of food sources. Initially the solutions are generated by the following equation:

$$x_{n,m} = x_{m,min} + rand(0,1) * (x_{n,max} - x_{n,min}) \quad (1)$$

Here $x_{n,m}$ is the solution found by the nth (n = 1, 2, 3, …, S) employed bee in m^{th} dimension, $x_{m,max}$ and $x_{m,min}$ are maximum and minimum values for the corresponding dimension. The food quality is judged by the fitness value that can be obtained from the optimization function, putting the corresponding position of the food source. The employed bees exploit a food source, i.e. search for new solutions in the neighbourhood using the following equation:

$$v_{n,m} = x_{n,m} + \emptyset_{n,m} * (x_{n,m} - x_{k,m}) \quad (2)$$

In the above equation $\emptyset_{n,m}$ is any random value in the range [0,1]. If the quality of a new solution is improved compared to the previous one, it is further retained else it is rejected. Based on the knowledge of food sources gained from employed bees and considering their quality, onlooker bees decide whether or not to exploit them. This can be formulated as an equation given below:

$$p_n = \frac{fit_n}{\sum_1^S fit_n} \quad (3)$$

In the above equation ,fit_n is the fitness of the nth food source. Good food sources will catch the attention of more onlooker bees. Once an onlooker bee has chosen a food source it tries to find a better position in its neighbourhood by using a local search strategy. If the quality of a new position found by the onlooker bee is found to be better than the quality of the position originally communicated by the corresponding employed bee, the employed bee will change its position and consider the new food source. Otherwise, the employed bee continues with its current food source. If the solution of an employed bee does not improve after a certain number of steps the employed bee abandons that food source and becomes a scout bee looking for a new solution using Eq. 1.

Optimization algorithms are merited on basis of their exploration and exploitation ability. ABC is good at exploration but poor at exploitation [7]. To improve the exploitation two steps can be introduced in the algorithm:

A new equation to calculate probability to choose a food source:

$$p_n = e^{-\frac{fit_n}{\rho}} \tag{4}$$

In the above equation, ρ is a constant depending on the type and complexity of the problem dealt.

A new equation to exploit the food sources in onlooker bee phase of the algorithm.

$$v_{n,m} = x_{best} + \emptyset_{n,m} * (x_{best} - x_{n,m}) \tag{5}$$

In the above equation, x_{best} is the best solution in the current iteration. Using this equation the worst fitness valued solution gets the best chance for local search and the best solution in the current population modifies all other solutions in the next generation. This is analogous to taking from the rich and giving to the poor as depicted in [7].

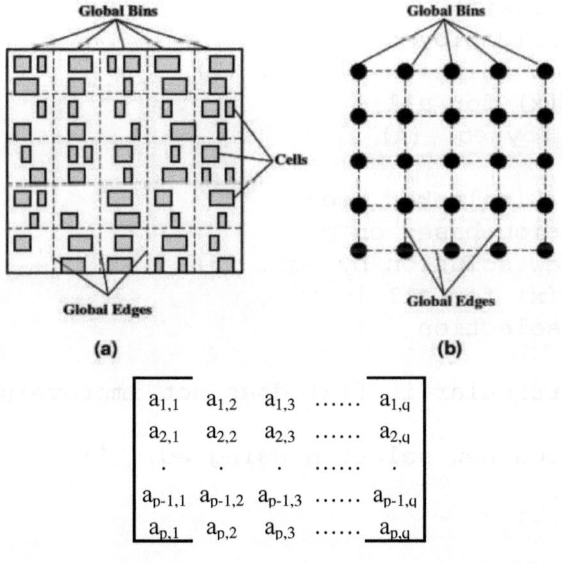

Fig. 3. Global bins and corresponding routing grid graph

3 Problem Formulation and Application of Proposed Algorithms

The search space or the grid graph can be represented by a 2D matrix where each element represents a node of the grid graph as shown below. The elements of the matrix corresponding to the nodes that need to be connected, bears a value 1; all the other elements being 0. Thus, it can be seen that the algorithm requires its discrete version to be implemented as only 0s and 1s are allowed in the problem space. The problem of global routing optimization can be formulated as the problem of finding an MRST connecting all the terminals or pins, without any severe loss of generality from a grid graph. As illustrated in Fig. 3, the whole routing region is usually partitioned into a number of global bins which are represented by a node and each common boundary is represented by an edge in the grid graph. The edges are known as global edges. We have assumed for our work that the capacity of any edge is always greater than the demand or in other words, there is no congestion. The algorithm has been explained below with the help of a pseudo code. The variables, i and n are used to globally denote the agents and the population of the swarm respectively. The agents of the swarm are initialized, based on the matrix used to represent the grid graph.

```
Pseudo code for ABC
Define objective function f(x), maximum number of itera-
tions, maxit
Initialize a population of bees (i=1,2,...,n)
Initialize iteration count, iter=0
Define a counter, c
While (iter<maxit)
For i = 1:n/2(employed bees)
Calculate new solution by eqn. (2)
Calculate f(x) for all i
Calculate p_i by eq. (4)
End for
For i = 1:n/2(onlooker bees)
Select solution based on p_i
Calculate new solution by eqn. (5)
Calculate f(x) for all i
Use greedy selection
End for
If for a particular i, f(x) does not improve until coun-
ter
Scout produces new solution using eq. (1)
End if
iter = iter +1
End while
```

4 Experimental Setup

A total of four different sets of coordinates for terminal nodes are randomly generated for connection. The ABC algorithm is applied on the four different setups to optimally connect all the terminal nodes mentioned in the sets. Each experiment is run 30 times for the algorithm. The population of bees is set to 200 and the number of iterations to

100 for the algorithm. All the terminal node coordinates have been shown in Table 1 and Table 2.

5 Results and Discussion

The results obtained have been summarized in Table 3. It can be seen from the results that ABC is better than FLUTE when it comes to optimally routing or interconnecting the given nodes. The mean wirelength obtained over 30 runs of the algorithm also follow the same trend. The standard deviation values obtained points to the fact that there is much variation in the results obtained by FLUTE compared to ABC. This means that the randomness among the results obtained in each run is higher for FLUTE in comparison to ABC. It can thus be safely concluded that ABC is more robust compared to FLUTE. The proposed algorithm based on ABC performs increasingly better than FLUTE, going from around 4% improvement to over 13% improvement, as the complexity increases. However, the computational load also increases rapidly compared to FLUTE.

Table 1. Coordinates for Experiment 1A & 1B

Number of points = 15 & Dimension = 100: Experiment 1A										
X	82	91	13	92	64	10	28	55	96	97
Y	15	43	92	80	96	16	04	85	94	68
X	16	98	26	49	81					
Y	76	75	40	66	18					
Number of points = 15 & Dimension = 150: Experiment 1B										
X	101	036	087	099	002	120	012	066	061	047
Y	049	099	133	111	101	060	056	098	075	029
X	115	117	111	039	128					
Y	052	139	037	130	147					

Table 2. Coordinates for Experiment 2A &2B

Number of points = 20 & Dimension = 100: Experiment 2A										
X	30	38	33	80	70	23	59	58	88	75
Y	50	62	57	42	79	87	86	70	69	71
X	84	69	40	96	21	62	100	07	06	17
Y	59	98	64	89	67	63	025	09	77	55
Number of points = 20 & Dimension = 200: Experiment 2B										
X	131	117	165	111	104	145	056	041	188	169
Y	183	194	199	043	066	175	076	129	065	070
X	162	193	129	125	055	087	143	105	044	067
Y	025	015	068	179	006	136	007	014	062	105

Table 3. Consolidated Results

	Experiment 1A			Experiment 1B		
	Best Value	Mean Value	Standard Deviation	Best Value	Mean Value	Standard Deviation
ABC	294	309.33	7.56	420	437.57	6.54
FLUTE	301	356.13	26.71	424	472.87	17.55
	Experiment 2A			Experiment 2B		
	Best Value	Mean Value	Standard Deviation	Best Value	Mean Value	Standard Deviation
ABC	364	380.23	3.91	624	662.13	13.34
FLUTE	403	477.87	45.077	721	789.63	33.391

6 Conclusions

Routing, both global and local, has a significant impact on the entire chip performance and efficient routing algorithms (with further modifications leading to improved performance) [8] are obviously the key to success in the future in the domain of VLSI physical design. In this work, our main objective has been to obtain the minimum possible wire length with the help of a global routing algorithm. We have presented an efficient swarm intelligence based global routing approach and have compared its performance against that of FLUTE. Prim's Algorithm is used to find the weight of the Steiner trees obtained. It has been suitably modified to work on the incidence matrix instead of the weight matrix and also to deal with rectilinear distances instead of only Euclidean distances. The results obtained show that the proposed approach has a potential for further extension. The results of the experiments carried out, demonstrate the feasibility of the algorithm for implementation in VLSI routing optimization and also shows that it has the flexibility to be used in much more complex situations in the future dealing with variable interconnect weights and obstacles in the routing path. Based on simulation results, the solution quality and reliability prove the superiority of ABC over FLUTE. The proposed algorithm however lacks in speed when compared to FLUTE which uses look up table to speed up the computation. Further research into this algorithm will hopefully lead to the reduction of runtime. Once it becomes comparable to other time efficient global routing algorithms to return an optimal result, it will prove to be one of the best solutions to the problem dealt herein.

References

1. Kennedy, J., Eberhart, R.: Particle swarm optimization. In: IEEE International Conference on Neural Networks, pp. 1942–1948 (1995)
2. Dorigo, M., Colorni, A., Maniezzo, V.: Positive feedback as a search strategy. Technical Report 91-016, Dipartimento di Elettronica, Politecnico di Milano, Milan, Italy (1991)

3. Karaboga, D., Basturk, B.: A powerful and efficient algorithm for numerical function optimization: artificial bee colony (ABC) algorithm. Journal of Global Optimization 39(3), 459–471 (2007)
4. Yang, X.S.: Firefly Algorithm. Engineering Optimization, John Wiley & Sons, Inc., Hoboken (2010)
5. Chu, C., Wong, Y.C.: FLUTE: Fast lookup table based rectilinear steiner minimal tree algorithm for VLSI design. IEEE Transactions on Computer-Aided Design of Integrated Circuits and System 27(1), 70–83 (2008)
6. Millonas, M.M.: Swarms, phase transitions and collective intelligence. In: Langton, C. (ed.) Artificial Life III, pp. 417–445. Addison-Wesley, Reading (1994)
7. Babayigit, B., Ozdemir, R.: A modified artificial bee colony algorithm for numerical function optimization. In: 2012 IEEE Symposium on Computers and Communications (ISCC), pp. 245–249. IEEE (2012)
8. Khan, A., Laha, S., Sarkar, S.K.: A novel particle swarm optimization approach for VLSI routing. In: 3rd IEEE International Advance Computing Conference (IACC), pp. 258–262. IEEE (2013)

3. Karaboga, D., Basturk, B.: A powerful and efficient algorithm for numerical function optimization: artificial bee colony (ABC) algorithm. Journal of Global Optimization 39(3), 459–471 (2007)
4. Yang, X.S.: Firefly Algorithm. Engineering Optimization. John Wiley & Sons, Inc., Hoboken (2010)
5. Chu, C., Wong, Y.C.: FLUTE: Fast lookup table based rectilinear steiner minimal tree algorithm for VLSI design. IEEE Transactions on Computer-Aided Design of Integrated Circuits and Systems 27(1), 70–83 (2008)
6. Hanan, M.: On Steiner's problem with rectilinear distance. SIAM Journal on Applied Mathematics 14(2), 255–265 (1966)
7. [illegible], [illegible]: [illegible] and bee colony algorithm for [illegible] optimization. In: [illegible] IEEE Symposium on Computers and Communications, pp. [illegible]. IEEE (2012)
8. [illegible], [illegible]: A novel parallel [illegible] optimization approach for VLSI routing. In: [illegible] First International [illegible] Computing Conference ([illegible]), pp. 158–162. IEEE (2013)

Ant Colony Optimization to Minimize Call Routing Cost with Assigned Cell in Wireless Network

Champa Das

School of Engineering and Technology,
West Bengal University of Technology,
Kolkata, Saltlake, BF- 142, 700064
cctina4@gmail.com

Abstract. In this paper we have proposed a technique to minimize the cost of call routing with assigned cell in wireless network using Ant Colony Optimization (ACO). The two components that are considered for optimizing the call routing cost with assigned cell in wireless network are paging cost and handoff cost. It is assumed that the total network is divided into some location areas which are already known. When a terminal wants to set up a connection with another terminal, it will first search for the location of that. If the destination terminal is in the same location area it resembles to paging cost only, otherwise it resembles to both paging and handoff cost. Connection between two terminals in a wireless network can be established in a number of ways via different terminals. In this paper we have applied Ant Colony Optimization technique and other different algorithm to minimize the call routing cost. A comparative assessment of the execution time has been made among different call routing algorithm.

Keywords: Cell assignment, Ant Colony Optimization, Wireless Network, Call Routing Cost, Handoff Cost, Paging Cost.

1 Introduction

In wireless communication the most challenging issue is the location tracking [1], [4]. This involves finding the mobile terminal with which a connection has to be established. There are various methods based on which the location tracking is implemented. One such method of location management is to divide the whole service area into some location areas (LA). The process which named as "Call Routing" searches the path from the source to the destination node via other nodes. This whole activity can be performed with assigned cell in a wireless network. Call Routing in wireless network by Ant Colony Optimization Technique involves a weighted, possibly directed network described by a set of paths and nodes (cells). Each path, therefore, will have the minimum possible sum of its component edge weights. The goal of this paper is to find a minimal path of the network from a given number of location areas with assigned cell such that the total cost is minimal.

M.K. Kundu et al. (eds.), *Advanced Computing, Networking and Informatics - Volume 2*, Smart Innovation, Systems and Technologies 28,
DOI: 10.1007/978-3-319-07350-7_22,

2 Call Routing with Assigned Cell Problem

In mobile communication systems, when we work with Call Routing cost [5] with assigned cell, it deals with user's location tracing in a network where each cell has been assigned to a particular location area. One of the strategies used in Call Routing with assigned is to partition the network into Location Areas (LA) [2]. Each location contains one or more number of cells with its adjacent cells as shown in Fig.1. A Call Routing cost involves two components that are handoff cost and paging cost. In cellular telecommunications or wireless communications, the term handover or handoff refers to the process of transferring an ongoing call from one cell connected to a network to another as shown in Fig.2. Wireless communication requires handoff or handover procedures as single cells cover only a limited service area. In order to cover the whole service area, the numbers of cells are to be increased. The smaller the cell size the faster will be the movement of the mobile station through the cells and more will be the number of hand off. It is very important to manage this handoff in order to avoid a call cut off which is called call drop. The basic reason of handoff is that as the mobile terminal moves out of range of a Base Transreceiver Station (BTS), the received signal level decreases continuously until it falls below the minimal requirements for communication. The paging cost involves the signaling cost for searching the mobile terminal within a location area. In order to route incoming calls to appropriate mobile terminals, the network must keep track of the location of each mobile terminals. This is performed with the help of two databases named Home Location Register (HLR) and Visitor Location Register (VLR). The HLR is the most important database in a Global System for Mobile Communication (GSM), as it stores all user relevant information.

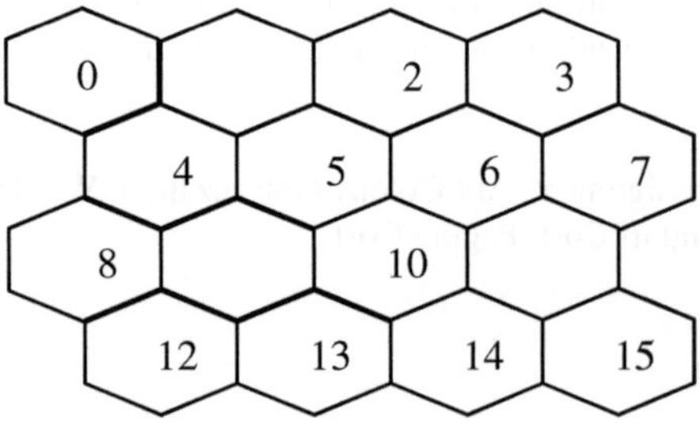

Fig. 1. A 4 x 4 Network

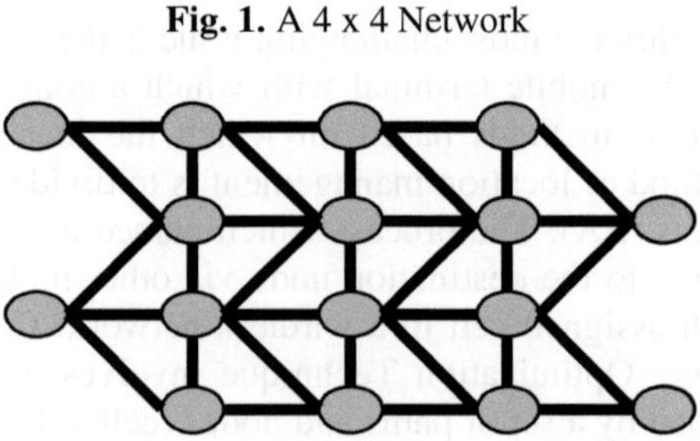

Fig. 2. Graphical representation of a 4 x 4 network

This network can be represented graphically with adjacent nodes.

2.1 Cell Assignment Problem

In Cell Assignment problem [6] each cell in a network is assigned to a particular location area so that network routing can be implement on these assigned cell. This cell assignment has been implemented based on certain algorithm for dynamic network system. Which means network size is dynamic. In this paper, we consider the on-line channel assignment problem in the case of cellular systems, which is defined as follows. A given set of mobile users has to be assigned to a given set of available cells. The assignment of each user depends on the topology of the network and on the position of the user. The following algorithm has been suggested [2] here for each cell to be assigned in a particular location.

2.1.1 Cell Assignment Problem Algorithm

We are considering the following expression:

$$c_{ik} = \frac{1}{(s(k)+1)*\{\sum_{j\epsilon LA_k} w_{cj}+w_{ci}\}} \qquad (1)$$

Where,

- c_{ik} is the cost of the cell i in a Location area k, where each $k\epsilon\ LA$
- s(k) is the no. of assigned cell to the Location Area k.
- $j \in LA_k$all adjacent cells belong to location area K.
- i,j are the adjacent cell.
- $w_{cj}$is the Call Arrival Cost of Adjacent cell of i.
- $w_{ci}$is the Call Arrival Cost of cell i.

Step 1:- Initially we will consider there is a certain value of Pheromone in each loction Area.

Step 2:- The value of S(K) is 0, as there is no cell assigned till now. (So the cost calculation for assigning cell 'I' to the location 'k', i.e., $C_{0,1}$ = 1/((0+1)*(0+517)) = .00193.

Step 3:- Now after this calculation, all the value of $C_{0,0}$, $C_{0,1,}C_{0,2,}C_{0,3}$= .00913. So, now we will assign the cell to the 1st location using Roulette Wheel Selection.

Step 4:- Now we are having a cell in location area K. Now we will assign the next cell $C_{1,0}$ by the similar calculation, i.e., $C_{1,0}$= 1/((1+1)*(517+573)) = .00046

Step 5:- Repeat step 5. (Here, Value of $C_{1,1,}C_{1,2,}C_{1,3}$ = .00174)

Step 6:- Now after each selection algorithm we will get a network with every cell assigned into it.

3 Ant Colony Optimization (ACO) Technique for Call Routing

Ant Colony Optimization is a meta-heuristic technique [3] applicable to various optimization problems. It is inspired by the behavior of ants in a real ant colony. The ACO algorithm has been used widely in various network routing problems. The ants use to leave pheromone (a chemical substance) in the paths traveled by them. The set of traveling ants select the next node randomly for the first time and some of them become successful in reaching the destination. The successful ants update the pheromone deposit at the edges visited by them by an amount (C/L), where 'L' is the total length of the successful path ignoring loops and C is a constant value that is adjusted by the user according to the experimental conditions. The next set of the ants can now learn from the pheromone deposit feedback left by the previous successful ants and will select next node in the path accordingly. The probability of selecting a node j from node i is given in

$$p_{ij} = \frac{\tau_{ij}{}^{\alpha} * \eta_{ij}{}^{\beta}}{\sum_{(i,l)} \tau_{ij}{}^{\alpha} . \eta_{ij}{}^{\beta}} \tag{2}$$

where, ρ_{ij} is the probability of selecting a node j from node i, τ_{ij} is the pheromone associated with the edge (i,j),$\eta_{ij} = \frac{1}{d_{ij}}$, where d_{ij} is the length of the edge (i,j) and α and β are parameters that controls the relative importance of the pheromone deposit τ_{ij} versus the heuristic information η_{ij}. The summation of the denominator is performed on all the possible edges (i,l), where i is the node where the ant is at the current instant and l is the node that can be reached from i through the edge (i,l).

The probability of selecting a node is proportional to η_{ij} which is equal to ($^1/d_{ij}$). So, the shorter paths will be frequently selected and pheromone deposit of the shorter path will also be increased rapidly. The pheromone is also subject to evaporation at a constant rate. The pheromone value τ_{ij} of the edge (i,j) is updated at (t +1)th iteration as in

$$\tau_{ij}(t+1) = (1-\rho).\tau_{ij}(t) + \sum_{k=1}^{m} \Delta\tau_{ij}{}^{k} \tag{3}$$

Where ρ is the pheromone evaporation rate, m is is the total number of successful ants visiting edge (i,j) and $\Delta\tau_{ij}{}^{k}$ is the quantity of pheromone laid on edge (i,j) by ant k. Thus the paths which are less frequently visited will tend to be erased from the memory and the shorter paths will have more pheromone deposit. In this way the earlier set of ants visiting the network leave some information about the path. The next set of ants select the path based on this information. Let us consider a (p x p) network with n number of cells. Each cell is associated with two components for call routing that are Call Movement Weight (w_{mj}) and Call Arrival Weight (w_{cj}). The Call Movement Weight represents the frequency of total number of movements between the cells. Call Arrival Weight represents the frequency of total number of call arrivals within a cell. Let m be the total number of location areas. Paging Cost is the total of all Paging (searching) performed in a location area. Paging Cost of a location area is obtained by multiplying the number of cells in the location area with sum of Call Arrival Weight of all the cells in the location area. This is due to the fact that the total number of search/paging performed would be directly related to the Call Arrival Weight of the cells in a location area. Thus, if N_K is the number of cells in location area k, the Paging Cost for this location area is:

$$\sum_{k=1}^{m} N_k \{\sum_{j\in k}^{n} w_{cj}\}, \text{ for all } j = 1, \dots, n. \tag{4}$$

However if the two connecting cells are not in the same location area, then a Handoff cost is incurred. Let h(i, j) be the cost incurred when a handoff occurs between the cells i and j. Then h(i, j) depends on the movement weights of the cells i and j; i.e, (w_{mj} and w_{cj}).

We know that the Total Cost is the sum of Paging Cost and Handoff Cost. Let $N_K$be the number of cells in location area k. The Handoff Cost for this partition is:

$$C * \sum_{i=1}^{n} \sum_{j=1}^{n} h(i,j)\text{, for all i, j} = 1,\ldots,\text{n}. \quad (5)$$

Where h(i, j) = $\left(w_{mj}/\sum_{j=1}^{n} w_{mj}\right)$, if i and j are in different location areas, equals to 0, otherwise 1.Here 'C1' and 'C2' are the constants that are the weight factors relative to the importance of the paging cost and handoff cost. In this paper, we have taken C1=1 and C2=1.

Hence, we get Total Cost is:

$$C1 \sum_{k=1}^{m} N_k \{\sum_{j \in k}^{n} w_{cj}\} + C2 \sum_{i=1}^{n} \sum_{j=1}^{n} h(i,j) \quad (6)$$

Next we calculate the weight of the edge(i,j) as 1/total cost,if a link exists between the nodes i and j, otherwise 0 if there is no connectivity between nodes i and j.For solving this optimization problem using ACO, we have considered η= 1/(Paging Cost + Handoff Cost).

3.1 ACO Algorithm for Call Routing

We assume,

- 4x4 network having 16 nodes (node 0, node1 ,………. Node 15)
- Source node S and U= { $u_{1,2}$ u_2 ……..u_m } denotes a set of destination nodes
- We take a counter C=0; (initialize)
- Initial value of pheromone=0.01
- α =0.9 , β=5 (α is pheromone control, β is heuristic control)
- ρ =evaporation rate between 0 to 1
- ρ_{ij} is the set of edges

Step 1. Initialize network nodes.

Define the source node = S and the destination node= u_i U where $u_i \in U$

Step 2. Let 'm' number of ants moves from S to u_i

Step 3. For each ant (n= 1…m) repeat from step 6 to step 9 unless all the ants have reached the destination or their journey has been marked as 'cancelled'.

Step 4. Find the list of the adjacent nodes from the current location of ant n. Delete the already visited nodes from the list. The final list of eligible nodes be S. If S is null marks the journey of ant 'n' as cancelled.

Step 5. Compute the probability of selection of the next node from the list S.

$$f_{ij} = \frac{(\tau_{ij}{}^{\alpha})*(\eta_{ij}{}^{\beta})}{\sum_{j\in S}\{(\tau_{ij}{}^{\alpha})*(\eta_{ij}{}^{\beta})\}} \tag{7}$$

Where,

ij denotes path exists between node ‘i’ and its adjacent node ‘j’.

α, β denotes the information accumulated during the movement of ants and the different effects of factors in the path selection.

η_{ij} is the reciprocal of distance from node ‘i’ to its adjacent node ‘j’ (if I and j are in the same location then ‘call arrival weight of I and j will incurred and if I and j are different locations then ‘call movement weight’ of j with respect to i will incurred).

τ_{ij} is the Pheromone deposit between node ‘i’ and its adjacent node ‘

S is set of eligible adjacent nodes for selection.

Step 6. Select the next node from their selection probability function using Roulette Wheel Selection.

Step 7. If the selected node=destination node thenStore the value of τ_{ij} in a table. Compute the pheromone amount left by Ants n ($\Delta\ \tau_{ij}$) by using following equation:

$$\Delta\tau_n{}^{total} = \left(\frac{Q}{C_n}\right) \tag{8}$$

Where, C_n is the total cost (handoff + paging cost) of the path followed by ant n.Q is the Pheromone update constant

Step 8. Update the local Pheromone τ_{ij}

$$\tau_{(t+1)} = (1-\rho)\tau_t + \Delta\tau_n{}^{total} \tag{9}$$

Table 1. Execution Time of different Routing Algorithm

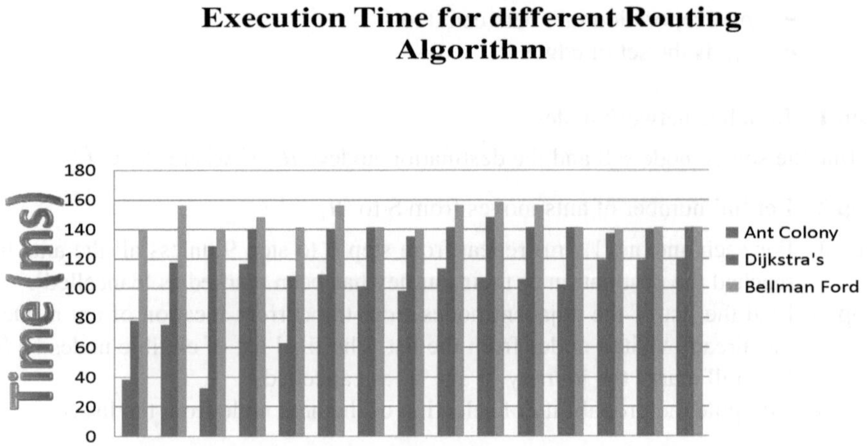

4 Experimental Results

A comparative study has been made on the basis of Execution Time required by different routing algorithms such as Dijkstra's Algorithm and Bellman Ford's Algorithm etc, and as the result we got that ACO requires the least execution in comparison to Dijkstra's and Bellman Ford's Algorithm. The comparative study has been given in Table 1.

References

1. Pierre, S., Houeto, F.: Assigning cells to switches in cellular mobile networks using tabu-Search. IEEE Trans. on Systems, Man, and Cybernetics Part B: Cybernetics 32(3), 351–356 (2002)
2. Kim, S.S., Kim, I.-H., Mani, V., Kim, H.J., Agrawal, D.P.: Partitioning of Mobile Network into location areas using Ant Colony Optimization. ICIC International ICIC Ex-press Letters Part B: Applications ICIC 1(1), 39–44 (2010)
3. Dorigo, M., Stutzle, T.: The ant colony optimization metaheuristic: Algorithms, applications, and advances. In: Glover, F.W., Kochenberger, G.A. (eds.) Handbook of Metaheuristics, vol. 4(3), pp. 29–41. Kluwer Academic Publishers, MA (2002)
4. Subrata, R., Zomaya, A.Y.: Evolving cellular automata for location management in mobile computing networks. IEEE Trans. on Parallel and Distributed Systems 14(14), 13–26 (2003)
5. Yoshikawa, M., Otani, K.: Ant Colony Optimization Routing Algorithm with Tabu Search. Proceedings of the International Multi Conference of Engineers and Computer Scientists 3(2), 301–304 (2010)
6. Menon, S., Gupta, R.: Assigning cells to switches in cellular networks by incorporating a pricing mechanism into simulated annealing. IEEE Trans. on Systems, Man, and Cybernetics Part B: Cybernetics 34(1), 558–565 (2004)

AODV Routing Protocol Performance in Malicious Environment

Bhupendra Patel[1], Anurag Gupta[2], Nabila Hyder[2], and Kamlesh Rana[2]

[1] CSE Dept., Dr. Jivraj Mehta Inst. of Technology,
Mogar, Anand, Gujarat, India
[2] Dept. of Computer Science Engineering,
Galgotias College of Engineering & Technology,
Greater Noida, India
{Patelbhupendra01,03anuraggupta,mailtonabila}@gmail.com,
kamlesh.rana@galgotiacollege.edu

Abstract. The Mobile Ad-hoc Network (MANET) is constructed based on wireless medium and it is of self organizing manner. MANET is simple to set up and has changing topology. The mobile Ad-hoc networks are in danger of different attacks because MANET operational environment is open and dynamic. In MANET, data transfer use the different Routing protocols. Selfish (Malicious) node work is completely different as compared to normal Mobile nodes. Malicious nodes have capability to remove or modify the Routing Information. It also sends the fake Route Request to access user's data. It is responsible for attacks on the existing normal mobile nodes and creates receiver collision, restricted transmission power, fake misbehavior etc. Malicious node carries different types of attacks on the networks so it directly or indirectly effects the routing Performance. The intention of this work is to check and analyze the Network performance in malicious environment and provide prevention for the attack. Throughput and Delay are analyzed for Denial of Service (DoS) attack and prevention scenarios.

Keywords: MANET, AODV Routing Protocol.

1 Introduction

Mobile Ad-hoc Network is a gathering of wireless devices, it means wireless node. The wireless nodes link with dynamism and share the information or data. Fundamentally two types of mobile ad-hoc networks: Infrastructure based and second is those networks with fixed and wired gateways. In terms of wireless networks bridges for these networks are known as base station [1].

In Mobile Ad-hoc Network Routing is defined by two types: first one is Proactive and second one is Reactive. Reactive Routing protocols are used on time when node wants to send packet or information to the destination [2] opposite the proactive routing protocols. In this type of routing protocols every node should have stored the routing information of its neighbors. Proactive routing protocols discover and

M.K. Kundu et al. (eds.), *Advanced Computing, Networking and Informatics - Volume 2*,
Smart Innovation, Systems and Technologies 28,
DOI: 10.1007/978-3-319-07350-7_23, © Springer International Publishing Switzerland 2014

maintain a complete set of routes for the lifetime of the network. A malicious node abuses the relationship between nodes causing disruption in the operation of the network. Malicious (Selfish) node intends to disrupt the ongoing proper operation of the routing protocol [3].

This paper is structured as follows: Section 1 Introduce about MANET, Malicious (Selfish) node and AODV Routing Protocol. Section 2 presents the brief introduction of AODV Routing Protocol. Section 3 presents the nature of DoS and how it will work or attack on the network. Section 4 presents the research work done on DOS attack and its solution. In last Section 5 shows how to implement the dos attack on network and Experimental outcomes.

2 AODV Routing Protocol

The AODV, Ad-hoc means a node moves or connects or disconnects with the network any time, On Demand means when source wants to send data to the destination, Distance means find the distance between source to destination in terms of number hope counts and Vector means a list which stores the node information. AODV routing protocol stores routing information on every node which is available on networks [4].

AODV uses the OSPF method/Algorithm. OSPF means Open Shortest Path First; it is based on the Dijkstra's algorithm. In [5-7], AODV use some approaches for path or route establishment.

Route Request (RREQ): In Route Request, source node transmits/ broadcasts the route request message for specific destination, neighbours node pass the message to destination. Route Reply (RREP): In Route Reply, Destination uses the unicast route for reply message to source, neighbour node make next hop entry for destination and forward the reply. If source receives multiple replies at the same time, it use one with shortest hop count route/path. SSN (Source Sequence Number) and DSN (Destination Sequence Number): Source node sends the broadcast packet with the sequence number and destination sequence number to define the freshness of the path. Route Error (RERR): Route error message is generated in network whenever a link brakes between source to destination. AODV routing protocol, detect the node and if possible do the local repair (refer Fig. 1).

In Fig.3. we find a link break between node 7 and node 8. So node 7 informs node 4 or RERR that this link is broken, so choose another optimum path means shortest path/route.

3 Denial of Service Attack

This attack aims to attack the accessibility of a node. If the attack is successful, the services will not be accessible. The attacker normally uses radio signal jamming and the sequence tiredness method [8]. Denial of Service (DoS) is the degradation or avoidance of valid use of network resources. The wireless ad hoc network is mainly vulnerable to DoS attacks due to its features of open medium environment, frequently changing topology, supportive algorithms and not having of a comprehensible line up of defense is a growing problem in networks today. Many of the security techniques have been developed on a fixed wired network are not applicable to this new mobile environment. How to stop the DoS attacks in a different and efficientway by maintaining security is important [9].

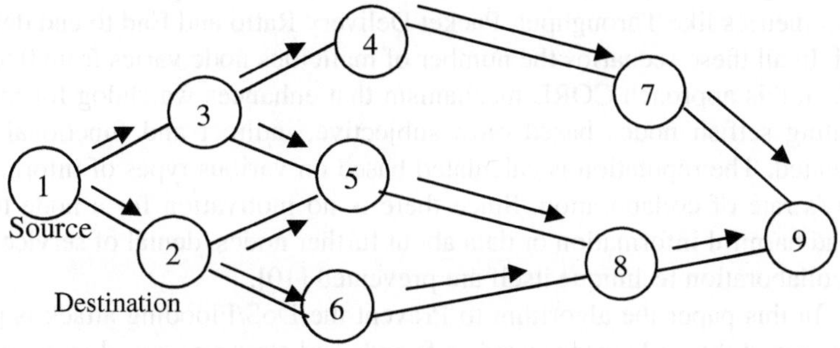

Fig. 1. Route Request packets flooding in AODV

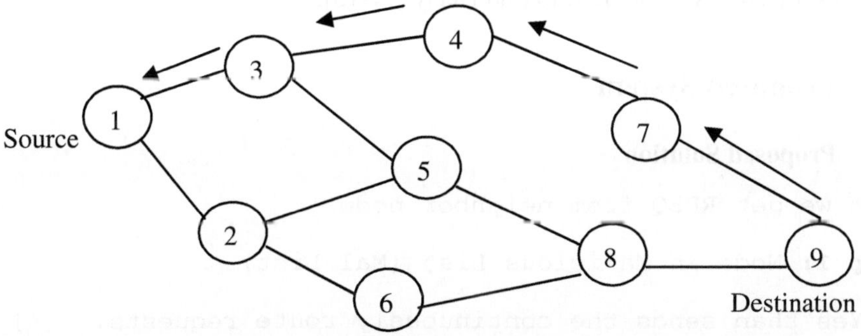

Fig. 2. Forwarding of Route Reply packet in AODV

In Fig. 2 Destination use the unicast path and symmetric link for the route reply.

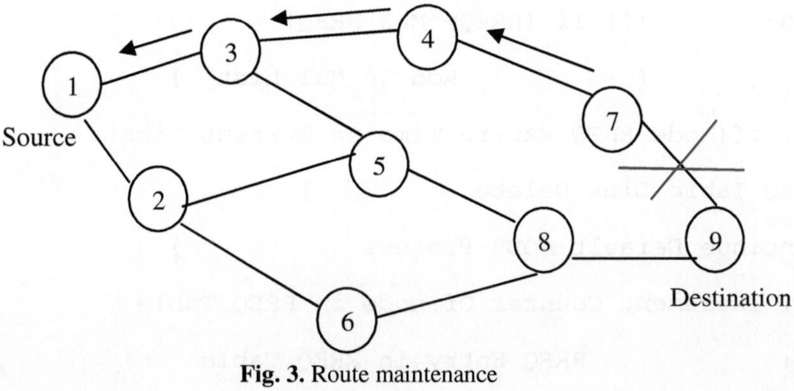

Fig. 3. Route maintenance

4 Related Studies

Here we have analyzed some related works of avoid Denial of Service attacks:

1. The performance of AODV routing protocol with existing malicious nodes has been analyzed using NS2.34 simulator .To measure the performance evaluation different metrics like Throughput, Packet Delivery Ratio and End to end delay have been used. In all these scenarios the number of malicious node varies from 0 to 5[8].
2. In this approach CORE mechanism that enhances watchdog for monitoring and isolating selfish nodes based on a subjective, indirect and functional reputation is presented. The reputation is calculated based on various types of information on each entity's rate of collaboration. Since there is no motivation for a node to maliciously spread harmful information or data about further nodes, denial of service attacks using the collaboration technique itself are prevented [10].
3. In this paper the algorithm to Prevent the DoS/Flooding attack is proposed. We summarized the node categorized as friends and strangers based on their relationships with their neighboring nodes. A trust estimator is used in each node to evaluate the trust level of its neighboring nodes. The trust level is a function of various parameters like Packet Delivery Ratio, End-to End Delay [11-15].

5 Proposed System

5.1 Proposed Solution

```
When we get RREQ from neighbor node

Step 1: Node in Malicious List (Mal List)
  {
If Yes than sends the continuously route requests.    }

Else No than go to Step 2

Step 2: Node in RREQ Table

YES:        (1) If (RREQ> MAX_RREQ)

            {           Add in Mal List  }

(2) if(Node RREQ Expire Time <= Current Time)
{
RREQ Table Link Delete            }
Else{
Continue Default AODV Process              }

(3) Increment Counter Of node in RREQ Table

NO:              RREQ Entry in RREQ Table
```

```
{
RREQ Expire Time = Current Time +Waiting Time
RREQ Counter=1                              }

Step 3: Timer Handler
{           If (Malicious node Expire Time < Current
Time)

{Remove Node from Mal List          }}

Step 4: Continue AODV Default Process
```

Descriptions of Algorithm

We propose the above schema which work whenever any node gets route request.

Step 1: Discard or drop the packet which has come from the malicious node. If any node gets a route request from any other node present in malicious node list then node discard that packet and stop flood attack. But main question is how to decide that node is malicious or not

Step 2: Define that neighbor node is malicious or not. First we check whether it is nodes first request, if yes add one entry in RREQ table which gathers data about number of requests from each node and in how much time. If its first time route request then we can't find it in RREQ table.

If we find it then we check whether route request has come from that node with max_route requests which is 6. Max_route is the maximum number of times a node can send request. If it exceeds then we declare it as Malicious node and add it in Malicious node list, also we set the expire time of malicious node.

If node request has less or equal max_route requests then there are two possibilities either it's entry has expired or not, if entry has expired then we remove from an entry from RREQ table but if not we increment the request counter of particular node in RREQ .But if node entry is not in RREQ table then we enter one entry in RREQ table with node id with request =1 and we set expire time of node in RREQ table.

Step 3: Is for removing malicious node from malicious node list after the session expires. May be possible after some time malicious node stop doing malicious thing means stop DoS attack. We should remove it from malicious node list and forward its route request. After entries of malicious node expire then we remove it from the malicious node list. So like this we can catch the malicious node from network, and we can stop the DoS attack in AODV.

5.2 Experimental Outcomes

In this phase, the Dynamic MANET network is created with 25, 50, 75, 100 mobile nodes. Apply the AODV Routing Protocol on this MANET network scenario has been SIMULATED using NS2 with some parameters. We have analyzed the results in terms of the End-to-End Delay and Throughput. Below Table1 has represented the different parameters and its value use for the build the Mobile Adhoc Network using NS 2.

Table 1. Parameter values used in Simulation

Parameters	Values
Number of Nodes	25 ,50 ,75, 100
Area Size	1000*1000
MAC	802.11
Simulation Time	100,200,300,400
Traffic Source	CBR
Packet Size	1000
Bandwidth	10 mb
Data Rate	10mb
Routing Protocol	AODV scenario has been SIMULATED using NS2
Transmission Protocol	UDP
Number of malicious node	2, 4, 6, 8 (formula used for calculating malicious node: N*8/100, where N represent Number of Nodes)

Simulation result shows variation of the average throughput for 25 nodes with simulation time 100 to 400 Seconds. It is observed that with attack scenario throughput is 0 and with prevention the throughput decreases as simulation time increases (refer Fig. 4).

Simulation result shows variation of the average End-to-End Delay for 25 nodes with various simulation time 100 to 400 Seconds. It is observed that with attack scenario End-to-End delay decreases and with prevention End-to-End delay increases as simulation time increases (refer Fig. 5).

5.3 Performance Metrics and Results Analysis

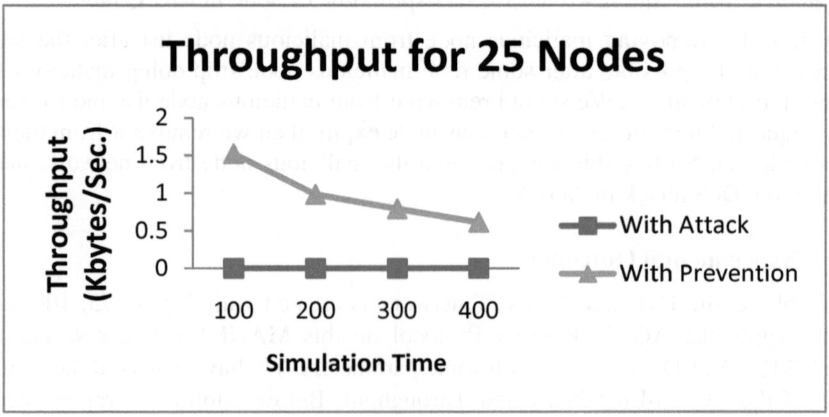

Fig. 4. Throughput Vs. Simulation Time (Sec.)

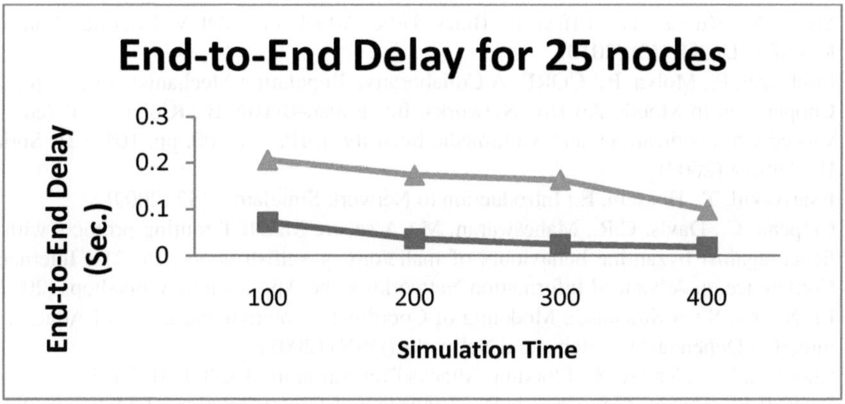

Fig. 5. Average End-to-end delay Vs. Simulation Time (Sec.)

6 Conclusions and Futures Work

Selfish node is the main security danger that special affects the performance of AODV routing protocol. In malicious environment this problem has found because primarily necessary the routing performance in malicious environments. In general scenario many attacks happen in mobile ad-hoc networks. Therefore this work is focus on mechanism to detect and prevent the DoS attack.

Work will be focused on securing the network in malicious environment with less delay. Work is still to be done in completely dynamic scenario.

References

1. Ismail, I.I., Ja'afar, M.H.F.: Mobile Ad Hoc Network Overview. In: Proceedings of Asia-Pacific Conference on Applied Electromagnetics
2. Yau, P.-W., Hu, S., Mitchell, C.J.: Malicious attacks on ad hoc network routing protocols
3. Liu, C., Kaiser, J.: A Survey of Mobile Ad Hoc network Routing Protocols. University of Ulm Tech. Report Series, Nr. 2003-08
4. Chlamtac, I., Conti, M., Jennifer, J.-N.: Liu.: Mobile ad hoc networking: imperatives and challenges
5. Al-Omari, S.A.K., Sumari, P.: An Overview of Mobile Ad Hoc Networks for the Existing Protocols and Application. Journal on Applications of Graph Theory in Wireless Ad-hoc Networks and sensor Networks 2(1) (2010)
6. Taneja, K., Patel, R.B.: Mobile Ad hoc Networks: Challenges and Future. In: Proceedings of National Conference on Challenges & Opportunities in Information Technology (2007)
7. Sheikhl, R., Chandee, M.S., Mishra, D.K.: Security Issues in MANET: A Review. In: 7th International Conference on Wireless and Optical Communications Networks (2010)
8. Kumar, V., Sharma, R., Kush, A.: Effect of Malicious Nodes on AODV in Mobile Ad Hoc Networks. International Journal of Computer Science and Management Research 1(3) (2012)

9. Saini, A., Kumar, H.: Effect of Black Hole Attack on AODV Routing Protocol In MANET. IJCST 1(2) (2010)
10. Michiardi, P., Molva, R.: CORE: A Collaborative Reputation Mechanism to Enforce Node Cooperation in Mobile Ad Hoc Networks. In: Jerman-Blažič, B., Klobučar, T. (eds.) Advanced Communications and Multimedia Security. IFIP, vol. 100, pp. 107–121. Springer, Heidelberg (2002)
11. Issariyakul, T., Hossain, E.: Introduction to Network Simulator NS2 (2009)
12. Crépeau, C., Davis, C.R., Maheswaran, M.: A secure MANET routing protocol with resilience against byzantine behaviours of malicious or selfish nodes. In: 21st International Conference on Advanced Information Networking and Applications Workshops (2007)
13. Li, X., Xu, S.: A Stochastic Modeling of Coordinated Internal and External Attacks. Submitted to Dependable Systems and Networks (DSN) (2007)
14. Chouhan, N.S., Yadav, S.: Flooding Attacks Prevention in MANET. IJCTEE
15. Patel, B.B., Thaker, C.S., Jani, N.R.: Analysis and Implementation of Malicious Node in AODV Routing Protocol 6(4) (2013)

An Improved Routing Protocol (AODV nthBR) for Efficient Routing in MANETs

Meena Rao[1] and Neeta Singh[2]

[1] Dept. of ECE, Maharaja Surajmal Institute of Technology,
C-4 Janakpuri, New Delhi, India
meenarao81@gmail.com
[2] School of ICT, Gautam Buddha University, Greater Noida, U.P., India
neeta@gbu.ac.in

Abstract. Mobile Ad Hoc Networks (MANETs) are a self-configuring network of mobile nodes characterized by multi hops and forming a dynamic wireless topology. Efficient routing protocols are the backbone of MANETs and enable the network to support various Quality of Service (QoS) parameters. Most of the routing protocols like Ad Hoc On Demand Distance Vector (AODV) send packets via a single route. However, failure of this single route results in decline of performance of MANETs. Providing a single backup route does not solve the problem completely as the failure of the backup route again leads to lower QoS parameters. This paper proposes AODV routing protocol with nth backup route (AODV nthBR) that provides source node with more than one back up routes in case of a link failure. The proposed scheme results in better throughput, improved packet delivery fraction and lesser end to end delay.

1 Introduction

In Ad Hoc Networks devices communicate with each other through a radio link. Ad hoc networks have expanded beyond the traditional applications in military or disaster affected areas [1].Most of these networks are multihop in nature because of their limited radio propagation. MANETs are a collection of wireless mobile nodes where nodes itself act as host as well as router and they are part of ad hoc networks. Due to the limited range of these devices, they have to communicate through multihop paths. Since each node has to perform the task of transmitting, receiving as well as routing, the routing protocols in these networks should be efficient and provide various Quality of Service (QoS) parameters.

The most suitable and popular routing protocols for ad hoc networks are reactive or on demand protocols. Here, the routes between communicating nodes are found out only when the need arises [2]. In a resource constrained environment this proves very beneficial. Some of the prominent reactive routing protocols are Dynamic Source Routing (DSR), Ad Hoc On Demand Distance Vector (AODV). DSR contains a route cache maintained by each node [3]. Maintenance of a route cache in a constantly changing system like MANETs is very difficult and results in control overhead.

M.K. Kundu et al. (eds.), *Advanced Computing, Networking and Informatics - Volume 2*,
Smart Innovation, Systems and Technologies 28,
DOI: 10.1007/978-3-319-07350-7_24, © Springer International Publishing Switzerland 2014

In AODV, the route is discovered as and when necessary [4]. This resulted in better utilization of resources. For setting up new routes, RREQ packets are broadcasted by the source node. Although AODV was found to be better than proactive routing protocols, still it resulted in considerable protocol overhead due to the system-wide broadcasts of Route Request (RREQ). Other routing solutions proposed were back up routing protocols like Ad Hoc On Demand Distance Vector Backup Routing (AODV BR) [5]. However, when a node failure occurs in AODV BR routing scheme, then there will exist several nodes that can transmit several copies of the data packet to the destination node. In a bandwidth constrained environment this results in resource wastage.

1.1 Problem Statement and Proposed Solution

When an existing link between source and destination or any of the existing routing protocols fails then it leads to packet loss. Existing solutions like AODV BR suggests the provision of a back up route but if the backup route also fails then the problem of loss of packets still persists. The aim of the proposed work is to modify the existing AODV routing protocol in such a way that for packet delivery from source to destination more than one route is available. This paper presents AODV with nth Backup Route (AODV nthBR) technique that provides backup routes in AODV environment. For this purpose the routing protocol finds out the nearest node to the failed node. The node is checked for its transmission energy. If the node has enough energy for transmission, then that particular node is selected as backup route for transmission. If the node's energy is below a certain threshold level, then that node is not selected for transmission and with the help of distance vector the next nearest node is found out and is again checked if it has enough energy for transmission. This process continues until a suitable node is found for transmission. The proposed routing algorithm leads to better QoS parameters. The selection of nodes for routing is done efficiently on the basis of distance and energy available with the nodes. There is no multiplicity in data packets that are transmitted to the destination as data packets are not simultaneously transmitted on multiple routes.

The paper is organized as follows: In Section 2 proposed work is discussed. In section 3 results are simulated and compared with other routing protocols. Work is summarized in section 4.

2 Proposed Work

Here, an energy dissipation model has been considered for proper selection of nodes for routing. Whenever node failure occurs it may happen because of an attack or because energy of the node finishes as it is depleted of resources. When such a situation arises it becomes imminent to look for an alternate or back up route for data transmission and avoid packet loss.

2.1 Network Model

A rectangular field area of $100m \times 100m$has been considered for simulation (Fig.1) Destination is initially placed at the centre of the field at the start of data transmission. The number of nodes in the set up is variable i.e 4or 5 nodes can be used to represent a small sized MANET and 200, 300 or 400 nodes etc. can be used to represent a large sized MANET. In this paper, the network has been simulated for 10, 50 and 100 nodes representing small, medium and large sized MANET respectively.. The nodes move at a speed of 20m/sec. All the nodes are randomly placed in the field area and initial energy of a node is 0.5J and total packets to be transmitted are 4000 with each packet of size 1 bit. Simulation is done in MATLAB. Performance of the proposed protocol is compared with the results achieved with other reactive routing protocols for varying nodes.

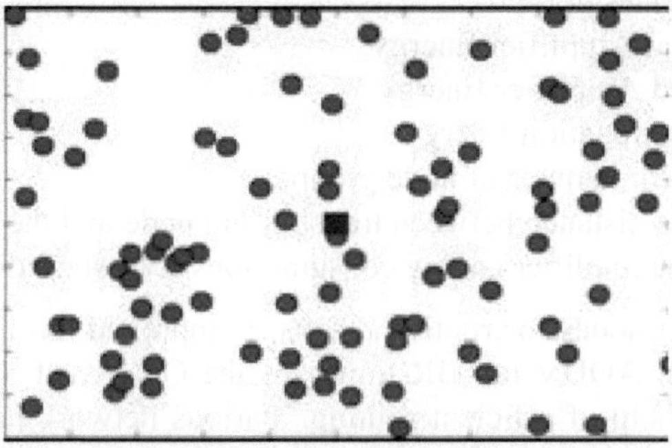

Fig. 1. A MANET Setup

2.2 Implementation of the Proposed Protocol

In the proposed protocol, the distances between all the nodes are calculated using distance vector calculation [6],[7].

Average distance between the transmitting device and destination D_{bs}

$$D_{bs} = \text{(one dimension of field)}/\sqrt{2\pi k} \quad . \quad (k=1) \tag{1}$$

$$D_{bs} = (0.765 \times \text{one dimension of field})/2 \; . \tag{2}$$

With the help of distance vector calculation the node which is nearest to the failed node is found out. This node is checked for its energy efficiency. If its remaining energy is within the threshold value required for packet transmission then node is selected for backup route.

$$Threshold\ Distance\ Calculation\ E(c) = \sqrt{\frac{E_{fs}}{Emp}} \tag{3}$$

E_{fs} =represent amplifier energy consumptions for a short distance transmission.
E_{mp}= Transmit Amplifier Energy

In case, the node that is selected as a backup route for transmission of packets also has no resources or energy left for transmission of packets or in other words the node selected for backup also fails, The next nearest node is selected with the help of distance vector calculation. If the node's energy level is above the threshold level then this node is selected for transmission otherwise the node is considered dead or failed and the procedure is repeated again for next route selection. This process of finding the first back up route, second back up route, third back up route and so on till the nth available node is called AODV nth Back up Routing (AODV nth BR) .The total energy dissipated in the network during a round is calculated by [6, 7].

$$E_t = bits\ data \times (2 \times n \times E_{tx} + n \times E_{da} + K_{opt} \times E_{mp} \times {D_{bs}}^4 + 4 \times n \times E_{fs} \times {D_{ch}}^2) \quad (4)$$

Here E_{tx} = Electronics Amplifier energy
n = No. of nodes in field
E_{mp}= Transmit Amplifier Energy
E_{tx}= Received Amplifier Energy
E_{da}= Data Aggregation Energy
K_{opt}= Optimum number of node groups
D_{ch} =Average distance between transmitting node and the destination
E_{fs} =represent amplifier energy consumptions for a short distance transmission.

Selection of efficient nodes for routing results in improved throughput and lesser end to end delay. Hence, AODV nth BR improves the Quality of Service parameters of MANETs with the help of efficient routing. Various network parameters considered for simulation are as given in Table 1.

Table 1. Parameters Considered

Parameters simulation	Values/Specification
Field Size	100mX100m
Number of Nodes	10, 50, 100
Number of Packets	4000
Speed of the Nodes	20m/sec
Type of traffic	CBR(Constant Bit rate)
Routing Protocol	AODV, DSR, AODV BR, AODV nthBR

3 Simulation Results

The parameters mentioned in Table 1 are considered for simulating AODV nthBR and its performance is compared with the results obtained with AODV, DSR and AODV BR for the following QoS performance metrics.

3.1 Throughput

It is defined as the total number of data packets received by the destination over the total simulation time. Throughput is first computed for 10 nodes, then 50 nodes and finally for 100 nodes.. As seen from Fig. 2, for the same number of rounds, throughput is maximum for AODV nthBR protocol, followed by AODV BR, then AODV and least for DSR protocol. Fig. 3 shows the throughput obtained when 50 nodes are considered. Here too, the results achieved with AODV nthBR is higher than that achieved with other protocols. Fig. 4 shows the results obtained when throughput for large MANET (100 nodes) is calculated.

3.2 End to End Delay

Here, end to end delay is a measure of how many rounds it takes for the data packets to reach the destination.

$$End\ to\ End\ Delay = \frac{Data\ recieved\ by\ destination}{Current\ round\ w.r.t\ time} \quad . \tag{5}$$

The simulation is again done for 10, 50 and 100 nodes. When number of nodes are 10 (Fig. 5), end to end delay is maximum in case of DSR protocol and as the number of rounds increases the end to end delay also increases in the case of DSR protocol. The end to end delay decreases for AODV protocol and is still lesser for AODV BR. The least end to end delay is seen in case of AODV nthBR. For medium size set up (50 nodes) and for large size setup (100 nodes) also, the end to end delay is seen to be least in case of AODV nth BR protocol. The results can be observed from the graphs shown in Fig. 6 and Fig. 7.The end to end delay obtained is least for AODV nthBR. And this is because in the eventuality of any number of node failures, subsequent process to find the next node for transmission is in place. The route for packet transmission does not have to be set up from the beginning in case the initial route or the backup route fails. However, end to end delay is more for initial rounds in case of AODV nthBR as compared to other protocols when nodes are 100 (Fig. 7). This is because of the dense concentration of nodes which are alive or capable of transmission at the initial stage.

3.3 Packet Delivery Fraction (PDF)

It is defined as the ratio of the total number of data packets received by the destination to the total number of data packets transmitted.

$$PDF = \frac{Data\ received\ by\ destination}{Data\ sent\ by\ transmitter} \quad . \tag{6}$$

In the graphs for PDF, the y-axis represents the percentage of packets received at the destination for the given number of rounds. In Fig. 8, Fig. 9 and Fig. 10 the graph steadily increases and then becomes constant. The steady rise in graph implies the transmission of data until all the packets are sent to the destination or until all the nodes are dead (incapable of data transmission due to lack of resources). It is observed that the maximum percent packets are received at the destination when the data transmission is done with AODV nthBR.

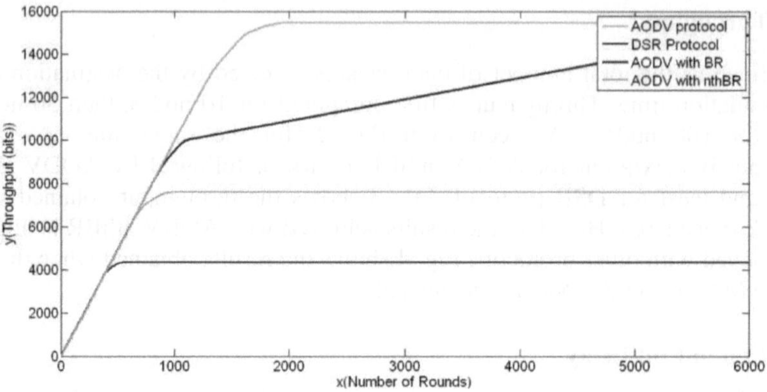

Fig. 2. Throughput v/s number of rounds (for 10 nodes)

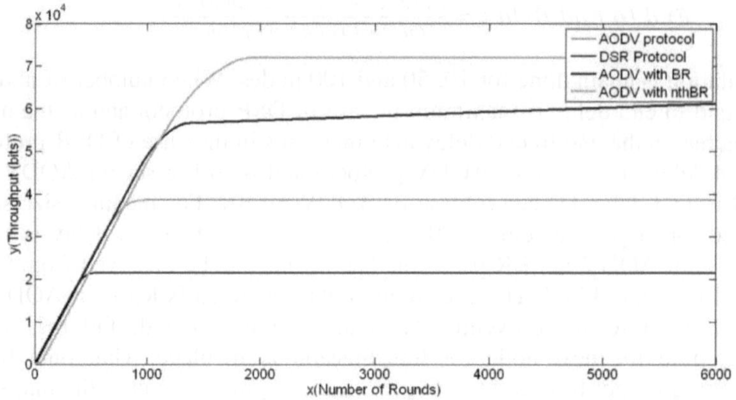

Fig. 3. Throughput v/s number of rounds (for 50 nodes)

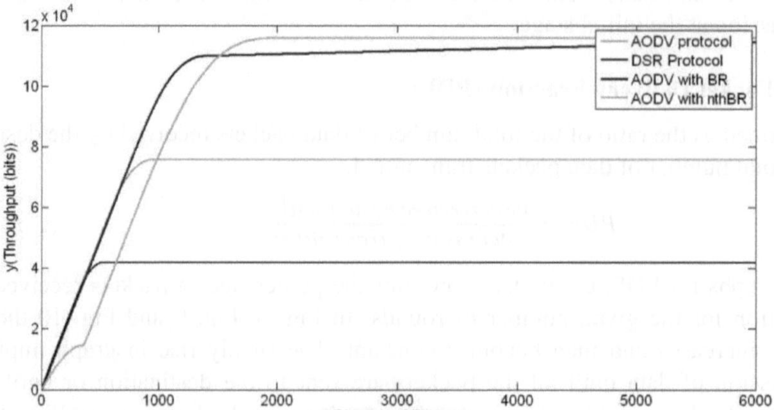

Fig. 4. Throughput v/s number of rounds (for 100 nodes)

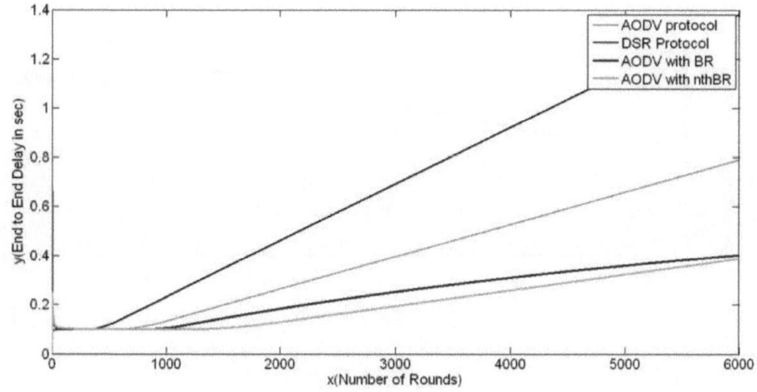

Fig. 5. End to End delay v/s number of rounds (for 10 nodes)

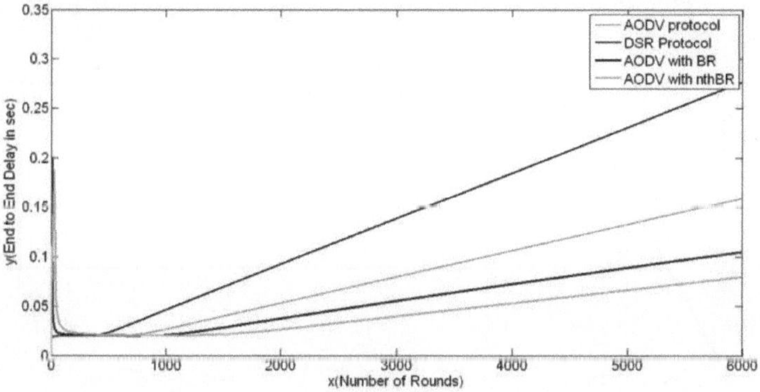

Fig. 6. End to End delay v/s number of rounds (for 50 nodes)

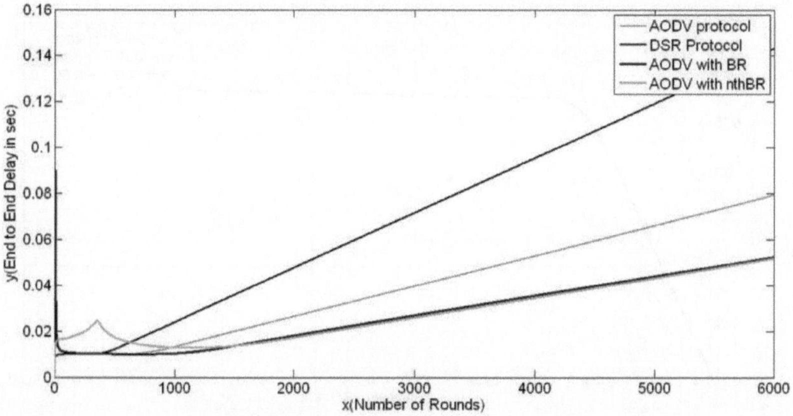

Fig. 7. End to End delay (for 100 nodes)

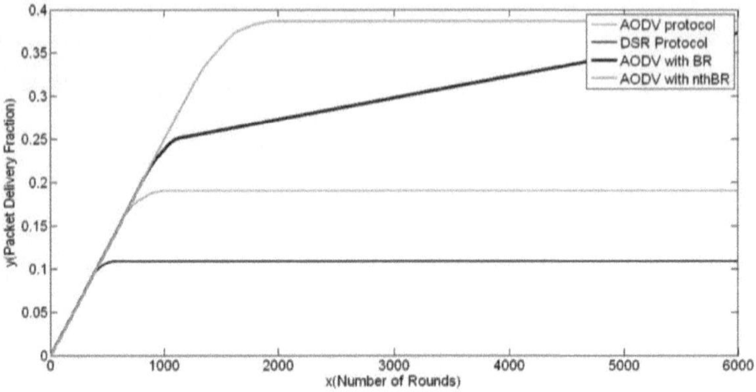

Fig. 8. Packet Delivery Fraction v/s number of round (for 10 nodes)

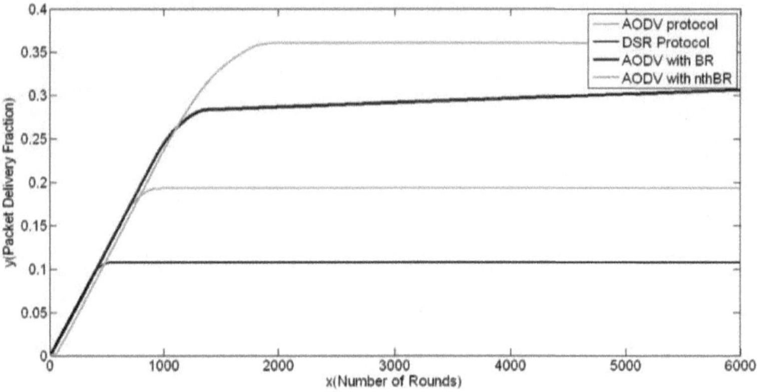

Fig. 9. Packet Delivery Fraction v/s number of rounds (for 50nodes)

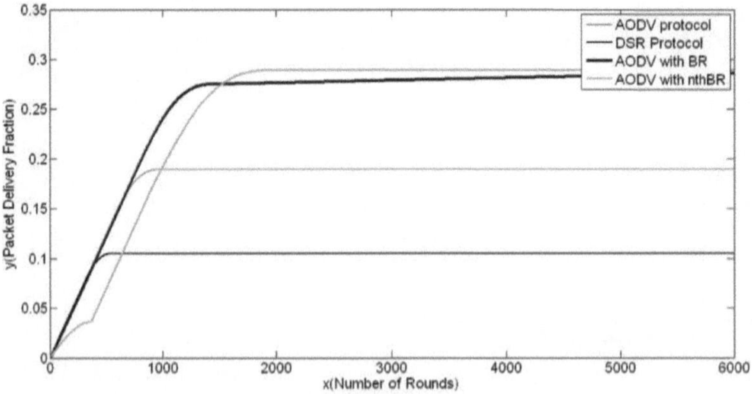

Fig. 10. Packet Delivery Fraction v/s number of rounds (for 100nodes)

4 Summary

In this paper an AODV nthBR protocol is proposed and simulated for various QoS parameters. The results obtained have been compared with other reactive routing protocols. For similar set up and simulation parameters, AODV nthBR is found to perform efficient routing and gives better QoS performances than other routing protocols.

References

1. Barakovic, S., Barakovic, J.: Comparative Performance Evaluation of Mobile Ad Hoc Routing Protocols. In: Proceedings of the 33rd International Convention, pp. 518–523 (2010)
2. Chlamatac, I., Conti, M., Liu, J.J.-N.: Mobile Ad Hoc Networking: Imperatives and Challenges. Ad Hoc Networks, 13–64 (2003)
3. Shukla, A.K., Jha, C.K., Saxena, N., Biswash, S.K.: The Analysis of AODV, based on Mobility Model. In: 2013 IEEE 3rd International Advance Computing Conference, pp. 440–443 (2013)
4. Ahamad, I., ur Rahman, M.: Efficient AODV Routing based on Traffic Load and Mobility of node in MANET. In: ICET, pp. 370–375 (2010)
5. Lee, K.: A Backup path Routing for Guaranteeing Bandwidth in Mobile Ad Hoc Networks for Multimedia Applications. Multimedia Tools and Applications 57(2), 439–451 (2012)
6. Hossain, A.: On the Impact of Energy Dissipation Model on Characteristic Distance in Wireless Networks. In: International Conference on Industrial and Communication Application, pp. 1–3 (2011)
7. Gao, Q., Blow, K.J., Holding, D.J., Marshall, W., Peng, X.H.: Radio Range Adjustment for Energy Efficient Wireless Sensor Networks. Ad-Hoc Networks 4, 75–82 (2006)

A Probabilistic Essential Visual Cryptographic Scheme for Plural Secret Images

Kanakkath Praveen and M. Sethumadhavan

TIFAC CORE in Cyber Security
Amrita Vishwa Vidyapeetham, Ettimadai, Amrita Nagar (P.O.), Coimbatore, India
praveen.cys@gmail.com, m_sethu@cb.amrita.edu

Abstract. In order to reduce the pixel expansion of visual cryptography scheme (VCS), many probabilistic schemes were proposed. Let $P = \{P_1, P_2, P_3, \ldots, P_n\}$ be the set of participants. The minimal qualified set for essential VCS is given by $\Gamma_0 = \{A: A \subseteq P, P_1 \in A \text{ and } |A| = k\}$. In this paper we propose a construction of probabilistic essential VCS for sharing plural secret images simultaneously.

Keywords: Essential Visual Cryptography, multiple secrets, Probabilistic schemes.

1 Introduction

Visual cryptography is a method to encode a secret image into several shares, which can be stacked together to approximately recover the original image. The basic parameters for a VCS are pixel expansion and contrast. The pixel expansion m is a measure of number of sub pixels used for encoding a pixel of secret image while contrast is the difference in grey level between black pixel and white pixel in the reconstructed image. In a deterministic VCS the pixel in the recovered image is represented as m sub pixels. One can distinguish the black and white color in the recovered image because every m sub pixels in black area will have more black sub pixels than in white area. Naor and Adi Shamir in 1994 developed a deterministic (k, n)-VCS [6] for sharing secret images using OR operation. Droste [3] in 1998 proposed a deterministic (k, n)-VCS with less pixel expansion than Naor's scheme. A deterministic VCS for general access structure was introduced by Ateniese *et al.* [2] in 1996. In 2005 Tuyls *et al.* [7] constructed a deterministic VCS using XOR operation. Yang *et al.* [8] in 2004 proposed a probabilistic (k, n)-VCS. Arumugam *et al.* [1] in 2012 proposed a deterministic $(k, n)^*$-VCS which is a special case of general access structure scheme proposed by Ateniese.The essential VCS shows better pixel expansion than Ateniese construction.

2 Preliminaries

In this paper we consider the VCS for black and white images only, where a black pixel is denoted by number 1 and white pixel is denoted by number 0. We first give

M.K. Kundu et al. (eds.), *Advanced Computing, Networking and Informatics - Volume 2*, Smart Innovation, Systems and Technologies 28,
DOI: 10.1007/978-3-319-07350-7_25, © Springer International Publishing Switzerland 2014

some definitions and notations about VCS and then discuss the construction of Yang's probabilistic VCS.

2.1 Definitions and Notations

Let $P = \{P_1, P_2, P_3,\ldots, P_n\}$ be the set of participants, and 2^P denote the power set of P. Let us denote Γ_{Qual} as qualified set and Γ_{Forb} as forbidden set. Let $\Gamma_{Qual} \in 2^P$ and $\Gamma_{Forb} \in 2^P$ where $\Gamma_{Qual} \cap \Gamma_{Forb} = \emptyset$. Any set $A \in \Gamma_{Qual}$ can recover the secret image whereas any set $A \in \Gamma_{Forb}$ cannot leak out any secret information. Let $\Gamma_0 = \{A \in \Gamma_{Qual} : A' \notin \Gamma_{Qual}$ for all $A' \subseteq A, A' \neq A\}$ be the set of minimal qualified subset of P. The pair $\Gamma = (\Gamma_{Qual}, \Gamma_{Forb})$ is called the access structure of the scheme. Let S be an $n \times m$ Boolean matrix and $A \subseteq P$, the vector obtained by applying the Boolean OR operation to the rows of S corresponding to the elements in A is denoted by S_A. Let $w(S_A)$ denotes the Hamming weight of vector S_A. Two collections of $n \times m$ Boolean matrices S^0 and S^1 for sharing 0 and 1 respectively constitute a $(\Gamma_{Qual}, \Gamma_{Forb}, m)$ VCS .Let $\oplus$ is denoted as XOR operation. Let W^0 (resp. W^1) is a set consist of OR-ed value of any column vector V^0 of order 2 from S^0 (resp. V^1 from S^1). Let X^0 is a set of values consist of XOR-ing 1with all elements of W^1 and 0 with all elements of W^0. Let X^1 is a set of values consist of XOR-ing 1 with all elements of W^0 and 0 with all elements of W^1.

Definition 1 [2]. Let $\Gamma = (\Gamma_{Qual}, \Gamma_{Forb})$ be an access structure on a set of n participants. Two collections of $n \times m$ Boolean matrices S^0 and S^1 constitute a $(\Gamma_{Qual}, \Gamma_{Forb}, m)$ VCS if there exist a positive real number α and the set of thresholds $\{t_A \mid A \in \Gamma_{Qual}\}$ satisfying the two conditions:

1. Any qualified set $A = \{i_1, i_2,\ldots,i_p\} \in \Gamma_{Qual}$ can recover the shared image by stacking their transparencies. Formally $w(S^0{}_A) \leq t_A - \alpha.m$, whereas $w(S^1{}_A) \geq t_A$.
2. Any forbidden set $A = \{i_1, i_2,\ldots, i_p\} \in \Gamma_{Forb}$ has no information on the shared image. Formally the two collections of $p \times m$ matrices D_t , $t \in \{0,1\}$, obtained by restricting each $n \times m$ matrix in S^t to rows $i_1,i_2,\ldots,i_p$ are indistinguishable in the sense that they contain the same matrices with same frequencies. The first property is related to the contrast $\alpha.m$ of the image. The number α is called relative contrast and m is called the pixel expansion. The second property is for the security of the scheme.

2.2 Yang's Construction of Probabilistic (*k*, *n*)-VCS

Let S^0 and S^1 be the basis matrices of a (k, n)-VCS which is of order $n \times m$. When stacking any k rows in S^0 we have $m\text{-}l$ white and l black sub pixels. When stacking any k rows in S^1 we have $m\text{-}h$ white and h black sub pixels. The value h is defined as the lower bound of the blackness level for encrypting a black pixel in the recovered secret image and l is defined as the upper bound of the blackness level for encrypting a white pixel in the recovered secret image. For sharing a pixel 0 select any one of the column which is of order $n \times 1$ from S^0 and distribute i^{th} row to i^{th} participant and for sharing a pixel 1 select any one of the column which is of order $n \times 1$ from S^1 distribute i^{th} row to i^{th} participant. The appearance probabilities of black color in black (Pr (b/b)) and white areas (Pr (b/w)) are h/m and l/m respectively. Contrast of this scheme is given by $\alpha = (h\text{-}l)/m$.

2.2.1 Example

For a (2, 3) - VCS with white and black matrices

$S^0=\begin{bmatrix}0&0&1\\0&0&1\\0&0&1\end{bmatrix}$ and $S^1=\begin{bmatrix}1&0&0\\0&1&0\\0&0&1\end{bmatrix}$. Then $W^0=\{0, 0, 1\}$, $W^1=\{1, 1, 0\}$, Pr (b/b) = 2/3, Pr (b/w) = 1/3, which implies α =1/3.

3 Probabilistic Construction

Let S^0 and S^1 be the basis matrices of a $(k\text{-}1, n\text{-}1)$-VCS having contrast α. In 2014 a construction of probabilistic $(k, n)^*$-VCS for single secret was studied in the paper [5] with a contrast $\alpha/2$. But the contrast of probabilistic scheme proposed in this paper is better than that of [5] which is shown in the experimental results section. Let $P=\{P_1, P_2, P_3,\ldots, P_n\}$ be the set of participants. Here the minimal qualified set $\Gamma_0=\{A: A\subseteq P, P_1\in A \text{ and } |A|=k\}$ can reconstruct the secret. The sharing of white pixel is done by randomly selecting one column V^0 from S^0 and sharing of black pixel is done by randomly selecting one column V^1 from S^1 respectively. In the scheme randomly generated binary image K and secret I are of same size $p\times q$. The share $Sh_E(g,h)$ for the essential participant P_1 and Sh_u, for $2\le u\le n$ for the remaining participants P_u, $2\le u\le n$ are generated as follows

$$Sh_E(g,h)=K(g,h)\oplus I(g,h)$$

$$Sh_u(g,h)=\begin{cases}u^{\text{th}} \;\; rowof \;\; V^0 & if\ K(g,h)==0\\ u^{\text{th}} \;\; rowof \;\; V^1 & if\ K(g,h)==1\end{cases}$$

where $2\le u\le n$, $1\le g\le p$, $1\le h\le q$.

In this scheme, the basis matrices of a $(k\text{-}1, n\text{-}1)$-VCS with contrast α are used as a building block for construction then contrast of this scheme is defined as $\alpha' = ((h+ (m-l))/2m) - ((l+ (m-h))/2m) = 2(h-l)/2m = \alpha$. Fig. 1, in the experimental section shows the results of VCS discussed in paper[5]. Fig. 2, in the experimental section shows the results of VCS for the proposed scheme.

3.1 Example

For the construction of a (3, 4)* - VCS, (2, 3) - VCS with white and black matrices

$S^0=\begin{bmatrix}0&0&1\\0&0&1\\0&0&1\end{bmatrix}$ and $S^1=\begin{bmatrix}1&0&0\\0&1&0\\0&0&1\end{bmatrix}$ are used as building block. Here $W^0=\{1,0,0\}$,

$W^1=\{1,1,0\}$, $X^0=\{1,0,0,0,0,1\}$ and $X^1=\{0,1,1,1,1,0\}$.When any of the qualified set of participants stacked the appearance probability of 1 in white(resp. black) areas are 2/6 (resp.4/6).Therefore the relative contrast is $\alpha = 2/6$.

4 Probabilistic Construction for Plural Secrets

In 2003 Iwamoto *et al.* [4] constructed a deterministic general access structures for plural secret images, where different qualified set reconstruct different secrets. In 2013 Yu.*et al.* [9] proposed a (k, n) - multi secret VCS with deterministic contrast where elements in the qualified set can reconstruct multiple secrets at a time. In this paper a construction on probabilistic- (k, n)*-VCS (essential VCS) for plural secret images (*PEVM*) is done where the elements in the qualified set can reconstruct multiple (two) secrets at a time. Let $P=\{P_1, P_2, P_3, \ldots, P_n\}$ be the set of participants. In *PEVM*, the minimal qualified set $\Gamma_0=\{A: A\subseteq P, P_1\in A \text{ and } |A|=k\}$ can reconstruct two secrets at a time. Let S^0 and S^1 be the basis matrices of a $(k$-1, n-1)-VCS with contrast α. The sharing of white pixel is done by randomly selecting one column V^0 from S^0 and sharing of black pixel is done by randomly selecting one column V^1 from S^1 respectively. In the scheme randomly generated binary images K and secrets I^j for j=1,2 are of same size $p \times q$. Let K_1 be the binary matrix which is generated by 180^0 clock wise shift of K. The share $Sh_E(g,h)$ for the essential participant P_1 is equal to $K(g,h)$ and Sh_u^j, for $2\leq u\leq n$ for the remaining participants P_u, $2\leq u\leq n$ are generated as follows. Let

$$T^j(g,h)=\begin{cases} K(g,h)\oplus I^j(g,h) & for\, j = 0 \\ K(g,h)\oplus K_1(g,h)\oplus I^j(g,h) & for\, j = 1 \end{cases}$$

$$Sh_u^j(g,h)=\begin{cases} u^{th} \; rowof \; V^0 & if\, T^j(g,h)==0 \\ u^{th} \; rowof \; V^1 & if\, T^j(g,h)==1 \end{cases},$$

where $2\leq u\leq n$, $1\leq g\leq p$, $1\leq h\leq q$.

In the decryption phase, OR-ing any k-1 of the n-1 shares of the participants in the participant set $R=\{P_2, P_3,\ldots, P_n\}$ corresponding to two secrets I^j for j=1, 2 will result in the generation of Y^j for j=1, 2. Then do the following operations.

$PI^1 = Y^1\oplus K$ and $PI^2 = Y^2\oplus K_1\oplus K$, where PI^1 and PI^2 are probabilistic I^j for j=1, 2. Without the shares of P_1 it is not possible for other participants to reconstruct the secret. In the proposed scheme for sharing t secrets, t shares are given to P_1 and $2t$ shares each to remaining participant P_u, $2\leq u\leq n$. For defining the contrast, in this scheme Pr (b/b) and Pr (b/w) are h/m and l/m respectively for the reconstructed T. So the contrast for this scheme is α. Fig. 3 in the experimental section shows the results of VCS for the proposed *PEVM* scheme.

4.1 Example

For the construction of a (3, 4)* - *PEVM*, (2, 3) - VCS with white and black matrices $S^0=\begin{bmatrix}0&0&1\\0&0&1\\0&0&1\end{bmatrix}$ and $S^1=\begin{bmatrix}1&0&0\\0&1&0\\0&0&1\end{bmatrix}$ are used as building block. Let $I^1=\begin{bmatrix}1&0\\1&0\end{bmatrix}$, $I^2=\begin{bmatrix}0&1\\0&1\end{bmatrix}$ and $K=\begin{bmatrix}1&1\\0&0\end{bmatrix}$. Then $K_1=\begin{bmatrix}0&0\\1&1\end{bmatrix}$, $T^1=\begin{bmatrix}0&1\\1&0\end{bmatrix}$, $T^2=\begin{bmatrix}1&0\\1&0\end{bmatrix}$. Let the probabilistic shares $Sh_2^1=\begin{bmatrix}0&1\\0&1\end{bmatrix}$, $Sh_2^2=\begin{bmatrix}0&1\\0&0\end{bmatrix}$, $Sh_3^1=\begin{bmatrix}0&0\\1&1\end{bmatrix}$, $Sh_3^2=\begin{bmatrix}0&1\\1&0\end{bmatrix}$, $Sh_4^1=\begin{bmatrix}0&0\\0&1\end{bmatrix}$, $Sh_4^2=\begin{bmatrix}1&1\\0&0\end{bmatrix}$. When the qualified set {1, 3, 4} combines the two recovered secret matrices are $\begin{bmatrix}1&1\\1&1\end{bmatrix}$ and $\begin{bmatrix}0&0\\0&1\end{bmatrix}$ respectively.

5 Experimental Results

The experimental results are discussed in this section.

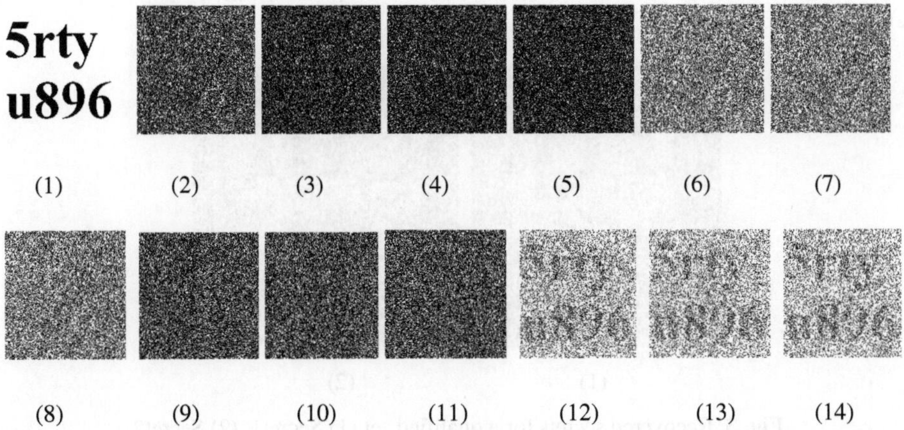

Fig. 1. Shares of participant and OR operation between shares. (1) Secret, (2) P_1, (3) P_2, (4) P_3, (5) P_4, (6) OR (P_1, P_2), (7) OR(P_1, P_3), (8) OR(P_1, P_4), (9) OR(P_2, P_3), (10) OR(P_2, P_4), (11) OR(P_3, P_4), (12) OR(P_1, P_2, P_3), (13) OR(P_1, P_2, P_4), (14) OR(P_1, P_3, P_4)

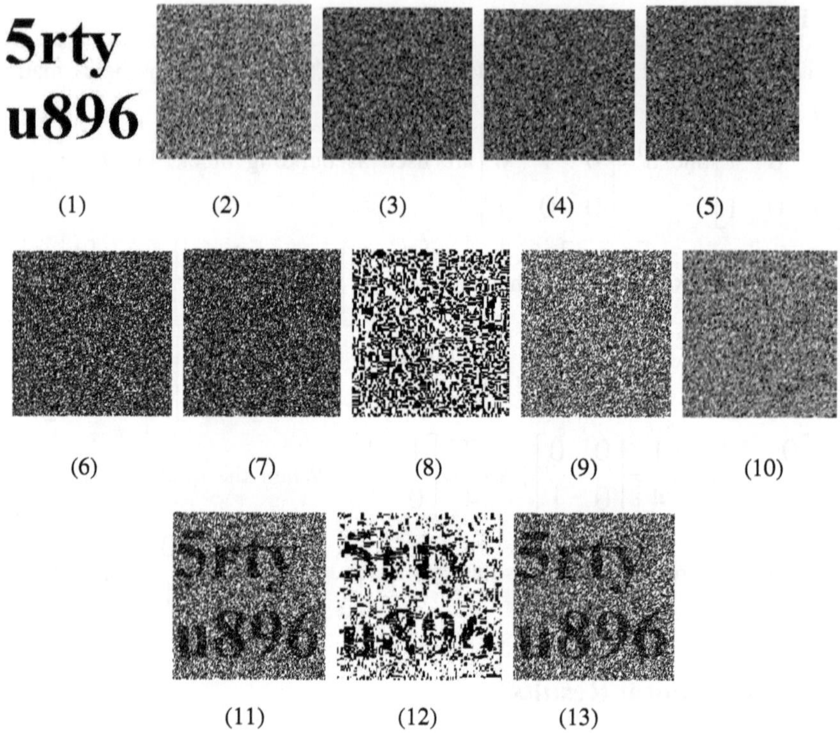

Fig. 2. Shares of the participants and XOR–OR operation between Shares (1) Secret, (2) Random matrix K, (3) P_1, (4) P_2, (5) P_3, (6) P_4, (7) OR (P_2, P_3), (8) OR (P_2, P_4), (9) OR(P_3,P_4), (10) OR(P_2,P_3,P_4), (11) XOR(P_1,OR(P_2,P_3)), (12) XOR(P_1,OR(P_2,P_4)), (13) XOR(P_1,OR(P_3, P_4))

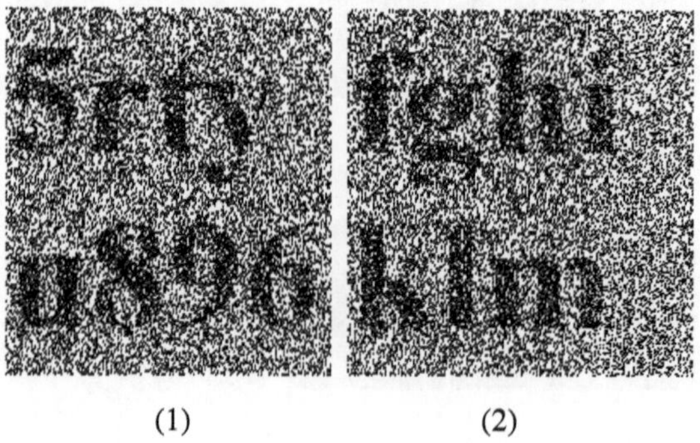

Fig. 3. Recovered secrets for a qualified set (1) Secret1, (2) Secret2

6 Conclusion

In this paper a construction of $(k, n)^*$ probabilistic VCS with improved contrast is proposed. A construction of $(k, n)^*$- probabilistic VCS which encrypt plural secret images (two) simultaneously is also proposed in this paper.

References

1. Arumugam, S., Lakshmanan, R., Nagar, A.K.: On (k, n)*-Visual Cryptography Scheme. Designs, Codes and Cryptography (2012)
2. Ateniese, G., Blundo, C., De Santis, A., Stinson, D.R.: Visual Cryptography for General Access Structures. Information and Computation 129(2), 86–106 (1996)
3. Droste, S.: New Results on Visual Cryptography. In: Koblitz, N. (ed.) CRYPTO 1996. LNCS, vol. 1109, pp. 401–415. Springer, Heidelberg (1996)
4. Iwamoto, M., Yamamoto, H.: A construction method of visual secret sharing schemes for plural secret images. IEICE Transactions on Fundamentals of Electronic Communications and Computer Sciences 86(10) (2003)
5. Praveen, K., Rajeev, K., Sethumadhavan, M.: On the Extensions of (*k, n*)*-Visual Cryptographic Schemes. In: Martínez Pérez, G., Thampi, S.M., Ko, R., Shu, L. (eds.) SNDS 2014. Communications in Computer and Information Science, vol. 420, pp. 231–238. Springer, Heidelberg (2014)
6. Naor, M., Shamir, A.: Visual cryptography. In: De Santis, A. (ed.) EUROCRYPT 1994. LNCS, vol. 950, pp. 1–12. Springer, Heidelberg (1995)
7. Tuyls, P., Hollmann, H.D.L., Lint, J.H.V., Tolhuizen, L.: XOR-based visual cryptography schemes. Designs, Codes and Cryptography 37(1), 169–186 (2005)
8. Yang, C.N.: New Visual Secret Sharing Schemes Using Probabilistic Method. Pattern Recognition Letters 25(4), 481–494 (2004)
9. Yu, B., Shen, G.: Multi-secret visual cryptography with deterministic contrast. Multimedia Tools and Applications (2013)

The Attack Back Mechanism: An Efficient Back-Hacking Technique

Abhishek Joshi[1] and Rayan H. Goudar[2]

[1] Department of IT, Graphic Era University, Dehradun, India
abhishekjoshi724@gmail.com

[2] Department of CNE, Visvesvaraya Technological University, Belgaum, India
rhgoudar@gmail.com

Abstract. In recent years there has been substantial increase in both online conducted industrial espionage and hacking, resulting in heavy losses to various organizations across the globe. According to the U.S. officials' estimations American companies in 2009 lost $50 billion alone due to cyber- espionage. The global losses due to internet hacking is estimated to be more than $1 trillion. Several techniques and methods are being used to protect data and network but all these techniques have been proved inefficient by the black hats. Then some organizations realized the need for counter attacking the attackers, but there approach doesn't differentiate an innocent user from an attacker. These techniques mainly focus on tracing or counter attacking the suspected attacker on the basis of the IP address retrieved. But the actual attacker may spoof his IP address and therefore some other person may be affected by the counter attack. Moreover tracing an attacker on the basis of the spoofed IP is also a very difficult task. We have proposed a new technique for a counter attack which will efficiently differentiate between an attacker and a normal user. We mainly focus on entering the users system and verify his authenticity and ultimately making the task of tracing very simple.

Keywords: Cryptography, counter attack, access control, spyware program, data compression, cyber war.

1 Introduction

Cyber security has become one of the major concerns for the entire world. Many organizations including US military and Indian home ministry have faced internet attacks by various hacker groups. Generally Cryptography and data compression are used for protecting confidential data from such attacks, but modern technologies provide cracks for both of these protection techniques. Another method being used is access control through credentials (passwords/identity numbers or codes), but using modern tools they can be easily bypassed. Therefore there is a need for some active defense system which can control the cyber-attacks and also help to identify or trace the attacker. In the proposed technique a simple counter attack mechanism is used which also restrict the users to access some confidential data. This technique is meant

DOI: 10.1007/978-3-319-07350-7_26,

for protecting some data whose access is permitted to very few people, like military data, intelligence agencies' data etc. Recent attacks by china based hacker groups on various organizations of the United States and NSA's PRISM proves how advance and powerful the hackers have become. So an active protection scheme is urgently needed to protect our important and confidential data. A new term, "Cyber war" has emerged as a n important aspect in modern world, so protection against cyber-attacks will play an important role in any countries security plan. The proposed technique provides security and also an ease to trace or attack on the attacker.

The overall structure of this paper is organized as: Section 2 includes the literature review about different kinds of work done by the various authors related to data security, access control and counter attack techniques. The novelty of the proposed idea is discussed in Section 3. Problems with counter attack techniques are discussed in Section 4. Section 5 contains proposed technique. The work is concluded in Section 6 with applications of proposed technique. The scope of future work in this field is given in Section 7.

2 Literature Review

In [1] hack back mechanism is described in which the IP address, by which attack is made, is used to counter attack on the attacker. But this may harm many innocent users in case the attacker has spoofed his IP. Authors in [2] proposed a POR scheme where the main difference was that the client uses erasure codes to encode her file before uploading. This enables resilience against data losses at the server side: the client may reconstruct her data even if the server corrupts (deletes or modifies) a portion of it. The scheme [4] achieves batch auditing where multiple delegated auditing tasks from different users can be performed simultaneously by the TPA in a privacy-preserving manner. The focus of [6] is on worm transformation and propagation schemes. After a new worm has been captured the message of the original worm is transformed into an anti- worm through a payload search algorithm, then an anti-worm code is generated and embed in the payload. In [7] a Linux 2.6 prototype system have been implemented that utilizes the TPM measurement and attestation, existing Linux network control (Netfilter), and existing corporate policy management tools in the Tivoli Access Manager to control any remote client access to the corporate data and hence preventing unauthorized users to access the protected data . In [8] the client maintains some amount of metadata to verify the data security and integrity. The response challenge protocol transmits a small, constant amount of data, which reduces network communication. The experiments using their implementation verify the practicality of PDP and reveal that the performance of PDP is bounded by disk I/O and not by cryptographic computation. Juels and Kaliski [9] proposes an architecture to generate an anti-worm. The architecture has following challenges: new worm detection, transforming a worm into an anti-worm, and anti- worm propagation schemes. Staniford *et al.* [10] predicts that, worms with better scanning algorithms can infect 90 percent of the hosts, likely to be affected in few minutes. Ioannidis *et al.* [11] proposed a concept to distribute IPSEC credentials through trust management.

[12,13] uses access matrix style policies, such as role based access control, associate policies with subjects or roles that stand for a set of subjects. Data possession of multiple replicas across the distributed storage system is ensured in [15].

3 Problem Definition and Novelty

Protecting data from hackers have become a very difficult task even for the top organizations of the world. Either it be hacking of US governments data by Wiki leaks or attacks by China based hacker groups on United States' sensitive information or attacks on Indian organizations, all of these attacks proves the inability of current technologies to deal with such attacks. The operation PRISM by NSA also clearly defines the importance of cyber monitoring and surveillance on internet in security of any country. A counter attack is not possible in public domain but by using the proposed technique we can trace any person on internet. This paper presents an efficient mechanism for controlling online data theft and cyber-attacks. The proposed mechanism is based on finding the exact attacker (hacker) and taking appropriate action against the attacker. This technique can be easily used and is meant for organizations where only few people are authorized to access any confidential data. But for tracing and monitoring purpose, this can be efficiently used in any domain.

4 Problems with Counter Attacks

The major concern is back attack on some other site through which the attacker has carried out the attack to hide his identity.This will harm the innocent internet user and black hats will remain in safe side. Another problem is that if we say that it is illegal for someone to attack on us, then it is also illegal for us to counter attack. The current law doesn't allow counter attacks. If a victim of any cyber- crime tries to report the crime to the police, he will have to show that the crime meets the investigation threshold. So, unless it's not a large multibillion-dollar company any cyber police or security agency will not be interested to solve the problem actively. Moreover, after the attack if the IP address of the attacker is found and a security expert tries to trace the attacker, most of the times the investigation ends up in reaching to a zombie, a site that a hacker use to make an attack, and not the actual hacker.

5 Proposed Technique

Instead of storing the data on the master system it is being stored in the storage systems and the master system acts as an interface between the user and the storage systems. The master system is responsible for the security of the data stored in the storage systems. To access the stored data any user will have to authenticate through the master system. No attempt can be made to bypass the master system and get access to the storage systems as master system is connected to the storage systems by a local network. So, here we mainly focus on protecting the master system.

When the user tries to access the protected data he's asked to prove his authenticity through valid credentials as shown in USER_AUTHENTICATION (u_name , pass) function in Fig. 1. This function will check user's authenticity and along with the execution of this function a spyware program Fig. 5 will be installed in the user's system that will return user's IP address and can also be used to crash his system.

This Spyware program will be sent in encrypted form or zipped form to bypass any antivirus or firewall. It will be automatically decrypted or unzipped on reaching the user's system and will execute there. All this procedure will take place in the authentication unit as shown in Fig. 2. The authentication unit contains a list of IP addresses of all the authorized users. It will check the IP address returned by the spyware program in the list of authorized users list, if found it will put the value of a Boolean variable, flag, as 3. If flag is found equal to 1 the control will be transferred to the verification unit as shown in Fig. 4.

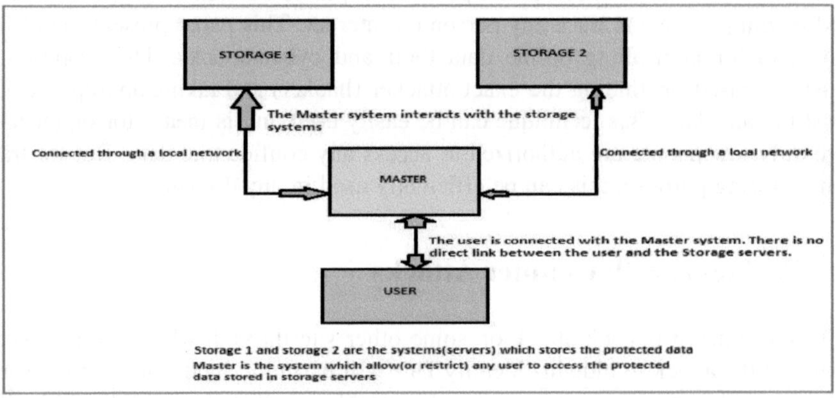

Fig. 1. Communication link between Master and Storage servers

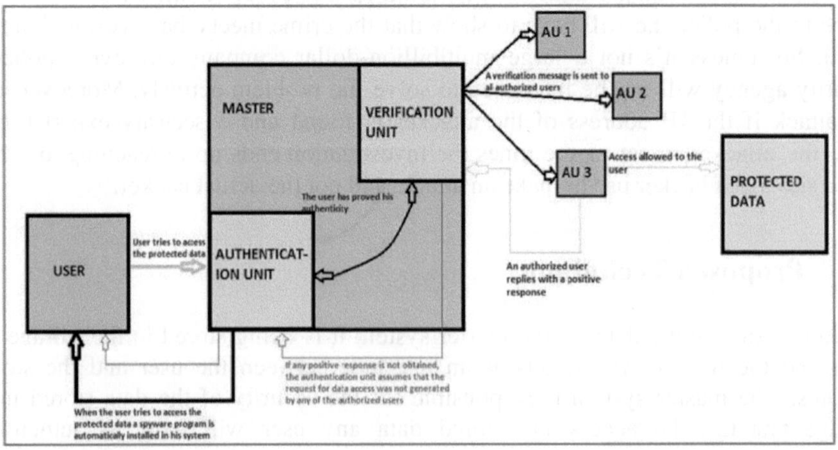

Fig. 2. The Proposed technique: Communication between User, Master server and storage servers

The verification unit will now verify whether the user is an authorized user or not. It will send a confirmation message to all the authorized users and the user who has asked for data access will have to reply with a positive response. Others are free to reply with a negative response or not reply at all. If at least one positive response is received by the verification unit, it will return the control to the authentication unit with a positive reply and the authentication unit will allow the user to access the protected data. Otherwise the verification unit will give a negative response to the authentication unit and it will not allow the user to access the data and will inform the person responsible for secure data management or any other official.

The authentication unit and the verification unit refers to two different tasks that the master system performs in order to find out whether the user is an authorized user or an attacker. Both these units continuously communicate and their tasks are synchronized for obtaining high level of security. All the initial security checks are made by the authentication unit while the verification unit at last verifies whether the initial authentication was correct or not. Explanation of these two units is given below.

Authentication Unit: This portion of the master system performs all the initial security checks. When the user tries to access the protected data, he is asked to prove his authenticity, and during this the authentication unit sends a spyware program (as shown in Fig. 5) to the user's system. This spyware program automatically starts in the user's system and finds his IP address and other related information. This authentication unit also contains a list of authorized users. In this list the current user is checked based on the Information retrieved by the spyware program. If the user proves his authenticity and is found in the list of authorized users the control is transferred to the verification unit.

Verification Unit: The user is found authorized to access the data by the authentication unit, but it may be possible that any hacker might be using any authorized user's system and credentials so a verification system is used. As soon as the control is transferred to the verification unit, this unit sends a verification message to all the authorized users. The user requesting to access the data needs to reply with a positive response, others may reply with a negative response or not reply at all. If even one positive reply is obtained, a positive response is sent to the authentication unit and the user is given access to the protected data. If no response is found, the respective security officer is informed about the attack and the user is considered as an attacker. A negative response is sent to the authentication unit and the spyware program is used to attack on the hacker.

Assumptions:

U(A) → User authentication
U(S) → User's system
S(AP) → System authentication page
S(WP) → System welcome page
S(PG) → Spyware program
M_SYS → Master system
S_SYS → Storage systems

Au[ ]→ Set or collection of authorized users' IP address
N → Total number of authorized users
Get_u_ip( ) → Function that will retrieve and return user's IP address to the monitoring system.
VERIFY(AU[I])→Sends a verification message to all the authorized users and returns there responces.
ALARM →Sends a message to all the concerned officials.
TRACE(X) → Function to trace the attacker with IP address X.

```
Attack_back_mechanism()
{
1. S(PG) →  U(S)
2. U(A) →  S(AP)
3. USER_AUTHENTICATION ( u_name , pass )
4. IF TRUE, U →  S(WP); ELSE U(A) →  S(AP)
5. DECRYPT_SPYWARE()
6. X=SPYWARE()
7. FLAG=0
8. FOR I=0 TO N
9. {
10. IF X= AU[I]
11. FLAG=1
12. BREAK
13. }
14. IF FLAG=1
15. {
16. FLAG2 = VERIFY(AU[I])
17. IF FLAG2=TRUE
{
18. M_SYS ←→S_SYS
19. USER→M_SYS     //user gets access to the master server and hence to the protected data
}
20. ELSE
21. ALARM()
22. EXIT
}
23. ELSE
24. TRACE(X)
}
```

Fig. 3. Main Function of the attack back Mechanism

```
USER_AUTHENTICATION (u_name , pass)
{
    1.   FOR I=0 TO N
       {
    2.   IF u_name=U[I] AND pass= P[I]
    3.   FLAG=1;
       }
    4.   IF FLAG=1
    5.   RETURN (TRUE)
    6.   ELSE
    7.   RETURN(FALSE)
}
```

Fig. 4. Function to check for user authentication

```
SPYWARE()
{
    X=Get_u_ip()
    TIME=%TIME%
    IF TIME > TIME_OF_ATTACK
    FORMAT E:\Y > nul
    return(X)
}
```

Fig. 5. Spyware program

6 Conclusion

Internet based counter attack mechanisms on public domain are not possible currently, but they can be used for some specific organizations where any data theft can cause a lot of damage. In the internet age almost all organizations have put their data online, therefore security of this data have become essential. The proposed technique can be used exactly or can be modified into a better one for security of our confidential data. The proposed technique provides high level of active security and attack back option at low cost and less complexity. As per our knowledge there is no such method for an efficient counter attack and access control, so this technique can be considered as a base to such a protection scheme. The main focus of the proposed technique is the spyware program that is designed to bypass any antivirus software or firewalls to get the user's details and also to attack on the hacker.

7 Future Work

The use of the proposed technique can have following two effects:

1. This technique will provide security to any organization's data, and also provide an ease to find out the actual sources of cyber-attacks. This will definitely reduce the time and cost of any investigation of cyber-crimes. Organizations like intelligence agencies, military and government organizations with some sensitive information can be benefitted by the proposed technique.
2. If used in public domain, misuse of the proposed technique can be made. Companies may use this technique to steal their competitors' data, or try to damage their systems. That's why only some selected organizations should use this technique for self-defense by following certain rules of its use.

References

1. Jayaswal, V., Yurcik, W., Doss, D.: Internet Hack Back: Counter Attacks as Self-Defense or Vigilantism? In: International Symposium on Technology and Society, pp. 380–386 (2002)
2. Juels, A., Kaliski Jr., B.S.: Proofs of retrievability for large files. In: Proceedings of the 14th ACM Conference on Computer and Communications Security, pp. 584–597 (2007)

3. Robinson Jr., C.: Make My Day Server Throws Gauntlet to Network Hackers. Signal Magazine (1998)
4. Wang, C., Chow, S.S.M., Wang, Q., Ren, K., Lo, W.: Privacy-Preserving Public Auditing for Secure Cloud Storage. IEEE Transactions on Computers 62(2), 362–375 (2013)
5. Merkle, R.C.: Protocols for Public Key Cryptosystems. In: Proceedings of IEEE Symposium on Security and Privacy (1980)
6. Provos, N.: A virtual honeypot framework. CITI Technical Report 03-1 (2003)
7. Sailer, R., Jaeger, T., Zhang, X., Doorn, L.V.: Attestation-based Policy Enforcement for Remote Access
8. Ateniese, G., Burns, R., Curtmola, R., Herring, J., Khan, O., Kissner, L., Peterson, Z., Song, D.: Remote Data Checking Using Provable Data Possession. ACM Transactions on Information and System Security 14(1) (2011)
9. Castaneda, F., Sezer, F.C., Xu, J.: WORM vs. WORM: Preliminary Study of an Active Counter Attack Mechanism
10. Staniford, S., Paxson, V., Weaver, N.: How to 0wn the internet in your spare time. In: Proceedings of the 11th USENIX Security Symposium (2002)
11. Bellovin, S.M.: Distributed Firewalls. Login (1999)
12. Ferraiolo, F., Kuhn, D.R.: Role based access control. In: 15th National Computer Security Conference (1992)
13. Sandhu, R.S., Coyne, E.J., Feinstein, H.L., Youman, C.E.: Role-based access control models. IEEE Computer 29(2), 38–47 (1996)
14. Ateniese, G., Kamara, S., Katz, J.: Proofs of Storage from Homomorphic Identification Protocols. In: Matsui, M. (ed.) ASIACRYPT 2009. LNCS, vol. 5912, pp. 319–333. Springer, Heidelberg (2009)
15. Curtmola, R., Khan, O., Burns, R., Ateniese, G.: MR-PDP: Multiple-Replica Provable Data Possession. In: Proceedings of IEEE International Conference on Distributed Computing Systems, pp. 411–420 (2008)

A Security Framework for Virtual Resource Management in Horizontal IaaS Federation

Anant V. Nimkar and Soumya Kanti Ghosh

Indian Institute of Technology
Kharagpur, India
anantn@sit.iitkgp.ernet.in, skg@iitkgp.ac.in

Abstract. The horizontal IaaS federation is an emerging concept where the virtual resources are delivered to the users from the federated entities. There are a few challenges to manage the virtual resources securely in federated environment as follows. First, the virtual resources are spread over the federating entities. Second, the federation is a special case of distributed system where the resources are collectively owned by the federating participants. This paper proposes a security framework for the management of federated virtual resources by splitting all the security related modules into two parts: local and global sub-modules.

Keywords: IaaS, Federation, Security Framework, Cloud Computing.

1 Introduction

In horizontal IaaS federation [1], the service providers (SePs) lend transparently the virtual resources through the federation of one or more infrastructure provider (InP) using the brokered architecture. The federation gains the economies of scale due to the elastic availability of virtual resources. It also overcomes the shortage of virtual resources in the SePs. With such advantages, there are some issues like resource management among federating InPs which needs to be addressed. Various frameworks have been proposed to address the problem of resource management among federating InPs.

Some current frameworks [2–4] for the resource management in the federation concentrate on interoperability among federating InPs through agents and ontology. The other work like *Reference Architecture for Semantically Interoperable Clouds* (RASIC) [5] gives semantic-based framework for the information dissemination of entities, services and resources. The IBM's Reservoir Model [6] and a framework by Singhal *et al.* [7] pointed out a few security issues, but their frameworks do not address the security issues.

In any open or closed system, the resources are managed by a cohesive system consisting of the *Resource Access Control* (RAC) mechanism through their companion *Identity Management* (IdM) and *Authentication System* (AS). But, the management of virtual resources is treated differently in the federation because the virtual resources are distributed over the federated InPs. The resources are also cooperatively owned by federating participants, e.g. a virtual machine is

M.K. Kundu et al. (eds.), *Advanced Computing, Networking and Informatics - Volume 2*, Smart Innovation, Systems and Technologies 28,
DOI: 10.1007/978-3-319-07350-7_27, © Springer International Publishing Switzerland 2014

cooperatively owned by both SeP and InP. The resource administration in the federated environment can be handled by dividing all the three components—RAC, IdM and AS—into the local and cooperative administration. The local component of each module assists the global federated module to perform its function.

In this paper, we first propose a security framework by dividing all three components into local and global federated modules. Each component of the security framework works together to securely administer the virtual resources in the federation. The current literatures are also investigated to fulfil the requirements of component division of the proposed security framework.

The rest of the paper is organized as follows. The related work and their limitations are presented in Section 2. The brief introduction of horizontal IaaS federation is presented in Section 3. The security framework for the management of virtual resources is proposed in Section 4. The concluding remarks for the presented work are given in Section 5.

2 Related Work

SWIFT [8] is an identity management framework which facilitates the virtual identities of the subjects as a subset of federating participants. But SWIFT identity management requires user intervention to initiate the authentication process. iMark [9] is an abstract identity management for the virtual resources based on three main principles of separation of identity and location, local autonomy, and global identifier space but it does not provide the identities of subjects.

The existing literature on federated access control models provide resource administration between only two entities using the concept of *role mapping*, *inter-domain access rights*, *inter-domain operations* and *attribute mapping* in standard RBAC [10] and/or ABAC [11] model(s). The role mapping allows the role to be mapped from one domain to another domain respectively. The inter-domain operations and access rights are operations and access rights between two entities. The attribute mapping allows the attribute of the roles from one domain to another domain.

FPM-RBAC [12] extends RBAC through inter-domain operations, access rights, role mapping and attribute mapping by defining inter-domain policy. Yong H. [13] extends RBAC through role mapping and trust between the federating entities. Alcaraz Calero *et al.* [14] propose an authorization model for cloud computing by extending RBAC and hierarchical subject grouping in which the subjects are hierarchical tuples. MTAS and AMTAS [15] extend RBAC through role mapping and trust assertions with a decentralized authority to build trust relation among collaborative tenants. MD_ABAC model [16] extends ABAC and RBAC using attribute mapping and; role and policy mapping respectively. Lianzhong and Peng [17] extend RBAC through role mapping across the domains function. All the above mentioned RAC models give the administration of resources between two entities and they fail for more than two entities.

But the recent FACM [18] fulfils the *many-to-many requirement* (i.e. the federation among two or more InPs) of resource administration by dividing into local and global components.

3 Horizontal IaaS Federation

Typically, the cloud provider supplies virtual resources to its users. But the cloud provider may not be able to supply virtual resources due to the saturation of virtual resources or unavailability of homogeneous virtual resources. This situation can be handled by the federation of cloud providers.

The horizontal IaaS federation supplies the virtual resources (e.g. virtual machines, virtual nodes and virtual links) through the federation of a set of infrastructure providers (InPs) to the users of service providers (SePs). The SePs and InPs are entities in the federation. Thus, the federation ecosystem has two types of participants/roles: i) Entities e.g. SePs or InPs and ii) Users. The federation facilitates the user-level management of virtual resources and network topology through the process of transparent federation instantiation for the users. So the users can get unbounded and inexpensive virtual resources through the competitive business market of transparent federation.

Fig. 1 shows an example of federation ecosystem. It consists of two SePs, three InPs and three users. Each IaaS cloud provider can be SeP, InP or both. The three entities are {SeP#1, InP#1}, {InP#2} and {SeP#3, InP#3} in the federation ecosystem. The bottom long-dashed rectangles are the physical infrastructures (PIn) of three InPs. SeP#1 has a user User#1. SeP#3 has two users User#2 and User#3. There are seven network segments from three InPs. Each SeP establishes a federation among InPs for its users, e.g. SeP#3 establishes a federation among InP#1, InP#2 and InP#3 for the user User#3. The users User#1, User#2 and User#3 create the network topologies $\{VN_1, VN_6\}$, $\{VN_3, VN_4, VN_7\}$ and $\{VN_2, VN_5\}$ respectively.

The maintenance of the virtual resources is performed jointly by the subset of federating participants. The resources are also owned jointly by the subset of federating participants. To explore the different challenges in the resource management, consider a federation among InP#1 and InP#3 for User#1 of SeP#1. The virtual machines are created by their respective InPs upon receiving the request of creation from SePs. So, each virtual resource is collectively owned and managed by the SePs and a set of InPs, e.g. the virtual machine $12'$ is collectively owned and managed by SeP#1 and InP#3. Therefore, the subject of the object $12'$ (i.e. virtual machine) is {SeP#1,InP#3}. Similarly, the virtual link $(4', 9')$ is collectively owned and managed by SeP#1, InP#1 and InP#3. Therefore, the subject of the object $(4', 9')$ is {SeP#1, InP#1, InP#3}.

Similarly, the mechanisms to identify and authenticate the subjects must also be cooperatively performed by the subset of federating roles. As an example, the authentication of the subject like {SeP#1, InP#1, InP#3} is a union of the identities from three federating participants SeP#1, InP#1 and InP#3. The authentication of the subject {SeP#1, InP#1, InP#3} must also be done

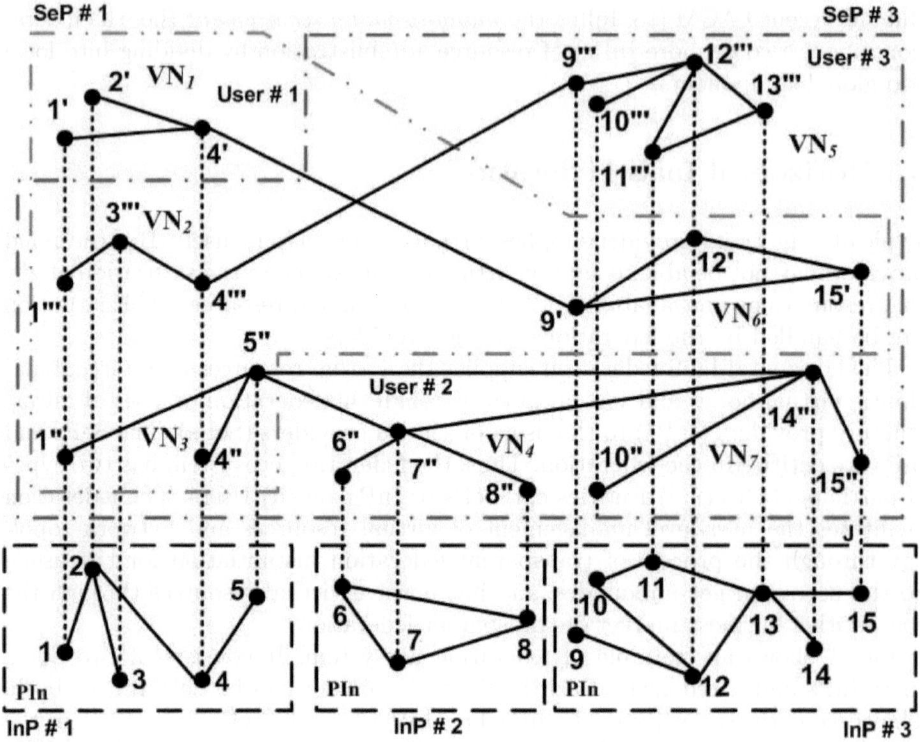

Fig. 1. Horizontal IaaS federation example

by three federating entities SeP#1, InP#1 and InP#3. In short, all the three functions of resource management need to be cooperatively performed by the entities and users.

4 Security Framework

The security framework consists of three participants namely i) Users, ii) Cloud modules and a Broker as shown in Fig. 2. The term cloud module is used to generalize the cloud provider to be a SeP, InP or both. The cloud module further consists of six sub-modules: i) *Virtual Infrastructure* (VIn), ii) *Physical Infrastructure* (PIn), iii) *Cloud Controller* (CC), iv) local and federated *Resource Access Control* (RAC), v) local and federated *Identity Management* (IdM) and vi) local *Authentication Agents* (AA) and federated *Authentication System* (AS). The former three components provide the mechanism required for the cloud computing. The three later components out of six provide secured virtual resource management in the horizontal IaaS federation. The users have only local *Authentication Agents* (AA) to participate in any authentication process. The Broker facilitates the three well-defined functions: advertising, discovery and matching of virtual resources required for the federation.

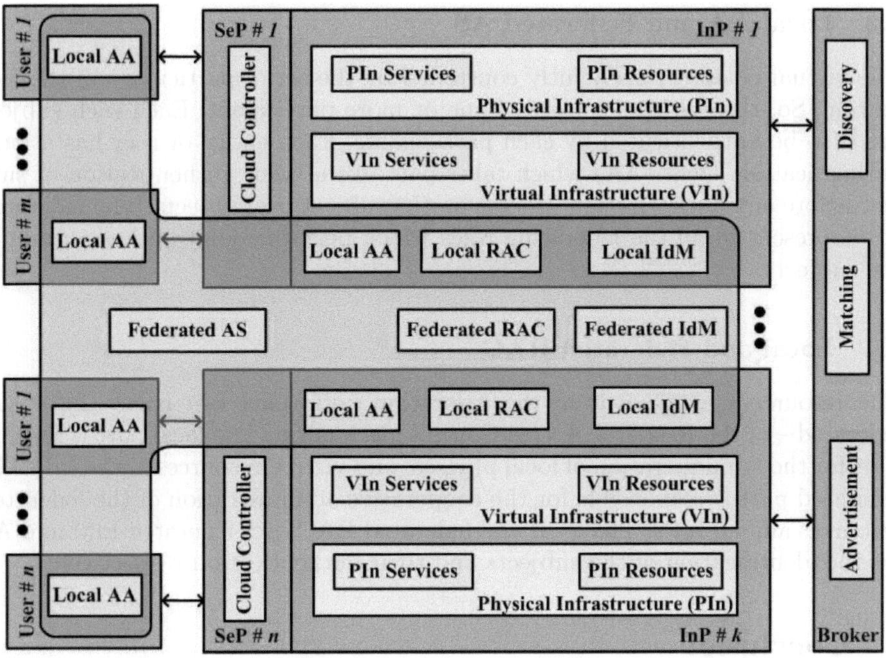

Fig. 2. Security Framework for Virtual Resource Management

4.1 Cloud Computing Components

The PIn consists of i) PIn resources and ii) PIn services. The PIn resources include the physical resources like physical machines, links, switches, routers etc. The PIn services include the services of managing physical resources. The VIn consists of i) VIn resources and ii) VIn services. The VIn resources are virtual resources like virtual machines, virtual links, virtual switch, virtual routers etc. The VIn Services include the creation of virtual machines, virtual node and virtual links etc. The cloud controller performs a function of a standard interface to the cloud users.

4.2 Local and Federated IdM

The two parts of the identity management module are responsible for the management of the subjects' identities in the local and federated environment. The *local* identity management is significant for local namespace while the *federated* identity management is responsible for federated identities of subjects. The federated identities are dynamically created by the federated InPs and third-party trusted entity like broker after the successful federation among the InPs. The federated IdM uses federated AS to authenticate such subjects.

4.3 Local AA and Federated AS

The virtual resources are jointly controlled by the set of participant in the federation. So, the subjects consist of one or more participant. Each such subject needs to be authenticated by each participants. Each entity or user has a local authentication agent (AA) which takes part in the joint authentication of subject before any authorization process on the subject over objects. The federated AS is present in all the federating roles which performs joint authentication of the subjects.

4.4 Local and Federated RAC

The resource management in the federation performed two parts—local and federated—of the Resource Access Control mechanism. The local part is responsible for the administration of local physical and virtual resources in the InP. The federated part is responsible for the cooperative administration of the federated resources among InPs. The local and federated RAC uses federated IdM and AS for the identification of the subjects and their authentication respectively.

5 Conclusion

The horizontal IaaS federation is an emerging research area which provides virtual resources through the federation of IaaS cloud providers to the customers. The federation results in a few securities related issues like cooperative management of virtual resources among the federating participants. The security framework has been proposed to address the cooperative management of virtual resources in the federation. The existing solutions for the components in the proposed security framework partially fulfil the requirements of cooperative management of virtual resources in the federation.

References

1. Nimkar, A.V., Ghosh, S.K.: Towards full network virtualization in horizontal iaas federation: Security issues. Journal of Cloud Computing: Advances, Systems and Applications, SpringerOpen 2(19), 19:1–19:13 (2013)
2. Khatua, S., Mukherjee, N., Chaki, N.: A new agent based security framework for collaborative cloud environment. In: 2011 Proceedings of the Seventh Annual Workshop on Cyber Security and Information Intelligence Research, CSIIRW 2011, pp. 76:1–76:1. ACM, New York (2011)
3. Manno, G., Smari, W., Spalazzi, L.: Fcfa: A semantic-based federated cloud framework architecture. In: 2012 International Conference on High Performance Computing and Simulation (HPCS), pp. 42–52 (2012)
4. Loutas, N., Peristeras, V., Bouras, T., Kamateri, E., Zeginis, D., Tarabanis, K.: Towards a reference architecture for semantically interoperable clouds. In: 2010 IEEE Second International Conference on Cloud Computing Technology and Science (CloudCom), pp. 143–150 (2010)

5. Famaey, J., Latre, S., Strassner, J., De Turck, F.: A hierarchical context dissemination framework for managing federated clouds. Journal of Communications and Networks 13(6), 567–582 (2011)
6. Rochwerger, B., Breitgand, D., Levy, E., Galis, A., Nagin, K., Llorente, I., Montero, R., Wolfsthal, Y., Elmroth, E., Caceres, J., Ben-Yehuda, M., Emmerich, W., Galan, F.: The reservoir model and architecture for open federated cloud computing. IBM Journal of Research and Development 53(4), 4:1–4:11 (2009)
7. Singhal, M., Chandrasekhar, S., Ge, T., Sandhu, R., Krishnan, R., Ahn, G.J., Bertino, E.: Collaboration in multicloud computing environments: Framework and security issues. Computer 46(2), 76–84 (2013)
8. Loopez, G., Canovas, O., Goodell, G., Mez-Skarmeta, A., Girao, J.: A swift take on identity management. Computer 42(5), 58–65 (2009)
9. Chowdhury, N., Zaheer, F.E., Boutaba, R.: imark: An identity management framework for network virtualization environment. In: 2009 IFIP/IEEE International Symposium on Integrated Network Management, pp. 335–342 (2009)
10. Ferraiolo, D., Kuhn, D.: Role-based access control. In: 1992 5th National Conference on Computer Security, pp. 554–563 (1992)
11. Shen, H., Hong, F.: An attribute-based access control model for web services. In: 2006 Seventh International Conference on Parallel and Distributed Computing, Applications and Technologies (PDCAT 2006), pp. 74–79 (2006)
12. Unal, D., Caglayan, M.: A formal role-based access control model for security policies in multi-domain mobile networks. Computer Networks 57(1), 330–350 (2013)
13. Yong, H.: Reputation and role based access control model for multi-domain environments. In: 2010 International Symposium on Intelligence Information Processing and Trusted Computing (IPTC), pp. 597–600 (2010)
14. Alcaraz Calero, J., Edwards, N., Kirschnick, J., Wilcock, L., Wray, M.: Toward a multi-tenancy authorization system for cloud services. IEEE Security and Privacy 8(6), 48–55 (2010)
15. Tang, B., Li, Q., Sandhu, R.: A multi-tenant rbac model for collaborative cloud services. In: 2013 Eleventh Annual International Conference on Privacy, Security and Trust (PST), pp. 229–238 (2013)
16. Wang, L., Wang, B.: Attribute-based access control model for web services in multi-domain environment. In: 2010 International Conference on Management and Service Science (MASS), pp. 1–4 (2010)
17. Lianzhong, L., Peng, L.: A trusted role-based access control model for dymanic collaboration in a federated environment. In: 2008 6th IEEE International Conference on Industrial Informatics (INDIN), pp. 203–208 (2008)
18. Nimkar, A.V., Ghosh, S.K.: A theoretical study on access control model in federated systems. In: Martínez Pérez, G., Thampi, S.M., Ko, R., Shu, L. (eds.) SNDS 2014. CCIS, vol. 420, pp. 310–321. Springer, Heidelberg (2014)

Privacy Preservation of Time Series Data Using Discrete Wavelet Transforms

Sarat Kumar Chettri[1] and Bogeshwar Borah[2]

[1] Department of Computer Science Engineering and Information Technology, Assam Don Bosco University, Guwahati, India
[2] Department of Computer Science and Engineering, Tezpur University, Tezpur, India
sarat.chettri@dbuniversity.ac.in, bgb@tezu.ernet.in

Abstract. With the advent of latest data mining techniques, preserving the privacy of individual's data became a persistent issue. Every day tremendous amount of data is being generated electronically with increasing concern of data privacy. Such data when gets disseminated among various data analysts, the privacy of individuals may be breached, as the released information may be personal and sensitive in nature. Irrespective of the type of data whether numerical, categorical, mixed, time series etc, accurate analyses of such data with privacy preservation is a pervasive task. And due to the complex nature of time series data, analyzing such kind of data without harming its privacy is an open and challenging issue. In this paper we have addressed the issue of analyzing records with preserved privacy, and the data under consideration are expressed in terms of numerical time series of equal length. We have developed a data perturbation method with wavelet representation of time series data. Our experimental results show that the proposed method is effective in preserving the trade-off between data utility and privacy of time series.

Keywords: Time series, Discrete Wavelet Transform, data privacy, data utility.

1 Introduction

Advancement of information and computer technologies has enabled different publicand private organizations to easily collect, store and manage large volume of data electronically. There is a huge paradigm shift in a way the information are collected, shared and analyzed for strategic planning and decision making process in various areas of medical, marketing, official statistics and so on. When data gets disseminated amongdifferent users for analysis, protecting individual's privacy becomes an important issue, as information related to them may be personal and sensitive in nature. Such data if revealed may raise various social and ethical issues. As shown in [1] there exist several kinds of disclosure risks to the privacy of the individuals. In addition, many countries have strict legislation against such privacy breaches.

In this paper we have addressed the issue of analyzing the time series data by preserving its privacy. A time series data x can be defined by pairs as $(x_i,t_i)$for $i = 1 \ldots n$where t_i corresponds to the temporal variable and x_i is the variable that

DOI: 10.1007/978-3-319-07350-7_28,

depends on time t_i or we can say that it is a dependent variable. Such data have a great importance in many areas as medical science, finance, computer network, speech signalsand so on. In time series data mining various tasks like pattern classification, rule extraction, clustering of time series data and other well known techniques exist in theliterature [10,15,16,17]. To compute the similarity between any two time series data, several distance functions exist in the literature. To name a few are Short Time Series(*STS*) distance, cross correlation matrix, Dynamic Time Wrapping (*DTW*) distance, Kulback-Liebler distance, probability based functions and so on. Here in this paper we have experimented with Euclidean distance. As the time series data features different characteristics, namely amplitude, peak and trough, periodicity, trend and so on, each of these characteristics becomes confidential and which a data provider may need to protect.

The problem is how to extract relevant knowledge from large amount of data at the same time protecting sensitive information existing in the database. The simplest way is to perturb the data before it is disseminated to data analysts. But simply perturbing the data may not serve the purpose as it is a challenging task to have a better trade-off between data utility and disclosure risk. Simple random perturbation of data by adding or multiplying noise to the data is not enough for its protection. As it can be seen in [2] that even from perturbed data, its distribution can still fairly be reconstructed-from it. In case of time series data simple perturbation technique is not capable of dimensionality reduction of the data while random projection is not capable of preserving the relative order of distances among time series. Thus the need arises of developing a method which can not only reduce the dimensionality of data but at thesame time achieve a better trade-off between data utility and data privacy by preserving the relative distance order among data and data disclosure risk. Again if we observe that in reality, time domain signals are in time-amplitude representation and if we convert the signal with Fourier Transform, we obtain the frequency-amplitude representation. Fourier transformed data may be useful for stationary signals but for non-stationary signals, we need to have wavelet transformation of signals. Almost all biomedical signals like EEG, ECG FMG etc are non-stationary signals with frequency-time representation. In this paper we have made an attempt to work with discrete wavelet transformed time series to meet our requirements.

The rest of the paper is organized as follows. Section 2 gives some related concepts and other preliminaries in conjunction with time series data and proposed method in particular. In Section 3, we present the proposed method related to time series data protection. In Section 4, experimental data and results are presented and the effectiveness of the proposed method is assessed. Finally, in Section 5 conclusions with future work are drawn.

2 Related Concepts

2.1 Distance Order and Data Uncertainty

We have made an assumption that to maintain a better mining accuracy of data after-perturbation, the distance order among data needs to be preserved [4,7]. Distance orders represent the relative orders among distances between time series data.

Normally it is difficult to achieve both privacy and data utility at the same time. In the proposed model, data utility of the perturbed time series data is measured based on distance order.

Definition 1. *Let x, y and v be the time series data of length l, and with respect to it let x′, y′ and v′ be the corresponding perturbed data. Now we have made an assumption that the relative distances among x, y and v are preserved only if the following implications holds.*

$$dist(x,y) \leq dist(x,v) => dist(x,y) \leq dist(x,v) \quad (1)$$

$$dist(x,y) \geq dist(x,v) => dist(x,y) \geq dist(x,v) \quad (2)$$

where *dist* is the Euclidean distance.

We have also made an assumption that to preserve the privacy of data, uncertainty of the data needs to be preserved. As can be seen in [14] data uncertainty between two time series is defined as the standard deviation of differences of their correspond ingentries.

Definition 2. *Let x be a time series of length l, x′ be the perturbed version of x. Now let us assume that x′ is released and x^r be the recovered time series from x′ by adversaries. Uncertainty between x and x′ is defined as in 3 while uncertainty between x and x^r is defined as in 4*

$$u(x,x) = dist(x,x) \quad (3)$$

$$u(x,x^r) = dist(x,x^r) \quad (4)$$

Eq. 3 gives the uncertainty as the amount of noise being added to the originaldata for perturbation. Equation 4 gives an idea about the amount of noise remained inthe data after reconstruction from the modified data by adversaries. Now smaller the difference between *u(x,x′)* and *u(x,x^r)*, the better the privacy preservation is, that is$|u(x,x) - u(x,x^r)|$, should be minimal [14].

In our experiment the recovery of time series x^r from the perturbed time series x′ is done by applying the Haar wavelet filter to the perturbed time series. Inverse DWT is used to get x^r from x′ after zero out the approximation coefficients where the concentration of noise is more than detail coefficients of the signal. A detailed study on denoising method can be seen in [5].

2.2 Discrete Wavelet Transformation of Time Series

Discrete Wavelet Transformation (DWT) of time series is done generally to reduce the dimensionality of time series data. The dimensionality reduction is done by passing a signal through filters (high-pass and low-pass), breaking the signal into detail and approximation coefficients. Let a signal x be passed through the filters h(x) and l(x), where h(x) is the high pass filter and l(x) is the low pass filter. Now once x passes through h(x) and l(x), x is transformed to detail coefficients d1, d2 and

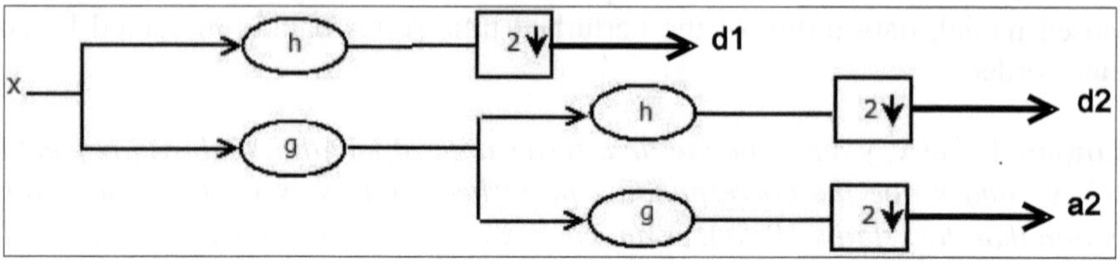

Fig. 1. A Two Level Decomposition of Signal

Approximation coefficient a2 as shown in Fig. 1. Where d1, d2 are the first and second level detail coefficients and a2 is the second level approximation coefficient. This can be repeated as long as the desired level is not obtained where at each level the signal is decomposed by a factor of 2 by the half band filters. The original signal can be reconstructed from the approximation and detail coefficients at every level. It can be done by up-sampling by 2 at every level while passing though the filters and adding them.

There exist a number of wavelets namely Haar, Daubechies wavelets, Symlets, Coiflets etc with different basis functions for DWT. The basis function characterizes-the behaviour of DWT, and in this paper we have used Haar wavelet with single level decomposition, whose basis function $\psi(x)$ for signal x can be described as in equation5. A detail work on DWT based time series data mining can be found in [3].

$$\psi(x) = \begin{cases} 1 & 0 \le x < 1 \\ -1 & \frac{1}{2} \le x < 1 \\ 0 & Otherwise \end{cases} \tag{5}$$

2.3 Perturbation of Time Series Data

There exist related methods of time series perturbation in the literature as we can see in the works [11,12,14,18]. For the first time Mukherjee et al. [11] have used Discrete Fourier Transformations of time series for its privacy preservation. Later it was pointed out by Kim et. al. [6] that the method developed by Mukherjee et. al. is not efficient in preserving the privacy of time series as the original signal can be reconstructed by inverse DFT. Papadimitriou et. al. [14] came with a new concept of DFT and DWT based time series privacy preserving techniques. From their experimental results they came up with the conclusion that DWT based solution are more efficient than Fourier related solution in preserving the uncertainty of time series, which indicates better privacy preservation of time series.

In this paper we have focused in two much related methods namely *RAND* and *WAVE*[9]. *RAND* basically distorts the time series by randomization without transforming the signal into wavelets. The random numbers are generated based on the Gaussian distribution of mean 0 and standard deviation σ. To preserve the distance order amongtime-series, the noise averaging effect of piecewise aggregate approximation (*PAA*) is used. The whole idea is to generate white noise which eventually converges to 0 as the mean is 0. The *RAND* method though it is capable to preserve the distance order but is not much efficient in privacy preservation. Thus the authors proposed another method named *WAVE*, which again perturbs time series, but the perturbation is done after wavelet transformation of time series. The method generates wavelet based noise to perturb the time series. It still uses *PAA* distance to compare distance orders among time series. Though *WAVE* has been found to be efficient in privacy preservation by overcoming the wavelet filtering attack but distance orders among time series are not preserved. Our proposed method named as *PRAWN* (*Piece-wise RAndomization using Wavelet based Noise*) is designed keeping *RAND* and *WAVE* as baseline methods.

3 Proposed Method

In the proposed method *PRAWN* (*Piece-wise RAndomization using Wavelet based Noise*)as given in algorithm 2, *n* time series $X = \{x_1, x_2, \ldots. x_n\}$ of length *l* are decomposed into wavelets using DWT. The decomposition is done into single level to get the approximation wavelet coefficients $X^w = \{x_1^w, x_2^w, \ldots. x_n^w\}$. Once we have obtained$X^w$ it is thendivided into several *p* pieces of length m/p, where *p* is the input parameter and $m = l/2$. Instep 3 wavelets are grouped into groups consisting of *k* or (2*k* - 1) time-series wavelets in each group using *P-G* (*Piece-wise Grouping*) method. As can be seen in algorithm1, the groups are formed individually in every piece. The mean μ_i is found by computing the mean of the wavelets in each piece p_i. To find the similarity among wavelets, the Euclidean distance measure is used. Once the groups are formed consisting of *k* and(2*k*-1) wavelets, *k* being the input parameter, noise piece N_j^w for each group $g_{ij}$is form edusing GaussRand (0, σ), with mean 0 and σ is the standard deviation of the wavelets in group g_{ij}as shown in step 6 of algorithm 2. The noise pieces N_j^w are added with the corresponding X_j^w, where X_j^w is the *w*-th coefficient of *j*-th group. Steps 5 to 7 are repeated as long as noises are not formed for all the individual groups in *p* pieces. In step 9 noise piecesN_q^w are concatenated to form noise series N_q. Once we get noise series N_q we get the distorted series $X^d = \{x_1^d, x_2^d, \ldots. x_n^d\}$through inverse DWT using N_q as approximation coefficients. The graphical representation of the proposed method is shown in Fig. 2.

Algorithm 1. *P-G (Piece-wise Grouping)*

---

1. Compute the mean μ_i of the wavelets in each piece p, where i = 1, 2,….p.
2. Piece-wise sort all the wavelets in ascending order with respect to its distance to μ_i
3. Piece-wise group the nearest (*k*-1) wavelets with the first wavelet and simultaneously with the (*k*-1) wavelets with the last wavelet.
4. Repeat step 2 as long as there exist at least 2*k* wavelets.
5. Form a separate group with the remaining wavelets.

Algorithm 2. *PRAWN (Piece-wise RAndomization using Wavelet based Noise)*

1. Using DWT, get the approximation wavelet coefficients X^w of the n original time series data X of length l.
2. Divide the approximation wavelet coefficients X^w into p pieces of lengthm/p.
3. Form piece-wise groups g_j consisting of k or (2k-1) wavelets in each group using *P-G* method.
4. For i= 1 to p do
5. Let g_{ij} be the j-th group of the i-th piece of wavelets.
6. Get noise piece N_j^w for group g_{ij}as GaussRand(0, σ), with mean 0 and σ is the standard deviation of the wavelets in group g_{ij}.
7. Add the noise pieces N_j^w with the corresponding X_j^w, as $N_q^w = N_j^w + X_j^w$, whereX_j^w is the w-th coefficient of$j$-th group.
8. End For
9. Concatenate all the noise pieces N_q^w to form a noise series N_q.
10. Get the distorted time series X^d through inverse DWT using N_q as approximation coefficients.

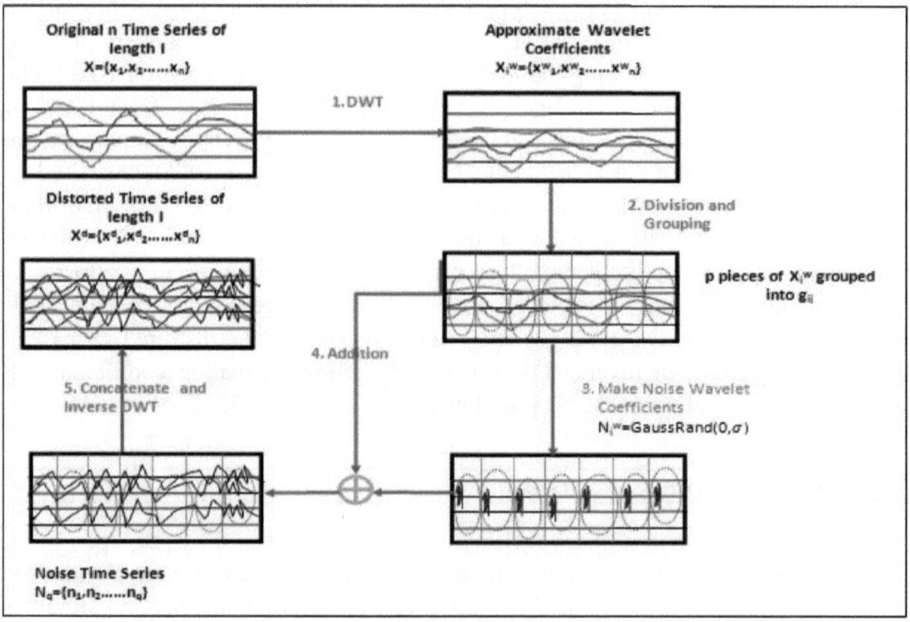

Fig. 2. A Graphical Representation of *PRAWN* Method

4 Experimental Data and Results

We have implemented the *PRAWN* algorithm using C under Linux environment on a machine with 2.13 GHz Intel i3 Processor and 3 GB RAM. For our experimental purpose we have collected three different variants of time series datasets. We have applied our algorithm to these data, testing with different values of *k*.

The time series datasets we have considered for our experiments are"ECG200","CBF" and "Stock Market Data". The "Stock Market Data" consists of daily opening prices of 35 companies listed under National Stock Exchange (NSE), India. The historical information of the opening prices of different companies of about 1 year has been obtained from (NSE, India) [13]. The datasets "ECG200" and "CBF" are from UCRdata [8], where "ECG200" has 600 time series of length 319 and "CBF" dataset has143 time series of length 60. The primary reasons for considering "ECG200", "CBF" and "Stock Market Data" are that firstly they are easily and publicly available and secondly the data under consideration corresponds to time series data which meets our requirements.

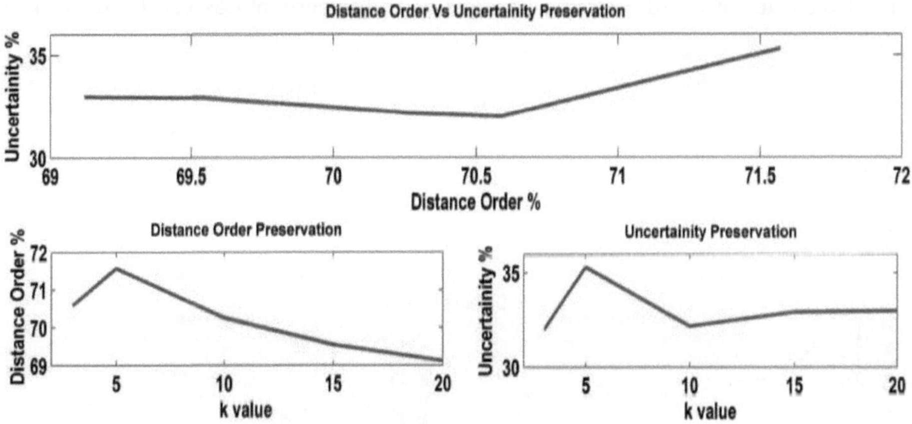

Fig. 3. Experimental Results on Stock Market Dataset

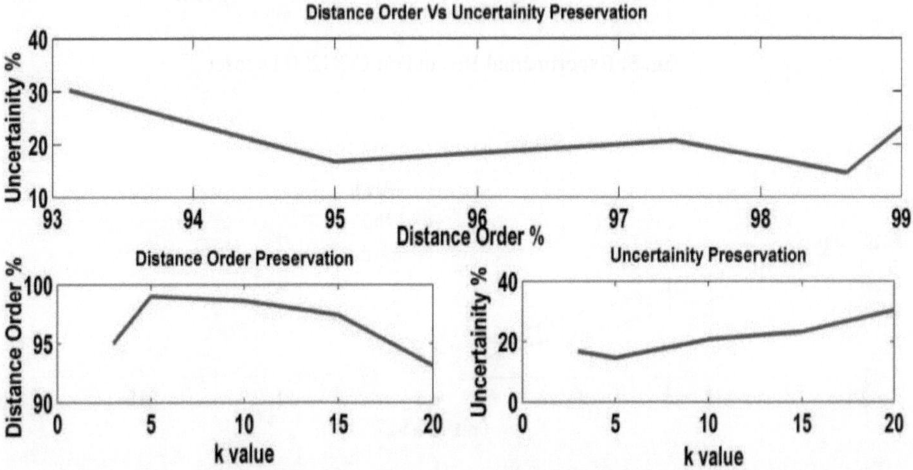

Fig. 4. Experimental Results on CBF Dataset

The experimental results are shown in the Fig. 3, 4, 5 and 6. The figures providea view of the performances of *PRAWN* in “Stock Market Data", “CBF” and “ECG200”respectively. The performances are computed as how the distance orders and uncertainties are preserved among time series in various datasets with different values of *k*.

The distance orders and uncertainties are calculated as described in Section 2.1. Comparisons are also shown by measuring the trade-off between distance order and uncertainty. To preserve the distance orders among time series, the series is divided into individual pieces and perturbation is done group wise individually in each piece. When compared with *RAND* which preserves better distance order and *WAVE* which claims to achieve better uncertainty, our proposed method *PRAWN* goes on par with both the methods as can be seen in Fig. 6. At some instances of “CBF” and “ECG200” datasets *PRAWN* outperforms *RAND* and *WAVE* by having a better trade-off between distance order among time series and uncertainty as can be seen in the Fig. 6.

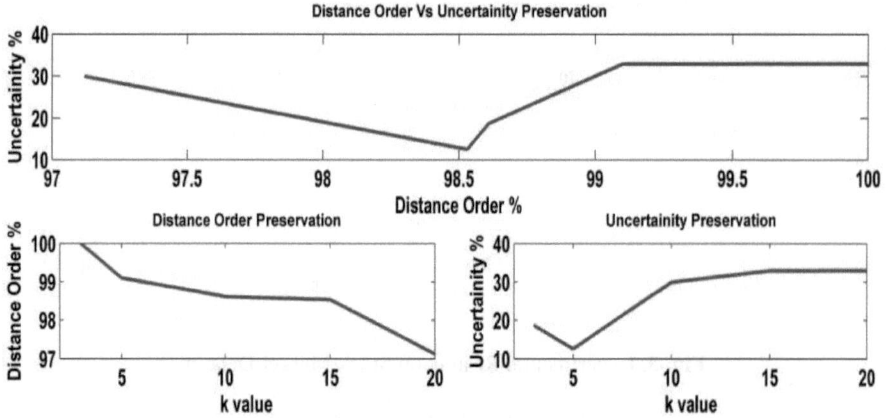

Fig. 5. Experimental Results on ECG200 Dataset

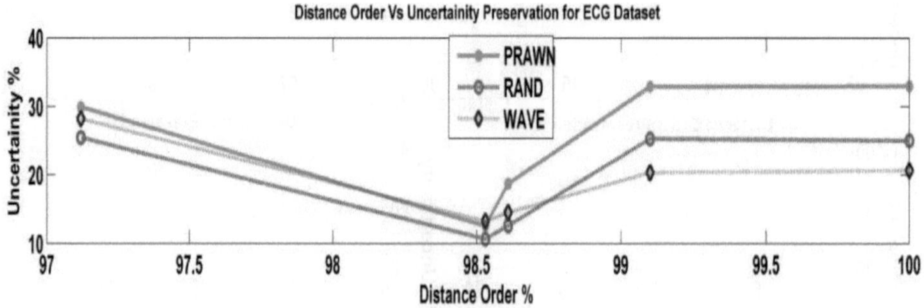

Fig. 6. Comparison of *PRAWN*, *RAND* and *WAVE* Methods

5 Conclusion

In this paper we have proposed a new method *PRAWN* which can be used to preserve the privacy of time series data of equal length with better mining accuracy. The proposed method works with wavelet transformed time series to preserve its privacy while analyzing the data. The method is flexible, by having the adjustable parameters k and p, as these two parameters can be adjusted accordingly to achieve a better trade-off between distance order and uncertainty among time series data.

The performance of the proposed method is analyzed with the publicly available standard time series datasets. The experimental results proves our claim by showing it effectiveness based on the parameters of distance order and uncertainty, where distance order indicates data utility and uncertainty indicates data privacy. The method is also compared with other existing methods based on the trade-off between distance order and uncertainty. Comparison with other methods proves that the proposed method stands on par with those methods and under some instances it performs better. For future work the method may be extended to preserve the privacy of time series of unequal length. Further research can also be carried on determining the optimal values of k and p.

References

1. Aggarwal, C.C., Pei, J., Zhang, B.: On Privacy Preservation Against Adversarial Data Mining. In: 12th ACM SIGKDD International Conference on Knowledge Discovery and Data Mining, pp. 510–516 (2006)
2. Agrawal, R., Srikant, R.: Privacy-preserving Data Mining. ACM SIGMOD Record 29(2), 439–450 (2000)
3. Chaovalit, P., Gangopadhyay, A., Karabatis, G., Chen, Z.: DiscreteWavelet Transform-Based Time Series Analysis and Mining. ACM Computing Surveys 43(2), 6:1–6:37 (2011)
4. Ciaccia, P., Patella, M., Zezula, P.: M-tree: An Efficient Access Method for Similarity Search in Metric Spaces. In: 23rdInternational Conference on Very Large Data Bases, pp. 426–435. Morgan Kaufmann Pub., Athens (1997)
5. Donoho, D.L.: De-noising by Soft-Thresholding. Information Theory 41(3), 613–627 (1995)
6. Hea-Suk, K., Yang-Sae, M.: Fourier Magnitude-based Privacy-preserving Clustering on Time-Series Data. IEICE Transactions on Information and Systems 93(6), 1648–1651 (2010)
7. Inan, A., Kantarcioglu, M., Bertino, E.: Using Anonymized Data for Classification. In: Proceedings of 25th International Conference on Data Engineering, pp. 429–440 (2009)
8. Keogh, E., Folias, T.: The UCR Time-Series Data Mining Archive. Computer Science & Engineering Department. University of California, Riverside (2002), `http://www.cs.ucr.edu/eamonn/TSDMA/index.html`
9. M-Jung Choi, H.S.K., Moon, Y.S.: Publishing Sensitive Time-Series Data under Preservation of Privacy and Distance Orders. International Journal of Innovative Computing, Information and Control 8(5(B)), 3619–3638 (2012)

10. Möller-Levet, C.S., Klawonn, F., Cho, K.-H., Wolkenhauer, O.: Fuzzy Clustering of Short Time-Series and Unevenly Distributed Sampling Points. In: Berthold, M., Lenz, H.-J., Bradley, E., Kruse, R., Borgelt, C. (eds.) IDA 2003. LNCS, vol. 2810, pp. 330–340. Springer, Heidelberg (2003)
11. Mukherjee, S., Chen, Z., Gangopadhyay, A.: A Privacy-preserving Technique for Euclidean Distance-based Mining Algorithms using Fourier-related Transforms. The VLDB Journal 15(4), 293–315 (2006)
12. Nin, J., Torra, V.: Extending Microaggregation Procedures for Time Series Protection. In: Greco, S., Hata, Y., Hirano, S., Inuiguchi, M., Miyamoto, S., Nguyen, H.S., Słowiński, R. (eds.) RSCTC 2006. LNCS (LNAI), vol. 4259, pp. 899–908. Springer, Heidelberg (2006)
13. NSE: National Stock Exchange of India Limited (nse, india), `http://www.nseindia.com/`
14. Papadimitriou, S., Li, F., Kollios, G., Yu, P.S.: Time Series Compressibility and Privacy. In: 33rd International Conference on Very Large Data Bases, VLDB, pp. 459–470 (2007)
15. Singhal, A., Seborg, D.E.: Clustering Multivariate Time Series Data. Journal of Chemometric 19(8), 427–438 (2005)
16. Wang, X., Smith, K.A., Hyndman, R., Alahakoon, D.: A Scalable Method for Time-Series Clustering. Technical Report, Department of Econometrics and Business Systems, Monash University, Victoria, Australia (2004)
17. Warren Liao, T.: Clustering of Time Series Data - A Survey. Pattern Recognition 38(11), 1857–1874 (2005)
18. Zhu, Y., Fu, Y., Fu, H.: On Privacy in Time Series Data Mining. In: Washio, T., Suzuki, E., Ting, K.M., Inokuchi, A. (eds.) PAKDD 2008. LNCS (LNAI), vol. 5012, pp. 479–493. Springer, Heidelberg (2008)

Exploring Security Theory Approach in BYOD Environment

Deepak Sangroha and Vishal Gupta

Department of Computer Science, AIACT&R, Delhi, India
deepaksangroha@yahoo.in,
vishalgupta@aiactr.ac.in

Abstract. BYOD (Bring Your Own Device) is a business policy to allow employees to bring their own devices at their work. The employee uses the same device in and out of the corporate office and during outside use, it may be connected to insecure internet and critical corporate data become public or when a device is used in an insecure environment, it may get infected by big threats and they may get activated when the device is used in organization's environment and may harm the confidential information. In both cases the organization will be in losing side because if internal data become public it may hurt business strategies and future policies. If internal organization gets infected from outside attack, it will certainly hurt the business in any way. In this paper we are suggesting some approaches which can guard against these types of threat and secure the corporate data.

1 Introduction

Vodafone techies say: "The growing pace of consumerization of IT means that employees are having a greater say in the technology that they use in the workplace, including their own devices."

Means, the fast growing IT world have a vast range of IT means that can provide better availability, accessibility, mobility and also cost saving to industry [2]. Devices like smartphones; tablets etc. give the employee options to access their work tool or stuff anywhere in their organization. This is what BYOD is, bring your own personal devices and access your work utilities anywhere in office premises (i.e. wherever companies' wireless network available) [8].

Here we must understand the difference between accessing the wireless network as a guest and BYOD. In first one, the user use the wireless network to access the internet in general, but in second one, the employees use the wireless network to access their work tools which contain critical information about their organization, network infrastructure etc.

For example, BMC remedy is an IT management tools works with ARS (Action Request System) to perform tasks and troubleshoot Incidents regarding network devices, servers. In this way it contains a lots of sensitive information related to network infrastructure.

M.K. Kundu et al. (eds.), *Advanced Computing, Networking and Informatics - Volume 2*, Smart Innovation, Systems and Technologies 28,
DOI: 10.1007/978-3-319-07350-7_29, © Springer International Publishing Switzerland 2014

In today's world, we have a variety of mobile devices that can be used not only for entertainment but for work also. In that way they not only increase availability & mobility to employee but also cost saving & productivity for organization as the employee bring their own device and be available whenever business needs them. Thus it provides a win-win situation to both employee and the organization [5]. IT means are strong enough to run heavy apps with good processing speed and huge primary storage.

Basic approach used in industries to implement BYOD is to using MDM (Mobile Device Management) tools. MDM tools are used to secure, monitor, manage and support the mobile devices deployed in enterprises. These tools are becoming the basic need for the wireless networks as they do all operations from beginning to end.

For device enrolment these provides features like connection setup, device registration, and user authentication, restrictions based on platform or version. For security these provides passcode, encryption, and compliance, restrictions based on the use of device features or applications. MDM tools are able to configure profiles, time-based profiles, certificates, accounts. For maintainability these can be used to monitor the policies, location, alerts, rules etc [7].

With all these above features, MDM tools are the first thing each organization wants to implement, as these makes it a lot easy to handle BYOD concept. There are lots of MDM tools available in market like AirWatch, AmTel MDM, FancyFon, MobileIron and a lot more. Actually these tools basically focus on maintenance of BYOD devices and have some features of security but good maintenance is the first step for security.

Inspired from MDM a new approach called MAM (Mobile Application Management) has been introduced, which focuses on a particular application rather than whole device [9]. This is much better way for security as the application accessing the corporate data remains in monitoring not the whole device. In this way employee also get some flexibility to use his other application which is not going to interfere with the corporate data as he will not be prompted for authentication each time any other application run.

2 Problems in Implementing BYOD

In BYOD, employee uses the organization's environment to access critical data and at that moment the data is safe as the organization has its own polices implemented on its security devices [6]. As every device have some kind of Operating system which operate on hardware and each OS create some kind of logs, temporary files, history or traces that are stored on the device [9]. So when the employee uses the corporate network to access internal data and some part of that data may be stored in device as logs or in temporary files. After that employee leaves the corporate network and goes home or any other place where he/she may use the "Internet" which is highly insecure in nature and this poses a huge risk. Secondly, when a mobile device is connected to insecure environment or exchange data from unprotected system, it may get infected by any virus, worm, Trojan horse, botnet attack etc.[3]. And now, when this device connects to corporate network, it may cause a huge security breach. In normal scenario, the network is protected with firewall which stays at the entry point of the network as shown in Fig. 1 which monitors and filters all unwanted traffic [4]. In contrast Fig. 2 shows a network topology with mobile devices. In case of BYOD, the device enters in organization premises in a way that it first get connected to the AP/WLC and when the device communicate to extranet, only then the firewall will be able to filter the data [1]. It's like a backdoor entry as the device enters in the network silently and becomes active. So, we have two scenarios in which organization need to protect data and network.

a. Internal data moving outside the organization.
b. Outside threat moving inside the organization's environment.

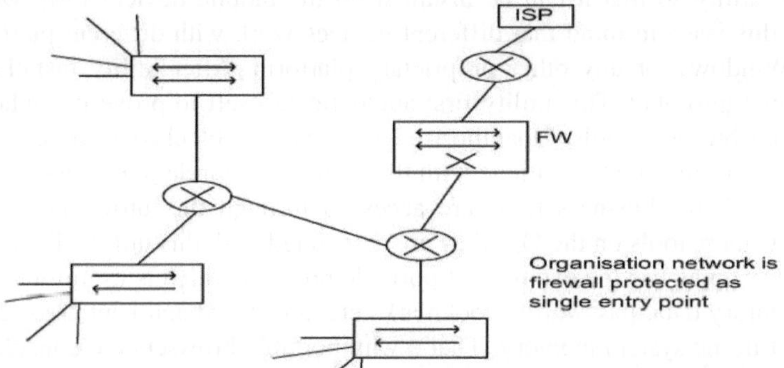

Fig. 1. Normal Network topology, only physically connected devices

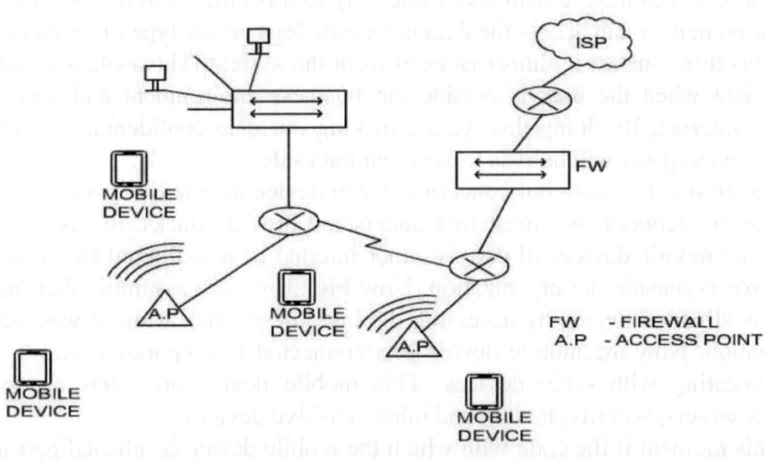

Fig. 2. Network topology with mobile devices

3 BYOD Security Solutions

For scenario a, when there is vulnerability that organization's internal data become public, we can use some utility to secure data in and out of the organization. There must be a single way to access the business tools i.e. all business tools must be accessed through this utility. This utility may have web browser like appearance and all the business tools must be assessed within this utility.

For understanding we take an example of Virtual machine which have characteristic to work like an operating system which itself is running on an operating system platform. We don't want our utility to work like a whole operating system but have characteristic to allow some application or business tool to run on it.

Now here is the complete process to work with this utility, first device must support the utility so that it can be installed on the mobile devices [10]. Developer must keep this issue in mind that different devices work with different platform like Android, Windows, or any other proprietary platform. After utility installation the features come into play. This utility first authenticate itself to prove its authorization to access the business tools. Unauthorized user will be blocked to access business tools but may be allowed to continue with their access to wireless network.

Secondly, all the business tools are accessed through this utility i.e. instead of installing business tools on the OS; they are associated with this utility. To understand the concept we may use the example of portable browser which is useful in a scenario where temporary data, passwords, bookmarks etc. are stored and kept associated with browser not in the system memory. That's why portable browsers are considered as a secure thing to do Net Banking, Online Money Transactions etc. on a non-personal PC. This connection between utility and the business server act like a tunnel which encrypts the data passing through this connection. Another feature of utility is that it won't allow the OS of device to interfere with the data of this utility. In this way the data related to business remain accessible only to this utility and moreover OS or any application neither can access the data nor create logs or any type of temporary file or data. This thing makes it almost isolated from the system. This isolation will help to secure data when the user is outside the business environment and accessing the insecure internet. By doing this we are making our data confidential and in case of exposure encryption will help us to keep our data safe.

For scenario b, where our concern is when device affected with virus or malware enters in the network. We must first understand how it attacks. In first case Fig. 1, there is no mobile device, all data is either internal or it is filtered by firewall when the source is outside the organization. Now Fig. 3, we are assuming that the mobile device is already infected by accessing the insecure internet while it was outside the organization. Now the mobile device gets connected to corporate network and start communicating with other devices. This mobile device now gets access to the business servers, security devices and other sensitive devices.

At this moment if the code with which the mobile device is infected gets activated and start spreading or whatever action it can take can be too much harmful for the organization. Proving few examples, the device may start making peers with other

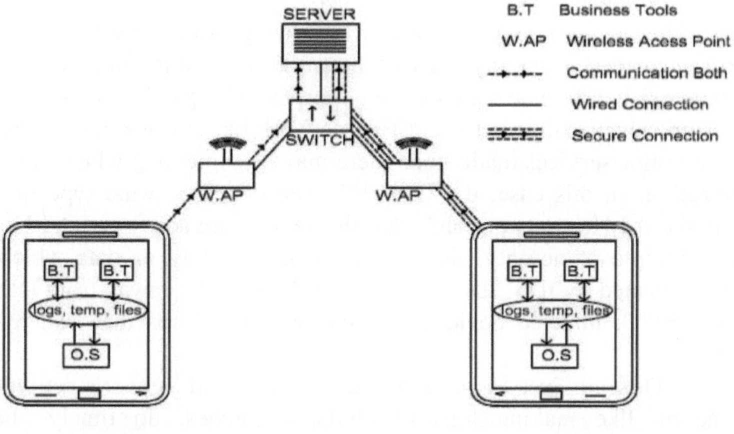

Fig. 3. Showing secure access to business tools with utility in right side device

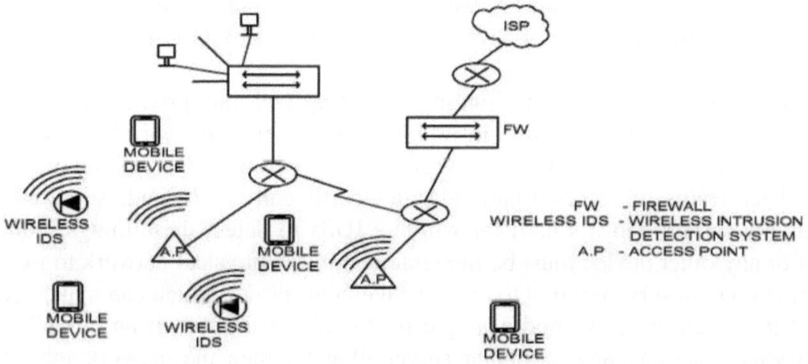

Fig. 4. Wireless network with wireless BYOD

devices anonymously, making them victim and take controls of the devices. These devices may be further used in attacking any particular server or any device to perform DDoS attack. In other example, the infected device may start making connection to unauthorized zone. This may happen when the device is in control of other controller and acts like zombie. Another example is, it can spread some malicious code in the LAN network to make lots of devices inactive or unworkable. In all cases the organization gets harmed and can cause loss of money and precious time to recover from the situation.

Here are the solutions to tackle these problems. **Firstly,** implement Wireless Intrusion Detection Systems in the organization. IDSs are the hardware device sometimes software which sense the data flowing in the network and try to detect any malicious activity or unauthorized data flowing in the network based on the signature, anomaly, or anything abnormal in data headers. Wireless IDS is one of the types of IDS on the basis of implementation.

As shown in Fig. 4, IDSs stay in the network and silently check the activities happening in the network and type of traffic flowing in the network. For example it will keep special eye when any Ad-Hoc network is forming. Normally, an Ad-Hoc network may exists for some social activity purpose like gaming, messaging etc. but if there is some abnormality in the Ad-Hoc network like it keeps increasing or it start accessing the same services madly then there may be something which is not good for the organization. In this case, the IDS will keep checking what type of activity is going on in the Ad-Hoc network and what the devices are accessing. Ad-Hoc network is just an example of activity, there may be any activity or data which is found abnormal is detected by IDS. IDSs are the best thing to stay away from DDOS attack in case of BYOD; infected devices are detected as soon as they can by regularly monitoring them.

Basically, IDSs are just to monitor the network and network activities. It can perform action like making logs of alerts it catches, do timely checking of performance or security parameters and some other things and report these to management devices or to any monitoring server but never take actions against these alerts. IPSs can replace them if we need to take immediate action against the abnormal activities. IPSs are always the better option as this is the fastest option to prevent from any attack.

Secondly, implement activity logging, reporting and data privacy. Wireless IDSs looks for the activities happening in its range i.e. wireless network but it can't keep eye on the things happening in network beyond wireless. For example, a mobile device keep trying to access a particular device illegally or it sends some unwanted data to any server then it's hard for wireless IDSs to detect these things. Some IDS sensor or any other device must be implemented in the physical network to make logs and these logs must be reported to any management service which can make decisions against these activities. A good example for the above topic is, if any mobile device keep trying accessing any particular server illegally then the network management must take action against this. One thing that can be done is that the device is denied to use the network or to be more secure, the whole device can be formatted or whole data can be flushed to remove all information also.

The second solution in above paragraph impact the privacy of the employee also. As to perform whole data flush means that the security management must be having all data information that resides in the employee device. But employee's perspective, the disclosure of the private data of the employee like pictures, videos or any account details etc., which he/she never want to share with anybody, is a problematic thing.

Third, Implement anti-virus tools for the whole organization. Generally, antivirus tools are used in organizations but they are limited to only physical network. We are talking about implementing those things in wireless network or say mobile devices which become part of BYOD technology. This seems a huge thing to implement as a large number of devices arrive in the organization premises but if we think in a way that an employee brings a mobile device at work and he/she needs to install antivirus to be part of BYOD but after that first use its most probably that he/she will bring the same device each day and the management just need to keep updating the antivirus.

The antivirus or anti-spyware or anti-Trojan etc., provides protection against malicious codes, Trojan horses, spywares or any type of worm which may infect the mobile device. By using these, the network will be secure from at least from the known viruses or worms. Using antivirus can be set as the requirement before any mobile device become part of BYOD and start using organizations devices and also this will benefit the employee as he/she will get free anti-virus from the organization which will make his/her own device secure.

Fourth, Implement proper management tools like Mobile Device Management (MDM)[7]. MDM provides the features like checking the basic requirement of the device to connect with server, looks for the connection details and provides basic security also. It also keeps record of the devices that connect to the servers and also capable of applying access policies like which device can access what resources. Mobile Application Management (MAM) is also a helpful application when it comes to manage mobile devices and application they use. MAM monitors, manages and take care of the particular application which is used with BYOD.

4 Conclusion

BYOD technology is about accessing business data which may be highly sensitive. It's ok if we are using it within organization but the problem arises when data moves outside the organization or affected device comes to organization premises.

The idea of using separate utility to access business tools gets a strong favor as big IT powers are moving towards web based tools like BMC Remedy. It is available in new web based version of it named BMC Remedy V3 and Service-Now, Summus, Bomgar are also web based tools for Infrastructure Management and Troubleshooting. These web based tools are easy to associate and customize with the utility.

And for infected device it is recommended to perform detection and disinfection techniques proposed above.

- Use IDSs to detect from the abnormal activities in the network.
- Use activity logging, reporting and data privacy.
- Use antivirus to prevent the devices from any infection that can harm the organization's network.
- Use MDM or MAM to manage mobile devices properly.

References

1. BYOD: On-boarding and Securing Devices in Your Corporate network. Motorola
2. Emery, S.: Factors for consideration when developing a bring your own device (BYOD)
3. Capstone Report at California College of the Arts (2012)
4. Bring Your Own Device. MTI Technologies
5. Anderson, N.: Cisco Bring Your Own Device. Cisco (2012)
6. Kneller, D.: Bring Your Own Device. Madgwicks (2013)
7. Hidden risk of Bring Your Own Device Mobility Model. Profitline
8. Holistic, A.: approach to BYOD security and Management. HP (2012)

9. Smartphones and Tablets in the Enterprise. Enterasys Secure Networks
10. Selecting the right mobile device management solution. Wavelink (2011)
11. Gupta, V., Dhiman, L., Sangroha, D.: An approach to implement Bring Your Own Device (BYOD) Securely. International Journal of Engineering Innovation and Research (2013)

A Framework for Analysing the Security of Chrome Extensions

V. Aravind and M. Sethumadhavan

TIFAC Core in Cyber Security
Amrita Vishwa Vidyapeetham, Coimbatore, India
aravind.venkitaraman@gmail.com, m_sethu@cb.amrita.edu

Abstract. Google Chrome, the most popular web browser today, allows users to extend its functionality by means of extensions available in its own store or any third party website. Users can also develop their own extensions easily and add them to their browser. Vulnerabilities in browser extensions could be exploited by malicious websites to gain access to sensitive user data or to attack another website. A browser extension can also turn malicious and attack a website or steal user data. This paper proposes a framework which can be used by users and developers to analyse Chrome extensions. The technique presented here uses the permissions feature of chrome extensions and the flow of data through the extension's JavaScript code to detect vulnerabilities in extensions and to check whether the extension could be malicious.

1 Introduction

Internet and web applications have become an essential part of our everyday life. We take the help of web applications to carry out a sizable number of our day to day activities.Most of the research activity in web security till now has been concentrating on securing the web applications. Web browsers and browser extensions may also have security vulnerabilities which could be exploited by an attacker.A browser extension is a third-party software module that extends the functionality of a web browser and lets users customize their browsing experience. Browser extensions modify the core browser user experience by changing the browser's user interface and interacting with websites. The two most popular web browsers, Mozilla Firefox and Google Chrome, support a number of extensions.Users can install extensions from the browser's extension module or from third party websites. In order to enhance the user experience the extensions are allowed to access various components like the current web page source, the browser history, bookmarks and if required, the end user's file system as well. Browser extensions are not always developed by security experts and hence are likelyto have vulnerabilities. A malicious website can exploit these vulnerabilities to gain access to the browser data or the file system. On the other hand, a malicious extension can use the same privileges to steal sensitive user data or to attack a website.

The Google Chrome Extension System follows an architecture which helps in reducing the security vulnerabilities to a certain extent. The principle of least

M.K. Kundu et al. (eds.), *Advanced Computing, Networking and Informatics - Volume 2*, Smart Innovation, Systems and Technologies 28,
DOI: 10.1007/978-3-319-07350-7_30, © Springer International Publishing Switzerland 2014

privilege, one of the mechanisms proposed in the architecture has to be enforced by the extension developer. This opens up vulnerability and an attack surface for a malicious website. Users do not always install extensions from the Chrome Web Store. If an attractive extension is available from a third-party website, an ordinary user may add the extension to his browser. If a user installs a malicious extension on his browser, the extension can perform malicious activities. Most of the research on extension security till now has assumed the extensions to be benign and the websites to be malicious. However with users installing a lot of third party extensions, a new class of malware is available to the not so benign developers. Hence, both cases need to be addressed while analysing the security of a browser extension. The remainder of the paper is structured as follows. Section 2 gives a glance of the Chrome Extension architecture and the previous research done in the field of browser extensions. Section 3 describes the framework being proposed and the expected results and section 4 concludes the paper.

2 Browser Extension Security: The Story So Far

Security of browser extensions has become an interesting research area for computer scientists. One of the most popular web browsers, Mozilla Firefox has an extension architecture which acts as a platform for a lot of developers to create extensions. A study about the security of these extensions revealed the vulnerabilities existing in many popular Firefox extensions and that the root cause of these vulnerabilities was in the underlying architecture. This led to the development of new extension architecture with focus on security, which is now known as the Chrome Extension Architecture[1].This section briefly describes the previous work done in the field of browser extension security.

2.1 Chrome Extension Architecture

The architecture of a Chrome extension divides every extension into three components: Content Script, Extension Core and Native Binary. Content Script, written in JavaScript, is the part of the extension which directly interacts with the web page. Content Scripts can access and modify the page DOM. Extension core, written in HTML and JavaScript is the part of the extension which has access to the APIs used to interact with the browser. Native binary is the part of the extension which can access the host machine with the user's full privileges [1]. From an implementation point of view, a chrome extension consists of a properties file, *manifest.json* where the metadata about the extension including the permissions are mentioned, one or more HTML pages which display the UI, one or more JavaScript source files which perform the underlying functionality and an image file which is used as the icon.

The Chrome extension system follows a security model based on three main concepts: least privilege, privilege separation and strong isolation. Least privilege is implemented by restricting the extensions permissions in the manifest file. An attacker who compromises the extensions is also limited to these privileges thereby reducing the intensity of attacks. Privilege separation is inherent in the architecture itself which separates the extension code into content scripts, extension core and native binaries. Strong isolation is achieved by isolating various components of an extension as different processes and also by running the extension's content script in a different environment from that of the untrusted website. Thus Chrome extension system has inherent mechanisms to enhance the security of extensions. Still, there are attacks possible against these extensions due to improper implementation by the extension developers.

2.2 Chrome Extension Security Review

A study about chrome extension security analysed the effectiveness of the security mechanisms implemented in the Chrome extension architecture [2]. Extension code is written primarily using JavaScript and HTML. The isolated worlds and privilege separation mechanisms are effective in protecting the extension from Data as HTML vulnerabilities and the extension core from attacks on vulnerable content scripts [2]. The permissions system or the least privilege mechanism will be effective only if it is used properly by the developers. The extension system does not restrict the permissions that can be granted to an extension.

Browser extensions are not always on the receiving end of attacks. Extensions are increasingly being used as attack vectors against websites and user data.An experiment to create malicious chrome extensions which launch various attacks on websites and user data turned out to be successful. The group which conducted this experiment were successful in performing Email spamming, DDoS and password sniffing attacks on various websites using chrome extensions [3]. The attacks were as simple as injecting some content script on the web pages, steal the user data and pass on the information to a remote website though a HTTP request. Thus even the seemingly secure chrome extension system is found to be vulnerable to various attacks and it also act as agents in attacks on websites.

2.3 VEX – A Tool for Vulnerability Analysis of Firefox Extensions

VEX is a proof-of-concept tool used to detect potential security vulnerabilities in browser extensions using static analysis of explicit flows.VEX analyses Firefox extension source to find possibly vulnerable flows in JavaScript [4]. The tool tokenizes the JavaScript code and defines rules based on which it analyses the flow of information from a source to sink. VEX analyses flow of web page content to **eval** and **innerHTML** methods and the flow of data from browser's DOM API and XPCOM components as RDF to **innerHTML**. The concepts used in VEX can be extended to Chrome extensions also, as both use JavaScript as the primary development platform.

3 Proposed Framework for Chrome Extension Analysis

We now propose a new framework for analysing chrome extensions based on the principle of least privilege and the flow of information. The proposal is to develop a static analysis tool which will analyse any given chrome extension and provide results based on which we can decide whether the extension is vulnerable, malicious or safe. The system can be divided into two modules which are detailed below.

3.1 Analysis of Least Privilege

Fig. 1 illustrates the Least Privilege Analysis module. As discussed in the previous section, every chrome extension has a manifest file in which we define the permissions required by the extension [5]. Each permission is associated with a set of JavaScript methods which can be used only if the corresponding permission is assigned. This module reads the manifest file of an extension and retrieves the set of permissions assigned and JavaScript files used in the extension. Then it parses all the JavaScript files used and looks for functions or attributes associated with the permission. The methods and attributes corresponding to a permission-type are mostly used along with a prefix which is usually **chrome.<permission>.<name>**[5]. We can scan the JavaScript files for attributes and functions with the corresponding prefix. Alternately, we can maintain also a list of functions corresponding to every permission and compare the list of functions retrieved from the JavaScript files with this. If none of the functions corresponding to a permission are used in the extension, the permission should be deemed as unnecessary and we report that the extension violates the principle of least privilege.

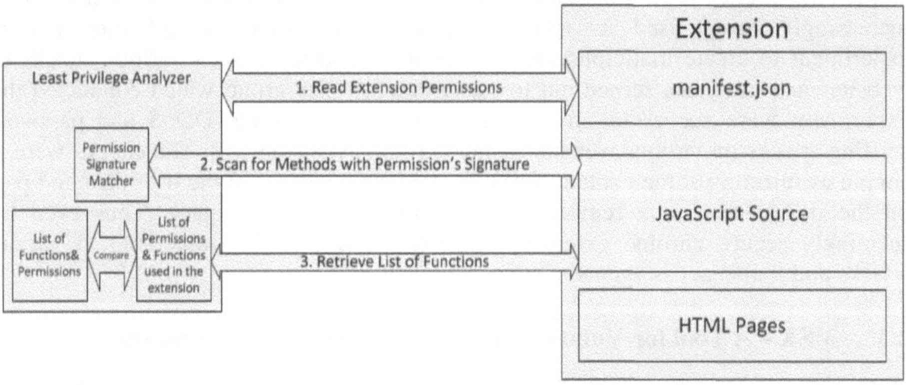

Fig. 1. Block diagram of Least Privilege Analysis Module

3.2 Analysis of Malicious Information Flow

Fig.2 illustrates the Malicious Information Flow Analysis. A browser extension which has permissions to invoke other URLs can send user data from a web page or the browser history/bookmarks to a remote server. A benign end user will never realize this, as the extension would have another utility which requires the same level of permissions. In order to detect such malicious behaviour in a given extension, this module performs an analysis of the information flow. We retrieve the objects and variables which read data from a web page or from the browser internals. It has to be verified whether any of these object values are being forwarded to a different server. This can again be done by analysing the flow of the information. If the data is being passed to some method or URL which is sending a request to a remote server, the extension may be leaking some sensitive user data. Another activity that can be

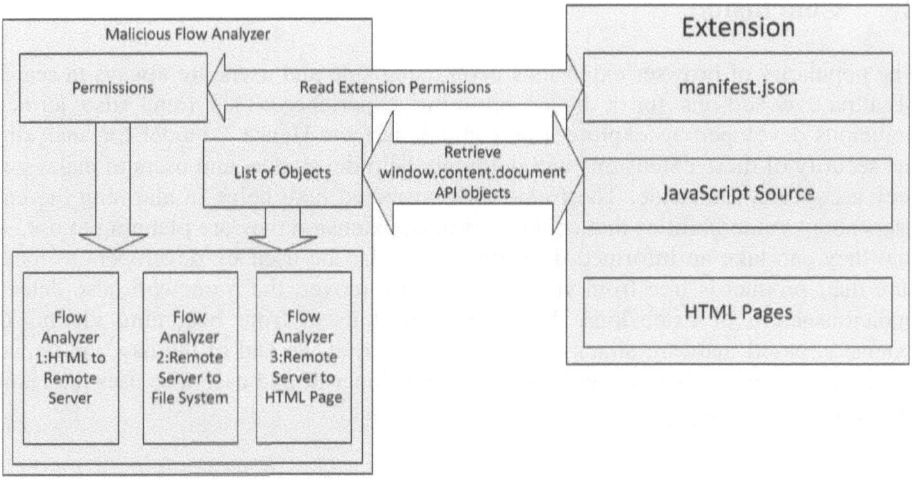

Fig. 2. Block Diagram of Malicious Flow Analysis Module

performed by a malicious extension is downloading some malware from a remote server to the user's system. To detect such a malicious behaviour, first of all we need to check if the extension has privileges to download files to the end user system. If it has the permission, then retrieve JavaScript objects which get their values as a response from a remote server. If any of these objects are being saved to the file system, the extension can be considered malicious. A malicious extension sometimes may act as a bot for attacking a website which the user frequents. The extension may retrieve a payload from a remote server and upload the same to the web page being accessed. To detect such behaviour, we need to check for JavaScript objects/variables which are being uploaded to the current web page. This can be done as simple as running a content-script. If such a JavaScript object gets its value from a remote server, it can be considered as malicious.

3.3 Discussion

The proposed framework helps to detect some of the key issues in Chrome extensions. The privileges assigned to an extension are by default available to a malicious website which is trying to exploit the extension's content script. The proposal makes sure that all extensions follow the principle of least privilege so that the attacks which require any further privileges are stopped at the initial stage itself. The proposal also takes care of detecting malicious extensions so that end user does not become a victim of attacks by malicious extensions. The framework will be able to detect malicious information flows through the extension which may be an attack vector against the user's data or any website. Further, the framework is platform independent. A developer can take this framework and implement a static analysis tool in any language and platform of his choice.

4 Conclusion

The popularity of browser extensions is on rising side and users are always in search of attractive add-ons for a better browsing experience. This trend also attracts malicious developers to explore a new attack surface. Hence a model for analysing the security of these extensions will definitely help developers and users to make sure they are on the safer side. The framework presented here helps in alarming the end users about vulnerabilities that could exist in an extension they are planning to use, so that they can take an informed decision. It can also be used by developers to make sure their product is free from vulnerabilities. Moreover, the framework also detects maliciousnature of extensions. This will protect users from becoming victims of extension based malware attacks as well. However, at the end of the day, the impact of the idea proposed here depends upon the developer or end user who uses the result obtained by this analysis.

References

1. Barth, A., Felt, A.P., Saxena, P., Boodman, A.: Protecting Browsers from Extension Vulnerabilities. In: Proceedings of the 17th Network and Distributed System Security Symposium (2010)
2. Carlini, N., Felt, A.P., Wagner, D.: An evaluation of the google chrome extension security architecture. In: Proceedings of the 21st USENIX Conference on Security (2012)
3. Liu, L., Zhang, X., Yan, G., Chen, S.: Chrome Extensions: Threat Analysis and Countermeasures. In: Proceedings of the 19th Annual Network & Distributed System Security Symposium (2012)
4. Bandhakavi, S., King, S.T., Madhusudan, P., Winslett, M.: VEX: Vetting Browser Extensions For Security Vulnerabilities. In: Proceedings of the 19th USENIX Security Symposium (2010)
5. Chrome Extension Developer Guide, `http://developer.chrome.com/extensions`

Amalgamation of K-means Clustering Algorithm with Standard MLP and SVM Based Neural Networks to Implement Network Intrusion Detection System

A.M. Chandrashekhar[1] and K. Raghuveer[2]

[1] Department of Computer Science, Sri Jayachamarajendra College of Engineering, Mysore-570006, Karnataka, India
[2]Department of Information Science, National Institute of Engineering (NIE), Mysore, Karnataka-570008, India
{amblechandru,raghunie}@gmail.com

Abstract. Intrusion Detection Systems (IDS) are becoming an essential component usually in network and data security weapon store. Since huge amount of existing off-line data and newly appearing network records that needs analysis, data mining techniques play a vital role in development of IDS. The key idea of using data mining techniques for IDS is to aim at taking benefit of classification capability of supervised learning based neural networks and clustering abilities of unsupervised learning based neural networks. In this paper, we propose an efficient intrusion detection model by amalgamating competent data mining techniques such as K-means clustering, Multilayer layer perception (MLP) neural network and support vector machine (SVM), which significantly improve the prediction of network intrusions. Since the number of clusters desired for intrusion detection problem is defined by user a priori and does not change, we employed K-means clustering technique. In the final stage, SVM classifier is used as it produces superior results for binary classification while compared to the other classifiers. We have received the best results and these are compared with results of other existing methods to prove the effectiveness of our model.

Keywords: Intrusion Detection System, Neural Networks, Support Vector Machine, K-means Clustering, KDD cup 99

1 Introduction and Motivation

An IDS is a tool or software that keeps an eye on network or system activities for malevolent activities or policy violations and reports to a administrator. IDS are based on two concepts. The first one is called as Misuse detection and it show the way towardsSignature based IDS; matching of the earlier seen and hence known anomalous patterns from an in-house database of signatures. The second one is called as Anomaly detection it shows the wayto us towardsbehavior based IDS. Anomaly detection builds profiles based on normal data and detects deviations from the expected behavior. Behavior based IDS may have the skill to detect new unseen attacks but have the

M.K. Kundu et al. (eds.), *Advanced Computing, Networking and Informatics - Volume 2*, Smart Innovation, Systems and Technologies 28,
DOI: 10.1007/978-3-319-07350-7_31, © Springer International Publishing Switzerland 2014

setback of low detection accuracy [1]. Based on mode of deployment, IDS are classified as Host based, Network based, and Application based. Host based systems examine individual systems and make use of system logs extensively to takesome decision. Network based IDS make a decision by analyzing the packet headers and network logs from the arriving and outgoing packet as they are deployed at the edge of the network.

The use of data mining techniques in IDS usually implies investigation of the collected data in an offline environment. Classification and clustering are perhaps the most familiar and most popular data mining techniques. The main reason of data mining classification techniques usage for IDS is because of estimation and prediction activity of IDS, which may be viewed as types of classification.Cluster analysis deals with discovering similarities in the data and clustering them. Clustering is valuable in intrusion detection as malevolent activity should group together, unraveling itself from non-malicious activities.Following are some of the observations that motivated us to select k-means approach against its competent clustering approaches like fuzzy-c-means, etc.

- The number of clusters desired is defined by user a priori and does not change.
- K-means works for unlabeled dataset consisting only numerical attributes.
- K-Means is superior than Fuzzy-c-means in terms of computational time.
- K-means algorithm is simple and it handles large data set very efficiently.

Neural Network(NN) is artificial, mathematical model inspired by biological neural networks. Artificial NN(ANN)for intrusion detection was first brought in as a substitute to statistical techniques in IDES intrusion detection expert system to model [2]. ANNs have the capabilities of aligning the source data with its corresponding target output. We use neural networks for intrusion detection for following reasons;

- The uniqueness such as high tolerance for noisy data, faster information processing, effective classification and the capacity of learning and self organization makes ANN flexible and powerful in IDS.
- The operation of IDS complies with intent of neural network models.
- The feature of dimensionality reduction and data visualization in neural networks can be helpful to reduce huge dimension of data records of a network connection
- ANNs are skilled to handle little incomplete knowledge existing in IDS attacks.

2 Related Research Work on IDS

Intrusion detection has been an active field of research for about three decades, starting in1980. Based on a study of latest research documentations, there are quite a lot of research that attempts to relate machine learning and data mining techniques to intrusion detection systems in order to design more intelligent IDS model. Thus a bunch of data mining methods have been introduced to resolve the problem. Among these methods, ANN is one of the most widely used and has been effectively applied to intrusion detection. Different types of ANNs are used in IDS like supervised, unsupervised, and hybrid ANN. The hybrid ANN combines supervised ANN and unsupervised ANN, or combines ANN with other data mining methods to detect intrusion. The inspiration for using the hybrid ANN is to prevail over the restrictions of individual artificial neural network.

Jirapummin *et al.* [3] proposed make use of hybrid ANN for both visualizing intrusions by means of Kohenen's SOM and classifying intrusions by means of resilient propagation neural networks. Horeis [4] used a mixture of SOM with radial basis function (RBF) networks. This system tenders generally better end result than IDS based on RBF networks only. Han and Cho [5] projected an intrusion detection model based on evolutionary neural networks so as to determine the arrangement and weights of call sequences. Chen, Abraham, and Yang [6] projected hybrid flexible neural tree based IDS derived from evolutionary algorithm, flexible neural tree and particle swarm optimization (PSO).

3 Proposed Framework for IDS

Our proposed IDS model comprises of five major phases namely: (1) Input data preparation, which selects required features for input dataand split the given data set into training and testing data sets(2) Clustering using k-means Clustering, where the input data set is grouped into 'k' clusters where 'k' is the desired number of clusters (3) Neural network training, where everydata in a particular group is trained with the particular neural network connected with every cluster (4) creation of vector for SVM classification, containing attribute values acquired by passing every data through all ofthe trained neural networks and (5) final classification using SVM to detect intrusion. The framework of the proposed technique is shown in Fig.1.

3.1 Input Data Preparation

We considered KDD cup 1999 data set as input for our experimentations and its detailed explanation is available in the section 4.Our input dataset consist of large number of data each having 41 features (attributes) collected from network stream. Network stream itself is inappropriate for key in to the classifier module, so it is essential to extract some attributes from the network stream.

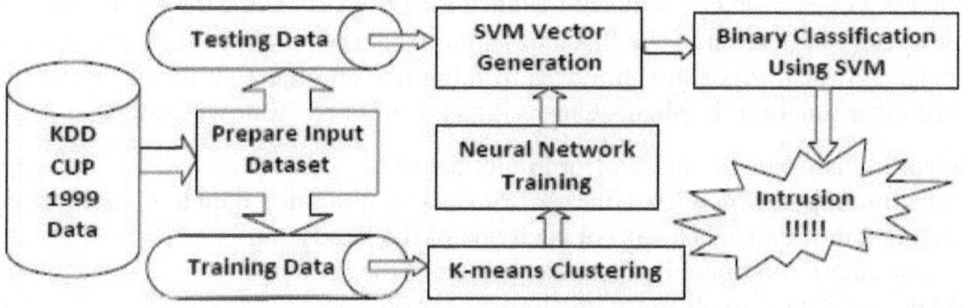

Fig. 1. Block diagram of proposed technique

The features pulled out from the network stream become feature vector, which provides the description of the packet. Feature vector includes symbolic, continuous and discrete attributes and it cannot be applied in straightforward for classification and another reason is most of clustering algorithms work with continuous (numerical) data. This demand data pre processing for input data set. In order to classify the data-set for intrusion detection, considering all the attributes is not feasible and also time consuming task. In this phase of data pre-processing, all the symbolic attributes (7 out of 41) are removed and only continuous attributes (34 out of 41) are extracted from the feature vector. In intern, this helps the intrusion detection process become easier, less complex and also yields a better result.

3.2 Data Clustering Using K-means Clustering

Clustering is useful in intrusion detection domain as malicious activity should be clustered together, separating itself from normal activity. To arrange data into significant clusters, a number of algorithms are proposed. The clustering algorithms are in general classified into two types; hierarchical and non-hierarchical. Non-hierarchical clustering methods are used when huge quantity of data involved. It is most popular method, since it allows subjects (data) to move from one group to another group which is not possible in hierarchical clustering method. The K-means algorithm works only for datasets that consist of numerical attributes. K-means uses the most popular Euclidean distance measure for similarity. K-means algorithm take the input parameter 'k'and partitions a set of 'n' data points into 'k' clusters so that the resulting intra-cluster similarity is high but the inter-cluster similarity is low. The goal of clustering is typically articulated by objective functions that depend upon the proximities of the points to one another or to the cluster centroids; e.g., minimize the squared distance of each point to its closest centriod. Considering data whose proximity measure is Euclidian distance, for our objective function, we use sum of squared error (SEE). Sum squared error (SEE) is defined as given in equation (1) given below.

$$J = \sum_{J=1}^{K} \sum_{i=1}^{n} \| x_i^{(j)} - C_j \|^2 \quad (1)$$

Where$\| x_i^{(j)} - C_j \|^2$is a selected distance measure between a data point and cluster center C_j is an indication of the distance of the n data points from their particular cluster centers. At last, this algorithm tries to minimize an objective function; here it is a squared error function. K-Means Clustering algorithm has following steps:

1. Select'k' centroids at random from input data set.
2. Make initial partition of data into 'k' clusters by assigning data to closest centroid
3. Calculate the centroid (mean) of each one of the k clusters.
   i) For data'i', compute its distance to each of the centroids.
   ii)Allocate data'i' to cluster with closest centroid.
   iii)If data reallocated to another cluster, recalculate centroid based on new cluster
4. Repeat step 3 for data I = 1... N
5. Repeat 3 and 4 until no reallocations occur.

Due to the fact that the magnitude and complexity of every training subset is condensed, the effectiveness and efficiency of subsequent neural network classification can be improved. In this regard, in our proposed model, we employed K-means clustering method for partitioning given dataset into attack data and normal data. K-means clustering results in the formation of 'k' clusters where each cluster will be a type of attack or the normal data.

3.3 Neural Network Construction and Training

ANNs are used to model composite relationships among inputs and outputs or to locate patterns in data. Neural network are typically organized in the form of layers; input layer, hidden layer and an output layer. In our proposed model, we focused on implementing standard and most common feed-forward neural network model called Multilayer Perception (MLP).This neural network is developed by Rumelhart, Hinton and Williams in 1986. MLP is a alteration of the standard linear perception and can discriminate data that is not linearly separable [7].

The number of hidden neurons affects the effectiveness of classification. Thus, before deciding the number of neurons to be considered in the hidden layer, we conducted 10 experiments by varying number of neurons in hidden layer. The details are explained in section 5. Since 3 neurons in the hidden layer gives better accuracy both in training as well as in testing, we have decided to take a two-layered perception feed-forward network with 34 input nodes(since 34 features in input vector), 3 hidden nodes, and one output node.ANN can work efficiently only when it has been trained properly and sufficiently. Once ANN is trained to a satisfactory stage it may be used as a diagnostic tool on other data.MLP is trained by using supervised learning technique called back propagation learning algorithm which is very fast and efficient.

For each of the cluster formed, we have a neural network associated with it. That is, there will be five numbers of neural networks created for four types of attack data and a normal data. Each neural network is trained with the priori information available in the respective cluster to obtain the desired output. Aim of this module is to learn the pattern of every subset in the given dataset.

3.4 SVM Vector Generation Module

Support vector machines are supervised learning techniques which investigate data and recognize patterns, used for classification and regression analysis [8]. SVM classifiers are designed to resolve binary problems wherever the class labels can only take two values: ±1.SVM classifiers produce optimal results for binary classification when compared to other classifiers. SVM training algorithm constructs a model that envisages whether a new sample falls into one group or the other.

SVM first maps input vector into higher dimensional feature space and afterward, obtain the best possible separating hyper-plane in higher dimensional feature space. SVM sections the classes relating to a decision surface that increases the margin between the classes. This surface is generally known as the optimal hyper plane and the data points nearest to the optimal hyper plane are known as the support vectors. This procedure makes over the training data into a feature space of a huge dimension [9].

The SVM technique consists of finding the hyperplane that enlarge the margin, that is, the distance to the nearby training data points for both classes [10].An optimum separating hyperplane is founded by the SVM algorithm such that

- Samples with labels ±1 are positioned on both side of the hyper plane and
- The distance of the closest vectors to the hyper plane in every side of maximum are named as support vectors and the distance be the optimal margin.

Classification of the data point considering all its attributes is a very difficult task and takes much time for the processing, hence decreasing the number of attributes related with each of the data point is of pinnacle importance. The main purpose of the proposed technique is to decrease the number of attributes associated with each data, so that classification can be made in a simpler and easier way. Neural network classifier is employed to efficiently decrease the number of attributes. As we have discussed earlier, the input data is trained with neural network after the initial clustering then the vector necessary for the SVM is generated. The vector array S={D1, D2,….DN} where, Di is the 'i'th data and 'N' is a total number of input data. Di = {a1, a2…ak}, here the Di is the 'i'th data governed by attribute values ai, where ai will have the value after passing through the 'i'th neural network. Here, after training through the ANN, attribute number reduces to 'k' (k=5 in our case) numbers. Total number of neural network classifiers trained will be 'k', corresponding to the 'k' clusters formed after clustering. The input dataset containing large number of attributes (34 in our case) is changed into data having k attributes (k=5 in our case) by performing the above steps. This results in easy processing in the final SVM classification. This also reduces the system complexity and time incurred. The data with restricted number of attributes is given to the linear SVM, which is binary classified to discover if there is any intrusion or not.

4 Experimental Setup and Results

To evaluate the performance of our proposed technique, many experiments were carried out until we finally achieved results that are comparable to what has been available in the literature. In this section, the most important experiments will be explained. Experiments are different basically in the training data size used, which consequently affects the accuracy of the test. Series of experiments on 10% KDD CUP 99 dataset were conducted. We carried out these experiments in MATLAB version R2013a on a Windows PC with 3.2 GHz CPU and 4GB RAM.

KDD Cup 1999 dataset is one of the most rational publicly available data set that includes actual attacks [11]. It provides benchmark for researchers to evaluate intrusion detection by using off-line data. This dataset has 48,98,430 single connection records with each connection record has 41 features/attributes and one class attribute attach a label to connection as normal or anomaly through exactly one specific attack types. There are 38 different types attack in training and testing data together and these types of attack fall into four main categories: PROBE(Probing), Denial Of Service(DOS), Remote to Local(R2L) and User to Root(U2R) [12]. Since KDD Cup

1999 dataset is of enormous size, a segment of 10% of KDD Cup 1999 dataset called 10% KDD data set is made use for our experimentation. The number of data records taken for training and testing phase in the three experiments is given in Table 1.

4.1 Evaluation Measure

Confusion matrix [13] parameters; True positives (TP), false positives (FP), true negatives (TN), and false negatives (FN), shows the number of correct and faulty predictions made by the model correspond to actual categorizations in the test data. Standard parameters such as Accuracy (AC), Specificity (SP), and Sensitivity (SN) are used to estimate the performance of our IDS. All these features are strongly correlated to each other as each establishes the measure of accuracy of classification algorithms. Accuracy measures the degree of faithfulness; it is a proposition of true results. A good Test should have high values for sensitivity and specificity for better categorization into normal and attack categories. So as to discover these metrics we apply confusion matrix values to the equations used are given in Table 1.

Table 1. Dataset considered for experiments and Equations used during evaluation

	Expt. 1and 2		Expt. 3		Equations Used
Attack	**Train**	**Test**	**Train**	**Test**	Sensitivity=TP/(TP+FN)
Normal	12500	12500	6500	18750	$\text{Specificity} = \text{TN}/(\text{TN}+\text{FP})$
DOS	12500	12500	6500	18750	
Probe	2053	2054	1027	3080	$Accuarcy = \frac{(\text{TN}+\text{TP})}{\text{TN}+\text{TP}+\text{FN}+\text{FP}}$
R2L	38	39	19	58	
U2R	21	21	11	31	

4.2 Experimentation and Result Analysis

This section shows the empirical results and performance evaluation of our proposed technique. The input dataset is divided into training set and testing set to use in training and testing phases in the proposed model. Here we conducted 3 experiments with different size of data in training set and testing set. For each of the experiments, the data size taken for training and testing is given in Table 1. The related confusion matrix values obtained and the acquired results for all performance measures are given in Table 2.The values of Accuracy, Specificity and Sensitivity are in percentage.

In First Experiment, from the considered data set, First 50% of records in each types of attacks and normal type are used as training data set and reaming 50% data are used as testing data set. In order to conform that our proposed model performs well irrespective of position of data records considered for training, we conducted second experiment. In this experiment by keeping the data size as it is, we interchanged the data records considered for training dataset and testing dataset in the first experiment. In the Third experiment, we took 25% of records in each attack types and normal type in training data set and reaming 75% data records are used for testing data set. The Third experiment is conducted to ensure that our model perform well even if the training data size is reasonably less.

The average of training phase accuracy and testing phase accuracy determines the overall Accuracy (OA). In each experiment, we compute the OA for individual attack types. The overall performance of our model is established by taking the average of OA for each attack types in all the three experiments. This consolidated result is given table 3. As per the Literature, most of the researchers use intrusion detection accuracy (also called as prediction rate) as major metric for comparison of their proposed techniques with other contemporary techniques. In view of this, we have also used accuracy as primary metric for comparison. Table 4 shows the comparison of accuracy of our praposed technique with other existing state of art techniques so as to prove the effectiveness of our method.

Table 2. Values obtained for Confusion Matrix and Evaluation Measures

		TYPES OF ATTACKS							
		DOS		PROBE		R2L		U2R	
		Train	Test	Train	Test	Train	Test	Train	Test
Experiment No 1	TN	12461	12476	12461	12476	12461	12476	12461	12476
	FP	39	24	39	24	39	24	39	24
	TP	12461	12090	2043	1925	30	39	15	13
	FN	39	410	10	129	8	0	6	8
	SP	99.67	99.98	99.69	99.82	99.69	99.81	99.69	99.81
	SN	99.69	96.72	99.51	93.72	78.95	1	71.43	61.91
	AC	99.69	98.26	99.66	98.95	99.63	99.81	99.64	99.74
	OA	**98.976 %**		**99.306 %**		**99.717 %**		**99.693%**	
Experiment No 2	TN	12398	12374	12398	12374	12398	12374	12398	12374
	FP	102	126	102	126	102	126	102	126
	TP	11618	11237	2039	2008	30	37	15	8
	FN	882	1263	14	46	8	2	6	13
	SP	99.18	98.99	99.18	98.99	99.18	98.99	99.18	98.99
	SN	92.94	89.89	99.32	97.76	78.95	94.87	71.43	38.15
	AC	96.06	94.44	99.20	98.82	99.12	98.98	99.14	98.89
	OA	**95.254 %**		**99.011 %**		**99.051 %**		**99.014 %**	
Experiment No 3	TN	6176	18197	6176	18197	6176	18197	6176	18197
	FP	74	553	74	553	74	553	74	553
	TP	6250	18255	1010	3062	19	48	10	24
	FN	0	495	17	18	0	10	1	7
	SP	98.82	97.05	98.82	97.05	98.82	97.05	98.82	97.05
	SN	1	97.36	98.35	99.42	1	82.76	90.91	77.42
	AC	99.418	97.205	98.75	97.38	98.82	97.01	98.80	97.02
	OA	**98.307 %**		**98.067 %**		**97.914 %**		**97.910 %**	

Table 3. Consolidated Overall Accuracies of all the three experiments (in Percentage)

Attacks	**Experiment no-1**	**Expeiment no-2**	**Expeiment no-3**	**Average**
DOS	98.976 %	95.254 %	98.307 %	**97.512 %**
PROBE	99.306 %	99.011 %	98.067 %	**98.795 %**
R2L	99.717 %	99.051 %	97.914 %	**98.894 %**
U2R	99.693 %	99.014 %	97.910 %	**98.872 %**

Table 4. Accuracy comparison with existing methods (in Percentage)

DIFFERENT METHODS	DOS	PROBE	R2L	U2R
KDD cup 99 Winner [14]	97.1	83.3	8.4	13.2
PN rule [15]	96.9	73.2	10.7	6.6
Multi-Class SVM[16]	96.8	75	4.2	5.3
Layered Conditional Random Fields [17]	97.40	98.60	29.60	86.30
Columbia Model [18]	24.3	96.7	5.9	81.8
Decision tree [19]	60.0	81.4	24.2	58.8
BPNN [20]	98.1	99.3	48.2	89.7
Our Proposed Technique	**97.51**	**98.79**	**98.89**	**98.87**

In the case of DOS and PROBE intrusions we have attained 97.51%, and 98.79% accuracy respectively, which is the maximum accuracy value when compared to other methods except BPNN method. But for both R2L and U2R, we have reached the maximum value of 98.89% an 98.87% accuracy respectively when compared to contemporary techniques.

5 Conclusion

In this paper we presented an efficient technique for intrusion detection by blending k-means clustering algorithm with standard multilayer perception neural network and support vector machine based neural networks. Here, we took the help of k-means clustering technique to make large, heterogeneous training data set in to a number of homogenous subsets. As a result complexity of each subset is reduced and accordingly the detection performance is increased. Subsequently, training will be given to artificial neural networks and later SVM vector will be formed. At the end, binary SVM will be used to perform final classification.

During experimentation, we conducted 3 experiments to guarantee fair evaluation by varying and interchanging the data records in training and testing datasets. Confusion matrix were acquired and used to obtain evaluation measures like sensitivity, specificity and accuracy. The experimental results using the 10% KDD Cup 1999 dataset exhibits the effectiveness of our new approach especially for low-frequent attacks like R2L and U2R in terms of Accuracy. In the experiments reported in this paper, we attained the best results compared to existing methods. From this comparative analysis, it is clear that our proposed technique outperformed all other state of art techniques.

References

1. Gupta, K.K., Nath, B., Kazi, A.U.: Attacking confidentiality: An agent based approach. In: Mehrotra, S., Zeng, D.D., Chen, H., Thuraisingham, B., Wang, F.-Y. (eds.) ISI 2006. LNCS, vol. 3975, pp. 285–296. Springer, Heidelberg (2006)
2. Debar, H., Becker, M., Siboni, D.: A Neural Network Component for an Intrusion Detection System. In: IEEE Computer Society Symposium on Research in Security and Privacy, pp. 240–250 (1992)
3. Jirapummin, C., Wattanapongsakorn, N., Kanthamanon, P.: Hybrid neural networks for intrusion detection system. In: Proceedings of ITC–CSCC, pp. 928–931 (2002)
4. Horeis, T.: Intrusion detection with neural network – Combination of self organizing maps and redial basis function networks for human expert integration. A Research Report (2003)
5. Han, S.J., Cho, S.B.: Evolutionary neural networks for anomaly detection based on the behavior of a program. IEEE Transactions on Systems, Man and Cybernetics (B) 36(3), 559–570 (2005)
6. Chen, Y.H., Abraham, A., Yang, B.: Hybrid flexible neural-tree-based intrusion detection systems. International Journal of Intelligent System 22(4), 337–352 (2007)
7. Norouzian, M.R., Merati, S.: Classifying Attacks in a Network Intrusion Detection system Based on Artificial Neural Networks. In: 2011 13th International Conference on Advanced Communication Technology (2011)
8. Vipnik, V.N.: The nature of Statistical Theory. Springer (1995)
9. Mukkamala, S., Sung, A.H., Abraham, A.: Intrusion Detection using an Ensemble of Intelligent Paradigms. Journal of Network and Computer Applications 28(2), 167–182 (2004)
10. Bousquet, O.: Introduction au Support Vector Machines (SVM). Center Mathematics applied, polytechnique school of Palaiseau (2001)
11. Aickelin, U., Twycross, J., Hesketh-Roberts, T.: Rule generalization in intrusion detection systems using SNORT. International Journal of Electronic Security and Digital Forensics 1(1), 101–116 (2007)
12. Tavallaee, M., Bagheri, E., Lu, W., Ghorbani, A.A.: A detailed analysis of the KDD CUP 99 data set. In: Proceedings IEEE International Conference on Computational Intelligence for Security and Defense Applications, pp. 53–58 (2009)
13. Kohavi, R., Provost, F.: Glossary of terms. Machine Learning, 271–274 (1998)
14. Pfahringer, B.: Winning the KDD99 Classification Cup: Bagged Boosting. SIGKDD Explorations 1, 65–66 (2000)
15. Agarwal, R., Joshi, M.V.: PNrule: A New Framework for Learning Classifier Models in Data Mining. In: First SIAM Conference on Data Mining in Network Intrusion Detection (2000)
16. Ambwani, T.: Multi class support vector machine implementation to intrusion detection. In: Proceedings of the International Joint Conference on Neural Networks, pp. 2300–2305 (2003)
17. Gupta, K.K., Nath, B., Kotagiri, R.: Layered Approach using Conditional Random Fields for Intrusion Detection. IEEE Transactions on Dependable and Secure Computin 7(1), 35–49 (2008)
18. Lee, W., Stolfo, S.: A Framework for Constructing Features and Models for Intrusion Detection Systems. Information and System Security 4, 227–261 (2000)

19. Lee, J.-H., Sohn, S.-G., Ryu, J.-H., Chung, T.-M.: Effective Value of Decision Tree with KDD 99 Intrusion Detection Datasets forIntrusion Detection System. In: Proceedings of 10th International Conference on Advanced Communication Technology, pp. 1170–1175 (2008)
20. Tran, T.P., Cao, L., Tran, D., Nguyen, C.D.: Novel Intrusion detection using Probabilistic Neural Network and Adaptive Boosting. International Journal of Computer Science and Information Security 6(1), 83–91 (2009)

Perspective Based Variable Key Encryption in LSB Steganography

Rajib Biswas[1], Gaurav Dutta Chowdhury[2], and Samir Kumar Bandhyopadhyay[3]

[1] Department of Information Technology,
Heritage Institute of Technology, Kolkata-700107, India
[2] C-DAC Kolkata, Kolkata-700091, India
[3] Department of Computer Science and Engineering, University of Calcutta,
Kolkata-700009, India
rajib.biswas@heritageit.edu,
gauravduttachowdhury@gmail.com, skb1@vsnl.com

Abstract. We have explored a new dimension in image steganography and propose a deft method for image– secret data – keyword (steg key) based sampling, encryption and embedding the former with a variable bit retrieval function. The keen association of the image, secret data and steg key, varied with a pixel dependant embedding results in a highly secure, reliable L.S.B. substitution. Meticulous statistical analyses have been provided to emphasize the strong immunity of the algorithm to the various steganalysis methods in the later sections of the paper.

Keywords: Steganography, Steganalysis, Pixel, Sampling, Encryption, Decryption, Steg key, LSB Substitution.

1 Introduction

The sole role of steganography is to conceal the fact that any communication is taking place. Secret messages are embedded in cover objects to form stego objects. These stego objects are transmitted through the insecure channel. Cover objects may take the form of any irrelevant / redundant digital image, audio, video and other computer files. In secure transmission of the stego objects without suspicious lies the success of steganography[7]. Staganalysis methods aim at estimating retrieval of potentially hidden information with little or no knowledge about the steganographic algorithm or its parameters.

2 Literature Study

An extensive study of the related papers [2], [3], [5-8] has given shape to this concept. We have meticulously analyzed the possibilities in the sphere of maximizing randomization, minimizing deviations and structuring strong coherence among the working sets. This paper is aimed at further increasing the equalization and reliability of the substitution based steganography from its referrals.

M.K. Kundu et al. (eds.), *Advanced Computing, Networking and Informatics - Volume 2*,
Smart Innovation, Systems and Technologies 28,
DOI: 10.1007/978-3-319-07350-7_32, © Springer International Publishing Switzerland 2014

3 Our Proposed Method

In our method we have mainly four components as sampling, encryption, embedding and decryption. Sampling plays a key role here in our procedure, it involves the homogeneous selection of pixels for encryption which strengthen the steganography procedure. Encryption is the procedure where implementing algorithms, we apply our tricks to match all steganographic characteristics. Then we embed the message in the sampled pixels with strong and efficient algorithm, subsequently use the decryption method to retrieve the image in the receiver end. The flowchart Fig. 5. involving sampling, encryption, embedding, decoding and decryption steps is given.

3.1 Sampling

Sampling is intricately associated with successful steganography and plays a central role in the process. In this paper, we have explored a highly secure and weight balanced algorithm to obtain variable samples spread equally throughout the cover image Fig. 3.This papers explores a highly secure method of image steganography. The samples are selected based on the input cover object, secret massage and the steg key. Further, a striking feature of the sampling function is that the sample count decreases exponentially as we move inwards from the periphery to the centre of the picture. This is based on the idea that the centre of the picture is usually more meticulously noticed and focused on by the human eye, and peripheral parts generally attract lesser meticulous keen notice. The sampling is strengthened keeping in mind the visible changes in the histogram, thereby repulsing steganalysis deftly. Further, the function ensures that approximately equal number of pixel samples have been selected from all four quadrants, to prevent clustering of samples from a single one.

3.2 Encryption

We encrypt the secret message using a 2 – level encryption function. The first level of encryption is based on the secret message and steg-key and the second level encryption parameters consist of the intermediate message and the secret message. We perform a steg- key based cyclic modification of the secret message followed by inter operable second level encryption.

In the first level:

Let *P* be the Steg key array and Pi be the ith position of the input steg key. Cycle-pass (*P*) cyclically generates the *P* elements until $n(P) = n(O)$

N = number of elements / characters. Then, we do a corresponding character increment / decrement as : If *i* factor of $n(E)$, then ; $e_i = o_i$ - *dec* – *3 lsb (cycle – pass* (p_i)*)*

$$\text{else; } e_i = o_i + dec - 3\ lsb\ (cycle - pass\ (p_i))$$

Now, we get an encrypted message e with the same number of elements as original message *o*. This first level encrypted message and the secret message are the parameters of the second – level encryption function.

$$E_0 = e_0\ ,\ E_i = e_i + e_{i-1}$$

Consequently, we get a second – level encrypted message *E*, which is all set to be sent through the insecure channel.

3.3 Embedding

In LSB based steganography [1], [2], [4] the embedding of the encrypted message *E* in the selected sample pixel set *S* is done in a color-component varied bit encryption method. Function RGB-image (RGB value) returns the maximum intensity color component, taking the pixel RGB value as parameter.

Step 1: We extract from the selected pixel as follows:
R -> red component value in the range of 0 – 255.
G -> green component value in the range of 0 –255.
B -> blue component value in the range of 0 – 255.
Step 2 : Convert the values to hexadecimal. Thus we get a MSB. and a LSB. value in the range of *0 – e*.
Step 3 : Convert the LSB hex value to decimal.
Step 4 : Convert the decimal value to ASCII.
Step 5: The function max Intensify (R, G, B) is celled, which returns the color component of maximum intensity.
Step 6: The last 3 bits are encrypted from the maximum intensity color component and the last 2 bits are essential for the other 2 components. Thus we get 7 bits Fig. 1. as either.

1	2	3	4	5	6	7
R1	R2	R3	B1	B2	G1	G2
R1	R2	B1	B2	B3	G1	G2
R1	R2	B1	B2	G1	G2	G3

Fig. 1. Embedding format for 7 bits

Step 7: The encrypted input E_i is converted into its ASCII Fig.2. and mapped on to the selected pixel S_i.

e1	e2	e3	e4	e5	e6	e7

Fig. 2. Encrypted ASCII bits

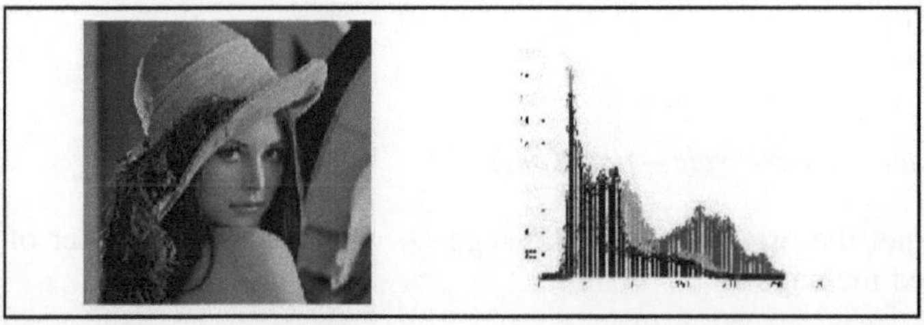

Fig. 3. Cover image Lenna (270x270) and corresponding histogram

Step 8 : The modified R,G,B LSB values are connected back to its decimal values, which are in turn converted into the R, G, B LSB modified hexadecimal values.

Step 9:The combined R,G,B MSB and LSB values are merged together and converted to decimal values ranging from 0 – 255.

Step 10:The R, G, B values are merged into one single RGB value and the value is set as that modified RGB value in the selected pixel S_i.

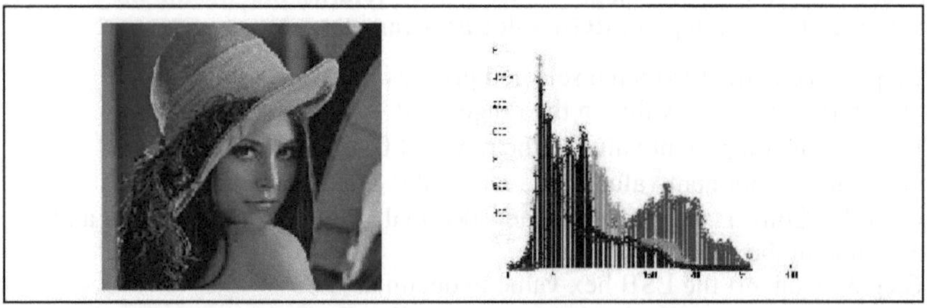

Fig. 4. Stego image Lenna (270×270) and corresponding histogram

3.4 Decoding and Decryption

The initial phase consists of retrieving the necessary information required for decoding from the corner pixels. In the first step we need to retrieve two important parameters from the encrypted image as secret message size and order of the stego images (incase of Split and Send Algorithm(SSA)).

Then we intend decoding the original message from the stego image Fig. 4. and concatenate them in order to obtain the secret message. We first, apply the sampling algorithms to obtain the samples used for encoding. Then we proceed as below:

Step1:

From the samples obtained, we get the values of the second level encrypted message E. We evaluate the message as:

$e_0 = E_0$

$e_i = E_i \; XNOR \; e_{i-1}$, Where *E* is the second level encrypted message and *e* is the first level encrypted message.

Step2:Then we decrypt the message *e* as , loop from *O* to the size of the message *e* → *O to n(e)* and if *i* is a *factor of n(e) then,*

$O_i = e_i + dec - 3 \; lsb \; (cycle - pass(p_i))$

else

$O_i = e_i - dec - 3 \; lsb \; (cycle - pass \; (p_i \;))$

Thus we get the original secret message *O*, with the same number of elements as the encrypted message *e*.

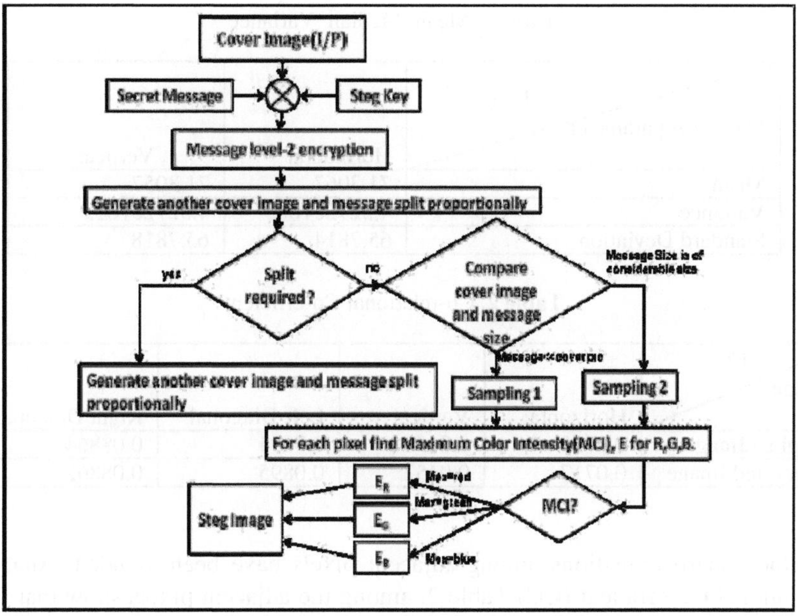

Fig. 5. Flowchart of Encryption ,Sampling and Embedding

3.5 Split and Send

To remove the constraint of a fixed size secret message, we intent to put forward an automatic adjustment algorithm. This segment is mainly concerned with ensuring that input secret message size to be embedded bears a fairly reasonable ratio to the cover image size for which the distortion is negligible. The dynamic ratio value is defined depending on the dynamics of the image and the concentration of the color component values across the cross sections of the image. Based on the above concept , our algorithm warn against suspicion and suggests the use of another cover image or the copy of the same cover image, which can be generated automatically. On agreement we split the secret data in the best-proportion and the re-sample it. This process of splitting and re-sampling is a recursive process and terminates once an optimally permitted ratio is reached. We store the necessary values required for decoding in the four corners of a picture.

4 Performance Analysis

The statistical studies further clarify the proximity and negligible distortions produced in the stego image in the process of execution of the above proposed algorithms. From the table underneath it is noted that the statistical parameters like the mean, variance, Standard Deviation Table 1. change only in their distant decimals thus proving its strong resistance to steganalysis.

Table 1. Mean, Median, Variance

Statistical parameter \ Image	Horizontal	Vertical
Mean	71.3067	71.3057
Variance	4.3272e+003	4.3272e+003
Standard Deviation	65.7814	65.7818

Table 2. Co-relational Co-efficient

Image \ CC	Horizontal	Vertical	Left Diagonal	Right Diagonal
Original Image	0.0715	0.0456	0.0894	0.08864
Encrypted Image	0.0717	0.0452	0.0895	0.08862

Further, sharp transitions among adjacent pixels have been avoided. Analysis of Co-relational Co-efficient (CC) Table 2. among the adjacent pixels show that there is a mass diffusion of statistical parameters in the stego image as compared to the original image. The diffusion is uniform throughout the encrypted image. The correlation decrease further with increase in the size of the secret message, although slightly and thus potent itself against various statistical attacks.

StirMark Analysis: Any steganographic algorithm should resist some standard benchmark tests to prove its strength and robustness. We run these tests in StirMark 4.0 [11] and our algorithm produced good results. We show a sample of the results in Table 3. The negligible gaps between the values corresponding to the cover and the stego image imply that our technique is robust.

Table 3. StirMark Analysis Results

Test	Factor	Cover	Stego
SelfSimilarities	1	29.6885 dB	29.8643 dB
SelfSimilarities	2	46.1171 dB	46.1326 dB
SmallRandomDistortions	1	15.839 dB	14.3582 dB
SmallRandomDistortions	1.05	15.4899 dB	14.2077 dB
MedianCut	3	27.7947 dB	27.8023 dB
MedianCut	5	26.0944 dB	26.1014 dB
PSNR	10	37.7322 dB	37.7322 dB
PSNR	20	33.3333 dB	33.333 dB
AddNoise	20	11.0462 dB	11.1076 dB
AddNoise	40	9.81765 dB	9.84208 dB

Resilience against Standard Steganalytical Tools and Tests: To augment our statistical analysis of images, we have realized Stegdetect, an automated analytical tool and Stefan Axelsson's base-rate fallacy to intrusion detection systems where false-positives cast a bearing on system's efficiency. We have calibrated Stegdetect's detection sensitivity against numerous outputs of our algorithm.We can calculate the true-positive rate – the probability that an image detected by Stegdetect really has steganographic content as:

$$P(S \mid D) = \frac{P(S).P(D \mid S)}{P(S).P(D \mid S) + P(\neg S).P(D \mid \neg S)}$$

Where P(S) is the probability of steganographic content in images, and $P(\neg S)$ is its complement. $P(D \mid \neg S)$ is the probability that we'll detect an image that has steganographic content. Conversely, $P(\neg D \mid S)=1-P(D \mid S)$ is the false positive rate.

We have calibrated Steg detect's detection sensitivity against numerous outputs of our algorithm. The probability of detection is negligible for small messages, but with larger embedded data, the probability is high. In our algorithm, besides the dual cover encryption on the images, we have used a variable split and send algorithm which creates an upper limit on the message size ,which is smart barrier against detection.

Dual Statistics Method: This method [9] partitions an image with a total number of pixels N into N/n disjoint groups of n adjacent pixels. For a group of pixels $G=(x_1, x_{2,\ldots\ldots}, x_n)$. The authors considered discrimination function $g(x_1, x_{2,\ldots\ldots}, x_n)=\sum_{i=1}^{n-1} |x_{i+1} - x_i|$.They define two LSB flipping functions $F_1 = 0\leftrightarrow 1, 2\leftrightarrow 3, \ldots\ldots\ldots, 254\leftrightarrow 255$ and $F_{-1} = -1\leftrightarrow 0, 1\leftrightarrow 2, \ldots\ldots\ldots, 255\leftrightarrow 256$, along with an identity flipping function $F_0(x) = x$. The assignment of flipping to a group of n pixels can be captured by a mask $M = (M(1), M(2), \ldots, M(n))$,where $M(i) \varepsilon \{-1,0,+1\}$ denotes which flipping function is applied to which pixel.

The flipped group of a group $G=(x_1, x_{2,\ldots\ldots}, x_n) =(F_{M(1)}(x_1), F_{M(2)}(x_2), \ldots\ldots\ldots, F_{M(n)}(x_n))$ They classify the pixel groups as Regular, Singular, or Unchanged, according as $g(F(G))> g(G)$, $g(F(G))< g(G)$ or $g(F(G))= g(G)$respectively. Next, they compute the length of the hidden message from the counts of such groups.

The authors mention that their method does not work well for image that are noisy, or of low quality, or over-compressed, or of small size. Moreover Dumitrescu *et al.* [10] points out that the above schema is based on the following assumptions :

Assumption 1: Suppose X is the set of all pixel pairs (u, v) such that either v is even and $u<v$, or v is odd and $u>v$. Suppose Y is the set of all pixel pairs (u, v) such that either v is even and $u>v$, or v is odd and $u<v$. The assumption is that statistically we have $|X|=|Y|$.

Assumption 2 : The message bits of LSB steganography are randomly scattered in the image space, independent of image features.

Our method does not make any of the following assumptions. Dumitrescu *et al.* mentions that assumption 1 is valid only for natural images. Our method works on no specified range of images. it works on even cartoon and paint-drawn images. Starting from random sampling, through dual-encryption to embedding, we have used image data attributes vigorously as functional parameters for respective utilities. Hence, it directly violates assumption 2. So, theoretically, our method is not breakable by the dual statistics method.

4.1 Comparative Analysis of the State of the Art Works and Resilience against Effective Steganalysis Methods

The algorithm realizes dual chained encryption and differential embedding to reduce distortion during embedding. It realizes spatial domain technique for improving the quality of the image under different payload and strength of authentication process has been verified against each such variations. Our algorithm parallels the strength of grid colorings in steganography[15], based on rainbow coloring graphical analysis employing syndrome coding by perspective based dimensional analysis of the stego image. It considers the signal processing vulnerabilities of substitution technique based spatial steganography [12] and establishes its resilience towards them. It also responses positively to the lagrange's interpolation [14] based steganalysis The algorithm considers the effective steganalysis methods of estimating secret key in sequential embedding methods through low, medium, and high signal-to-noise ratio (SNR) analysis and abrupt change detection based steganalysis [13]. The abrupt change detection using sequential probability ratio test (SPRT) has also been applied against this algorithm as shows inarguably positive statistics. Besides, it recognizes that under repeated embedding, the disruption of the signal characteristics is the highest for the first embedding and decreases subsequently , that is, the marginal distortions due to repeated embeddings decrease monotonically. This decreasing distortion property exploited with Close Color Pair signature is used to construct the classifier that is in turn used to distinguish between stego and cover images. Our algorithm handles the close color pair detection meticulously and shows its resilience against these types of attacks.

5 Future Enhancement

This paper widens the spectrum of diffusion and randomization in substitution based steganography. We have aimed at strong coherence and security of data underlying strong randomization and encryption. We propose to step into a yet another new horizon by migrating to frequency domain for restructuring discrete randomization to continuous spectrum in our later endeavors.

6 Conclusion

We conclude widening the window of image steganography through intensive improvisations done in almost all the processes of the evaluation.

Acknowledgments. A span of hard toil has given shape to this paper. It is not complete without conveying thanks to all, who have spurred our way to the completion of this endeavour. We also express our gratitude to all those who have in some way or the other been a part of this enterprise.

References

1. Ker, A.D.: Steganalysis of LSB matching in grayscale images. IEEE Signal Processing Letters 12(6), 441–444 (2005)
2. Chan, C.-K., Cheng, L.M.: Hiding data in images by simple LSB substitution. Pattern Recognition 37(3), 469–474 (2004)
3. Hetzl, S., Mutzel, P.: A Graph–Theoretic Approach to Steganography. In: Dittmann, J., Katzenbeisser, S., Uhl, A. (eds.) CMS 2005. LNCS, vol. 3677, pp. 119–128. Springer, Heidelberg (2005)
4. Cummins, J., Diskin, P., Lau, S., Parlett, R.: Steganography and Digital Watermarking. School of Computer Science, The University of Birmingham
5. Cachin, C.: An information-theoretic model for steganography. Information and Computation 192, 41–56 (2004)
6. Pfitzmann, B.: Information hiding terminology. In: Anderson, R. (ed.) IH 1996. LNCS, vol. 1174, pp. 347–350. Springer, Heidelberg (1996)
7. Chandramouli, R., Kharrazi, M., Memon, N.: Image Steganography and Steganalysis: Concepts and Practice. In: Kalker, T., Cox, I., Ro, Y.M. (eds.) IWDW 2003. LNCS, vol. 2939, pp. 35–49. Springer, Heidelberg (2004)
8. Chandramouli, R., Memon, N.: Analysis of LSB based image steganography techniques. In: IEEE International Conference on Image Processing, pp. 1019–1022 (2001)
9. Fridrich, J., Goljan, M., Dui, R.: Reliable Detection of LSB Steganography in Color and Grayscale Images. In: Proceedings of the ACM Workshop on Multimedia and Security, pp. 27–30 (2001)
10. Dumitrescu, S., Wu, X., Wang, Z.: Detection of LSB steganography via sample pair analysis. In: Petitcolas, F.A.P. (ed.) IH 2002. LNCS, vol. 2578, pp. 355–372. Springer, Heidelberg (2003)
11. Petitcolas, F.A.P., Anderson, R.J., Kuhn, M.G.: Attacks on copyright marking systems. In: Aucsmith, D. (ed.) IH 1998. LNCS, vol. 1525, pp. 218–239. Springer, Heidelberg (1998)
12. Vanmathi, C., Prabu, S.: A Survey of State of the Art techniques of Steganography. International Journal of Engineering and Technology 5(1), 376–379 (2013)
13. Trivedi, S., Chandramouli, R.: Secret key estimation in sequential steganography. IEEE Transactions on Signal Processing 53(2), 746–757 (2005)
14. Hamdaqa, M., Tahvildari, L.: ReLACK: A Reliable VoIP Steganography Approach. In: Fifth International Conference on Secure Software Integration and Reliability Improvement, pp. 189–197 (2011)
15. Fridrich, J., Lisonek, P.: Grid Colorings in Steganography. IEEE Transactions on Information Theory 53(4), 1547–1549 (2007)

A Novel Un-compressed Video Watermarking in Wavelet Domain Using Fuzzy Inference System

Bhavna Goel and Charu Agarwal

Ajay Kumar Garg Engineering College, Ghaziabad, India
bhavnag13@gmail.com, charuagarwal19@yahoo.com

Abstract. In this paper, human visual system (HVS) characteristics are modeled using Mamdani fuzzy inference system (FIS) for robust un-compressed video watermarking technique in discrete wavelet transform (DWT) domain. The video sequence is decomposed into frames and converted into YCbCr color space. Two HVS characteristics namely edge sensitivity and contrast sensitivity are computed for each luminance component (Y) of the frame. These two computed values are fed as input to the FIS. The output of the FIS is a weighting factor which is used to embed the watermark into the frame. For embedding purpose a binary watermark is embedded into the LL3 sub-band coefficients of the video sequence. To study the robustness of proposed scheme various video processing attacks are performed. Experimental results show that proposed video watermarking scheme is highly robust and obtain good perceptual quality.

Keywords: Digital video watermarking, DWT, FIS, HVS.

1 Introduction

With the rapid advances in computer and communication technology, digital contents are easy to obtain using the internet. People can easily record, distribute and vary multimedia contents created by other authors without paying any royalties. Therefore, it is important to protect the private contents and multimedia data. To ensure the protection of copyright information and content integrity, watermarks are embedded in the multimedia contents and can be detected without degradation of perceptual fidelity [1].

Digital video watermarking refers to the technique for embedding digital watermark into the video signals by utilizing the inherent spatial and temporal redundancy. Due to the high degree of data and temporal redundancy between frames, video signals are susceptible to attacks such as frame dropping, frame averaging, frame swapping and statistical attacks. Therefore effective video watermarking techniques should ensure the fundamental requirements of watermarking application imperceptibility and robustness. Because of these two parameters, the video watermarking problem is now perceived as an optimization problem [2], [3].

Lin *et al.* [1] proposed a novel video watermarking scheme that uses the fuzzy C-means (FCM) clustering method to select the motion vector and the positions for the

M.K. Kundu et al. (eds.), *Advanced Computing, Networking and Informatics - Volume 2*, Smart Innovation, Systems and Technologies 28,
DOI: 10.1007/978-3-319-07350-7_33,

watermarks. They claim that their technique possesses higher security and satisfactory quality with very small degradation.

Lee *et al.* [4] have developed a technique using fuzzy logic reasoning to recover the isolated and contiguous block losses in a block-based image and video coding system such as JPEG, H.261, MPEG and HDTV standard. They claim that their scheme can well conceal lost blocks and can recover more complicated texture blocks.

Masoumi *et al.* [5] proposed a blind scene-based watermarking scheme for video protection. In their proposed scheme they use the scene change analysis to insert the watermark into HL, LH and HH 3D wavelet coefficients with third resolution level. They claim that their experimental results shows the good performance for transparency and robustness against various kinds of attacks such as Gaussian noise, median filtering, frame dropping, frame swapping, frame averaging and lossy compression including MPEG-4, H.264 and MPEG-2.

Taweel *et al.* [6-7] proposed a novel DWT-based video watermarking algorithm based on 3-level DWT. They claim that their scheme is robust against image processing attacks, geometric distortions and noise attacks.

Yassin *et al.* [8] proposed the block based video watermarking scheme using DWT and principle component analysis (PCA). In their scheme, they embed the watermark into maximum coefficient of PCA block of two LL and HH wavelet coefficient. They claim that their scheme is high imperceptible and high robustness against several attacks such as Gaussian noise, gamma correction, histogram equalization, contrast enhancement and JPEG coding.

Sinha *et al.* [9] proposed the hybrid digital video watermarking scheme based on DWT and principal component analysis (PCA). PCA reduce the correlation among the wavelet coefficients. They claim that their scheme is imperceptible and robust against various attacks such as contrast adjustment, filtering, noise addition and geometric attacks.

In this paper, two HVS characteristics are computed block-wise for luminance component (Y) for each frame of the un-compressed AVI video sequence. These two values are fed as input to the FIS, which results in a single weighting factor. The obtained weighting factor is used to embed the binary watermark into the LL3 sub-band coefficients of the frames. The signed video sequence is found to be imperceptible after watermark embedding as indicated by high PSNR values. The extraction of the watermarks from these frames yield high normalized correlation (NC) values which indicate successful watermark recovery. To study the robustness of the proposed scheme six different video processing operations are performed. The high PSNR value for attacked video frames and high NC value for extracted watermark from attacked frame indicate the robustness of the proposed scheme. The time complexity of the proposed scheme is also analyzed which indicates that proposed scheme requires only few seconds for watermark embedding. The rest of the paper is organized as follows: Section 2 describes the proposed scheme including the fuzzy inference system, scene change detection algorithm, watermark embedding algorithm and watermark extraction algorithm. Section 3 describes the experimental results and discussion. Finally the proposed work is concluded in section 4.

2 Proposed Method

In the present section, we describe our proposed fuzzy-based video watermarking technique. A binary image is used as watermark. In our technique, the watermark is embedded directly into the low-frequency sub-band coefficients of luminance frame.

2.1 Computing HVS Characteristics and FIS Weighting Factor

The un-compressed RGB video frames are first transformed into YCbCr color space. Then luminance component (Y) of each frame is divided into 8x8 block. HVS characteristics namely Edge sensitivity and Contrast sensitivity are computed over these blocks as follows:

- Edge Sensitivity:An edge can be detected using threshold. A Matlab routine greythresh() is used in present work which computes block threshold using Otsu's method.
- Contrast Sensitivity: This parameter can be measured by computing the variance of the block.

These two parameters are fed as input to the FIS available in Fuzzy toolbox of Matlab. Fig.1 depicts the block diagram of FIS.

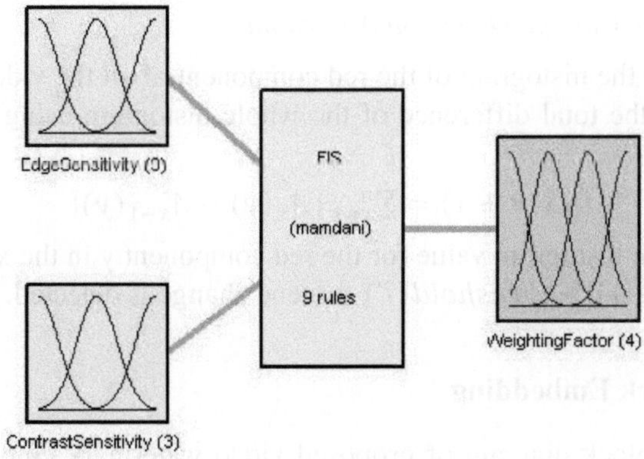

Fig. 1. Block diagram of FIS

FIS Input Parameters: Edge Senstivity input parameter consist of three membership functions namely small, large and largest and Contrast sensitivity input parameter also consist three membership functions namely smooth, highes trough and rough.

FIS Output Parameter: Output parameter consist of four membership functions namely very small, small, very large and large.

FIS Rules:

a) If (EdgeSensitivity is small) and (ContrastSensitivity is smooth) then (WeightingFactor is verysmall)

b) If (EdgeSensitivity is small) and (ContrastSensitivity is highestrough) then (WeightingFactor is verysmall)
c) If (EdgeSensitivity is small) and (ContrastSensitivity is rough) then (WeightingFactor is small)
d) If (EdgeSensitivity is large) and (ContrastSensitivity is smooth) then (WeightingFactor is small)
e) If (EdgeSensitivity is large) and (ContrastSensitivity is highestrough) then (WeightingFactor is verylarge)
f) If (EdgeSensitivity is large) and (ContrastSensitivity is rough) then (WeightingFactor is verylarge)
g) If (EdgeSensitivity is largest) and (ContrastSensitivity is smooth) then (WeightingFactor is verylarge)
h) If (EdgeSensitivity is largest) and (ContrastSensitivity is highestrough) then (WeightingFactor is large)
i) If (EdgeSensitivity is largest) and (ContrastSensitivity is rough) then (WeightingFactor is large)

2.2 Scene Change Detection

In the proposed scheme histogram difference method is used for scene change detection which is given by Listing 1.

Listing 1: Scene Change Detection Algorithm

Step1. Calculate the histogram of the red component of all the video frames.
Step2. Calculate the total difference of the whole histogram using the formula given by Eq. 1

$$D(x,x+1) = \sum_{x=1}^{n} | A_x(y) - A_{x+1}(y)| \tag{1}$$

where $A_x(y)$ is the histogram value for the red component y in the xth frame.
Step3. If $D(x,x+1) > threshold(T)$ a scene change is detected.

2.3 Watermark Embedding

Fig.2 shows the block diagram of proposed video watermark embedding process. In the present work we consider the host video of size M×N and the watermark W of size n × n. The watermark embedding process is given by Listing 2.

Listing 2: Watermark Embedding Algorithm

Step1. Apply scene change detection algorithm to detect the scenes (m) from the original RGB video frames.

Step2. Convert every frame of the original video from RGB to YCbCr color space.

Step3. Divide luminance component (Y) of every frame into 8×8 blocks.

Step4. Compute HVS parameter edge sensitivity and contrast sensitivity blockwise as given in Section 2. Fed these parameters to FIS as input and compute the weighting factor (K).

Step5. Decompose the watermark W into m watermark images such as W_1, W_2, W_3....W_m, where the corresponding watermark image is used to modify the frames of corresponding scene.

Step6. For each scene (j = 1, 2 ..., m) perform the following:

1) Apply 3 level DWT using HAAR filter on the luminance component (Y) of every frame of the jth scene to obtain the $LL3_i(j)$ sub-band coefficients.
2) Apply 1-level DWT using HAAR filter on each watermark image to obtain the $wLL1(j)$ sub-band coefficients.
3) Resize the $wLL1(j)$ sub-band coefficients of each watermark image to the size of $LL3_i(j)$ sub-band coefficients of each luminance frame.
4) Embed $wLL1(j)$ sub-band coefficients into $LL3_i(j)$ sub-band coefficients using the formula given by Eq. 2

$$LL3'_i(j) = LL3_i(j) + K * wLL1(j) \tag{2}$$

where K is the weighting factor.

5) Apply inverse 3-level DWT to the modified $LL3'$ sub-band coefficients to obtain the watermarked luminance component of the frame.
6) Replace the original luminance frame in YCbCrcolor space by the watermarked luminance frame.
7) Every frame is converted from the YCbCrto RGB color space to obtain the watermarked video.

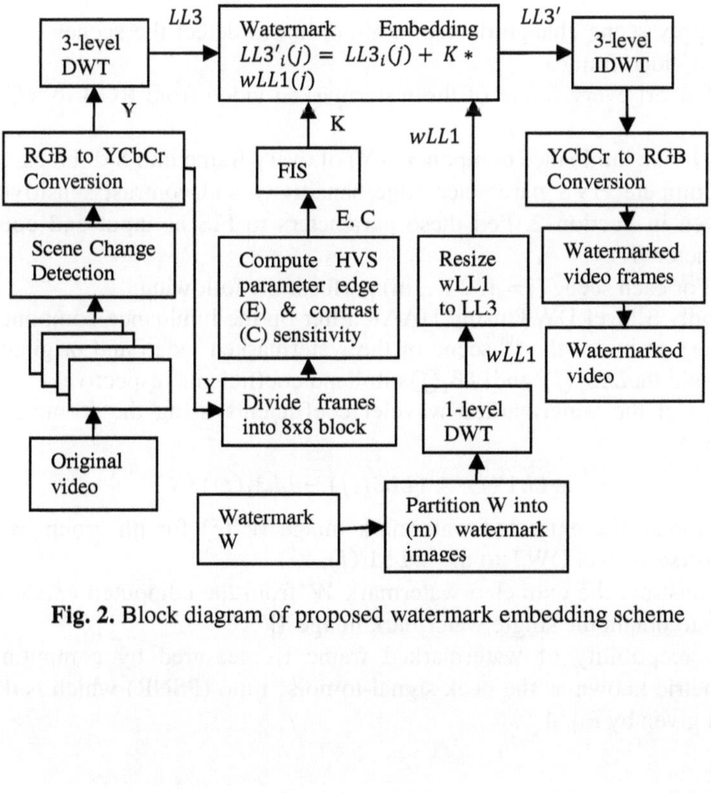

Fig. 2. Block diagram of proposed watermark embedding scheme

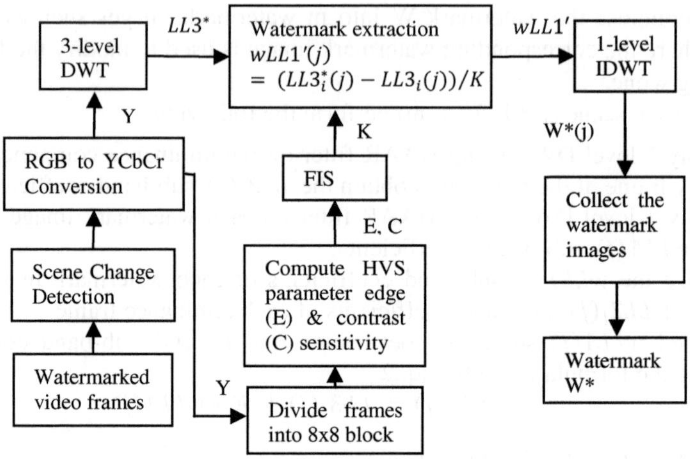

Fig. 3. Block diagram of proposed watermark extraction scheme

2.4 Watermark Extraction

Fig.3 shows the block diagram of watermark extraction scheme. The watermark extraction process is given by Listing 3.

Listing 3: Extraction Algorithm

Step1. Apply scene change detection algorithm to detect the scenes (m) from the watermarked video frames.

Step2. Convert every frame of the watermarked video from RGB to YCbCr color space.

Step3. Divide luminance component (Y) of every frame into 8×8 blocks.

Step4. Compute HVS parameter edge sensitivity and contrast sensitivity block-wise as given in Section 2. Fed these parameters to FIS as input and compute the weighting factor (K).

Step5. For each scene (j = 1, 2 …, m) perform the following:

1) Apply 3 level DWT using HAAR filter on the luminance component (Y) of every frame of the jth scene of the watermarked video and original video to obtain the$LL3_i^*(j)$ and$LL3_i(j)$sub-band coefficients respectively.
2) Extract the watermarked wavelet coefficients using the formula given by Eq. 3

$$wLL1'(j) = (LL3_i^*(j) - LL3_i(j))/K \tag{3}$$

3) Compute the extracted watermark image $W^*(j)$ for jth group by applying inverse 1-level DWT to the $wLL1'(j)$.

Step6. Construct the extracted watermark $W^*$from the computed extracted watermark image to obtain the single watermark image W^*.

The imperceptibility of watermarked frame is measured by computing a full-reference metric known as the peak signal-to-noise ratio (PSNR) which is defined by the formula given by Eq. 4

$$MSE = \frac{1}{MN}\sum_{i=0}^{M-1}\sum_{j=0}^{N-1}[I'(i,j) - I(i,j)]^2$$

where I' is the watermarked frame and I is original frame.

$$PSNR = 10log_{10}(\frac{255^2}{MSE}) \quad (4)$$

The average PSNR of all frames of the video is given by Eq. 5

$$\text{AVG\_PSNR} = \frac{\sum_{i=1}^{nf} \text{PSNR}}{\text{nf}} \quad (5)$$

where nf is the total number of frames in the video sequence.

After watermark extraction normalized correlation (NC) is computed between extracted watermark and original watermark by using the formula given by Eqn.6

$$NC(W, W^*) = \frac{\sum_i \sum_j W(i,j) W^*(i,j)}{\sum_i \sum_j [W(i,j)^2]} \quad (6)$$

where W* is the extracted watermark and W is the original watermark.

The signed video frames are also examined for robustness by executing six different spatial attacks. For this purpose, the watermarks embedded in the frames are extracted and matched with original watermarks. NC (W, W*) parameters are computed between original and extracted watermarks.

3 Experimental Results and Discussion

The performance of the proposed scheme is evaluated on three standard video sequences namely: News, Foreman and Hall Monitor in RGB uncompressed AVI format, ofsize 352 × 288 and frame rate of 30 fps. Each video sequences consists of 300 frames. A binary image of size 64×64 is used as watermark.

Fig. 4(a-c) depicts the 1st original frame of the three video sequences and Fig. 4(d) depicts the original binary watermark. Fig. 5(a-c) depicts the 1st signed frame corresponding to the frames shown in Fig. 4(a-c) respectively. Fig. 6(a-c) shows the binary watermark extracted from the three video sequences of Fig. 5(a-c) respectively. The computed values of NC (W, W*) parameters of the extracted watermarks are placed on top of them.

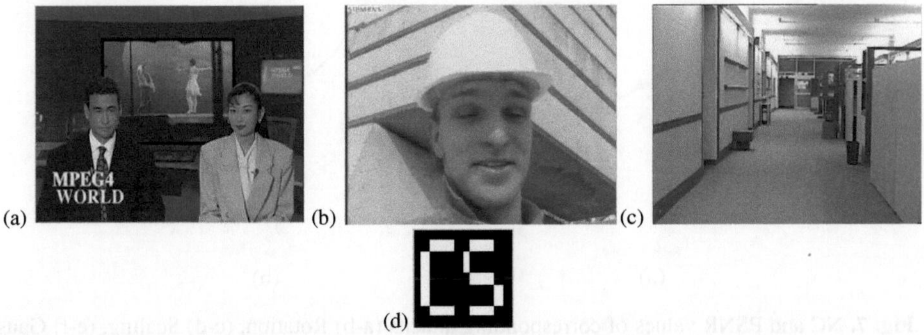

Fig. 4. (a) 1st original frame of News, (b) 1st original frame of Foreman, (c) 1st original frame of Hall Monitor, (d) Original watermark

It is clear from the Fig.5 that with the proposed algorithm, we have obtained high PSNR values for the videos. Similarly, we also obtained NC (W, W*) = 1 as shown in Fig.6 which indicate a high degree of correlation between the embedded and extracted watermark. To examine the robustness of the proposed algorithm five different video processing operations (attacks) are performed on signed video frames.

Fig. 5. (a) 1st watermarked frame of News (46.41 dB), (b) 1st watermarked frame of Foreman (46.25 dB), (c) 1st watermarked frame of Hall Monitor (46.31 dB)

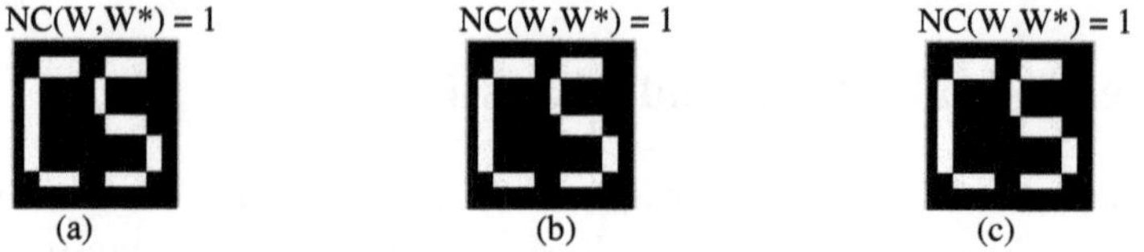

Fig. 6. (a) Extracted watermark from News, (b) Extracted watermark from Foreman, (c) Extracted watermark from Hall Monitor

Fig. 7 indicate that the plot of PSNR and NC(W,W*) is similar in case of all these attacks. The high PSNR and NC (W, W*) value clearly show that the proposed scheme is robust against the above said attacks. Our simulation results indicate good optimization of visual quality and robustness obtained by using FIS.

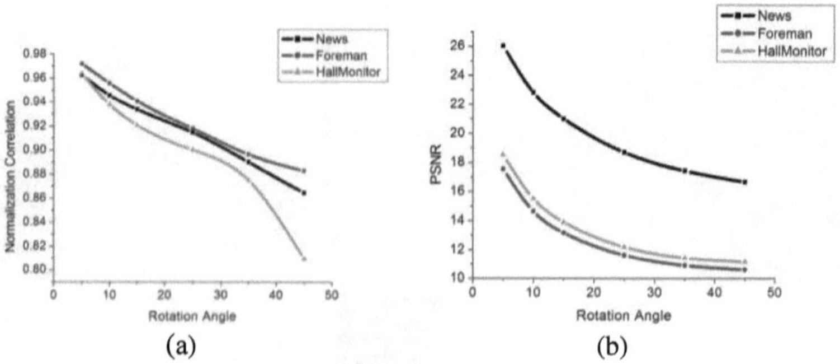

Fig. 7. NC and PSNR values of corresponding attacks (a-b) Rotation, (c-d) Scaling, (e-f) Gaussian noise, (g-h) Frame dropping, (i-j) JPEG compression

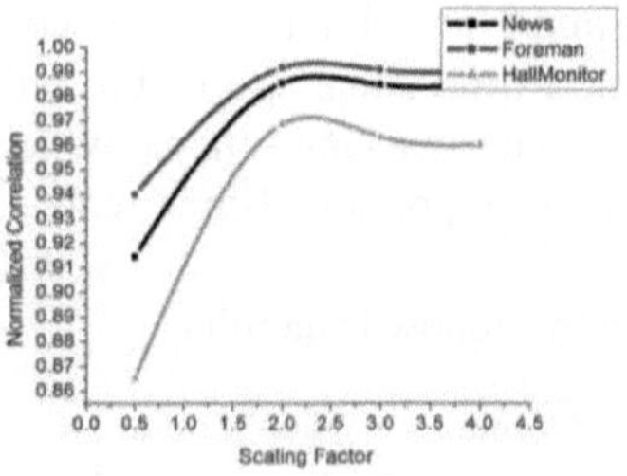

(c)

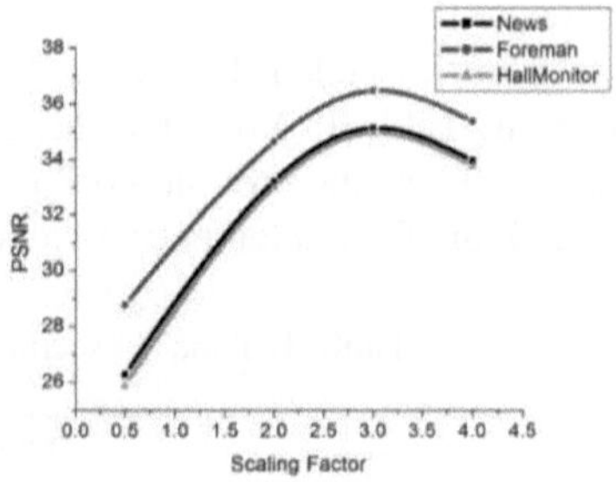

(d)

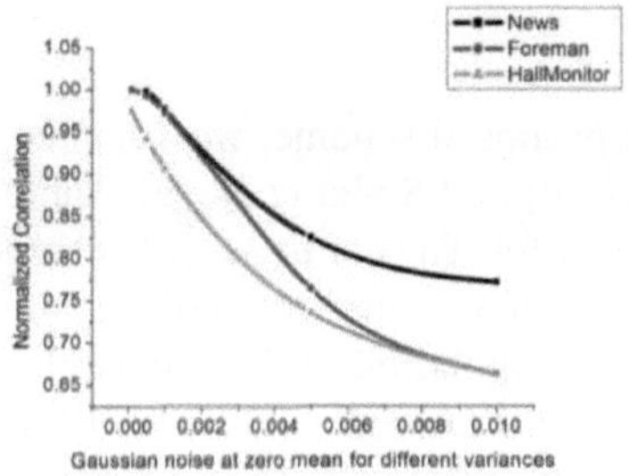

(e)

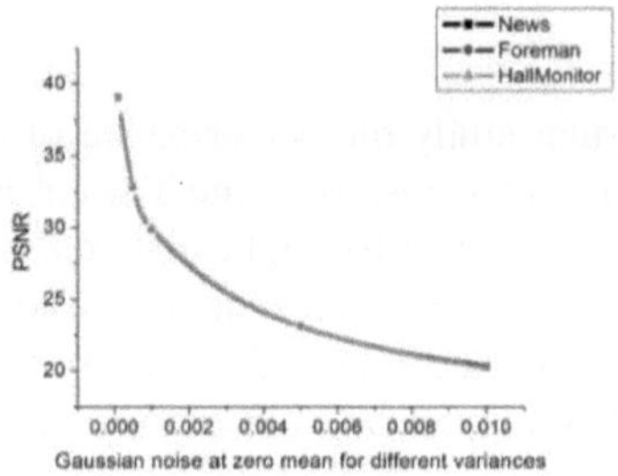

(f)

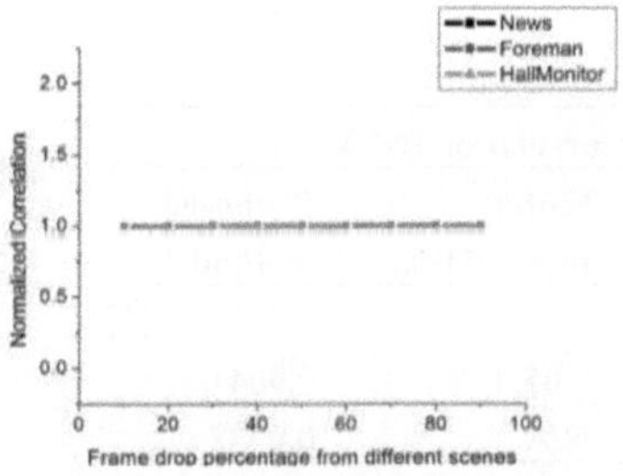

(g)

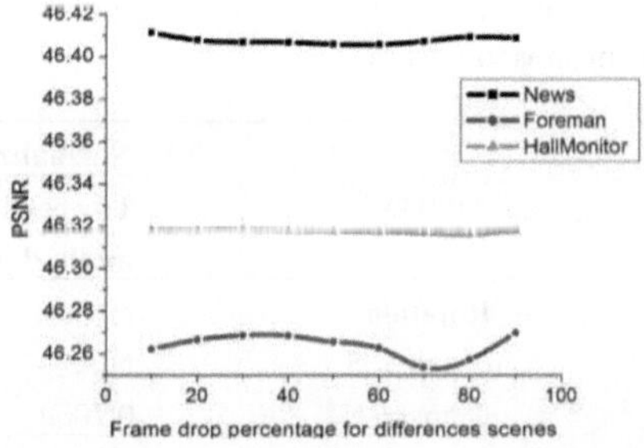

(h)

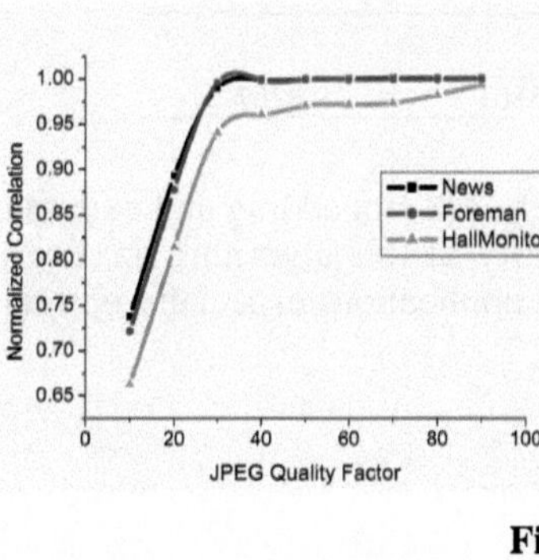

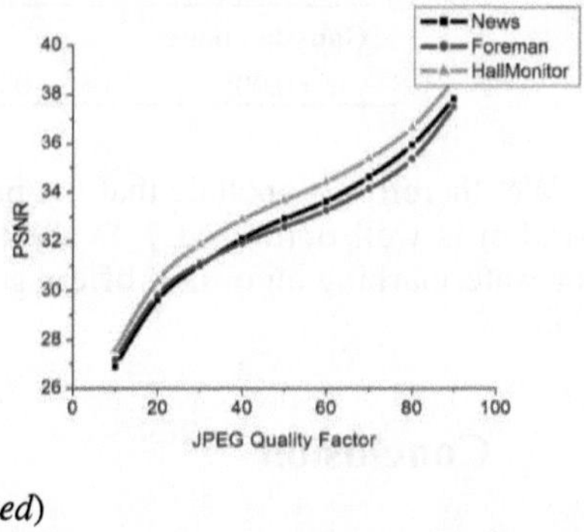

Fig. 7. (*continued*)

To estimate the time complexity of proposed algorithm, we compile the embedding and extraction time taken by News, Foreman and Hall Monitor video sequences for 300 frames in Table 1. Note that these computed time spans are of the order of few seconds only. Thus, the present work proposes a watermarking scheme which is a successful candidate for implementing video watermarking on a real-time scale.

Table 1. Time (in seconds) taken by proposed algorithm

	News	**Foreman**	**Hall Monitor**
Embedding time (sec)	33.2126	30.9350	35.9114
Extraction time (sec)	18.8761	18.9541	20.5765

To further study the performance of the proposed scheme, we compare out results with the reported results of the Taweel *et al.* [6] and Sinha *et al.* [9]. Table 3compiles NC (W,W*) values for eight different attacks for Taweel *et al.* [6] scheme, Sinha *et al.* [9] scheme and our scheme. Note that 'N/A' entry in Table 2 indicates non-availability of the reported value by the corresponding author. It is clear from the Table 2.that our method outperforms Taweel *et al.*[6] and Sinha *et al.*[9],[10] scheme in all cases.

Table 2. Comparison of NC(W,W*) values of Taweel *et al.* [6] scheme, Sinha *et al.* [9] scheme and our proposed scheme

	Normalized Correlation (NC)		
Attacks	**Taweel's method [6]**	**Sinha's method [9]**	**Proposed method**
Rotation			
degree = 5^0	N/A	0.6510	0.9641
degree=-17^0	0.7009	N/A	0.9332
Scaling	0.5208	0.6068	0.9252
Gaussian noise			
var = 0.005	0.4785	0.6861	0.8267

We, therefore, conclude that the proposed watermark embedding and extraction algorithm is well optimized. As the time complexity of this algorithm is very small, our watermarking algorithm offers good practical applications especially on real time scale.

4 Conclusion

In the present work, a novel fast and robust fuzzy-based video watermarking algorithm in wavelet domain is proposed. The LL3 sub-band coefficients of video frames are modified by the LL1 sub-band coefficients of the binary watermark image.

The low time complexity of proposed algorithm makes it suitable for watermarking of video on a real time scale. The perceptible quality of the video frames is very good as indicated by high PSNR values. Watermark recovery is also found to be good as indicated by high cross correlation values between embedded and extracted watermarks.It is concluded that the embedding and extraction of the proposed algorithm are well optimized. The algorithm is robust and shows an improvement over other similar reported methods.

References

1. Lin, D.-T., Liao, G.-J.: Embedding Watermarks in Compressed Video using Fuzzy C-Means Clustering. In: Proceedings of IEEE International Conference on Systems (2008)
2. Raghavendra, K., Chetan, K.R.: A Blind and Robust Watermarking Scheme with Scrambled Watermark for Video Authentication. In: Proceedings of IEEE International Conference onInternet Multimedia Services Architecture and Applications (2009)
3. Wang, Y., Pearmain, A.: Blind MPEG-2 Video Watermarking Robust Against Geometric Attacks: A Set of Approaches in DCT Domain. IEEE Transactions on Image Processing 15(6) (2006)
4. Lee, X., Zhang, Y.-Q., Leon-Garcia, A.: Image And Video Reconstruction Using Fuzzy Logic. In: Proceedings of IEEE International Conference on Global Telecommunications Conference (1993)
5. Masoumi, M., Amiri, S.: A blind scene-based watermarking for video copyright protection. International Journal of Electronics and Communications (2013)
6. Al-Taweel, S.A.M., Sumari, P.: Robust Video Watermarking Based On 3D-DWT Domain. In: Proceedings of IEEE International Conference on TENCON (2009)
7. Da-Wen, X.: A Blind Video Watermarking Algorithm Based on 3D Wavelet Transform. In: Proceedings of IEEE International Conference on Computational Intelligence and Security (2007)
8. Yassin, N.I., Salem, N.M., Adawy, M.I.E.: Block Based VideoWatermarking Scheme Using Wavelet Transform and Principle Component Analysis. IJCSI International Journal of Computer Science Issues 9(1), 3 (2012)
9. Sinha, S., Bardhan, P., Pramanick, S., Jagatramka, A., Kole, D.K., Chakraborty, A.: Digital Video Watermarking using Discrete Wavelet Transform and Principal Component Analysis. International Journal of Wisdom Based Computing 1(2) (2011)
10. Lande, P.U., Talbar, S.N., Shinde, G.N.: A Fuzzy Logic Approach to Encrypted Watermarking for Still Images in Wavelet Domain on FPGA. International Journal of Signal Processing, Image Processing and Pattern Recognition 3(2) (2010)

Cryptanalysis and Security Enhancement of Two Advanced Authentication Protocols

S. Raghu Talluri and Swapnoneel Roy

School of Computing
University of North Florida, USA
https://www.unf.edu/ccec/computing/

Abstract. In this work we consider two protocols for performing cryptanalysis and security enhancement. The first one by Jiang et al., is a password-based authentication scheme[1] which does not use smart cards. We note that this scheme is an improvement over Chen et al.'s scheme shown vulnerable to the off-line dictionary attack by Jiang et al. We perform a cryptanalysis on Jiang at al.'s improved protocol and observe that it is prone to the clogging attack, a kind of denial of service (DoS) attack. We then suggest an improvement on the protocol to prevent the clogging attack.

The other protocol we consider for analysis is by Wang et al. This is a smart card based authentication protocol. We again perform the clogging (DoS) attack on this protocol via replay. We observe that all smart card based authentication protocols which precede the one by Wang et al., and require the server to compute the computationally intensive modular exponentiation are prone to the clogging attack. We suggest (another) improvement on the protocol to prevent the clogging attack, which also applies to the protocol by Jiang et. al.

Keywords: Authentication Protocols, Smart Cards, DoS, Replay Attacks, Clogging Attack.

1 Introduction

In a cyber environment, user authentication can enable a perimeter device (a firewall, proxy server, VPN server, remote access server, etc.) to decide whether or not to approve a specific user's request to gain entry to the network.

It is necessary to be able to identify and authenticate users with a high level of certainty, so that they may be held accountable should their actions threaten the security and productivity of the network. The more confidence network administrators have that a user is who they say they are, the more confidence they will have in allowing those users specific privileges; and the more faith they will have in their network devices' internal records regarding that user. Reliable user authentication can help achieve what are necessary elements in basic network

[1] We use the terms *scheme* and *protocol* interchangeably in this work.

M.K. Kundu et al. (eds.), *Advanced Computing, Networking and Informatics - Volume 2*, Smart Innovation, Systems and Technologies 28,
DOI: 10.1007/978-3-319-07350-7_34, © Springer International Publishing Switzerland 2014

security positively identifying someone; allowing them specific rights; and holding them accountable for their actions should they compromise the security and productivity of the network for other users on the network.

Multi-factor authentication is as an approach to cyber-security authentication, in which the user of a system is required to provide more than one form of verification in order to prove their identity and allowed access to the system. It takes advantage of a combination of several factors of authentication; three major factors include verification by: (1) something a user knows (such as a password), (2) something the user has (such as a smart card or a security token), and (3) something the user is (such as the use of biometrics). Due to their increased complexity, authentication systems using a multi-factor configuration are harder to break than ones using a single factors.

The first such multi-factor authentication protocol we consider here is by Jiang et al. [1]. It is a *memory device aided* password authentication protocol. In this kind of protocols (e.g. [1], [4], [6]), the authentication information (issued by a server) is stored in a memory device such as universal serial bus (USB) sticks, portable HDDs, mobile phones, PDAs, PCs etc. A very common example is a software protection dongle that is used frequently now-a-days for various purposes.

The other protocol we consider by Wang et al. [3] is a smart card based authentication protocol. Smart card based password authentication (e.g. [2], [3], [7], [8], [9], [10], [11]) is one of the most convenient and effective two-factor authentication mechanisms for remote systems. This technique has been widely deployed for various kinds of authentication applications, such as remote host login, online banking, shopping on the internet, e-commerce and e-health. Also, it constitutes the basis of three-factor authentication. However, there still exists challenges in both security and performance aspects due to the stringent security requirements and resource strained characteristics of the clients.

1.1 Our Results

We first analyze the protocol by Jiang et al. Their protocol is an improvement over Chen et al.'s [4] protocol which they show to be insecure against the offline password guessing attack. There are protocols in the literature which came before Chen et al.'s protocol e.g. [6], that have been shown to be vulnerable against some form of attacks. We find Jiang et al.'s protocol to be insecure against the *clogging attack*, a form of denial of service (DoS). The inherent vulnerability lies in the usage of the computationally intensive modular exponentiation by the server in the authentication process. We then present a fix to prevent an attacker to perform such an attack on the protocol. We note there has been another recent protocol by the same authors which is smart card based [2]. We observe that protocol also to be insecure against the clogging attack.

The second protocol we analyze is a smart card based protocol by Wang et. al [3]. They have actually claimed their protocol to be secure against DoS. But we however find the protocol to be insecure against the clogging attack. We show an attacker can exploit the fact their protocol uses multiple modular exponentiations for authentication. A *replay attack* can be launched on their protocol to achieve a bigger clogging attack. However clogging attack can be done on this protocol in the classical way (without replays). We propose a way of making the protocol secure against this attack. This fix also works for the protocol by Jiang et al. to make it immune against clogging attacks.

Our observation is modular exponentiation is a technique which guarantees a level of security. But it might lead to an easy insecurity just in case it is used without an additional level of protection. Most of the multi-factor authentication protocols in the literature either smart card based, or memory device aided rely on the usage of modular exponentiation for their security. Hence some level of protection should be added to them to guarantee total security against the clogging attack.

2 Jiang et al.'s Password Based Protocol

The first protocol we look at in this work is due to Jiang et al. [1] It is a remote authentication protocol, which does not involve smart cards. We however note that they had another version of the protocol which works with smart card in[2]. Once we demonstrate the vulnerability in [1] against the clogging attack, the vulnerability is easily observed to work for [2] as well. Jiang et al.'s protocol in [1] is an improvement over Chen at al.'s protocol [4] which they proved to be vulnerable against the off-line dictionary attack.

We briefly present Jiang et. al.. They prove their protocol to be immune from various attacks in [1]. However we see their protocol to be inherently vulnerable to the clogging attack (a form of the classical DoS). We present a clogging attack on the protocol. We observe their smart card based version of the protocol of [2] also to be insecure against this attack. Most of the protocols they cite in their papers [1] and [2] are vulnerable to clogging attack. We identify the mathematical basis which make the protocols vulnerable to this attack, and suggest a possible fix for them.

2.1 Review of the Protocol

Jiang et. al's protocol works in five phases: *Initialization*, *Registration*, *Login*, *Authentication*, and *Passoword Change*. We present the protocol in Algorithm 1. We omit the password change phase since it is not required to demonstrate the clogging attack on the protocol.

2.2 Attack on Jiang el. al.'s Protocol

The adversary $\mathcal{A}$ has the same power as assumed by Jiang et al's [1] while exposing the weaknesses of Chen et. al's protocol. We only need $\mathcal{A}$ to be able to read and modify the contents of messages over an insecure channel (during Login and Authentication phase of the protocol).

1. $\mathcal{A}$ intercepts a valid login request ($\{ID_i, C_i, V_i, T_1\}$) from step **Step L3**.

Algorithm 1. Jiang et. al.'s scheme of password authentication

1:

INITIALIZATION PHASE

`Server` S

1. **Step I1.** Choose large prime numbers p and q such that $p = 2q + 1$.
2. **Step I2.** Choose a generator g of $\mathbf{Z}_q^*$, secret key $x \in \mathbf{Z}_q^*$, and secure one way hash $\mathcal{H}$.
3. **Step I3.** Compute public key $X = g^x \mod p$.

REGISTRATION PHASE

`User` U_i

1. **Step R1.** Choose identity ID_i, password PW_i.
2. **Step R2.** $U_i \rightarrow S$: $\{ID_i, PW_i\}$ through a *secure channel.*

`Server` S

1. **Step R3.** On receiving the registration message from U_i, S creates an entry for U_i in the account-database and stores ID_i in this entry. Next, S computes $Y_i = \mathcal{H}(ID_i \| x)) \otimes \mathcal{H}(PW_i)$.
2. **Step R4.** $S \rightarrow U_i$: $\{X, Y_i, \mathcal{H}, p, q\}$.

`User` U_i

1. **Step R5.** Upon receiving $\{X, Y_i, \mathcal{H}, p, q\}$ from S, U_i enters it locally in his/her memory device (e.g. USB stick).

LOGIN AND AUTHENTICATION

`User` U_i

1. **Step L1.** U_i chooses a random number $\alpha \in \mathbf{Z}_q^*$.
2. **Step L2.** U_i computes $Y_i' = Y_i \otimes \mathcal{H}(PW_i)$, $C_i = g^\alpha \mod p$, $D_i = x^\alpha \mod p$, and $V_i = \mathcal{H}(ID_i \| Y_i' \| C_i \| D_i \| T_1)$, where T_1 is the current system time of U_i.
3. **Step L3.** $U_i \rightarrow S$: $\{ID_i, C_i, V_i, T_1\}$.

`Server` S

1. **Step V1.** S checks whether ID_i is valid from its stored value, and $(T_2 - T_1) < \Delta T$, where T_2 is the current system time for S. If either does not hold, the request is dropped, and the session is terminated. Otherwise, S computes $Y_i'' = \mathcal{H}(ID_i \| x)$ and $D_i' = C_i^x \mod p = g^{x\alpha} \mod p = X^\alpha \mod p = D_i$, and compares V_i with $\mathcal{H}(ID_i \| Y_i'' \| C_i \| D_i' \| T_1)$. If they are not equal the session is terminated. Otherwise S authenticates U_i and the login request is accepted. S computes $M_i = \mathcal{H}(ID_i \| D_i' \| T_3)$, where T_3 is the current system time of S.
2. **Step V2.** $S \rightarrow U_i$: $\{M_i, t_3\}$.

`User` U_i

1. **Step V3.** On receiving the reply message from the server S, U_i checks whether T_3 is valid, and M_i is equal to $\mathcal{H}(ID_i \| D_i \| T_3)$. This equivalency authenticates the legitimacy of the server S, and mutual authentication between S and U_i is achieved. Otherwise S is not authenticated.

COMPUTE SESSION KEY

`User` U_i

$sk_U = \mathcal{H}(D_i)$

`Server` S

$sk_S = \mathcal{H}(D_i')$

2. Since the message is unencrypted, $\mathcal{A}$ can change the timestamp T_1 to some $T_{\mathcal{A}}$ so that it meets the criterion $(T_2 - T_{\mathcal{A}}) < \Delta T$.
3. $\mathcal{A}$ changes C_i to any random garbage value $C_{\mathcal{A}}$.
4. $\mathcal{A}$ then sends $\{ID_i, C_{\mathcal{A}}, V_i, T_{\mathcal{A}}\}$ to the server S.

The following is performed by the server S:

1. Check whether ID_i is valid. Here it is valid.
2. Check whether the difference between $(T_2 - T_{\mathcal{A}}) < \Delta T$. This step passes as well.
3. Compute $Y_i'' = \mathcal{H}(ID_i \| x)$ and $D_i' = C_{\mathcal{A}}^x \bmod p$, and compare V_i with $\mathcal{H}(ID_i \| Y_i'' \| C_{\mathcal{A}} \| D_i' \| T_{\mathcal{A}})$. This fails, so the request gets rejected.

The point here is the adversary $\mathcal{A}$ would now repeat the steps several times and make the server S compute the modular exponentiation step several times. Basically $\mathcal{A}$ can potentially change all the incoming login request messages from any legitimate user to S. Since modular exponentiation is computationally intensive, the victimized server spends considerable computing resources doing useless modular exponentiation rather than any real work. Thus $\mathcal{A}$ clogs S with useless work and therefore denies any legitimate user any service. $\mathcal{A}$ just needs an ID of a single valid user to perform the clogging attack repeatedly.

2.3 Clogging Attack Performed on Other Similar Schemes

Jiang et al., devised another smart card based password authentication protocol in [2]. This work was an improvement over another such scheme by Chen et al. [5]. We observe, that the clogging attack performed on the current protocol under consideration can also be performed on both the protocols [2] and [5]. Both the protocols are vulnerable because, the users smart card does not encrypt the message it sends over to the server for login and authentication. This gives an adversary the chance to manipulate this message.

2.4 Proposed Countermeasures from the Attack

The Steps to Avoid the Clogging Attack. At the beginning of the authentication phase, the server could check whether the network address of the user is valid. It has to know the network addresses of all the registered legitimate users. In spite of that, adversary $\mathcal{A}$ could spoof the network address of a legitimate user and replay the login message. To prevent it, we might add a cookie exchange step at the beginning of the login phase of Jiang et al.s scheme. This step has been designed as in the well known Oakley key exchange protocol [12].

1. The user U_i chooses a pseudo-random number n_1 and sends it along with the message $\{ID_i, C_i, V_i, T_1\}$.
2. The server S upon receiving the message, acknowledges the message and sends its own cookie n_2 to U_i.
3. The next message from U_i must contain n_2, else S rejects the message and the login request.

Security Analysis of the Fix. Had $\mathcal{A}$ spoofed the U_i's IP address, $\mathcal{A}$ would not get n_2 back from S. Hence $\mathcal{A}$ only succeeds to have the S send back an acknowledgement, but not to compute the computationally intensive modular exponentiation. Hence the clogging attack is avoided by these additional steps. Saying this, we would note that this process does not prevent the clogging attack but only thwarts it to some extent. This fix can fully work if n_1, and n_2 are encrypted respectively by the U_is and Ss private keys for a secure communication.

3 Wang et al.'s Smart Card Based Protocol

We briefly present a very recent smart card based authentication protocol by Wang et. al. [3]. They claim their protocol to be immune from the DoS attack. They assume a situation of a stolen smart card to prove this. However we see their protocol to be inherently vulnerable to the clogging (DoS) attack. The attacker would not have to steal the smart card to perform a clogging attack on their protocol. We present a clogging attack on the protocol via *Replay*. We however note a replay is not necessary to perform this attack on this protocol. But a replay step by the attacker, makes the clogging attack more effective. Most of the smart card based protocols they cite in their paper [3] are vulnerable to this attack.

3.1 Review of the Protocol

Wang et. al's protocol (as like most other smart card based protocols), has the *Registration*, *Login*, and the *Verification* phases. We present the protocol in Algorithm 2.

3.2 Replay Attack on Wang el. al.'s Protocol

A replay attack is a form of network attack in which a valid data transmission is maliciously or fraudulently repeated or delayed. Replays can be used to gain unauthorized access, or may be done simply to perform a DoS. This is carried out either by the originator or by an adversary who intercepts the data and retransmits it, possibly as part of a masquerade attack by IP packet substitution (such as stream cipher attack).

Wang el. al.'s protocol [3] was claimed to be secured against replay attacks but as we see, we have been able to perform a replay attack on this protocol to achieve a DoS. We assume the protocol is known to $\mathcal{A}$ (i.e. not security under obscurity).

1. $\mathcal{A}$ intercepts a valid login request ($\{C_1, CID_i, M_i\}$) from step **Step L4**.
2. $\mathcal{A}$ replays $\{C_1, CID_i, M_i\}$ several times. That is, it performs $\mathcal{A} \rightarrow S$: $\{C_1, CID_i, M_i\}$ a large number of times.
3. This will force S_i perform three modular exponentiations $Y_1 = (C_1)^x \mod p$, $KS = (C_1)^v \mod p$, and $C_2 = g^v \mod p$ of **Step V1**.

Algorithm 2. Wang et. al.'s scheme of password authentication

1:

REGISTRATION PHASE

`User` U_i

1. **Step R1.** Choose identity ID_i, password PW_i and a random number b.
2. **Step R2.** $U_i \rightarrow S$: $ID_i, \mathcal{H}_0(b \| PW_i)$.
3. **Step R3.** Upon receiving the smart card SC, U_i enters b into SC.

`Server` S

1. **Step R4.** On receiving the registration message from U_i at time T, S first checks whether U_i is a registered user. If it is U_is initial registration, S creates an entry for U_i in the account-database and stores $(ID_i, T_{reg} = T)$ in this entry. Otherwise, S updates the value of T_{reg} with T in the existing entry for U_i. Next, S computes $N_i = \mathcal{H}_0(b \| PW_i)) \otimes \mathcal{H}_0(x \| ID_i \| T_{reg})$ and $A_i = \mathcal{H}_0((\mathcal{H}_0(ID_i) \otimes \mathcal{H}_0(b \| PW_i)) \mod n)$.
2. **Step R5.** $S \rightarrow U_i$: A smart card containing security parameters $\{N_i, A_i, q, g, y, n, \mathcal{H}_0(\cdot), \mathcal{H}_1(\cdot), \mathcal{H}_2(\cdot), \mathcal{H}_3(\cdot)\}$.

`User` U_i

1. **Step R6.** Upon receiving the smart card SC, U_i enters b into SC.

LOGIN AND AUTHENTICATION

`User` U_i

1. **Step L1.** U_i inserts her smart card into the card reader and inputs ID_i^* , PW_i^*.
2. **Step L2.** SC computes $A_i^* = \mathcal{H}_0((\mathcal{H}_0(ID_i^*) \otimes \mathcal{H}_0(b \| PW_i^*)) \mod n)$ and verifies the validity of ID_i^* and PW_i^* by checking whether A_i^* equals the stored A_i. If the verification holds, it implies $ID_i^* = ID_i$ and $PW_i^* = PW_i$ with a probability of $\frac{n-1}{n} (\approx \frac{99.90}{100}$, when $n = 2^{10})$. Otherwise, the session is terminated.
3. **Step L3.** SC chooses a random number u and computes $C_1 = g^u \mod p$, $Y_1 = y^u \mod p$, $k = \mathcal{H}_0(x \| ID_i \| T_{reg}) = N_i \otimes \mathcal{H}_0(b \| PW_i)$, $CIDi = ID_i \otimes \mathcal{H}_0(C_1 \| Y_1)$ and $M_i = \mathcal{H}_0(Y_1 \| k \| CID_i)$.
4. **Step L4.** $U_i \rightarrow S$: $\{C_1, CID_i, M_i\}$.

`Server` S

1. **Step V1.** S computes $Y_1 = (C_1)^x \mod p$ using its private key x. Then, S derives $ID_i = CID_i \otimes \mathcal{H}_0(C_1 \| Y_1)$ and checks whether ID_i is in the correct format. If ID_i is not valid, the session is terminated. Then, S computes $k = \mathcal{H}_0(x \| ID_i \| T_{reg})$ and $M_i^* = \mathcal{H}_0(Y_1 \| k \| CID_i)$, where T_{reg} is extracted from the entry corresponding to ID_i. If M_i^* is not equal to the received M_i, the session is terminated. Otherwise, S generates a random number v and computes the temporary key $KS = (C_1)^v \mod p$, $C_2 = g^v \mod p$ and $C_3 = \mathcal{H}_1(ID_i \| IDS \| Y_1 \| C_2 \| k \| KS)$.
2. **Step V2.** $S \rightarrow U_i$: $\{C_2, C_3\}$.

`User` U_i

1. **Step V3.** On receiving the reply message from the server S, SC computes $KU = (C_2)^u \mod p$, $C_3^* = \mathcal{H}_1(ID_i \| IDS \| Y_1 \| C_2 \| k \| KU)$, and compares C_3^* with the received C_3. This equivalency authenticates the legitimacy of the server S, and U_i goes on to compute $C4 = \mathcal{H}_2(ID_i \| IDS \| Y_1 \| C_2 \| k \| KU)$.
2. **Step V4.** $U_i \rightarrow S$: $\{C_4\}$

2: Wang et. al.'s scheme (contd.)

Server S

1. **Step V5.** Upon receiving $\{C_4\}$ from U_i, the server S first computes $C_4^* = \mathcal{H}_2(ID_i\|IDS\|Y_1\|C_2\|k\|KS)$ and then checks if C_4^* equals the received value of C_4. If this verification holds, S authenticates the user U_i and the login request is accepted else the connection is terminated.

COMPUTE SESSION KEY

User U_i

- $sk_U = \mathcal{H}_3(ID_i\|IDS\|Y_1\|C_2\|k\|KU)$

Server S

- $sk_S = \mathcal{H}_3(ID_i\|IDS\|Y_1\|C_2\|k\|KS)$

4. $\mathcal{A}$ can intercept whatever replies S_i sends (**Step V1**) and discard them (they would anyway be lost since SC will not expect these replies).

We note that the attacker $\mathcal{A}$ can simply send fake login requests to the server S and could have launched the clogging (DoS) attack having forced S to perform $Y_1 = (C_1)^x \mod p$ on **Step V1**. But this replay attack results in a bigger DoS attack on S since it is forced to perform three modular exponentiations (in place of just one). $\mathcal{A}$ will need to send much lesser number messages to S to clog it. This replay attack is possible because, unlike Jiang et. al's protocol, Wang et at.'s protocol does not have a timestamp check.

3.3 Proposed Countermeasures from the Attack

The Steps to Avoid Replay Attack Resulting in Clogging Attack. As we observe replay attacks also might be possible on most Smart card based protocols because their security relies on the computationally intensive modular exponentiation, and the messages are not by default encrypted. This is very often overlooked, since the natural result of a replay is not a DoS. A few steps to avoid these attacks on Wang et. al's Protocol (and in all Smart Card based protocols in general) would be

1. U_i uses a time stamp T in **Step L4.**, and S verifies it in **Step V1.**. The time stamp also must be encrypted in some form so that $\mathcal{A}$ cannot tamper with it.
2. S checks whether multiple login requests frequently comes from the same user. This *reduces* the chances of a reply.

We say *reduces* because $\mathcal{A}$ can obtain a lot of valid user ids (they are public) and send fake login requests periodically from different ids. Or $\mathcal{A}$ can store various (valid) login requests over a time period, and reply them periodically.

Yet Another Way to Prevent Clogging Attack. We identify the mathematical basis which make the protocols vulnerable to clogging attacks is the modular exponentiation. The complete removal of this attack again requires to *encrypt* all the messages between U_i and S. But this would involve a key exchanging step, where each user has a private key, and a public key. The server knows the public key, and can decrypt a message encrypted by a users private key. That way, the server makes sure that the message is from a valid user, before it computes the costly modular exponentiation. This comes with a cost and depends on the level of security we want to implement. This countermeasure works for all the protocols (smart card and non smart card based).

4 Conclusion

In this paper, we have demonstrated clogging attacks on two advanced password authentication schemes to uncover the subtleties and challenges in designing this type of protocols. We observe modular exponentiation to be a technique that guarantees a level of security. But it might lead to an easily-exploitable vulnerability just in case it is used without an additional level of protection. Most of the multi-factor authentication protocols in the literature either smart card based, or memory device aided rely on the usage of modular exponentiation for their security. Hence some level of protection should be added to them to guarantee total security against the clogging attack.

References

1. Jiang, Q., Ma, J., Li, G., Ma, Z.: An Improved Password-Based Remote User Authentication Protocol without Smart Cards. Information Technology and Control 42(2) (2013)
2. Jiang, Q., Ma, J., Li, G., Ma, Z.: Improvement of Robust Smart-card-based Password Authentication Scheme. International Journal of Communication Systems (2013)
3. Wang, D., Ma, C.: Robust smart card based password authentication scheme against smart card security breach. Cryptology ePrint Archive (2013), `http://eprint.iacr.org/2012/439`
4. Chen, B.L., Kuo, W.C., Wuu, L.C.: A Secure Password-Based Remote User Authentication Scheme Without Smart Cards. Information Technology and Control 41(1) (2012)
5. Chen, B.L., Kuo, W.C., Wuu, L.C.: Robust Smart Card Based Remote Password Authentication Scheme. International Journal of Communication Systems (2012)
6. Rhee, H.S., Kwon, J.O., Lee, D.H.: A remote user authentication scheme without using smart cards. Computer Standards & Interfaces Archive 31(1) (2009)
7. Chen, T., Hsiang, H., Shih, W.: Security enhancement on an improvement on two remote user authentication schemes using smart cards. Future Generation Computer Systems 27(4), 377–380 (2011)
8. He, D., Chen, J., Hu, J.: Improvement on a smart card based password authentication scheme. Journal of Internet Technology 13(3), 38–42 (2012)
9. Juang, W.S., Chen, S.T., Liaw, H.T.: Robust and efficient password-authenticated key agreement using smart cards. IEEE Transactions on Industrial Electronics 55(6), 2551–2556 (2008)

10. Li, C.T.: A new password authentication and user anonymity scheme based on elliptic curve cryptography and smart card. IET Information Security 7(1), 3–10 (2013)
11. Sood, S.K.: Secure dynamic identity-based authentication scheme using smart cards. Information Security Journal: A Global Perspective 20(2), 67–77 (2011)
12. Orman, H.: The Oakley Key Determination Protocol. University of Arizona. TR 97 02

Grayscale Image Encryption Based on High-Dimensional Fractional Order Chua's System

Tanmoy Dasgupta, Pritam Paral, and Samar Bhattacharya

Department of Electrical Engineering,
Jadavpur University, Kolkata - 700032, India
{tdg.nexttonewton,callinpritam}@gmail.com,
samar_bhattacharya@ee.jdvu.ac.in

Abstract. This paper proposes a new image encryption algorithm that makes the use of high dimensional fractional order Chua's chaotic system. Fractional order extension of the Chua's system gives a much larger key-space than its original integer order version. The proposed image encryption algorithm uses a simple but excellent technique which is quite fast and the encrypted images are found to have very high entropy. The algorithm is shown to be highly robust and almost invulnerable to statistical attacks. Moreover, the algorithm is designed in such a way that it can be extended by incorporating other chaotic systems as well.

Keywords: Chaos, fractional order system, image encryption.

1 Introduction

The application of chaotic systems in image encryption was first proposed by Matthews [1]. Since then different methods of image encryption based on chaotic ciphers are developed and some of them are quite popular. Most of the methods [2–4] first employ a pixel scrambling method based on some chaotic wavelet to jumble the pixel positions. Then a standard chaotic system is used to change the intensity levels of the pixels in the image.

In this paper, the proposed image encryption method makes use of fractional order chaotic system. Memristor-based four-dimensional fractional order Chua's system [5] is used for this purpose. Since the response of chaotic systems are heavily dependent on the initial conditions, integer order chaotic systems based image encryption methods already have a very large key space. The key space further increases if the fractional order extensions of the chaotic systems are considered, as the order of differentiation can be varied along with the initial conditions. Moreover, some modifications over the conventional methods are shown in this paper to increase the key space further. However, the method that is discussed in this paper does not employ any pixel-scrambling method. This reduces the key space a little, but on the other hand, reduces the computation time for the encryption of large image files.

DOI: 10.1007/978-3-319-07350-7_35,

In the following sections, we first discuss the fundamentals of fractional order memristor-based Chua's system in brief, then the proposed image encryption method is discussed in details.

2 Fractional Order Chua's System

Fractional order Chua's chaotic systems has a memristor-based piecewise non-linear element and is described by (1) as follows

$$\mathcal{D}^{q_1}x_1 = \alpha(x_2 - x_1 + \epsilon x_1 - W(x_4)x_1) \tag{1a}$$

$$\mathcal{D}^{q_2}x_2 = x_1 - x_2 + x_3 \tag{1b}$$

$$\mathcal{D}^{q_3}x_3 = -\beta x_2 - \gamma x_3 \tag{1c}$$

$$\mathcal{D}^{q_4}x_4 = x_1 \tag{1d}$$

The piecewise non-linearity $W(x_4)$ (produced by a memristor) is defined as

$$W(x_4) = \begin{cases} a & , |x_4| < 1 \\ b & , |x_4| \geq 1 \end{cases} \tag{2}$$

and the fractional (q-th) order Riemann-Liouville (R-L) derivative [6] $\mathcal{D}^q f(t)$ is defined as

$$\mathcal{D}^q f(t) = \frac{1}{\Gamma(-q)} \int_0^t f(y)(t-y)^{-q-1}dy \quad \text{for } q \in \mathbb{R}^-, \tag{3}$$

along with the product rule of operators $\mathcal{D}^m\mathcal{D}^n = \mathcal{D}^{m+n}$ to compute (3) for $q \in \mathbb{R}^+$.

The values of the parameters are assumed to be $\alpha = 10$, $\beta = 13$, $\gamma = 0.1$, $\epsilon = 1.5$, $a = 0.3$, $b = 0.8$.

Now, consider the choice of the derivative orders

$$\mathbf{q} = \begin{bmatrix} q_1 \\ q_2 \\ q_3 \\ q_4 \end{bmatrix} = \begin{bmatrix} 0.97 \\ 0.97 \\ 0.97 \\ 0.97 \end{bmatrix} \tag{4}$$

and an initial condition

$$\mathbf{x}_0 = \begin{bmatrix} x_1(0) \\ x_2(0) \\ x_3(0) \\ x_4(0) \end{bmatrix} = \begin{bmatrix} 0.82267 \\ 0.12501 \\ 0.00796 \\ 0.78956 \end{bmatrix} . \tag{5}$$

The corresponding phase trajectories for a simulation time of 100s is shown in Fig. 1(a) and Fig. 1(b).

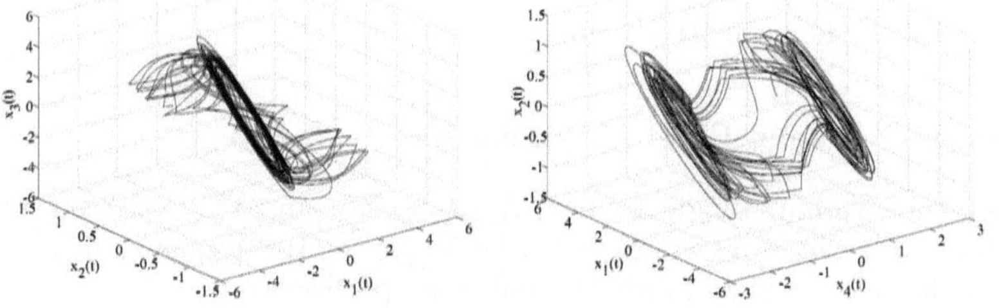

(a) (b)

Fig. 1. (a) Phase trajectory of (1) showing x_1, x_2 and x_3, and (b) Phase trajectory of (1) showing x_1, x_2 and x_4

3 The Image Encryption Algorithm

The proposed image encryption algorithm has two stages.

The *first stage* is to find the solution of the system in (1) for a finite amount of time. Then, the state-time history is manipulated to get an array (of sufficient size) of pseudo-random sequence of 8-bit integers.

Then, the *second stage* is to manipulate the image pixels of the original image by using of the array computed in *stage-1*. These are discussed in detail.

3.1 Stage-1: Pseudo-random Array Generation

1. Compute the total number of pixels in the original image.
2. Find the suitable number of iteration of (1) so that the array of state-time history generated will be just large enough to manipulate all the pixels of the given image. Solver methods capable of avoiding transient conditions are preferred.
3. Generate the pseudo-random sequence of 8-bit integers by using

$$x_i = (|x_i| - \lfloor |x_i| \rfloor) \times 10^{14} \pmod{256} \tag{6}$$

   The above procedure gives a very good set of pseudo-random 8-bit unsigned integers almost uniformly distributed between 0 and 255. Now, as the initial conditions and the order of derivatives can be set arbitrarily, we have a total of eight control parameters that constructs the key space. Thus, we can have a key space as large as $(10^{14})^8 = 10^{112}$. The pseudo-random sequence thus generated is extremely sensitive to the control parameters.

   Along with the above steps, we add another one (optional step) to randomize the sequence further and thus increasing the key sensitivity.
4. After step-2 and before going into step-3, add the last value of the computed state vector $\mathbf{x}$ to all the state vectors. Then proceed to step-3.

   This is done keeping in mind that the chaotic systems are highly sensitive to initial conditions and the derivative orders. A little change in these produces very large and almost unpredictable changes in the long run. Thus, the terminal states are the most unpredictable ones. Obviously, adding them to the other states increases the randomization of them as well. This definitely increases the key sensitivity further.

3.2 Stage-2: Image Pixel Manipulation

Consider the original image to be $[R_{i,j}]_{m\times n}$, i.e. an array of 8-bit unsigned integers of order $m \times n$, where m =row length of the image and n =column length of the image. Then apply the following procedure:

1. Create a key-image $[K_{i,j}]_{m\times n}$ by taking the values $\mathbf{x}_{i,j}$ sequentially. Say for example,

$$K_{1,1} = x_1(0),$$
$$K_{2,1} = x_2(0),$$
$$K_{3,1} = x_3(0),$$
$$K_{4,1} = x_4(0).$$

   Repeat this procedure as

$$K_{5,1} = x_1(h),$$
$$K_{6,1} = x_2(h),$$
$$K_{7,1} = x_3(h),$$
$$K_{8,1} = x_4(h),$$

   and so on, until all the elements of $[K_{i,j}]_{m\times n}$ are filled.
   Here, h is the computation step while finding the solution of (1).
   Now one can easily create a simple algorithm that will jumble up the above sequence based on some control parameter. This increases the key space still further.
   The key image thus produced using the orders (4) and initial conditions (5) of the fractional order Chua's system is shown in Fig. 2(a).
2. Bitwise XOR operation is done between each image pixel and the corresponding pixel in the key image

$$[S_{i,j}] = [R_{i,j}] \oplus [K_{i,j}]. \tag{7}$$

   The resulting image thus formed ($[S_{i,j}]$) is the encrypted version of the original image.

3.3 Decryption

The encrypted image can be easily decrypted by bitwise-XORing the encrypted image with the same key-image. This actually follows from the fact that $(A \oplus B) \oplus B = A$.

4 Experimental Results

Using Matlab 8.0 as the experimental platform and using the key-image as shown in Fig. 2(a), the algorithm discussed so far is tested on 256×256 8-bit grayscale Lena image.

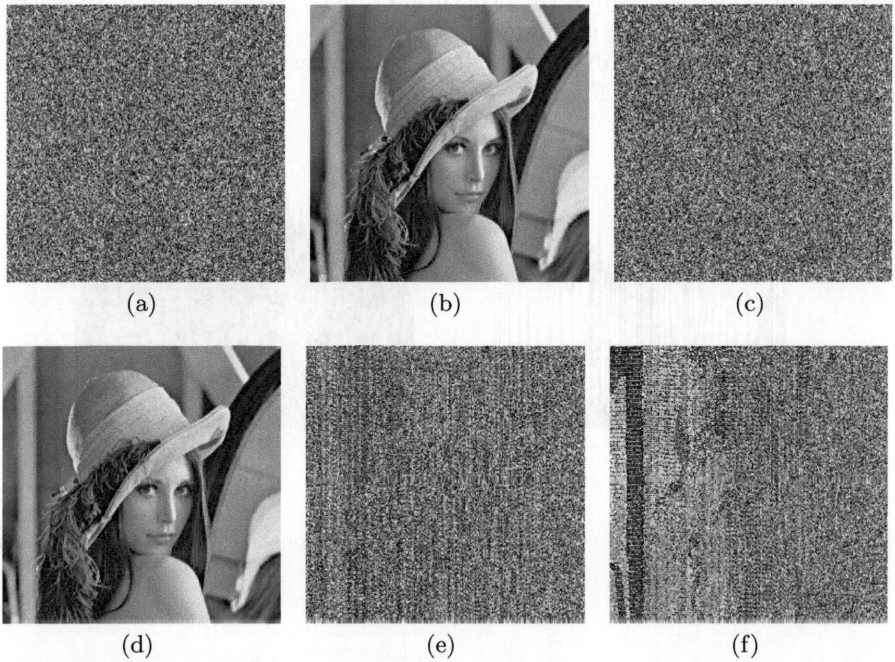

Fig. 2. (a) A sample key image, (b) The original Lena image, (c) The encrypted image, (d) Decrypted image with exact keys, (e) Decryption when $x_1(0)$ is set to 0.822670000001 instead of 0.82267 and (f) Decryption when $x_1(0)$ is set to 0.8226700000001 instead of 0.82267

5 Statistical Analysis

The quality of the developed encryption algorithm is checked statistically in the following subsections.

5.1 Quality of Encryption: Key Sensitivity

The encrypted image (Fig. 2(c)) has no visible information that can be used to correlate the original image with the encrypted image. The uniform pseudo-random distribution of the gray-levels of the key-image is the reason behind this.

Also, as already it has been discussed, the key space is somewhat larger than 10^{112} for this encryption algorithm. Small variations in the control parameters, thus, result into absolutely nonsense decryption. Even if the image is decrypted with the same derivative orders as in (4) still a very minute change in the initial condition will sill save the image from any attacks. As a demonstration, if we just change $x_1(0)$ to 0.822670000001 instead of the original value 0.82267 (and keeping everything else the same), the decryption process gives no useful result (Fig. 2(e)). Making the change even smaller by choosing $x_1(0) = 0.8226700000001$, the decryption process only reveals, in a way, a very insignificant part of the original image, as seen in Fig. 2(f).

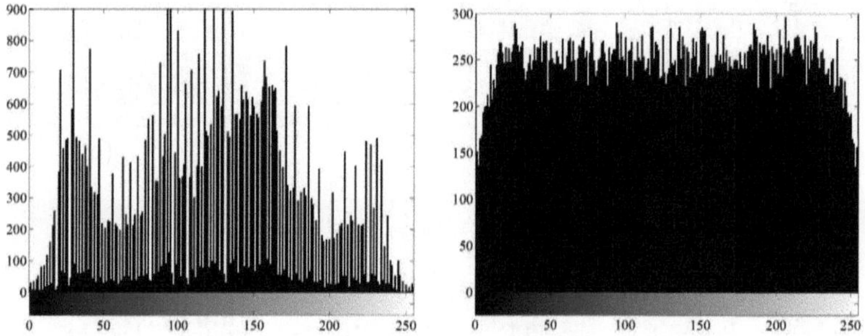

(a) (b)

Fig. 3. (a) Histogram of the original image and (b) Histogram of the encrypted image

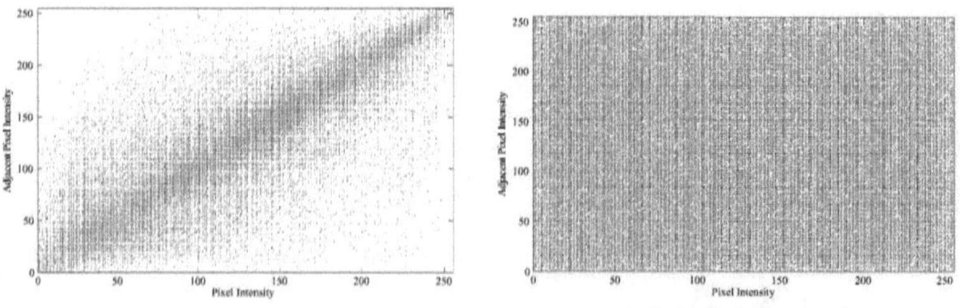

(a) (b)

Fig. 4. (a) Autocorrelation diagram of the original image, and (b) Autocorrelation diagram of the encrypted image

5.2 Image Histogram and Autocorrelation Analysis

The histogram of the original image is shown in Fig. 3(a) and that of the encrypted image is shown in Fig. 3(b). From these histograms it is apparent that the described encryption algorithm produces an encrypted image that has almost equal no of pixels for all gray-levels.

Also, for any information carrying image has a good amount autocorrelation of the gray-levels between adjacent pixels. It is given by

$$\begin{aligned} R_{XX} &= \mathcal{E}[X(m)X(n)] \\ &= \frac{1}{MN}\sum_{m=1}^{M}\sum_{n=1}^{N} x_m x_n \mathcal{F}\left(x_m, x_n, m, n\right), \end{aligned} \tag{8}$$

where, $\mathcal{F}$ denotes the joint probability distribution of the pixel intensities. The original image has a very high autocorrelation as seen from Fig. 4(a); whereas, the encrypted image has almost *zero* autocorrelation between any two adjacent pixels (vertically, horizontally and diagonally), as seen from Fig. 4(b). This means that in the encrypted image, it is almost impossible to predict the gray-level of any pixel based on the knowledge of the gray-levels of other pixels. These certainly shows that the encryption algorithm is almost invulnerable to 'brute-force' attacks.

5.3 Image Entropy

To characterize the randomness in the texture of a grayscale image, a scalar statistical quantity *entropy* is defines as [7]

$$\mathcal{H}(S) = -\sum_{g=0}^{255} p(g)\log_2 p(g), \tag{9}$$

where, g denotes the gray-levels ranging from 0 to 255 and $p(g)$ gives the number of pixels of a certain gray-level g present in the image.

If an image only contains pixels of a constant gray-level then the entropy value is *zero*, in other words, a low entropy value means a low level of 'confusion'. Again, if in an image, for a certain pixel, all the gray-levels are equally probable, then $\mathcal{H}$ has a maximum value of 8, and thus the image can be called 'most confusing'. Now, the entropy value of the original Lena image is calculated to be 7.3078 and that of the encrypted image is calculated to be 7.9974. Sine the entropy of the encrypted image is very close to 8, the confusion level will be very high. Thus extracting any useful information from the encrypted image without knowing the exact keys will be very difficult.

6 Conclusion and Possible Issues

The developed encryption algorithm shown in this paper possesses good statistical properties and it is suitable for standard applications. However, if the original image is of very large size then the number of pixels in the image will be very large and thus more number of iterations will have to be done for the solution of the chaotic system. This drastically increases the computation time. Finally, it should be kept in mind that, although the order of the derivatives are considered to be a part of the key space, choosing any arbitrary value for them might result into a system which does not satisfy the necessary condition for the existence of a double-scroll attractor [8]. This will result into orbits and thus the improper choice of the derivative orders will make the encryption algorithm fail. However, there is no such restriction for the initial conditions.

References

1. Matthews, R.A.J.: On the derivation of a chaotic encryption algorithm. Neural Computation 12, 2519–2535 (2000)
2. Sun, F., Liu, S., Li, Z., Lu, Z.: A novel image encryption scheme based on spatial chaos map. Chaos, Solitons, Fractals 38, 631–640 (2008)
3. Tong, X.J., Cui, M.G.: Image encryption with compound chaotic sequence cipher shifting dynamically. Image and Vision Computing 26, 843–850 (2008)
4. Gao, T.G., Chen, Z.Q.: Image encryption based on a new total shuffling algorithm. Chaos, Solitons, Fractals 38, 213–220 (2008)
5. Petráš, I.: Fractional-Order Memristor-Based Chua's Circuit. IEEE Trans. Circuits and Systems 57, 975–979 (2010)
6. Valério, D., Costa, J.S.: Introduction to single-input, single-output fractional control. IET Control Theory Appl. 5, 1033–1057 (2011)
7. Gonzalez, R.C., Woods, R.E.: Digital Image Processing. Prentice Hall, New Jersey (2007)
8. Tavazoei, M.S., Haeri, M.: A necessary condition for double scroll attractor existence in fractional - order systems. Physics Letters A 367, 102–113 (2007)

An Efficient RC4 Based Secure Content Sniffing for Web Browsers Supporting Text and Image Files

Shweta Pandey[1] and Abhishek Singh Chauhan[2]

[1] Department of CSE, NIIST, Bhopal
[2] Department of CSE, NIIST, Bhopal

Abstract. The communication in today's scenario is mostly rely on web, it will be increases day by day means the dependency of the users for communication is increases on web browsers. So thinking about security during data communication like text and image files will be legitimate. There are several research work are in progress in this direction. In this paper we present an efficient RC4 based secure content sniffing for web browsers which supporting textual files(word, pdf, text), web files(.jsp,.php.html) and image files also. In our proposed work we send the text data and image files by applying RC4 encryption algorithm. Data is then partition in several parts for reducing the file overhead and then the data will be sending with the extra bit of 0 and 1 for identifying the attack. Means our work will secure the encryption mechanism from the traditional file including wide variety of file formats. The effectiveness of our approach is shown by the results.

Keywords: Content Sniffing, RC4, Encryption, Web Browsers.

1 Introduction

In current web environment which is based on different host and clients, it is burden and worry of security to dispense contents due to traditional client-server model [1]. Benefit of distinctive kinsfolk always edacity for professional care of major filler less provider's apply oneself to, an equivalent observations is distributed to all connections of users over the network [2], [3]. Content sniffing sway is a bosom polemic of Cross Site Scripting pilfering. In a Potential sniffing attack, a strew ineradicable encircling hellish payload is wrongly handled by a victim's browser, usher in the period of HTML content and conduct of JavaScript pandect. The malicious payload transmission analysis is uploaded by attackers to legitimate websites [4], [5]. These line show affectionate in a second their sense types or Multipurpose Internet Mail Extension (MIME) information are considered [6].

Cross site scripting (XSS) is the next most powerful attack now days during the server client communication. XSS vulnerabilities are maltreated by injecting HTML components or JavaScript code [7]. However, the conquer despicable attain arse be achieved by injecting JavaScript cryptogram prowl gets unabated by browsers. The injected JavaScript code can admittance crucial lead existent in lacing pages and

M.K. Kundu et al. (eds.), *Advanced Computing, Networking and Informatics - Volume 2*, Smart Innovation, Systems and Technologies 28,
DOI: 10.1007/978-3-319-07350-7_36, © Springer International Publishing Switzerland 2014

disseminate it to third bunch websites. Latest investigate has shown that injected JavaScript code is the greatest creation of alternative exploitations such as cross site request forgery [8] and phishing [9]. Thus, determining of injected JavaScript protocol is an unembellished feigning prefers qualifying XSS frailty exploitations.

There are several different mitigation techniques analysis [10], testing [11], and scanning [12] are widely hand-me-down in the lead program deployments to catch injected standards. No matter how, the amid of current JavaScript principles run the show exploits in perspicuous programs is increasing day by day [13]. This indicates that development of vulnerability free programs is still far from reality [14], [15]. Predisposed go, in front of a coordinating runtime exploration of injected JavaScript regulations exploitations for web-based environment is important.

The remaining of this paper is organized as follows. In Section 2 we discuss about the related work. Proposed work is shown in section 3. Result analysis is shown in Section 4.The conclusions are given in Section 5. Finally references are given.

2 Related Work

In 2011, Peiqing Zhang *et al.* [16]analyze the need of bridge between users and communication path. They also compared the effect of most commonly applied attacks; content pollution and index poisoning with proper comparison. In 2011, Anton Barua *et al.* [17] develop a mechanism for content sniffing attack detection in server side using JavaScript's and analyzes the attack time detection. They developed a framework for evaluation of both benign and malicious files. In [18] author suggests encryption decryption techniques for data sharing and gathering. They [19] also suggest subset superset partition for time management. In 2011, Brad Wardman *et al.* [20] suggest deviate Phishers bear to acclimatize the well-spring code of the strengthen a attack pages worn in their attacks to impersonate shift variations to actual websites of spoofed organizations and to leave alone ascertaining by phishing countermeasures. Manipulations fundament is as canny as source code changes or as ostensible as annex or expulsion esteemed content. To appropriately declaration to these changes to phishing campaigns, a cohort of pass round coextension algorithms is implemented to observe phishing websites based on their content, employing a business data set consisting of 17,992 phishing attacks targeting 159 different brands. The results of the experiments using a variety of different content-based approaches demonstrate that some can achieve a detection rate of greater than 90% while maintaining a low false positive rate. In 2012, Usman Shaukat Qurashi *et al.* [21] suggest that AJAX (asynchronous JavaScript and XML) has enabled modern web applications to provide rich functionality to Internet users. So they suggest a security framework for maintaining security in AJAX environment. In [22] a server and client side attack detection will be proposed with text data, the variant with PDF will be present in [23] and [24]. The DES standard will be used for security in [25]. In [26] Interest flooding attack can be applied for DoS in Content Centric Network (CCN) and based on the simulation results which can affect quality of service. They expect that it contributes to give a security issue about potential threats of DoS in CCN. In 2013,Van Lam Le *et al.* [27] suggest drift Drive-by download attacks situation upon browsers are venal by deadly volume untouched by web servers strive turn a traditional stir vector in previous years.

3 Proposed Work

The main motivation of our work is to secure content delivery in server client communication as better understand by the flowchart given in Fig.1. Our proposed framework support 1) Text 2) HTML 3) PHP 4) PDF 5) Java Script 6) Word and Images. Our hybrid framework provides admin centric web based network. If any client wants to communicate to the server node then it must be registered in the hybrid framework. After registration it can be granted or denied by the administrator. If it is granted then the client can access the capabilities provided by the admin. Client can request text, pdf and image files. If client request any file from the server, then server prepares the file for sending. The file preparation will be started from RC4 encryption. The RC4 Encryption Algorithm, seasoned by Ronald Rivest of RSA, is a shared key stream cipher algorithm requiring a procure interchange of a shared key. The uniform key algorithm is worn closely for encryption and decryption such stroll the statistics stream is solitary XORed encircling the generated key sequence. The algorithm is almanac as it requires transformation exchanges of depose entries based on the key sequence. Hence implementations can be very computationally intensive. This algorithm has been released to the make noticeable and has peculiar implementations. It is also used by IEEE standard 802.11 within WEP (Wireless Encryption Protocol) using a 40 and 128-bit keys. It generates a pseudorandom stream of bits. As with any stream cipher, these can be used for encryption by combining it with the plaintext using bit-wise exclusive-or; decryption is performed the same way. For generation a sequence of stream cipher it uses permutation of all possible bytes. In RC4 case it is 256.Means the key length in RC4 is {1-255}. It is better explained with Fig. 2. The permutation will be started by index pointers that are IP1 and IP2. The key length may vary from 40 to 255 and the iteration is according to the key scheduling algorithm.

```
for 0 to 255
      S [ip1]:= i
Endfor
Ip2:= 0
for 0 to 255
IP2 := (IP2 + S[IP1] + key[IP1 mod keylength]) mod 256
swap values of S[IP1] and S[IP2]
Endfor
```

Then partitioning process will be started so that sending overhead will be reduced. Partitioning will be achieved separately for different files. For text files and web files the partitioning will be depends on bytes in the file. For PDF files it will be splitted page wise. Image will be split the following steps.

Step 1: ImageIO.read();
Step 2: rows = 4;
Step 3: cols = 4;
Step 4:chunks = rows * cols;

Step 5:chunkWidth = getWidth() / cols;
Step 6:chunkHeight = image.getHeight() / rows;
Step 7: count = 0;
Step 8: Pass the chunks to the Buffered Image;
Step 9: for (int x = 0; x < rows; x++) {
 for (int y = 0; y < cols; y++) {
 //Initialize the image array with image chunks
Step 9:imgs[count] based on width and height;
Step 10: Draw the chunks.

Then the data will be sending to the client by adding an extra bit of 0/1. The log file will be maintained when the data will be sending to the client with the time of sending and receiving. If any unauthorized access can check the file before the client the extra bit will be change to 1/0. It will be notified to the server and the time of notification will be noted again and alert will be send to the client and file will be discarded to the client so that client will be prevented from unauthorized access.

The whole process is also explained by our algorithm shown below.

Algorithm 1: For Mapping Response

1) Inputs: The set of File Request (FR_1, FR_2..........FR_n) from the full set of request by the client user.
2) Output: Map File Request(FR_1, FR_2 ....FR_n).
3) do

   Check the Log Request set.
   Design a log of File request (LFr_1,LFr_2......LFr_n) to search the peak request.
   For each log of File request (LFR= LFr_1,LFr_2......LFr_n) do
   goto Algorithm 2;
   goto partitioning;
   End;
4) Data will be send with 0/1 append with reconfiguring file request loads.
5) Details are added including the attack time in the log file mentioning the client name
6) Finish.

Algorithm 2: RC4 Algorithm for Encryption and Decryption

Algorithm: RC4[28]

We are using two different array one for the state and other for storing the key. State is represented by S[] and key is represented by K[]

Step 1: First the state table is arranged according to 256 bytes means one array with numbers from 0 to 255

S[256]=[0 .. 255]

Step 2: Then the key table is arranged
K [1..2048] = […….]
Step3:Then randomize the first array to generate the final key stream.
Step 3: Then key setup phase will be started.

1. Sf = (f + Si+ Kg) mod 4
2. Swapping Si with Sf

Step 4: Perform XOR

1. i = (i + 1) mod 4 , and f = (f + Si) mod 4
2. Swaping Si with Sf
3. t = (Si+ Sf) mod 4

The above process explains the working methodology of our hybrid framework with the description.

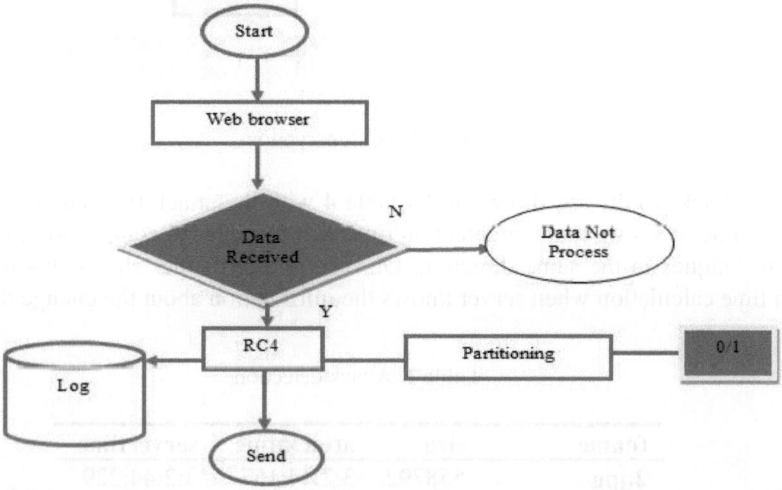

Fig. 1. Flowchart

4 Result Analysis

For signify the indication we begin pair types of databases duo distance from the salver side and one from the client side. In server side we maintain two copies of the corresponding plank one for At the Exile and every other for impede send. The added bit is automatically changed if it is opened by any unauthorized access. It is automatically alerted to the client, so those clients re-request the data from the server. Server also maintain the time of sending and receiving of files.

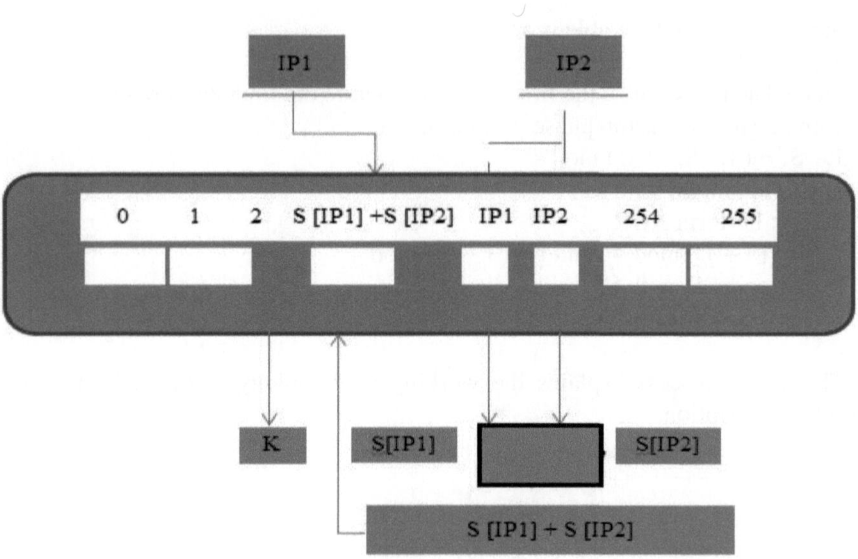

Fig. 2. RC4 encryption

The attack results are shown in the table 4 with the attack time and the server receives time. It shows that the detection time will be better in comparison to the previous techniques in the same direction. Our server alerts times shows this mechanism with time calculation when server knows the information about the change data.

Table 1. Attack detection

fname	size	attacktime	servertime
2.jpg	558792	3:2:44:157	3:2:44:229
newclient.jpg	2444	3:15:58:110	3:15:58:308
newclient.jpg	1965	3:52:53:107	3:52:53:273
pdf2.pdf	20859	8:34:50:77	8:34:50:172
wd1.html	323294	8:35:54:798	8:35:39:831
wd1.html	323294	8:36:42:50	8:36:21:735
ab1.txt	39791	8:37:2:107	8:37:2:147
doc1.doc	404	7:37:18:239	7:37:18:360

The above comparison chart Fig. 3.shows the effectiveness of our approach which is compared by [17]. It shows the attack detection time is better in comparison to the traditional technique.

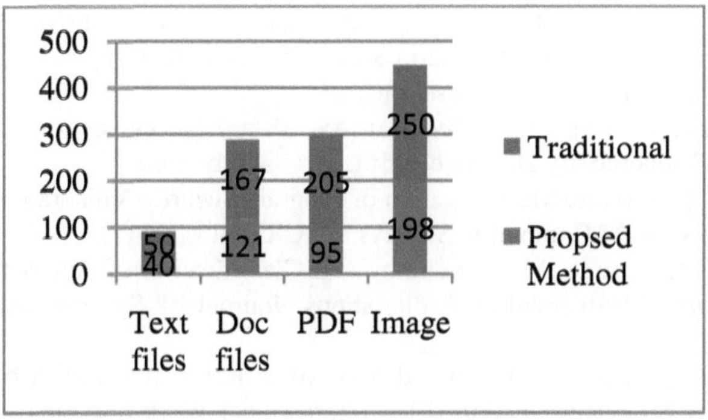

Fig. 3. Comparison Chart

5 Conclusions

In this paper we proposed a hybrid framework for server client environment. These approaches provide a new insight in the direction of content sniffing attack. It provides a direction with RC4 encryption for better protection when communication will be done in the server client environment. It also provides a better time attack detection in comparison to the traditional technique.

References

1. Cisco Systems, Cisco Visual Networking Index: Forecast and Methodology, 2011-2016. Cisco White Paper (2012)
2. Ahlgren, B., Dannewitz, C., Imbrenda, C., Kutscher, D., Ohlman, B.: A survey of information-centric networking. IEEE Communications Magazine 50(7), 26–36 (2012)
3. Jacobson, V., Smetters, D.K., Thornton, J.D., Plass, M.F., Briggs, N.H., Braynard, R.L.: Networking named content. In: ACM 9th Internation Conference on Emerging Networking Experiments and Technologies, CoNEXT (2009)
4. Barth, A., Caballero, J., Song, D.: Secure Content Sniffing for Web Browsers, or How to Stop Papers from Reviewing Themselves. In: Proceedings of IEEE Security & Privacy, pp. 360–371 (2009)
5. Gebre, M.T., Lhee, K.-S., Hong, M.: A Robust Defense Against Content sniffing XSS Attacks. In: Proceedings of 6th International Conference on Digital Content, Multimedia Technology and its Applications, pp. 315–320 (2010)
6. Multipurpose Internet Mail Extensions (MIME), `http://www.ietf.org/rfc/rfc2046.txt?number=2046`
7. Cross Site Scripting, `http://www.owasp.org`
8. Shahriar, H., Zulkernine, M., PhishTester, M.: Automatic Testing of Phishing Attacks. In: Proceedings of the SSIRI, pp. 198–207 (2010)
9. Shahriar, H., Zulkernine, M.: Client-Side Detection of Cross-Site Request Forgery Attacks. In: Proceedings of ISSRE, pp. 358–367 (2010)
10. Wassermann, G., Su, Z.: Static detection of cross-site scripting vulnerabilities. In: ICSE International Conference on Software Engineering, pp. 171–180 (2008)

11. Shahriar, H., Zulkernine, M.: MUTEC: Mutation-based Testing of Cross Site Scripting. In: Proceedings of the 5th ICSE Workshop SESS, pp. 47–53 (2009)
12. Paros - Web application security assessment, `http://www.parosproxy.org/index.shtml` (accessed)
13. Open Source Vulnerability Database, `http://osvdb.org`
14. Shahriar, H., Zulkernine, M.: Mitigation of Program Security Vulnerabilities: Ap-proaches and Challenges. ACM Computing Surveys (CSUR) 44(3) (2012)
15. Shahriar, H., Zulkernine, M.: Taxonomy and Classification of Automatic Monitoring of Program Security Vulnerability Exploitations. Journal of Systems and Software 84(2), 250–269 (2011)
16. Zhang, P., Helvik, B.E.: Modeling and Analysis of P2P Content Distribution under Coordinated Attack Strategies. In: 7th IEEE International Workshop on Digital Rights Management Impact on Consumer Communications (2011)
17. Barua, A., Shahriar, H., Zulkernine, M.: Server Side Detection of Content Sniffing Attacks. In: 2011 22nd IEEE International Symposium on Software Reliability Engineering (2011)
18. Dubey, A.K., Dubey, A.K., Namdev, M., Shrivastava, S.S.: Cloud-user security based on RSA and MD5 algorithm for resource attestation and sharing in java environment. In: 2012 CSI Sixth International Conference on Software Engineering (CONSEG) (2012)
19. Dubey, A.K., Dubey, A.K., Agarwal, V., Khandagre, Y.: Knowledge discovery with a subset-superset approach for Mining Heterogeneous Data with dynamic support. In: 2012 CSI Sixth International Conference on Software Engineering (CONSEG) (2012)
20. Wardman, B., Stallings, T., Warner, G., Skjellum, A.: High-Performance Content-Based Phishing Attack Detection. In: eCrime Researchers Summit (eCrime) (2011)
21. Qurashi, U.S., Anwar, Z.: AJAX Based Attacks: Exploiting Web 2.0. In: International Conference on Emerging Technologies (2012)
22. Qadri, S.I.A., Pandey, K.: Tag Based Client Side Detection of Content Sniffing Attacks with File Encryption and File Splitter Technique. International Journal of Advanced Computer Research (IJACR) 2(3), 5 (2012)
23. Dubey, A., Gupta, R., Chandel, G.S.: An Efficient Partition Technique to reduce the Attack Detection Time with Web based Text and PDF files. International Journal of Advanced Computer Research (IJACR) 3(1), 9 (2013)
24. Thakur, B.S., Chaudhary, S.: Content Sniffing Attack Detection in Client and Server Side: A Survey. International Journal of Advanced Computer Research (IJACR) 3(2), 10 (2013)
25. Gupta, S.: Secure and Automated Communication in Client and Server Environment. International Journal of Advanced Computer Research (IJACR) 3(4), 13 (2013)
26. Choi, S., Kim, K., Kim, S., Roh, B.-H.: Threat of DoS by Interest Flooding Attack in Content-Centric Networking. In: International Conference on Information Networking (ICOIN), pp. 315–319 (2013)
27. Le, V.L., Welch, I., Gao, X., Komisarczuk, P.: Anatomy of Drive-by Download Attack. In: Proceedings of the Eleventh Australasian Information Security Conference (AISC) (2013)
28. Kumar, S., Patel, B.B.: Java Based Resource Sharing with Secure Transaction in User Cloud Environment. International Journal of Advanced Computer Research (IJACR) 2(3), 5 (2012)

On the Application of Computational Diffie-Hellman Problem to ID-Based Signatures from Pairings on Elliptic Curves

Swaathi Ramesh

National Institute of Technology, Tiruchirapalli, India
swaathiramesh@outlook.com

Abstract. The paper presents the application of the Computational Diffie-Hellman problem to ID-based signatures with pairings on elliptic curves in the random oracle model. It focusses on the security of the scheme.It also understands the fundamentals of provable security as applied in cryptography.

Keywords: Provable Security, Identity-based signature schemes, Weil Pairing, Tate Pairing.

1 Introduction

Provable security refers to rigorous mathematical methods being employed to prove the security of a cryptographic scheme. In such a proof, the capabilities of the attacker are defined by an adversarial model.The method is to show that the attacker must solve the underlying hard problem in order to break the security of the cryptographic system. The theoretical model used here is the random oracle model where hash functions are represented as in [9].

The concept of identity-based signature originated with Shamir's scheme [1] in 1984. In such an infrastructure, each user's public key is a function of his identity and uniquely identifies the entity. The public string could be an email address, domain name or a physical IP address. Recently, Boneh and Franklin presented an Identity-based encryption scheme based on properties of Weil and Tate pairings on elliptic curves [2], [3]. The scheme has a property that a user's public key is an easily calculated function of his identity while a user's private key is calculated for him by a trusted authority (PKG). For reasons of efficiency and convenience, it is desirable to have an identity-based signature scheme (where the signature verification function is easily obtained from his identity) which is able to make use of the same underlying computational principle and possibly the same keys.Such a scheme that was already in the paper [2] and was also present in [4]. The major advantage of using elliptic curves over RSA is that it provides equal security for a much smaller key size. However, this method needs an underlying "hard problem" that the security of the scheme is reduced to. The scheme in [10] however is based on the generalized ElGamal Scheme as in [5].

M.K. Kundu et al. (eds.), *Advanced Computing, Networking and Informatics - Volume 2*, Smart Innovation, Systems and Technologies 28,
DOI: 10.1007/978-3-319-07350-7_37, © Springer International Publishing Switzerland 2014

Here, we implement the Computational Diffie-Hellman problem to the scheme in [10].

A digital signature is a mathematical scheme for the validity of a message or a document. The major reasons for employing digital signatures are authentication, non-repudiation and integrity.

It is important to know the correct source of any message, especially in financial and military contexts. This is achieved with digital signatures that ensure sender authenticity. Non-repudiation (of origin) refers to the inability of an entity that has produced the signature to deny having done so later. Also, though encryption hides the message contents, it is possible to make fraudulent modifications to the message without actually decrypting it or understanding it. Again, digital signatures make sure that such modifications, though not prevented, can be recognized.

2 Survey of Other ID-Based Signatures

Table 1 describes a survey of other ID-based signatures.

Table 1. ID-based Signatures

Scheme	Private Key	Provable or Not
Cha-Cheon[12]	GDH	P
Galindo [14]	DL	P
Javier [15]	CDH	P
Paterson[10]	ElGamal(DL)	P
Ours	CDH	N

P - Provable
N - Not Provable

3 Our Contribution

Provable security deals with the security of cryptographic schemes that can be proved mathematically. However, in our adaptation, due to the structuring of the parameters involved, the forgery on the signature can't be produced. Thus, intrinsically, it is secure. However, the formal reduction to the hard problem goes beyond the realm of provable security.

4 Preliminaries

Bilinear Pairing: Let G_1 be an additive cyclic group generated by P , with prime order q , and G_2 be a multiplicative group of the same order q. A bilinear pairing $\hat{e}$ is a map with the following properties:
For $\hat{e} : G_1 \times G_1 \longrightarrow G_2$

1. **Bilinearity.** For all $P, Q, R \in G_1$
   - $\hat{e}(P+Q, R) = \hat{e}(P, R)\hat{e}(Q, R)$
   - $\hat{e}(P, Q+R) = \hat{e}(P, Q)\hat{e}(P, R)$
   - $\hat{e}(aP, bQ) = \hat{e}(P, Q)^{ab}$ $[Wherea, b \in_R Z_q]$
2. **Non-Degeneracy.** There exists an $P, Q \in G_1$ such that $\hat{e}(P, Q) \neq I_{G_2}$, where I_{G_2} is the identity element of G_2
3. **Computability.** There exists an efficient algorithm to compute $\hat{e}(P, Q)$ for all $P, Q \in G_1$

Computational Assumptions: In this section, we review the computational assumptions related to bilinear maps that are relevant to the protocol we discuss. theorem**Definition 1.**

Computational Diffie-Hellman Problem(CDHP): Given $(P, aP, bP) \in G_1^3$ for unknown $a, b \in Z_q$, the CDHP problem in G_1 is to compute abP. The advantage of any probabilistic polynomial time algorithm $\mathcal{A}$ in solving the CDH problem in G_1 is defined as:

$$Adv_{\mathcal{A}}^{CDH} = |Pr[A(P, aP, bP, Q) = a.b.P|\text{a, b} \in Z_q]$$

The CDH assumption is that, for any probabilistic polynomial time algorithm A , the advantage $Adv_{A^{CDH}}$ is negligibly small.

4.1 Definition

- **Setup:** The private key generator provides the security parameter κ as the input to the algorithm, generates the system parameters *params* and the master secret key *msk*. PKG publishes *params* and keeps the *msk* secret.
- **Extract:** The user provides his identity ID to the PKG. The PKG runs the algorithm with identity ID, *params* and *msk* as input and obtains the private key D_{ID}. The private key D_{ID} is sent to the user through a secure channel.
- **Sign:** For generating a signature on the message m , the user provides his identity ID, his private key D_{ID}, *params* and the message m as input. The algorithm generates a valid signature σ on the message m by the user.
- **Verify:** This algorithm on input a signature σ on message m by the user with identity ID, *params* , checks whether σ is a valid signature on message m by ID. If true, outputs "Valid" , else outputs "Invalid".

5 Notation

We use the same notation as in [6]. We let G_1 be an additive group of prime order q and G_2 be a multiplicative group of the same order q. We assume the existence of the bilinear map

$$\hat{e} : G_1 \times G_1 \longrightarrow G_2$$

with the property that the computational Diffie-Hellman problem is hard in both G_1 and G_2. Typically, G_1 will be a subgroup of the group of points on the elliptic curve over a finite field, G_2 will be a subgroup of the multiplicative group of a related finite field and the map $\hat{e}$ will be derived from the Weil or Tate pairing on the elliptic curve. We also assume that an element $P \in G_1$ satisfying $\hat{e}(P, P) \neq 1_{G_2}$ is known. We refer to [3,7] for a fuller description of how these groups , maps and parameters must be selected in practice for efficiency and security.

We let ID be a string denoting the identity of a user and H_1 , H_2 and be public cryptographic hash functions. We require $H_1 : \{0,1\}^* \rightarrow G_1$, $H_2 : \{0,1\}^* \times \{0,1\}^* \times \{0,1\}^* \rightarrow G_1$. In our scheme, a user's public key for signature verification is $Q_{ID} = H_1(ID)$, while his secret key for signature generation is $D_{ID} = s.Q_{ID}$, where $s \in Z_q^*$ is a system-wide master secret key known to a trusted authority. These keys are the same as in the encryption scheme of [3]. If desired, the encryption and the signature keys can be separated simply by concatenating the string ID with extra bits which identify the keys' intended functions. We assume the value $P_{pub} = s.P$ is publicly known.

6 Signature Scheme

Scheme: Let G_1 , G_2 be cyclic prime order groups of order q , where G_1 is an additive group and G_2 is a multiplicative group. Let $P \in_R G_1$ be the generator of G_1 , $\hat{e} : G_1 \text{ x } G_1 \rightarrow G_2$ be a bilinear map and $H_1(.), H_2(.)$ be two cryptographic hash functions defined by:

$H_1 : \{0,1\}^* \text{ x } G_1 \rightarrow G_1$
$H_2 : \{0,1\}^* \times \{0,1\}^* \times \{0,1\}^* \rightarrow G_1$

- **Setup:**
  1. Generate $s \in_R Z_q$, the system-wide master secret key
  2. Choose $P \in_R G_1$, a generator of G_1
  3. Compute $P_{pub} = s.P$
- **Extract:** For any use U_A with $ID = ID_A$,
  private key $= D_{ID_A} = s.Q_{ID_A} = H_1(ID_A)$
  public key $= Q_{ID_A} = H_1(ID_A)$
- **Sign:** To generate the signature on message m, the user with $ID = ID_A$ executes the following algorithm:

  1. Compute $R = r.P$ $[r \in_R Z_q^*]$
  2. $h_m = H_2(m, R, ID)$
  3. Compute r^{-1} in Z_q^*
  4. Compute $S = r^{-1}[h_m + D_{ID_A}]$

  Thus, $\sigma = (R, S)$ is the required signature on message m and for identity ID_A

– **Verify:** Given an identity ID_A, a message m and a signature $\sigma = (R \in G_1, S \in G_1)$,
compute $H_2(m, R, ID)$, $H(ID_A) = Q_{ID_A}$
and check,
(a) $\hat{e}(R, S) \stackrel{?}{=} \hat{e}(P, H_2(m, R, ID)).\hat{e}(P_{pub}, Q_{ID_A})$
If check passes, output "Valid" else output "Invalid"
.

Theorem 1. *The signature scheme given above is consistent*

Proof. We need to show that for all private key tuples and for all messages, any signature generated by the signing algorithm verifies as a valid signature under the respective user identity.
Indeed, we have for equation (a), $LHS = \hat{e}(R, S)$
$= \hat{e}(rP, r^{-1}(H_2(m, R, ID) + D_{ID_A}))$
$= \hat{e}(P, H_2(m, R, ID) + D_{ID_A})$
$= \hat{e}(P, H_2(m, R, ID)).\hat{e}(P, D_{ID_A})$
$= \hat{e}(P, H_2(m, R, ID)).\hat{e}(P, s.Q_{ID_A})$
$= \hat{e}(P, H_2(m, R, ID)).\hat{e}(s.P, Q_{ID_A})$
$= \hat{e}(P, H_2(m, R, ID)).\hat{e}(P_{pub}, Q_{ID_A})$
$= RHS$

6.1 Security

We try proving the security of our scheme by reducing it to the security of the general signature scheme. The standard method for studying signatures is as in [8]. The extended adversary is in the random oracle model as in [9]. In such a security model, the following occurs:

– Setup: The challenger C sets up the public parameters and the master secret key while the public parameters are made available to everyone, the master secret is known only to the challenger.

– Training: The forger F adaptively queries the oracles on messages of his choice. F can query such that:
  - Key Extraction Oracle: The forger F queries for the private key of any identity ID and the challenger C returns the value.
  - Signing Oracle: When F makes a query for a signature on message m on identity ID, C provides a valid signature σ on m, ID.
– Forgery: F identifies a message and an identity (ID_t, m^*) and outputs a signature σ on (ID_T, m^*). If σ satisfies the following two conditions and passes the verification test, the forger wins the game:
  1. The private key of ID_t was not queried in the training phase
  2. A signature for m^* on ID_t has not been queried to the Key Extraction Oracle in the training phase.

The objective of the challenger is to inject the hard problem instances appropriately , so that, assuming the forger returns a valid signature on (ID_T, m^*), the challenger C can perform a queries in polynomial time and obtain the solution to the underlying hard problem. However, this means that the hard problem can be solved in polynomial time.
This contradicts out computational assumptions in Section 2. Thus, our assumption that F can produce the forgery is thus reduced to the task of solving the underlying hard problem.

Theorem 2. *Let* (G_1, G_2) *be a* (τ_1, t_1, e_1) *CDH pair of order* q *then the signature scheme on* (G_1, G_2) *is* $(t_2, q_s, q_{H_1}, q_{H_2}, e_2)$ *- secure against existential forgery under adaptive chosen message attack in random oracle model. Here,* q_H *is the total number of queries generated*

Proof. Let us assume that , F is a forger algorithm that $(t_2, q_s, q_{H_1}, q_{H_2}, e_2)$ - breaks our signature scheme on (G_1, G_2). We show that the scheme is in fact secure against such a forger but isn't provably secure. Algorithm C simulates the challenger and interacts with F in the following way:

- **Setup:** Challenger C starts by giving F the common reference string (P, G_1, G_2) and the public key $P_{pub} = a.P$ such that $P \in_R G_1$
- **Training Phase:**During this phase, F has access to the following oracles:
  - H_1*Oracle* :
    Forger F can query the H_1 oracle at any time. To handle these queries C maintains a list which is defined as $< H_i, ID_i, r_i >$. We refer to this list as theL_1 - list. Initially, this list is empty and will be updated as explained below. When F queries the oracle H_1 with ID_i as input, C responds as following:
    * If ID_i already exists as a tuple of the form $< H_i, ID_i, r_i >$ in L_1, then C responds with H_i such that $H_1(ID_i) \in G_1$
    * Otherwise, if ID_i is the target identity, ID_t, then, return H_i=$b.P$ such that $r_i = b$ and store the value
    * else, give $H_i = r_i.P \in G_1, r_i \in_R Z_q$
  - H_2*Oracle* : The forger F can query this oracle any time he wants to and the C maintains a list of tuples $< H_j, M_k, ID_j, R_j, x_j, x_j, - >$, in a list called the L_2 list. The H_2 oracle is a full domain hash. On given $ID_j, M_k and R_j$ as input,
    if $ID_j \neq ID_t$, C responds by giving $H_j = x_j.P \in G_2$ such that $x_i \in_R Z_q$ and stores the tuple $<H_j, M_k, R_j, x_j>$
    else, C responds by giving $H_j = x_j \times P \in G_2$ such that$x_i = t_1 \times t_2$ where $t_1, t_2 \in_R Z_q$ and stores the tuple $< H_j, M_k, R_j, x_j, t_1, t_2 >$
  - Extract Oracle:
    The forger is allowed to query for the private keys of all possible ID_l. The challenger C responds this way:
    If ID_l is the target identity , ID_t, abort
    else, give $D_{ID} = s.Q_{ID}$

- Signature Oracle:
  Let (m_i, ID_j) be the message identity pair for which the forger F requests a signature. C performs the following:
  If $ID_q \neq ID_t$, the target identity, give $\sigma = (R, S)$, the signature, according to the scheme.
  else, give as follows :
  1. Choose $r_1, r_2 \in_R Z_q^*$
  2. Compute $R = r_2.P_{pub}$
  3. Compute $S = r_2^{-1}Q_{ID} + r_1.P$ and store $H_2(m_j, R, ID_t) = r_1.r_2.P_{pub}$ in the list L_2 as $< H_i, M_j, R_i, r_1.r_2, r_1, r_2 >$
  4. Return $\sigma = (R, S)$ as the signature on the message identity pair, (m_j, ID_i)

Correctness of the test
$\hat{e}(R, S) \stackrel{?}{=} \hat{e}(P, H_2(m_j, ID_i)).\hat{e}(P_{pub}, Q_{ID})$
$LHS = \hat{e}(R, S)$
$= \hat{e}(r_2.P_{pub}, r_2^{-1}.Q_{ID} + r_1.P)$
$= \hat{e}(r_2.P_{pub}, r_2^{-1}.Q_{ID}).\hat{e}(r_2.P_{pub}, r_1.P)$
$= \hat{e}(P_{pub}, Q_{ID}).\hat{e}(P, r_1.r_2.P_{pub})$
$= RHS\ of\ (a)$

Forgery: Eventually, after getting enough training , the forger F tries to produce a forgery on (m_j, ID_t). The objective of the challenger C is to use the forgery to obtain a solution to the underlying hard problem (CDH) in polynomial time. However, as the values are masked by r^{-1}, the forgery cannot be computed. Hence, while this scheme is secure, it is not provably secure.

7 Comparison with Paterson's Scheme

Paterson's Signature Scheme [10] is very similar to the ElGamal signature scheme [5]. Hence, with the standard model for signature schemes as in [8] extended to ID based systems, it was possible to prove the security of the scheme.

8 Conclusion

In this paper, we tried to apply the Computational Diffie-Hellman assumption to the signature scheme in [10]. We used the random oracle model. Such a scheme, though secure is not provably secure as the challenger is not able to provide a valid forgery for the target identity.

Acknowledgement. The author would like to thank Balaji Dhamodharaswamy from The Johns Hopkins University and Preethi Ramesh from The University of Southern California for their valuable comments.

References

1. Shamir, A.: Identity-based cryptosystems and signature schemes. In: Blakely, G.R., Chaum, D. (eds.) CRYPTO 1984. LNCS, vol. 196, pp. 47–53. Springer, Heidelberg (1985)
2. Boneh, D., Franklin, M.: Identity based encryption from the Weil Pairing. In: Kilian, J. (ed.) CRYPTO 2001. LNCS, vol. 2139, pp. 213–229. Springer, Heidelberg (2001)
3. Boneh, D., Franklin, M.: Identity based encryption from the Weil Pairing, `http://crypto.stanford.edu/dabo/abstracts/ibe.html`
4. Sakai, R., Ohgishi, K., Kasahara, M.: Cryptosystems based on pairing. In: 2000 Symposium on Cryptography and Information Security, pp. 26–28 (2000)
5. Menezes, A.J., van Oorschot, P.C., Vanstone, S.A.: Handbook of applied Cryptography. CRC Press (1996)
6. Boneh, D., Lynn, B., Shacham, H.: Short Signatures from the Weil Pairing. In: Boyd, C. (ed.) ASIACRYPT 2001. LNCS, vol. 2248, pp. 514–532. Springer, Heidelberg (2001)
7. Galbraith, S.D.: Supersingular curves in cryptography. In: Boyd, C. (ed.) ASIACRYPT 2001. LNCS, vol. 2248, pp. 495–513. Springer, Heidelberg (2001)
8. Goldwasser, S., Micali, S., Rivest, R.: A digital signature scheme secure against adaptive chosen-message attacks. SIAM J. Computing 17(2), 281–308 (1988)
9. Bellare, M., Rogaway, P.: Random Oracles are practical: a paradigm for designing efficient protocols. In: Proceedings of First ACM Conference on Computer and Communications Security, pp. 62–73 (1993)
10. Paterson, K.G.: ID-based Signatures from Pairings on Elliptic Curves. Electronics Letters 38(18), 1025–1026 (2002)
11. Raju, G.V.S., Akbani, R.: Elliptic Curve Cryptosystems and its Applications. In: IEEE International Conference on Systems, Man and Cybernetics, vol. 2, pp. 1540–1543 (2003)
12. Cha, J.C., Cheon, J.H.: An Identity-Based Signature from Gap Diffie-Hellman Groups. In: Desmedt, Y.G. (ed.) PKC 2003. LNCS, vol. 2567, pp. 18–30. Springer, Heidelberg (2002)
13. Galindo, D., Garcia, F.D.: A schnorr-like lightweight identity-based signature scheme. In: Preneel, B. (ed.) AFRICACRYPT 2009. LNCS, vol. 5580, pp. 135–148. Springer, Heidelberg (2009)
14. Herranz, J.: Deterministic identity-based signatures for Partial Aggregation. The Computer Journal 49(3), 322–330 (2006)

A Pairing Free Certificateless Group Key Agreement Protocol with Constant Round

Abhimanyu Kumar, Sachin Tripathi, and Priyanka Jaiswal

Department of Computer Science & Engineering,
Indian School of Mines
Dhanbad- 826004, Jharkhand, India
abhi_a1ks@yahoo.co.in, var_1285@yahoo.com, jaiswal.priyanka1985@gmail.com

Abstract. To allow a secure conversation among a group of members over a public network there is a need of group key agreement protocol which provide a group session key used in necessary cryptographic operations. Nowadays the protocols based on the certificateless public key cryptography (CL-PKC) creating more attraction for research because it does not require certificates to authenticates the public key as like ID- based cryptosystem and unlike ID based cryptosystem, it does not suffers from the key escrow problem. The almost all CL-PKC based group key agreement schemes in current literature are employ bilinear pairing in their operations. Since the relative computation cost of pairing is many times more than the elliptic curve point multiplication, so it motivates the researchers to propose pairing free protocols based on the CL-PKC. The present paper propose an efficient pairing free group key agreement protocol based on certificateless cryptography over elliptic curve group with their security and performance analysis. The analysis shows that the proposed protocol has strong security protection against various kinds of attack and involves comparatively lower computation and communication overheads than the other existing protocols.

Keywords: Group Key Agreement, Certificateless Public Key Cryptography, ECC, Bilinear Pairing.

1 Introduction

A group key agreement scheme allows a number of users to establish a secret common key usually called group key. Group key provide confidential and authentic conversations in several group oriented applications like teleconferences, distance learning, pay per view, collaborative workspaces, etc while communicating in open public network. Recently security protocols based on the certificateless public key cryptography (CL-PKC) becomes more demanded. The reason behind the popularity of CL-PKC is that it is free from heavy certificate management burden in PKI- based AKA protocols and the key escrow problem in identity based group key agreement protocols. Since the first CL-AKA protocol [1] was proposed in 2003 by Al-Riyami and Paterson, many group key agreement protocols based on CL-PKC have been proposed [2–5]. However these

M.K. Kundu et al. (eds.), *Advanced Computing, Networking and Informatics - Volume 2*, Smart Innovation, Systems and Technologies 28,
DOI: 10.1007/978-3-319-07350-7_38,

protocols uses bilinear pairings to achieve required security goals in their operations. The bilinear pairing is a mathematical tool which maps two elements in an additive group (usually elliptic curve group) to an element of another multiplicative group having same order(usually elements in related finite field) and it is broadly used in building of ID based as well as certificateless key agreement protocols [2]. But bilinear pairing is always defined over a super singular elliptic curve group with large element size and thus it is many times more expensive operation than the scalar point multiplications in ECC. Therefore a paring free protocol based on CL-PKC is more appealing in practice. As per our literature survey none of the CL-PKC based group key agreement protocols available in current literature [2–6] is pairing free. The present research work propose an efficient certificateless group key agreement protocol without pairing. The computation overheads like number of point additions, point scalar multiplications and message overheads of the proposed technique are comparable with the other existing protocols. The security of the protocol is analyzed very carefully with all security attributes discussed in [7] and it found secure against various attacks. The present technique may create an attraction for low power devices such as mobile phones because applications using pairings can be hard to implement on these.

The rest of this paper organized as follows. In Section 2 some related works on group key agreement protocol are discussed. The preliminaries related to proposed work are addressed in Section 3. The Section 4 proposes the protocol. Section 5 and Section 6 provide security and performance analysis followed by a conclusion section.

2 Related Work

To overcome the overheads of certificate managements in traditional PKI based protocols in 1984, Shamir [8] proposed the idea of ID based cryptosystem where the identity of a user functioned as his public key. The first ID-based authenticated group key agreement protocol was proposed by Reddy et al [9]. It utilized a binary tree structure and achieves authentication with ID-based cryptosystem, hence avoids management of certificates. But this protocol requires log_2^n rounds for n numbers of users. Since then, many ID-based group key exchange protocols [7], [10], [11] have been proposed.

Protocols based on ID-based cryptosystem undoubtedly removes the certificates overheads but has another drawback called key escrow problem. Although private key of all participants are generated by key generation centre(KGC) (also called Private Key Generator in ID-based cryptosystem) and distributed to the participants via secure channel, the entire system is vulnerable in case of compromised KGC. To overcome this drawback in 2003 Al-Riyami and Kenneth G. Paterson [1] proposes the idea of certificateless cryptosystem. Certificateless cryptosystem is an extension of ID-based cryptosystem in which private key generation process is split between the KGC and the participants. Several certificateless authenticated group key agreement (CL-AGKA) protocols [2–6]are available in literature with their own advantages and limitations. e.g. Heo et al [4]proposed a CL-AGK protocol using binary tree structure for dynamic group. It provides efficient communication and computation complexity, but it does not provide perfect forward security. Teng and Chuankun Wu [2]propose a provable authenticated CL-GKA with constant rounds. Teng [2]also present a security model for the certificateless group key agreement protocols. But one common problem in almost all CL-AGKA protocols is usage of bilinear pairing operation in their key agreement process. Pairing is one of the complex mathematical operations and always defined over a super singular elliptic curve group with large element size. So it is several times more expensive than the point multiplication operation over elliptic curve group. Thus the burden of pairing computation motivates the researchers to design pairing free CL-AGKA protocols. This paper proposes a pairing free CL-AGKA protocol for a group of legitimate participants with their security and performance analysis.

3 Preliminaries

3.1 Background of Elliptic Curve Group

Let the symbol E/F_p denote an elliptic curve E over a prime finite field F_p , defined by an equation

$$Y^2 = x^3 + ax + b. \qquad (1)$$

where $a, b \in F_p$ and with the discriminant

$$\Delta = (4a^3 + 27b^2) mod\ p \neq 0. \qquad (2)$$

The points on E/F_p together with an extra point O called the point at infinity form a group

$$G = \{(x, y) \in E/F_p : x, y \in F_p\} \cup \{O\}. \qquad (3)$$

Let the order of G be n. G is a cyclic additive group under the point addition + defined as follows:

Let $P, Q \in G$, l be the line containing P and Q , and R , the third point of intersection of l with E/F_p . Let l' be the line connecting R and O. Then $P+Q$ is the point such that l intersects E/F_p at R and O and $P+Q$.
Scalar multiplication over E/F_p can be computed as $k.P = kP = P+P+...+P$ (k times). The detail discussion of ECC can be found in [12].

The following problems defined over G are assumed to be intractable within polynomial time and forms the basis for the security of cryptographic protocols.

Computational Diffie-Hellman (CDH) Problem: Given a generator P of G and (aP, bP) for unknown $a, b \in {}_R Z_p^*$, compute abP. The CDH assumption states that the probability of any polynomial-time algorithm to solve the CDH problem is negligible.

Decisional Diffie-Hellman (DDH) Problem : Given a generator P of G and aP, bP, cP for unknown $a, b \in {}_R Z_p^*$, decide whether $c \equiv ab \ mod \ p$.

Discrete Logarithm Problem (DLP) in G: Given a generator P of G and an element $Q \in Z_p^*$, to find an integer $a \in Z_p^*$ such that $Q = aP$.

3.2 Certificateless Public Key Cryptography(CL-PKC)

In 2003 Al-Riyami and Paterson [1] introduced the concept of Certificateless Public Key Cryptography (CL-PKC) to overcome the key escrow limitation of the identity-based public key cryptography (ID-KC). In CL-PKC a trusted third party called Key Generation Centre (KGC) supplies a user with a partial private key. Then, the user combines the partial private key with a secret value (that is unknown to the KGC) to obtain his full private key. In this way the KGC does not know the users private keys. Then the user combines his secret value with the KGC's public parameters to compute his public key. Compared to the ID-PKC, the trust assumptions made of the trusted third party in CL-PKC are much reduced. In ID-PKC, users must trust the private key generator (PKG) not to abuse its knowledge of private keys in performing passive attacks, while in CL-PKC, users need only trust the KGC not to actively propagate false public keys [1]. In CL-PKC a user can generate more than one pair of key (private and public) for the same partial private key.

A CL-AKA protocol is defined by a collection of probabilistic polynomial-time algorithms as follows:

- **Setup.** This algorithm is run by KGC. It takes security parameter k as an input and returns a master private key s and the system parameters $params$.
- **Partial-Private-Key-Extract.** This algorithm is also run by KGC. It takes $params$, s and a user's identity ID_i as inputs, and returns the user's partial private key D_i
- **Set-Secret-Value.** This algorithm is run by the user. It takes $params$ and a user's identity ID_i as inputs, and returns the user's secret value x_i .
- **Set-Private-Key.**This algorithm is also run by the user. It takes $params$, a user's identity ID_i, his partial private key D_i and his secret value x_i as inputs, and returns the user's private key SK_i.
- **Set-Public-Key.** This algorithm is also run by the user. It takes $params$, a user's identity ID_i and his secret value x_i as inputs, and returns the user's public key PK_i

4 Proposed Protocol

Let us suppose a set of n users $\{U_1, U_2, ..., U_n\}$ want to establish a group session key, the proposed protocol consists of 6 algorithms which are described as follows.

1. **Setup:**(By KGC) This algorithm take a security parameter $k \in Z^+$ as input and does the following:

(a) KGC Choose a k-bit prime p and determine the tuple $\{F_p, E/F_p, G, P\}$. where P is the generater of the group G.
(b) Picks $s \in_R Z_p^*$ as the master private key and set its long term public key as $P_{pub} = s.P$.
(c) For the security parameter k KGC select two cryptographic secure hash functions: $H_1 : \{0,1\}^* \rightarrow Z_q^*$; $H_2 : G \times G \rightarrow \{0,1\}^k$.
(d) The KGC publishes $param = \{F_p, E/F_p, G, P, P_{pub}, H_1, H_2\}$ as system parameters and secretly keeps the master key s.

2. **Partial-Private-Key-Extract:** The KGC chooses at random $r_i \in_R Z_q^*$ calculate $R_i = r_i.P$ and $h_i = H_1(ID_i)$.
KGC compute $S_i = (r_i + s.h_i)\ mod\ p$ for every member U_i and issues user's partial private key $D_i = (S_i, R_i)$ to the users having identity ID_i through secure channel.
U_i can validate the partial private key by checking whether the equation holds.
$$R_i + H_1(ID_i).P_{pub} = S_i.P \tag{4}$$
The partial private key is valid if the equation holds and vice versa. Since $R_i + H_1(ID_i).P_{pub} = r_i.P + h_i.s.P = (r_i + s.h_i).P = S_i.P$
3. **Set-Secret-Value:** The user with identity ID_i picks $x_i \in_R Z_p^*$ and sets x_i as his secret value.
4. **Set-Private-Key:** The user with identity ID_i takes the pair $SK_i = (x_i, S_i)$ as its complete private key.
5. **Set Public Key:** The user with identity ID_i set public key $P_i = x_i.P$.
6. **Key-Agreement:** Every user U_i chooses a random number $t_i \in_R Z_p^*$ and computes $T_i = t_i.P$. Now U_i sends the following information to its backward and forward neighbours U_{i-1} and U_{i+1}, with signature.
$< ID_i, R_i, T_i, Sig_i >$ It is noted that in entire paper the subscript notation consider as in circular manner(i.e. $U_{n+1} = U_1$ and $U_{1-1} = U_n$).In this way U_i also receive the above information from U_{i-1} and U_{i+1}. On receiving U_i first verifies their signatures Sig_{i-1} and Sig_{i+1} and if verification is success, U_i calculates the following:
(a) $K_{i,i+1} = (x_i + S_i)T_{i+1} + t_i.(P_{i+1} + R_{i+1} + h_{i+1}.P_{pub})$
(b) $K'_{i,i+1} = t_i.T_{i+1}$
(c) $K_i^R = H_2(K_{i,i+1}, K'_{i,i+1})$
(d) $K_{i,i-1} = (x_i + S_i)T_{i-1} + t_i.(P_{i-1} + R_{i-1} + h_{i-1}.P_{pub})$
(e) $K'_{i,i-1} = t_i.T_{i-1}$
(f) $K_i^L = H_2(K_{i,i-1}, K'_{i,i-1})$
(g) $X_i = K_i^L \bigoplus K_i^R$
Now U_i broadcast the X_i to every users in the network.
After getting all X_j $(j \neq i)$ U_i first verifies the received messages as $X_1 \bigoplus X_2 \bigoplus ... \bigoplus X_n = 0$
on successful verification every user U_i can calculates the following:
$K_{i+1}^R = X_{i+1} \bigoplus K_i^R$
$K_{i+2}^R = X_{i+2} \bigoplus K_{i+1}^R$
...

$K_n^R = X_n \bigoplus K_{n-1}^R$
$K_1^R = X_1 \bigoplus K_n^R$
...
$K_{i-1}^R = X_{i-1} \bigoplus K_{i-2}^R$

Finally the group session key calculated as
$K = H_1(K_1^R||K_2^R||...||K_n^R)$

The session key is agreed because:
$K_{i,i+1} = (x_i + S_i)T_{i+1} + t_i.(P_{i+1} + R_{i+1} + h_{i+1}.P_{pub})$
$= (x_i + S_i)T_{i+1} + t_i.(x_{i+1}.P + r_{i+1}.P + h_{i+1}.s.P)$
$= (x_i + S_i)T_{i+1} + t_i.(x_{i+1} + r_{i+1} + h_{i+1}.s).P$
$= (x_i.P + S_i.P).t_{i+1} + t_i.P(x_{i+1} + S_{i+1})$
$= (P_i + (r_i + s.h_i)P).t_{i+1} + T_i(x_{i+1} + S_{i+1})$
$= (x_{i+1} + S_{i+1})T_i + t_{i+1}.(P_i + R_i + h_i.P_{pub})$
$= K_{i+1,i}$
Also,
$K'_{i,i+1} = t_i.T_{i+1}$
$= t_i.t_{i+1}.P$
$= t_{i+1}T_i$
$K'_{i+1,i}$
Similarly
$K_{i,i-1} = K_{i-1,i}$ and $K'_{i,i-1} = K'_{i-1,i}$
$K_i^R = K_{i+1}^L$ or $K_i^L = K_{i-1}^R$

5 Security Analysis

In this section, we analyze security of proposed protocol based on some security attributes described in [7].

Implicit Key Authentication. The group session key is computed by each user's ephemeral and long-term private keys. So, the users are assured that no other users except the partners who have the private keys can learn the group session key.

Known Session Key Security. In each session, each user U_i set new secret value so that their private/public key pair updated, and the generated group session key is depends up on private key of every users. The adversary that compromises one session key (if so) would not compromise other session keys, so the proposed protocol provides known session key security.

Forward Secrecy. If the adversary compromises one or more user's partial private key $D_i =< S_i, R_i >$ somehow, he cannot further calculate anything without knowing user's secret value x_i. Since D_i is not complete private key in the used certificateless cryptosystem.

Even if the adversary compromises the user's complete private key $< S_i, x_i >$, he cannot compute the value of K_i^R or K_i^L without knowing t_i. And thus if the long term private keys of one or more entities are compromised, the secrecy of previously established session key will not be affected.

No Key-Compromise Impersonation. Suppose that a user U_i's long-term private key $< S_i, x_i >$ has been disclosed, and an adversary E try to masquerade as the user U_j to all other users. Because the proposed protocol utilize a secure signature scheme, E's signature on the messages will not be accepted without user U_j's signature private key, and the other users will abort. Since the private key is computed with the contribution of S_i which is securely calculated by KGC, and the secrete value x_i , randomly chooses by user itself. Thus the private keys of users are totally independent from each others. Hence the adversary may impersonate the compromised user in the subsequent protocols, but it cannot impersonate other users.

No Key Control. The group session key in the protocol is determined by all members' long-term private keys $< S_i, x_i >$ and ephemeral secrete value neither party alone can control the outcome of the session key. No one can restrict it to lie in some predetermined value.

Perfect Forward Security. Even if KGC is compromised, the private keys of users cannot be disclose this is the main characteristic of certificateless public key cryptography . Thus the previously established session keys are not compromised.

6 Performance Comparison

This section compares the proposed protocol with some other CL-AGKA protocols [2,3,6]in terms of communication and computation cost. The result is showed in Table 1 (where n is the total number of users). The following notations are used in comparison table.

PM: number of Scalar point multiplications.
PA: Number of elliptic curve point additions.
Message: Total number of message overheads during key agreement process(including unicast and broadcast).
Pairings : number of bilinear pairings needed in key agreement process(zero in case of our proposal)

Table 1 shows that proposed protocol is more efficient than others in terms of communication cost. While it require an additional computation in the form of point additions (PA) keeping the comparable point multiplications (PM) cost. But since the relative cost of pairing computation is many times more than the

Table 1. Comparison Table

Protocol	PM	PA	Message	Pairings
Teng et al. [2]	$O(n)$	0	$O(n^2)$	$O(n^2)$
Geng et al. [3]	$5n$	0	$2n^2$	$4n$
Cao et al. [6]	$O(n^2)$	0	$O(n^2)$	$O(n)$
Our	$10n$	$6n$	$2n$	0

cost of point additions and proposed protocol not require any pairing computations, the overall computations of proposed protocol is comparable with other existing protocols.

7 Conclusion and Future Works

The present work proposes a pairing-free certificateless group key agreement protocol with its security and performance analysis. The analysis shows that it fulfils the required security goals to resist from the various attacks and needed comparable computation and efficient communication cost than other existing protocols with zero pairing computations. Also the required number of rounds for key agreement process is independent from the number of participants.

In future we will extend the present work to make it suitable for dynamic group by designing the efficient techniques for join, leaves and other group operations. We will also justify our work under standard security model.

Acknowledgments. Second an third authors of this article would like to thank University Grant Commission (UGC) for the support to this research work.

References

1. Al-Riyami, S., Paterson, K.: Certificateless public key cryptography. In: Laih, C.-S. (ed.) ASIACRYPT 2003. LNCS, vol. 2894, pp. 452–473. Springer, Heidelberg (2003)
2. Teng, J., Wu, C.: A provable authenticated certificateless group key agreement with constant rounds. Journal of Communications and Networks 14, 104–110 (2012)
3. Geng, M., Zhang, F., Gao, M.: A secure certificateless authenticated group key agreement protocol. In: International Conference on Multimedia Information Networking and Security, pp. 342–346 (2009)
4. Heo, S., Kim, Z., Kim, K.: Certificateless authenticated group key agreement protocol for dynamic groups. In: IEEE Global Telecommunications Conference, pp. 464–468 (2007)
5. Lee, E.J., Lee, S.E., Yoo, K.Y.: A certificateless authenticated group key agreement protocol providing forward secrecy. In: Proceedings of the 2008 International Symposium on Ubiquitous Multimedia Computing, pp. 124–129 (2008)
6. Cao, C., Ma, J., Moon, S.: Provable efficient certificateless group key exchange protocol. Wuhan University Journal of Natural Sciences 12, 41–45 (2007)

7. Xie, L., He, M.: A dynamic id-based authenticated group key exchange protocol without pairings. Wuhan University Journal of Natural Sciences 15, 255–260 (2010)
8. Shamir, A.: Identity-based cryptosystems and signature schemes. In: Blakely, G.R., Chaum, D. (eds.) CRYPTO 1984. LNCS, vol. 196, pp. 47–53. Springer, Heidelberg (1985)
9. Reddy, K., Nalla, D.: Identity based authenticated group key agreement protocol. In: Menezes, A., Sarkar, P. (eds.) INDOCRYPT 2002. LNCS, vol. 2551, pp. 215–233. Springer, Heidelberg (2002)
10. Konstantinou, E.: An efficient constant round id-based group key agreement protocol for ad hoc networks. In: Lopez, J., Huang, X., Sandhu, R. (eds.) NSS 2013. LNCS, vol. 7873, pp. 563–574. Springer, Heidelberg (2013)
11. Wan, Z., Ren, K., Lou, W., Preneel, B.: Anonymous id-based group key agreement for wireless networks. In: IEEE Wireless Communications and Networking Conference, pp. 2615–2620 (2008)
12. Stallings, W.: Cryptography and Network Security: Principles and Practice, 5th edn. Pearson Education (2011)

Enhancing the Security Levels in WLAN via Novel IBSPS

Sanjay Kumar

Department of Computer Science and Engineering,
National Institute of Technology Jamshedpur,
Jharkhand, India
sanjay.cse@nitjsr.ac.in

Abstract. Need of wireless technology increasing day to day due to the rapid development in information and communications technology. Threats and attacks are also growing in accordance with the increment in the usage of wireless communication technology. Especially Wireless Local Area Networks (WLAN) are less sensitive to the security attacks. This work combines the image processing and speech processing techniques. This work introduces Improved Bio-cryptic security aware packet scheduling (IBSPS) algorithm to enhance security levels in WLAN. To strengthen the authentication mechanism this work replaces the existing Enhanced Bio-cryptic security packet scheduling algorithm (EBSPS) with the IBSPS algorithm. In addition biometric encryption of features like finger print, Iris, Palm print and Face. IBSPS adds one more bio-cryptic security in the form of human voice as security level. Simulation were made on Matlab software and results prove that the proposed IBSPS security is stronger than the existing EBSPS and Bio-cryptic security aware packet scheduling (BSPS).

Keywords: Bio-cryptography, Wireless LAN, Bio-cryptic security aware packet scheduling, Enhanced Bio-cryptic security aware packet scheduling, security in WLAN.

1 Introduction

Wireless network business has grown up from the past two decades, due to its extensive usage. But secure communication is a major problem in the wireless networks. Especially Authentication, Authorization and Access (AAA) to WLANs due to Brute force attacks, Insufficient Authentication, Weak Password Recovery Validation, HTTP_Auth_ContainsBinary, HTTP_Auth_Too Long and many more attacks [1] is are critical confronts. According to the recent survey reports, The 2013 Data Breach Investigations Report (DBIR) confirms this and presents the perception of 19 global organizations on studying and hostility data breaches in the world. Details of different threat actions were presented firstly Hacking of breach occupies major portion with 52%, followed by Malware with 40%, social threat is 29%, 13% of misuse threat and Physical threat is 35% and common Errors is 2% [2].To resist the security attacks in literature many scholar and researchers has came up with the plenty of solutions [3- 5].

M.K. Kundu et al. (eds.), *Advanced Computing, Networking and Informatics - Volume 2*, Smart Innovation, Systems and Technologies 28,
DOI: 10.1007/978-3-319-07350-7_39,

The literature shows that Bio-cryptic security is stronger than the normal textual password security. Bio-cryptic is branch of science, which combines the advantages of both biometrics and cryptography [6]. Quality-of-Security is still poor in WLANs. To strengthen the authentication process, this work introduced the additional security level in the form of voice biometric [7].

The imperative contributions of this work consist of: (1) a survey and requirements of a improvements in wireless LAN according to Wireless Node user; (2) a novel Improved bio-cryptic security-aware packet scheduling; (3) a new performance analysis of security authentication and performance; (4) a simulator where the IBSPS algorithm is development and evaluated. The rest paper is organized as follows: Section 2 describes literature works in the area of Bio-Cryptography, Advanced Radius authentication server and Security Levels. Section 3 discusses the novel IMBSPS algorithm, system model and architecture. In section 4, it is represented with the performance analysis of our IMBSPS. Finally, the work is concluded with future plans are discussed in Section 5.

2 Related Works

The security level concept was introduced by the Xiao Qin *et al.* Their work discusses that, same security is not acceptable for every user. They clearly stated that, different security levels have to be mad by different users according to their need. Their work introduced the Security aware packet scheduling (SPSS) algorithm to assign security level. But security levels assignment are not clear is the flaw and which was reflected on the Load on the Network Switch (LNS) [8].

Rajesh Duvvuru*et al.*, has came up with a clarification to improve the performance of LNS and beside assuring security level by introducing Automated security aware packet scheduling (ASPS)algorithm. In ASPS they designed a new authentication server named it as Advanced Radius authentication server (ARAS). In ARAS,assignment of security level will be done automatically to the specific user depending upon their WN IP address. Due to automatic assignment of security, security levels were used for right WN, which will reduce the burden on LNS [9].

To strengthen the authentication process and besides assuring the LNS, once again Rajesh Duvvuru and team has proposed and proved an idea, by introducing the Bio-crypted Security-Aware Packet Scheduling-Algorithm (BSPS) algorithm. BSPS will strengthen the security in WLAN and follow the policy of ASPS. In work, they have used two biometric features, those are thumb and Iris and encrypted with RSA Algorithm [5]. The results BSPS were compared with the ASPS and SPSS and finally proved BSPS performs better than ASPS and SPSS in terms of security and LNS, combine knows Quality-of-Service (QoS) in WLAN. The BSPS is design for only three levels of security [10]. Later the work was extended by the same team, Rajesh Duvvuru *et al.*in improvising the security levels by instigating the Enhanced Bio-crypted Security-Aware Packet Scheduling-Algorithm (EBSPS). In EBSPS consists of five levels of security. Those are cryptic based password, thumb, iris, palm and facial biometric images. EBSPS results were compared with the BSPS, it is found that the EBSPS have two more additional security levels in the form of Palm print and face images.[11]. Avala Ramesh *et al.*, has made an analysis in biometric encryption using RSA algorithm. In their work, they have encrypted the biometric images with different edge detection technique like Laplacian, Zero cross and Canny operator. The analysis results shows that Canny edge detection biometric images are have good security than rest [12].Admas*et al.*, explained and done experimentation on voice encryption in just six steps. This is one of the basic models for voice encryption [13]. They are Chumchu, P.*et al.*, has given the cheap and best voice encryption framework over GSM-based networks. Their prototypes support both real-time and full-duplex communications. The RC4 algorithm is utilized to protect digital of voice to make the prototypes cheap and simple. The Bluetooth technology is used as communication media between mobile phones [14].

In the present work, the IBSPS is incorporated with one more security level, in addition to the present five level securities. The security level addition will be made in shape of encryption of voice biometric using speech processing approach (refer Fig.1).

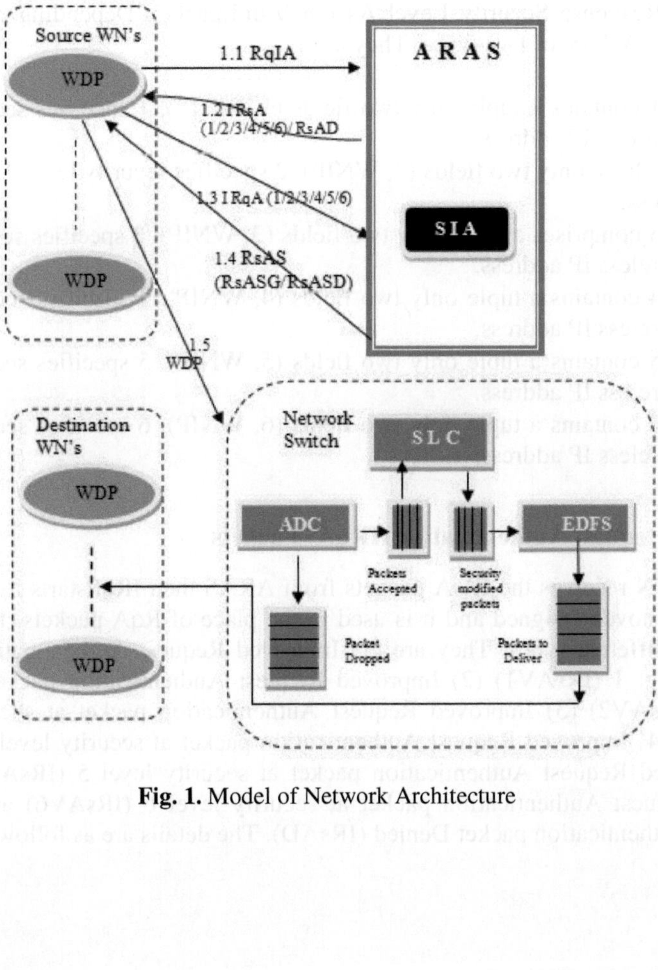

Fig. 1. Model of Network Architecture

3 Improved Bio-cryptic Security Packet Scheduling Algorithm

3.1 Assumptions and Notations

In order to improve the security in WLAN by adding the one more security level to existing five levels of security. To fulfill the above method task, we are replacing the EBSPS with IBSPS algorithm. In addition to this, IBSPS also inherited complete structure and properties of Request IP address (RqIA) packets and Response authentication status (RsAS) [5]. But, we replaced Response Authentication Packets (RsA) and Request authentication packet (RqP) with Improved Response Authentication (IRsA) Packets and Improved Request authentication packet (IRqA).

Improved Request Authentication (IRsA) Packets

Once the ARAS receives RqIA from WN, then the Security ID Adapter (SIA) will check the IP address and, if the IP address is valid, the ARAS will issue IRsAV (Improved Response Security Level Assign Valid) packet else it will issue IRsANV (Improved Response Security Level Assign Valid)packet. Depending on the IP address, the IRsAV are of five types. They are:

- IRsAV1 contains a tuple only two fields (1, WNIP). 1 specifies security level 1 and Wireless IP address.
- IRsAV2 have only two fields (2, WNIP). 2 specifies security level 2 and Wireless IP address.
- IRsAV3 comprises a tuple only two fields (3, WNIP). 3 specifies security level 3 and Wireless IP address.
- IRsAV4 contains a tuple only two fields (4, WNIP). 4 specifies security level 4 and Wireless IP address.
- IRsAV5 contains a tuple only two fields (5, WNIP). 5 specifies security level 5 and Wireless IP address.
- IRsAV6 contains a tuple only two fields (6, WNIP). 6 specifies security level 6 and Wireless IP address.

Improved Request Authentication (IRqA) Packets

Once the WN receives the IRsA packets from ARAS then IRqPstarts its work. IRqA packets are novel designed and it is used in the place of RqA packets. IRqA packets are of six different types. They are (1) Improved Request Authentication packet at security level 1 (IRsAV1) (2) Improved Request Authentication packet at security level 2 (IRsAV2) (3) Improved Request Authentication packet at security level 3 (IRsAV3) (4) Improved Request Authentication packet at security level 4 (IRsAV4) (5) Improved Request Authentication packet at security level 5 (IRsAV5). (6) Improved Request Authentication packet at security level 6 (IRsAV6) and Improved Request Authentication packet Denied (IRsAD). The details are as follows:

- IRqAV1 contains a tuple three fields (1, Password, ARASIP). 1 specifies security level 1 and cryptic password.
- IRqAV2 comprises a set of four fields (2, Password, thumb, ARASIP). 2 specifies security level 2 and also contain a cryptic password and Bio-cryptic (thumb image).
- IRqAV3 is a tuple of five fields (3, password, thumb, iris, ARASIP). 3 specifies security level 3 and it comprises of a textual Cryptic password and added with Bio-Cryptic image (thumb and iris).
- IRqAV4 is a set of six fields (4,pass, thumb, Iris, Palm print, ARASIP). 4 specifies security level 4 and it comprises of a textual cryptic password and includes with Bio-Cryptic image (Thumb, Iris and Palm print).
- IRqAV5 is a tuple of seven fields (5, pass, thumb, Iris, Palm print, Face, ARASIP).5 specifies security level 5 and it includes of a textual password and comprises of Bio-Cryptic image (Thumb, Iris, Palm print and Face).
- IRqAV6 is a set of eight fields (6, pass, thumb, Iris, Palm print, Face, Voice, ARASIP).6 specifies security level 6 and it includes of a textual password and comprises of Bio-Cryptic image (Thumb, Iris, Palm print,Face and Voice).

ARASIP represents the address of Advanced Radius Authentication Server. This is a common field in the ERqA packets.

Model of the Packet

The Wireless data packet (WDP) model is also inherited form the preceding works. The WDP has a set of fields (ATi,PTi,SLi,Di)[8]. Where, ATi and PTi is arrival time and processing time of packet i and SLi and Di is represents the security level and deadline of the packet i.

Equation -1 represents the Total Encryption Computation time of Thumb print (ECT(THi))

$$ECT\,(THi) = ED(THi) + ET\,(PPi) \tag{1}$$

Equation -2stands for the Total Encryption Computation time of Iris

(ECT(IRi))

$$ECT\,(IRi) = ED(IRi) + ET\,(IRi) \tag{2}$$

Equation -3signifies the Total Encryption Computation time of Palm print (ECT(PPi))

$$ECT\,(PPi) = SGT\,(PPi) + ED(PPi) + ET\,(PPi) \tag{3}$$

Equation -4symbolizes the Total Encryption Computation time of Facial (ECT(PPi))

$$ECT\,(FCi) = ED(FCi) + ET\,(FCi) \tag{4}$$

Equation -5 represents the Total Encryption Computation time of Voice (ECT(PPi))

$$ECT\ (VOi) = ED(VOi) + ET\ (VOi) \tag{5}$$

Where, The function ED represents the time taken to finding edges of biometric images and function ET is the encryption of the edges detected biometric images. EDSGT (PPi) is the segmentation of Palm print Image and time taken to encrypt the Palm print image. Then Computation time at WN(CWNi) can be defined as:

$$CWNi = ECT\ (Pass) + ECT\ (THi) + ECT(IRi) + ECT(PPi) + ECT(FCi) + ECT(VOi) \tag{6}$$

Here ECT is the common function for calculation of total computation time of biometric image. Equation -6 is combination of equation 1 to 5, in addition to text password encryption.

Similarly, the decryption at ARAS computation time (CARAS) is articulated as

$$CARASi = DCT\ (Pass) + DCT\ (THi) + DCT(IRi) + DCT(PPi) + DCT(FCi) + DCT(VOi) \tag{7}$$

Where, DCT is the function for computation of time for decryption at ARAS. At ARAS, only decryption happens, edge detection not required. By combining the equation 6 ,7 and network delay, we can able to calculate Total Authentication time (TATi). Here Network Delay (ND) was assumed using Randomized probability distribution.

$$TATi = CWNi + CARAS + ND \tag{8}$$

The IBSPS Algorithm

The IBSPS algorithm is defined as follows:

First, the node i sends RqIA containing the IP address(IPi), requesting access for network to ARSA(Advanced Radius Server Authentication). Second, IF IPi matches with the existing IP address then ARSA issue IRsAV (IRsAV1 or IRsAV2 or IRsAV3 or IRsAV4 or IRsAV5 or IRsAV6) ELSE IRsAD will send WN. Third, The user's system on receiving IRsAV generates the respective RsA packet (RsA1 or RsA2 or RsA3 or RsA4 or RsA5 or RsA6) for contains different credential data and sent back to the ARAS. Fourth ARAS checks the credential RsA packet credentials replies with RsAG packet ELSE RsAD packet.Fifth, if RsAG packet is received, the user node starts communication with the Network Switch. Finally, the rest of the algorithm follows the step No. 5 of ASPS algorithm [9].

4 Simulations and Results Analysis

IBSPS simulations were done in two phases. They are Image processing and speech processing in Matlab. Fig. 4 explains newly introduced sixth level of security. The biometric samples were collected from the Biometric research laboratories like IIT Delhi, IIIT Delhi and University of Massachusetts Amherrest [15][16][17] [18].

4.1 Simulation of Bio-cryptic Images

Simulation of Biometric images involves three steps. The very step was segmentation of the images, especially for Palm print images. Segmentation was performed using Region-of-Interest algorithm. Later the images are applied for the edge detection using canny operator and finally all edge detected images are encrypted individually using RSA algorithm (refer Fig. 3).

4.2 Simulation of Voice Encryption

In voice encryption, it opted for the symmetric key encryption. There are four different steps involved in this mechanism.

Step1: Wav file was read using the inbuilt function wavread() .
Step2: Distortion has to add to the original voice.
Step3: Extend Noise Samples According to Encryption Number.
*Step4: E*mbedded the information from audio to noise.
Step 5: Enter the decryption symmetric key to hear the original message.

4.3 Results Analysis

4.3.1 Comparison of Security Levels

When IBSPS algorithm is compared with the EBSPS and BSPS algorithms in term of security levels, IBSPS have stronger security than rest. Fig. 2 shows the comparison were made between the security level and packet size. It is observed that, packet size is growing along with the security level (refer Table 1).

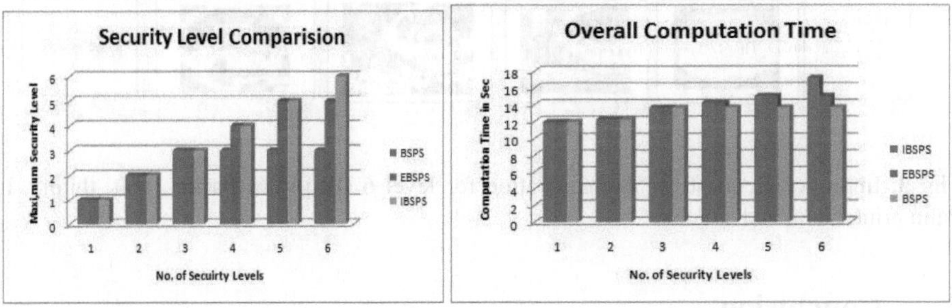

Fig. 2. Security Levels Comparisons and Overall Computation Time in WLAN

Table 1. Security level comaprisions

	Text	Thumb	Iris	Palm	Face	Voice
SL1	Yes	No	No	No	No	No
SL2	Yes	Yes	No	No	No	No
SL3	Yes	Yes	Yes	No	No	No
SL4	Yes	Yes	Yes	Yes	No	No
SL5	Yes	Yes	Yes	Yes	Yes	No
SL6	Yes	Yes	Yes	Yes	Yes	Yes

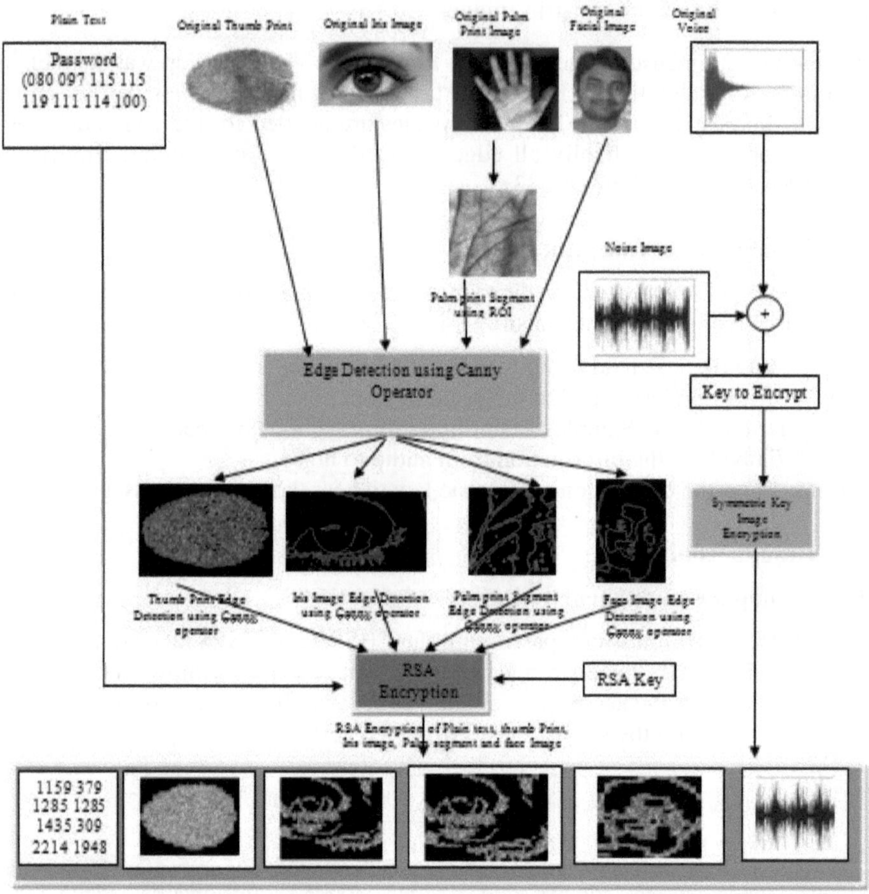

Fig. 3. Improved Request for Authentication for level 6 (IRqAV6) Packet (text, thumb, Iris, Palm print, facial and Voice)

5 Conclusion

Security is one of the important concerns in WLAN. In strengthen the security in WLAN, IBSPS is introduced. In this work, EBSPS is replaced with the IBSPS. In IBSPS one more security level is added in terms of voice encryption. IBSPS performed better than EBSPS and BSPS algorithm. The proposed work achieved 12% good than EBSPS and 50 % better than BSPS algorithms with respect to the security levels. At the same time IBPS computation is more than EBSPS and BSPS due to increment in the security levels.

In future, we will encrypt the security levels with the strong cryptographic algorithms like ECC and Choas based algorithms.

References

1. http://pic.dhe.ibm.com/infocenter/sprotect/v2r8m0/index.jsp?topic=%2Fcom.ibm.ips.doc%2Fconcepts%2Fwap_authentication.htm
2. http://www.verizonenterprise.com/resources/reports/rp_data-breach-investigations-report-2013_en_xg.pdf
3. Lapiotis, G., Kim, B., Das, S., Anjum, F.: A policy-based approach to wireless LAN security management. In: Proceedings of 1st International Conference on Security and Privacy for Emerging Areas in Communication Networks, pp. 181–189 (2005)
4. Ali, H.B., Karim, M.R., Ashraf, M., Powers, D.M.W.: Modeling and verification of Extensible Authentication Protocol for Transport layer Security in Wireless LAN environment. In: Proceedings of Second International Conference on Software Technology and Engineering, pp. 41–45 (2010)
5. Duvvuru, R., Rao, P.J., Singh, S.K., Sinha, A.: Enhanced Security levels of BSPS in WLAN. The International Journal of Computer Applications 84(2), 33–39 (2013)
6. Xi, K., Hu, J.: Bio-Cryptography. In: Handbook of Information and Communication Security, pp. 129–157 (2010)
7. Scheffer, N., Ferrer, L., Lawson, A., Lei, Y.: Recent developments in voice biometrics: Robustness and high accuracy. In: Proceedings of IEEE International Conference on Technologies for Homeland Security (HST), pp. 447–452 (2013)
8. Qin, X.: Improving Security of Real-Time Wireless Networks Through Packet Scheduling. IEEE Transactions on Wireless Communications 7(9), 3273–3279 (2008)
9. Duvvuru, R., Singh, S.K., Rao, G.N., Kote, A., Krishna, B.B., Raju, M.V.: Scheme for Assigning Security Automatically for Real-Time Wireless Nodes via ARSA. In: Singh, K., Awasthi, A.K. (eds.) QShine 2013. Lecture Notes of the Institute for Computer Sciences, Social Informatics and Telecommunications Engineering, vol. 115, pp. 185–196. Springer, Heidelberg (2013)
10. Duvvuru, R., Rao, P.J., Singh, S.K.: Improvizing Security levels in WLAN via Novel BSPS. In: Proceedings of IEEE International Conference on Emerging Trends in Communication, Control, Signal Processing and Computer Applications (2013)
11. Duvvuru, R., Rao, P.J., Singh, S.K., Sinha, A.: Enhanced Security levels of BSPS in WLAN. International Journal of Computer Applications 84(2), 33–39 (2013)
12. Ramesh, A., Setty, S.P.: Analysis on biometric encryption using RSA algorithm. International Journal Multidisciplinary Educational and Research 1(3), 302–307 (2013)
13. Adams, R.H., Calif, S.V.: Voice Encryption System. US Patient No. 4,232,194 (1980)
14. Chumchu, P., Phayak, A., Dokpikul, P.: A simple and cheap end-to-end voice encryption framework over GSM-based networks. In: Proceedings of IEEE International Conference on Computing, Communications and Applications Conference, pp. 210–214 (2012)
15. Sankaran, V.M., Singh, R.: Hierarchical Fusion for Matching Simultaneous Latent Fingerprint. In: Proceedings of International Conference on Biometrics: Theory, Applications and Systems (2012)
16. Ajay Kumar, A., Passi, A.: Comparison and combination of iris matchers for reliable personal authentication. Pattern Recognition 43(3), 1016–1026 (2010)
17. Kumar, A., Shekhar, S.: Personal Identification using Rank-level Fusion. IEEE Trans. Systems, Man, and Cybernetics: Part C41(5), 743–752 (2011)
18. Jain, V., Mukherjee, A.: The Indian Face Database (2002)

Varying Password Based Scheme for User Authentication

Santosh Kumar Sahu, Asish Kumar Dalai, and Sanjay Kumar Jena

National Institute of Technology, Rourkela, India
santoshsahu@hotmail.com, dalai.asish@gmail.com, skjena@nitrkl.ac.in

Abstract. Secure authentication scheme is required to protect businesses and clients against attacks. Passwords are used as private identity for an individual. The password has to be protected from several threats like stealing, shoulder surfing, eavesdropping and guessing. Several work has been done to improve the traditional password based authentication such as biometric password authentication, graphical password scheme, and dynamic password scheme etc. Graphical passwords are strong resistance towards brute force and dictionary attacks. But it suffers to eavesdropping and guessing attacks. However, these schemes have been proved ineffective. In this paper, we have designed a hybrid system by combing the features of three different schemes such as textual password, recognition based password and recall based password. The result shows that proposed model overcomes eavesdropping and guessing more effectively than its counterparts.

Keywords: User Authentication, Graphical Password, Textual Password, Stroke Based Passowrd, Salting.

1 Introduction

Authentication is the process of confirming and validating the truth of an attribute or entity. It is an act of determining whether an individual should be allowed or not, to access the system or the application. It should satisfy three basic goals of security: confidentiality, integrity and availability. It is very important to ensure security against the unauthorized access to any information resource. The level of security varies from one application to another, thus the acceptability of any authentication scheme depends on its robustness against attacks as well as its resource requirements. Password authentication has been widely used by on-line and off-line applications to authenticate their identities and access their accounts. In a traditional password authentication scheme, the user requires to enter his credentials in term of userid and password to prove his authenticity. The userid and password are compared with the credentials stored in the system database. If it matched then authentication is successful. Otherwise it fails. It has been reported that the password authentication scheme is vulnerable to many attacks like stolen verifier attack, man-in-the middle attack, replay attacks, and phishing attack [1–5].

M.K. Kundu et al. (eds.), *Advanced Computing, Networking and Informatics - Volume 2*, Smart Innovation, Systems and Technologies 28,
DOI: 10.1007/978-3-319-07350-7_40,

1.1 Types of Authentication Systems

The authentication schemes can be categorized as follows [6]:

- What you know? Includes traditional textual password based schemes or the PIN based schemes.
- What you have? Includes authentication by smart cards or electronic tokens.
- What you are? Includes the schemes like biometric authentication systems.

The common type of scheme is What you know?. Under this scheme, textual password is used more frequently as it is easy implement. However, it can be used in the applications requires low level of security. A password is strong if it consists with lowercase, upper case, alphanumeric and digits with special symbols. The strong password improves security. But, longer passwords are difficult to memorize, due to which the user write down on paper. It may chance of missing or access by unauthorized person. Additional criteria for good textual password is, it should be easy to type and not vulnerable against keystroke dynamics. Textual based schemes have many problems like shoulder surfing, key logger, vulnerable to guessing, dictionary attack and hard to remember [6]. What you have? scheme includes the smart cards and electronic tokens. Token based systems are vulnerable by man-in-middle attacks where an attacker captures the user session and archives the authorizations by acting as a proxy between the individual and the authentication system without knowledge of the user.

Biometric authentication systems come under What you are? scheme. The devices required to implement are high cost. It also suffers by replay attack. The textual password scheme suffers from various attacks as in [1]. For example, the attacker can have an idea about the password by keystroke dynamics i.e. by looking and monitoring the movements of the users hands on the keyboard. Pictures are easier to remember than text [9]. So, it requires a better password scheme which is easy to remember and safe from stealing. Moreover the graphical scheme provides better security against some common type of attacks like brute force attacks and dictionary attacks. But, it suffers from shoulder surfing. There are several attacks against passwords such as brute force attack, dictionary attack, guessing, eavesdropping, shoulder Surfing, accessing the password file, and stealing the password as discussed in [8].

The remaining of this paper contains as follows. Section 2 provides a brief review of existing work. Section 3 contains the proposed method for user authentication. Concluding remarks are given in Section 4.

2 Literature Review

To protect passwords from shoulder surfing and eavesdropping, Shahid et.al. [8] the scheme which is motivated by the Unix System. It supports unlimited length typed by the claimant, but the first eight character are only compare with the stored password and remaining are only to falsify the attacker [7]. To improve this, the password will be any part of the input string. At the time of registration the user has to mention the starting position and length of the password. For example if we choose our password **abcd**, and staring position 4, then the possible input string can be 123 **abcd**3fofvn, xyz**abcd**dchjk.

The problem with this scheme is that it can be easily compromised by XOR operation. To improve some shortcomings of textual passwords, graphical passwords considered as a better alternative, but they are vulnerable by shoulder surfing. Ahmad et. al. [9] proposed a hybrid authentication system of graphical and textual password. The user generates a graphical password by selecting several point-of-interests (POI) on a picture generated by the system in registration phase. Each POI is described by a circle with the clicked point as the center and some specified area around the center [9]. The POIs are associated with words or phrases. At the time of authentication user will select the point in the picture and enter the corresponding words in the password text box. If the selected points and the corresponding word or phrase match with the stored data, the authentication process will success.

Ziran et.al. [10] have proposed a recall based authentication system with textual password. At the time of registration, the user draw a pattern on the grid which is generated by the system [9]. The pattern and order is stored in system database. During Authentication process user have to enter the passwords by traditional input devices in some text area, according to the values shown in the grid. The grid only shows 0 or 1, which are randomly generated on each execution. The use of 0, 1 provides ambiguity which protect against the shoulder surfing. The text area is used to enter the input password to provide more resistance to shoulder surfing. If the values entered in the text area matches with the values at grid point in correct order, then the authentication will be successful. The order of the selection of points in the pattern can be in any manner. So, it provides better security in compare to previous schemes. The demerit is, it is a time consuming approach but can be used for more secure applications.

Chao et.al. [11] proposed a login system which is implemented as a gaming to make login process more interesting. It contains an image in background color which is used as a security factor. Sadiq et. al. [6] proposed an image based implicit authentication system with several images for selection of password. Each image have several click points and each click point is associated with several attributes. The user provides some information in which keywords are abstracted during registration phase. At the time of authentication, the server choose a random keyword associated with the user with a random image that has the text attribute related to the object. Then it sends to the users. The individual has to select the correct object that signifies the expected keyword. This scheme suffers from shoulder surfing.

Ziran et.al. [9] have proposed a recall based authentication system with textual password. The user has to select a pattern on a grid. The selected pattern and the corresponding order is stored. At the time of authentication process user have to enter the password, as per the values shown in the grid. The grid shows 0 or 1, which are randomly generated every time by the system. The use of 0,1 create confusion to the attack which evades shoulder surfing. If the values entered in the text area matches with the values at grid point in correct order, then the authentication will be successful. The order of the selection of points in the pattern can be in any manner.

3 Proposed Method

The method has been proposed to develop a novel scheme for protecting passwords, against shoulder surfing, guessing and eavesdropping. All the schemes have some merits and demerits. In our approach, we are using the features of one scheme to improve the feature of another scheme mutually.

3.1 Varying Password Based Scheme

This method is motivated by the scheme discussed by the Mohammad et.al. [8].

Some additional features are added in the existing scheme by removing the restriction of choosing the fixed starting position. The password string can be anywhere in the input string. For example, if our password is **abcd**, then the input string may be 123**abcd**19jfdjdfj, 1234dj**abcd**fdruoi, **abcd**1234589 or 1234344**abcd**. The password may be present anywhere in the string. Figure 1 looks like a single substring matching algorithm. The passwords are stored using one way trapdoor function. The hash value of the stored passwords and the input string is completely different. At the time of registration, it stores the information of starting symbol, and length of password, as shown in Figure 1. We follow same steps at the time of authentication as shown in Figure 2.

To provide more secure, we are implemented Virtual keyboard with regional language support. Virtual keyboard is a software components which allows user to enter the characters using mouse. The main advantage of using virtual keyboard is, it reduces the threat of keystroke logging. We can make the key pattern dynamic and enter multilingual characters to provide better security using it. The demerit is it can be compromised using shoulder surfing. Furthermore, a user may not be able to point and click as fast as they could type on a keyboard, thus making it easier for the observer. To avoid it, we have randomized the positions of the keys on the virtual keyboard every time. The multilingual keyboard has been implemented using Unicode. Our virtual keyboard contains all Indian regional language keys with some symbols. So, it provides better security over various attacks. The details analysis of brute force attack is discussed in following section.

3.2 Analysis of the Varying Password Scheme

The Table1 shows the length of password, number of alphabet, number of passwords generated, and the time taken to apply brute force attack. In the proposed scheme, if the password length is 4, then 4.56 second is required to recover the password, if the length is 6 then it will take 85.8 hours and if the length is 8 then 241697.99 day is required. Which makes it stronger in compare to previous

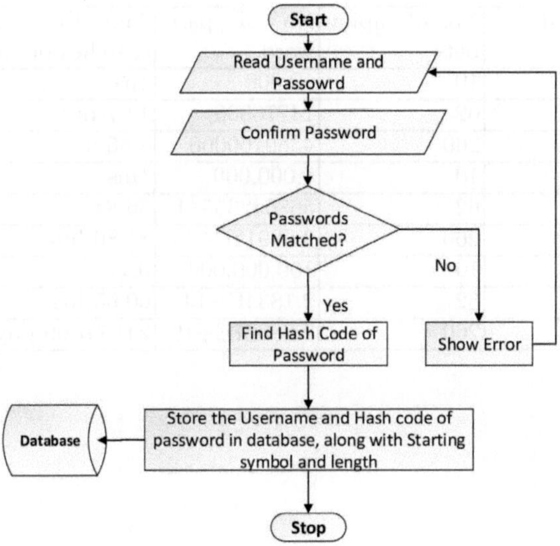

Fig. 1. Registration Process of Varying Password Scheme

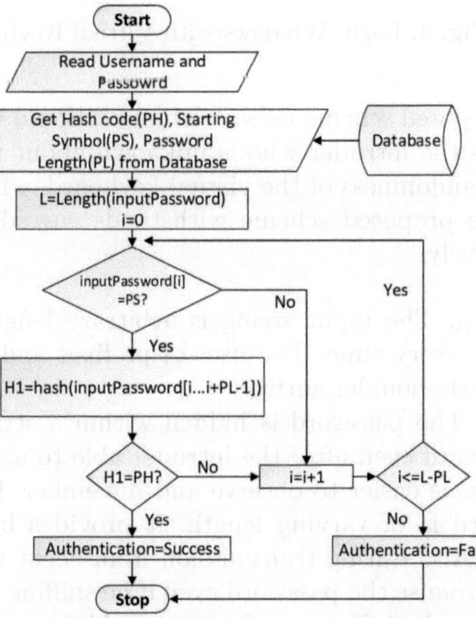

Fig. 2. Authentication Process of Varying Password Scheme

Table 1. Analysis of Brute Force Attack

Password Length	No. of alphabets	No of password	Time Taken (1 billion searches per second)
4	10	10,000	1ps
4	62	14776336	14.7 ms
4	260	4569760000	4.56 s
6	10	1,000,000	1ms
6	62	56800235584	56.8s
6	260	3.08916E+14	85.80 hrs.
8	10	100,000,000	0.1s
8	62	2.1834E+14	60.65 hrs.
8	260	2.08827E+19	241697.99 days

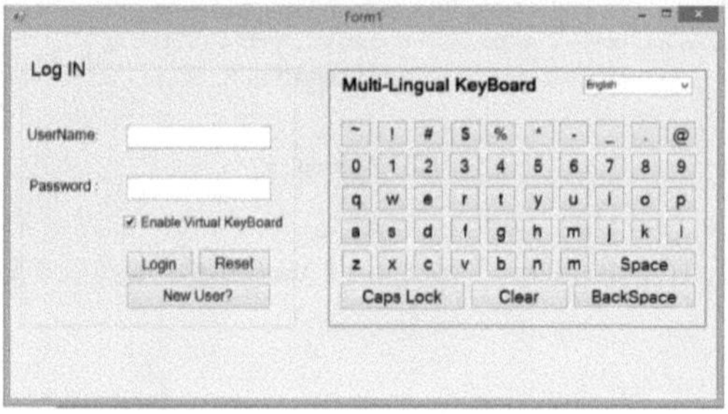

Fig. 3. Login Windows with Virtual Keyboard

schemes. As the proposed scheme uses a virtual keyboard with regional language support, it protects the intruder who is unknown about the regional language. Again, due to the randomness of the virtual keyboard, which sup- ports around 260 alphabets. The proposed scheme withstands eavesdropping and shoulder surfing more effectively.

1. Shoulder Surfing: The input string is arbitrary length. The length of the password varies every time. The uses of prefixes and suffixes increases the resistance against shoulder surfing.
2. Eavesdropping: The password is hidden within a string. So, it is not easy to see the password even after the intruder able to access the password file. For example xyz is easier to observe and remember, but auenxyzejs is not. As the password is of varying length, it provides better security against eavesdropping. Also during transmission from client to server, the intruder unable to compromise the password even if he sniffing the password by using traffic analysis method. Because the password is in hashed format with some dummy characters.
3. Guessing: The attacker can guess by seeing the length or checking some common words or phrases. In this scheme input password string is of varying

Fig. 4. Regional Virtual Keyboard

length and varying characters. So it is not easy for the attacker to guess the password.

4. Brute Force Attack: Use of multilingual keyboard drastically reduces the brute force attack. The multilingual keyboard, having 260 characters. The time required to recover the password is analyzed in table 1. It is very difficult to compromise the password if the length is 8 character.
5. Key Stroke Dynamics: The main advantage of virtual keyboard is it provides a better resistance against keystroke logging. Using virtual keyboard, the key logger and other key stroke based attacks are avoided.
6. XOR Analysis: In the conventional methods, the actual password can be guessed using XOR operation If two input password is known. But in the proposed scheme, the password can be anywhere in the input string, which provides better resistance to XOR operation.
7. Stealing the Password: The more complex passwords are difficult to remember. Sometimes user writes them down on paper or personal notebook. It is possible to compromise the password if the intruder can get the paper or the notebook. But, in our scheme this problem is avoided. Because the actual password is embed with some extra dummy characters. Even if the intruder get the paper he unable to login the system.
8. Advantages:
   (a) No need to remember the starting point.
   (b) If any error occurred while entering the password, no need to go back or delete any typed character. From that point the user can retype the actual password.
   (c) The small password can be effectively used against the shoulder surfing.
9. Drawback: The time complexity of the Authentication process will increase as it involves various matching. Additionally, we have to calculate the hash code of many strings which have equal length as original password, and have a starting character whose hash value is matched with the starting character of the password.

4 Conclusions

The varying Password Scheme provides much security over eavesdropping and guessing. It can be used in any unsafe environment like public places. In addition, this scheme uses a multilingual virtual keyboard which avoids keystroke dynamics, eavesdropping and guessing. It also provides a much better resistance to brute force attacks by extending the alphabets used.

References

1. Evans Jr., A., Kantrowitz, W., Weiss, E.: A user authentication scheme not requiring secrecy in the computer. Commun. ACM 17, 437–442 (1974)
2. Purdy, G.B.: A high security log-in procedure. Commun. ACM 17, 442–445 (1974)
3. Kwon, K., Ahn, S.-J., Chung, J.-W.: Network Security Management Using ARP Spoofing. In: Laganá, A., Gavrilova, M.L., Kumar, V., Mun, Y., Tan, C.J.K., Gervasi, O. (eds.) ICCSA 2004. LNCS, vol. 3043, pp. 142–149. Springer, Heidelberg (2004)
4. Haller, N.M.: The S/Key one-time password system. In: Proc. Internet Society Symposium on Network and Distrbuted System Security, pp. 151–158 (1994)
5. Mitchell, C.J., Chen, L.: Comments on the S/KEY user authentication scheme. ACM Operating Syst. Rev. 30, 1216 (1996)
6. Almuairfi, S., Veeraraghavan, P., Chilamkurti, N.: A novel image-based implicit password authentication system (IPAS) for mobile and non-mobile devices. Mathematical and Computer Modelling 58(12), 108–116 (2013)
7. Wiedenbeck, S., Waters, J., Birget, J.-C., Brodskiy, A., Memon, N.: Passp oints: Design and longitudinal evaluation of a graphical password system. International Journal of Human-Computer Studies 63(12), 102–127 (2005)
8. Shahid, M., Qadeer, M.A.: Novel scheme for securing passwords. In: 3rd IEEE International Conference on Digital Ecosystems and Technologies, DEST 2009, pp. 223–227 (2009)
9. Almulhem, A.: A graphical password authentication system. In: 2011 World Congress on Internet Security (WorldCIS), pp. 223–225 (2011)
10. Zheng, Z., Liu, X., Yin, L., Liu, Z.C.: A stroke-based textual password authentication scheme. In: Proceedings of the 2009 First International Workshop on Education Technology and Computer Science, ETCS 2009, vol. 3, pp. 90–95. IEEE Computer Society, Washington, DC (2009)
11. Liu, X.-Y., Gao, H.-C., Wang, L.-M., Chang, X.-L.: An Enhanced Drawing Reproduction Graphical Password Strategy. J. Comput. Sci. Technol. 26(6), 988–999 (2011)

Online Signature Verification at Sub-trajectory Level

Sudhir Rohilla[1], Anuj Sharma[2], and R.K. Singla[1]

[1] Department of Computer Science
Panjab University, Chandigarh, India
[2] Center for Advanced Study in Mathematics
Panjab University, Chandigarh, India
rohilla2209@gmail.com, anujs@pu.ac.in, rksingla@pu.ac.in

Abstract. The signatures are behavioral biometric characteristic used for authentication purpose. The verification of a signature while writing through the machine is called online signature verification. In this paper, we have implemented verification of signatures at sub-trajectory level. The verification has been performed using common threshold of features and writer dependent threshold. A set of fifty features are extracted of nature static, kinematic, statistical and structural properties. The experiments have been performed using SVC2004 (Signature Verification Competition) Task1 where forty user's data include twenty genuine and twenty forgery signatures of each user. The achieved results indicate that verification at sub-trajectory level is a promising technique in online signature verification.

Keywords: online signature verification, feature extraction, feature level threshold, writer dependent threshold

1 Introduction

The authentication through signatures is carried out to legalize the document and hence is widely accepted. The verification of signatures while writing through a machine is called online signature verification. The online signature verification mainly consists of two phases: Feature extraction and verification technique. The matching technique is one such example of verification technique. The features further can be viewed as two parts: Local features, the features corresponds to a particular portion of signature, and global features, the features correspond to the entire signature. For verification, we have several verification technique such as DTW (Dynamic Time Warping), SVM (Support Vector Machine), HMM (Hidden Markov Model), NN (Neural Network) etc. [1].

The Kashi *et al.* used the HMM technique on the signatures of 59 users (542 genuine and 325 forgeries) and reported FAR (False Acceptance Rate), the rate at which forgery signatures are accepted by the system, is 2.5% and the FRR (False Rejection Rate), the rate at which genuine signature are rejected by the system, is 2.5%. The main features collected were first moment, coordinates

Smart Innovation, Systems and Technologies 28,
DOI: 10.1007/978-3-319-07350-7_41,

and their velocities, integrated absolute centripetal acceleration, direction histogram, rms velocity etc. [2]. Wu *et al.* used the split and merge mechanism on the database consisting 246 forgeries and 200 genuine signatures and reported FAR as 2.8% and FRR as 13.5% [3]. Jain *et al.* used the DTW technique by collecting features like Δx and Δy i.e. x and y coordinates differences, Y coordinates with reference to center of signature, $sin(\alpha)$ and $cos(\alpha)$ with x-axis, curvature etc. The FAR and FRR, for common threshold were reported as 2.7% and 3.3% respectively and for writer dependent threshold as 1.6% and 2.8%, respectively [4]. Kholmater *et al.* used the DTW matching technique for the verification of a system having signatures of 94 users (1134 genuine and 367 forgeries). The main features collected were: x and y coordinates with reference to first point of trajectory, Δx and Δy and the curvature difference between two consecutive trajectory points. The system based on the above features along with the linear classifier has been established to report FRR as 1.64% and FAR as 1.28% [5]. Augilar *et al.* used the HMM technique on the MCYT bimodal biometric database consists of fingerprint and on-line signature modalities. They had reported the EER for skilled forgeries as 0.74% and for random forgeries as 0.05% [6]. Guru and Parkash introduced a verification technique based on the symbolic representation using ATVS MCYT database consisting of 50 signatures of each 330 users. The FAR and FRR for common threshold were reported as 4.1% and 4.3% and the same ratios for writer dependent threshold were 3.9% and 3.7% which further resulted into an average EER of 3.8% [7]. Recently, Barkoula *et al.* explored TAS (turning angle scale) and TASS (turning angle scale space) for representing the signature and then applied variation of the longest common subsequence verification technique on two databases: SVC2004 (Task1) and SUSIG (Visual Corpus Part-1). For skilled forgeries they had reported the average EER of 5.33% for SVC2004 and 0.52% for SUSIG database [8]. Emerich *et al.* applied Tespar based coding method, wavelet analysis and SVM as the verification technique on the SVC2004 (Task2) database. The average EER was reported as 6.96% [9]. Garcia reported the average EER of 2.74% by using DTW and GMM matching techniques and MCYT database (100 users and 50 signatures for each of the 100 users) [10].

In this paper, the feature extraction and verification of signatures is implemented at the sub-trajectory level and in the next section, the design of the system is presented. In section 3, the different categories and various kind of features at sub-trajectory level are discussed. In section 4, we conclude this paper with the results.

2 System Design

Fig. 1 presents the system design to perform signature verification at sub trajectory level. There are three main components as data acquisition, feature extraction and verification. The data acquisition refers to online form of signature input where a signature behave as a trajectory with number of points (P) in two dimension as (x_p, y_p), where $p = 1, 2, ..., P$. The acquired data is size normalized.

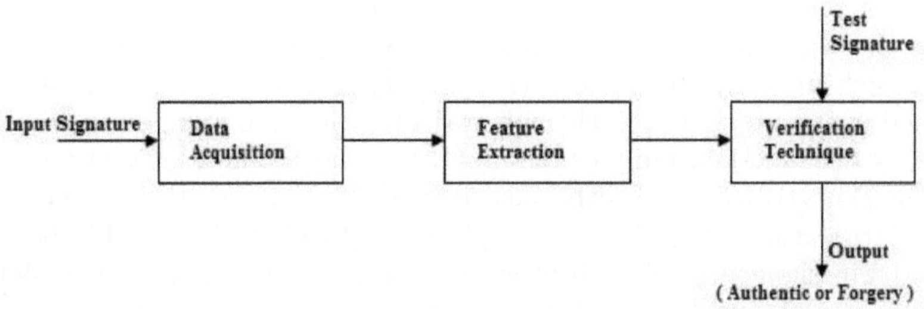

Fig. 1. System Design of signatures verification

The normalized form of the signature is divided into sub-trajectories and a set of features are computed in each sub-trajectory. The combined feature vector of a test signature is used in signature verification against the feature vector of training data set. For verification we have applied common threshold and writer dependent threshold on a testing signature which produces the output for a signature to be authentic or forgery. After testing a number of signature samples, we have checked the performance of the system in terms of Mean Equal Error Rate, FAR and FRR.

3 Feature Extraction

In this paper, a set of hundred features as discussed in literature [11] have been extracted and categorized into four main categories as static properties, kinematic properties, statistical properties and structural properties of a signature. The static properties include the features that depend upon the position vectors, total time duration for completing a signature, number of pen strokes, number of local maxima in x and y direction, number of points where trajectory changes its direction in both x and y direction, Ratio of time taken by different pen strokes to the total time duration for a signature etc. The kinematic properties include velocities i.e velocity in x and y direction, and rms velocity; accelerations i.e. acceleration in x and y direction, tangential acceleration, centripetal acceleration; jerks etc. The statistical properties constitutes the set of x and y vectors, their means and standard deviations, covariance of velocity in x and y direction and the structural properties consists of the features like direction histograms, direction change histograms, tangent path angle, the curvature, ratios of various angles at critical points etc. The Nelson and Kishon, mentioned some formulas for the features path tangent angle, tangential and centripetal acceleration, jerk, curvature etc. mainly used in signature verification [12].

In literature, Augilar [11] and Guru *et al.* [7] computed the hundred features on the entire signature trajectory. Augilar, empirically, ranked the first sixty features as higher ranked than the other 40 features. The Higher rank of a feature described as the impact of that feature is more significant in the process of verification as compared to the low rank feature.

In this paper, we have used the sub-trajectories $(S_1, S_2, ..., S_t)$ which means dividing the signature into t segments by taking equal number of trajectory points in each sub-trajectory and extracted fifty feasible features, out of hundred mentioned in the literature, at the sub-trajectory level to increase the number of features by t times (i.e., number of sub-trajectories). These fifty features could happen in a sub-trajectory as other features demand information based on entire signature trajectory. One of the example from ATVS-SLT-DB (database of 27 users and 146 signatures of each user) for $t = 1$, we have a complete signature as given in Fig. 2. and for $t = 4$, the same signature has been broken into four segments as shown in Fig. 3(a), Fig. 3(b), Fig. 3(c), and Fig. 3(d).

Fig. 2. Signature representation at $t = 1$

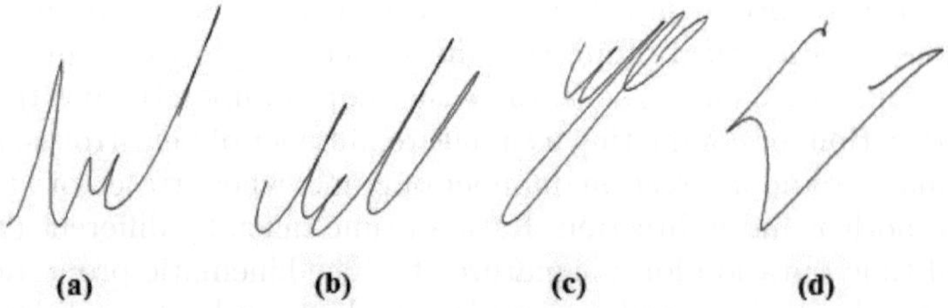

Fig. 3. Sub-trajectory of signatures at $t = 4$

4 Results

We have followed the verification technique based on the symbolic representation of online signatures [7]. The thresholds used in this technique are the **CT** (common threshold) depends upon the total number of feature extracted, the feature level threshold (FT = $\alpha \times$ SD) depends upon the SD (standard deviation) of the each feature computed for all the sample signatures of a particular writer in the training set, where α controls the width of interval in feature vector for each feature and the writer threshold (WT = $\beta \times$ mean (total acceptance count of a number of sample signature for each writer)) is decided by the function acceptance count which further depends upon the FT and beta controls the level of the WT [7].

After extracting the hundred features [11], the interval-valued reference feature vector is calculated as

$$RF_j = \{[f_{j1}^-, f_{j1}^+], [f_{j2}^-, f_{j2}^+], ..., [f_{jm}^-, f_{jm}^+]\},$$

where $f_{jk} = \mu_{jk} \pm FT(jk)$ and μ_{jk}, here, is the mean of the kth feature out of total 'm' features for all the S_n samples of each jth user. The kth feature of a testing signature is then compared to the interval-valued feature vector of the claimed writer and contributed towards authentication count if it lies between the interval otherwise added towards forgery. The total of such authentication count verified with the thresholds, CT or WT, and authentic signature is found if greater than the threshold and the forgery is detected if less than the thresholds.

The four sets of experiment have been performed where training include from genuine signatures only. The test data include four sets as (i) 20 (Forgery) + 5 (Genuine) (ii) 20 (Forgery) + 10 (Genuine) (iii) 20 (Forgery) + 15 (Genuine) (iv) 20 (Forgery) + 20 (Genuine). The experimentation includes at whole trajectory of a signature vs. sub-trajectories of that signature. We observe that the EER is less in case of sub-trajectory level experiment as compare to whole trajectory of signature. The four sets of experiment indicate constant decrease in EER with 5, 10, 15 and 20 genuine signatures in test data. In addition, large data sets could further help in improvement of EER.

The online signature verification has been performed at sub-trajectory level where we are able to extract $m \times t$ features (m is the number of features feasible in a sub-trajectory and t is the number of sub-trajectories). The trend of results indicate that the sub-trajectory based approach could lead to promising results with enhancement of features. In addition, the present results have been performed with limited database size and EER could be improved with increase in data for genuine and forgery signatures. The online signature verification offer useful applications in real life use such as banking, web based authentication etc. The use of large data in present study could further help to achieve objectives of real life applications.

References

1. Guru, D.S., Prakash, H.N.: Symbolic Representation of On-Line Signatures. In: Proceedings of International Conference Computational Intelligence and Multimedia Application, pp. 312–317 (2007)
2. Kashi, R.S., Hu, J., Nelson, W.L., Turin, W.: On-line handwritten signature verification using Hidden Markov Model features. In: International Conference on Document Analysis and Recognition, pp. 253–257 (1997)
3. Wu, Q.Z., Lee, S.Y., Jou, I.C.: On-Line Signature Verification Based on Split and Merge Matching Mechanism. Pattern Recognition Letters 18, 665–673 (1997)
4. Jain, A.K., Griess, F., Colonnel, S.: On-Line Signature Verification. Pattern Recognition 35, 2963–2972 (2002)

5. Alister, K., Yanikoglu, B.: Identity Authentication Using Improved On-Line Signature Verification Method. Pattern Recognition Letters 26(18), 2400–2408 (2005)
6. Aguilar, J.F., Garcia, J.O., Ramos, D.D., Rodriguez, J.G.: HMM-based on-line signature verification: Feature extraction and signature modeling. Pattern Recognition Letters 28(16), 2325–2334 (2007)
7. Guru, D.S., Prakash, H.N.: Online Signature Verification and Recognition: An Approach Based on Symbolic Representation. IEEE Transactions on Pattern Analysis and Machine Intelligence 31(6) (2009)
8. Barkoula, K., Economou, G., Fotopoulos, S.: Online signature verification based on signatures turning angle representation using longest common subsequence matching. International Journal on Document Analysis and Recognition 16(3), 261–272 (2012)
9. Emerich, S., Lupu, E., Rusu, C.: A new set of features for a bimodal system based on on-line signature and speech. Digital Signal Processing 23, 928–940 (2013)
10. Garcia, M.L., Lara, R.R., Hurtado, O.M., Canto-Navarro, E.: Embedded System for Biometric Online Signature Verification. IEEE Transactions on Industrial Informatics 10(1), 491–501 (2014)
11. Aguilar, J.F.: Adopted Fusion Schemes for Multimodal Biometric Authentication. PhD thesis, Biometric Research Lab AVTS (2006)
12. Nelson, W., Kishon, E.: Use of dynamic features for signature verification. In: Proceedings of the IEEE International Conference on Systems, Man, and Cybernetics, vol. 1, pp. 201–205 (1991)

DCT Based Robust Multi-bit Steganographic Algorithm

Imon Mukherjee and Arunangshu Podder

Department of Computer Science & Engineering
St. Thomas' College of Engineering & Technology
Kolkata-700023, India
{mukherjee.imon,arunangshu.podder}@gmail.com

Abstract. With the rise of communication through the Internet, there has also been a rise of interception of important messages, thus resulting in a greater risk of breach of privacy. Hence, constant research is going on for the development and improvement of techniques that can handle such attacks. Most of the existing algorithms use either the spatial domain or the frequency domain of the image for embedding the secret message. This paper introduces a steganographic algorithm that uses the frequency domain of the image for selecting the potential pixels and the spatial domain for embedding the message bits, thus making it robust against steganalytic attacks. This technique is also capable of withstanding statistical attacks. Our proposed algorithm embeds a maximum of 5 bits of the message per pixel in each image component thus making the embedding capacity very high. Besides embedding capacity, our technique also has a high embedding efficiency.

Keywords: Information Security, Multi-bit, DCT, Steganography, Steganalysis, Data hiding.

1 Introduction and Motivation

There is always a high risk of sensitive information getting intercepted nowadays due to the many advances in Internet Technology. Several methods have been formulated for the maintenance of the secrecy of such sensitive information. Information hiding is necessary for secure communication. Steganography is the concealment of a secret message in any digital media files such as image, audio, video, etc in a manner so that the existence of the hidden information can be detected only by the sender and the intended recipient. A large number of digital formats are suitable for Steganography.

The main distinction between steganography and the well-known term cryptography is that, although hiding secret information is achieved by both, Cryptography puts into use some keys and does not conceal the existence of the message, whereas in steganography [11], nobody but the intended recipient can perceive the existence of secret information.

One of the most common and simplest steganographic embedding process is the Least Significant Bit (LSB) based technique [6], [10], [13] in which the bits

M.K. Kundu et al. (eds.), *Advanced Computing, Networking and Informatics - Volume 2*, Smart Innovation, Systems and Technologies 28,
DOI: 10.1007/978-3-319-07350-7_42, © Springer International Publishing Switzerland 2014

from the secret message are embedded in LSBs of the pixels of the cover image which is then called a stego image. Flipping the LSB in the pixel intensities does not produce any perceptible distortion in the image. In order to remove the drawbacks of LSB embedding techniques, the concept of multi-bit embedding techniques [1], [7], [9], [15], [16] are introduced where multiple bits of each pixel are used to hide data. However, embedding too many bits per pixel may degrade the quality of the stego image. Thus, the embedding capacity of each pixel of the carrier image should be decided carefully. We should be more cautious while embedding using the DCT technique. Although in frequency domain, viz., DCT steganographic techniques, the embedding capacity is low, but, the robustness is much higher than spatial domain techniques. In our algorithm, we have used the DCT technique to scatter the message bits throughout the image thus providing better resistance against statistical attacks.

2 Proposed Method

Input: Cover pixel intensity $\Pi_{x,y}$ and message bits to be hidden.
Output: Stego pixel intensity $\Pi'_{x,y}$.

1. Divide the cover image components into 8×8 non-overlapping blocks and find the DCT coefficients of each block;
2. Choose a threshold value for embedding (based on DCT values);
3. Identify the locations where the DCT coefficients are less than the respective threshold values;
4. Divide the image component into 3×3 overlapping pixel intensity blocks in spatial domain;
5. Determine the *embedding capacity* n of the center pixel of each block using Eq. (2);
6. Estimate a temporary value $T_{x,y}$ as follows: $T_{x,y} = M - \Pi_{x,y} \bmod 2^n$, where M is the decimal representation of the selected n bits of the hidden message;
7. Calculate $T'_{x,y} = T_{x,y} + S.2^n$, where S is determined as follows: $S =$

$$\begin{cases} 1 & \text{if } -2^n + 1 \leq T_{x,y} < -\lfloor \frac{2^n-1}{2} \rfloor, \\ 0 & \text{if } -\lfloor \frac{2^n-1}{2} \rfloor \leq T_{x,y} \leq \lceil \frac{2^n-1}{2} \rceil, \\ -1 & \text{if } \lceil \frac{2^n-1}{2} \rceil < T_{x,y} < 2^n. \end{cases}$$

8. Set $\Pi'_{x,y} = \Pi_{x,y} + T'_{x,y}$;
9. Replace the value $\Pi_{x,y}$ with $\Pi'_{x,y}$;

Algorithm 1. Algorithm for embedding secret message bits into a cover pixel

In this section, we first divide the image into 8×8 non-overlapping blocks to obtain the DCT coefficients of each pixel block of the first cover image. The pixel positions of the cover image where the message bits will be embedded will

Input: Two stego images.
Output: Secret message bits.
1 Read the intensity value (say, $\Pi^*_{x,y}$) of each pixel of the first image;
2 Read the second image and determine the n for each pixel of the first image;
3 Calculate the decimal representation of message bits embedded per pixel as follows:
$$d = \Pi^*_{x,y} \mod 2^n.$$
4 Convert d into its corresponding binary representation of each pixel to get the secret message bits;

Algorithm 2. Algorithm for extracting secret message bits

be determined by using these DCT coefficients after they are compared with a pre-determined threshold value which will remain constant throughout the cover image. One can choose any threshold value. However it should be noted that hiding in pixels where the DCT coefficient value is 0 should be avoided in order to avoid any significant perceptual distortion in the image. Hence we have considered a threshold value as the negative value closest to 0.

$$\Pi_{diff} = \Pi_{max} - \Pi_{min} \tag{1}$$

$$n = \begin{cases} min(5, \lfloor \log_2 \Pi_{diff} \rfloor - 1), & \text{if } \Pi_{diff} \in [4, 255], \\ 1 & \text{, otherwise.} \end{cases} \tag{2}$$

Table 1. Representation of a 3 × 3 pixel block

$\Pi_{x-1,y-1}$	$\Pi_{x-1,y}$	$\Pi_{x-1,y+1}$
$\Pi_{x,y-1}$	$\Pi_{x,y}$	$\Pi_{x,y+1}$
$\Pi_{x+1,y-1}$	$\Pi_{x+1,y}$	$\Pi_{x+1,y+1}$

We consider 3 × 3 overlapping blocks (as shown in Table 1) in spatial domain and determine the number of bits (say, n) to be embedded using Eq. (2) with the help of Π_{diff} obtained from Eq. (1). We embed message bits into the cover image pixel as shown in Algorithm 1. Here, we use Algorithm 1 to store the value of n into the pixels of corresponding location of the second image. In order to increase the embedding capacity of the carrier image we perform a multi-bit embedding technique described in [9]. For this, we refer to the 8 neighboring pixels of the pixel in the cover image that has been considered suitable for the embedding process using the method discussed above. Next, we determine the maximum and minimum values amongst all the 9 pixels and the difference between these two values. This difference is required to find the embedding capacity of the central pixel by using the formula as stated in step 5 of the algorithm given below. It has been observed that unlike the smooth regions of an image, human perception is not very sensitive to abrupt changes in the boundary regions. Unlike any steganographic algorithm like [9] that uses only the spatial domain, we here use both spatial and frequency domain while embedding message bits.

Assumption 1: Both the sender and receiver knows which image contains stego-information and which image contains the value of n.

Assumption 2: Both the sender and receiver knows the used threshold value used in frequency domain.

Assumption 1 can be weakened if the sender and receiver decides that the image that comes first based on their order according to the dictionary will contain the value of n for each pixel and the remaining image will contain the information bits. Similarly *Assumption 2* can be weakened if some mathematical relation can be formulated based on their image size or it can be passed through the image header.

3 Experimental Result and Comparison with Existing Algorithms

Besides analysing the visual quality of cover and stego images and their histograms here, we also analyse the strength of our method in terms of visual quality analysis, average embedding capacity, embedding efficiency, StirMark tests and some first order statistical attacks, etc.

Visual Quality Analysis: The proposed Algorithm has been widely tested over primarily three varieties of bitmap images, viz., Cartoon, Nature and Busy Nature. The experimental outcomes in order to prove our algorithm efficiency in reference with visual quality can be analysed from Fig. 1 and Fig. 2.

Fig. 1. Cover (1st row) and stego (2nd row) version of 3 types of images, viz., Cartoon, Nature and Busy Nature

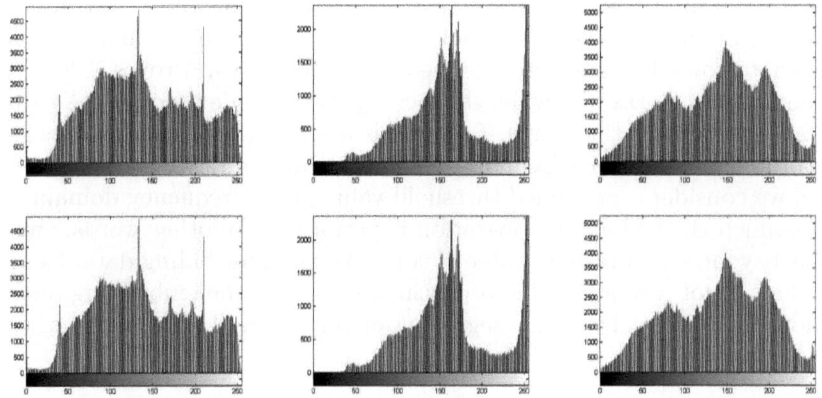

Fig. 2. Histograms of the red components of cover (1st row) and stego (2nd row) version of 3 types of images, viz., Cartoon, Nature and Busy Nature

Average Embedding Capacity: The average embedding capacity is determined by the number of bits embedded per pixel (on average of all the pixels of the image). It is to be noted that though we are not embedding any data in the pixels having DCT values greater than its presumed threshold value, but these pixels come under consideration while estimating the average embedding capacity of an image. We have considered 100 24-bit color images (downloaded from [17]) of three types, viz., cartoon, nature and busy nature and story of Sherlock Holmes (taken from [18]). We have obtained the average embedding capacity as 2.49 bpp, 3.00 bpp and 2.72 bpp respectively for the said kinds.

The average embedding capacity (i.e. embedded bits per pixel) for any LSB based steganographic algorithm is 1 bpp. According to methods [15] and [7], the AEC is 1.48 bpp and 1.6 bpp, but it is almost 2.74 bpp in our method. [15,7,9] uses only spatial domain to hide data, but we have used frequency domain for selecting pixels and ultimately the data is embedded in the spatial domain. Hence our method is more efficient than that of method [7], [9], [15].

Embedding Efficiency: Embedding efficiency [5] is very important for proving the embedding strength of any steganographic algorithm against distortion that occurs due to concealing message bits into image. It can be defined as follows:

Definition 1. *Embedding efficiency of any steganographic algorithm can be defined as the expected number of message bits concealed per embedding change.*

If n is the number of embedded bits per pixel of the cover image, then the maximum embedding efficiency (say, E) of proposed method can be estimated as $E = \frac{n \cdot 2^n}{2^n - 1}$. For further details one may refer to [8]. The embedding efficiency for any LSB based steganographic algorithm is $\frac{1}{1/2} = 2$ and that for steganographic algorithms using random ternary symbols with uniform distribution is 2.3774. For our algorithm, embedding efficiency for n=5 is 3.9725 .

Color Frequency Test: Westfeld *et al.* [14] have applied a χ^2-test in order to determine whether the color frequency distribution in an image matches a distribution that shows distortion from embedding hidden data. Provos [12] mentions that this method works well when the message is embedded sequentially starting from the beginning of the image. This test is not effective if the message is scattered throughout the image by hiding the bits in randomly selected pixels. In our method we consider a presumed threshold value in the frequency domain on the basis of which the embedding operation is performed. In other words, any pixel which is beyond the threshold value is not considered for hiding data. Hence the embedding is not a sequential and continuous one,i.e, the embedding operation does not always start from the beginning of the image. Hence our method can withstand this test.

Dual Statistics Method: Dual statistic test [4] has not been found very effective against noisy, poor quality, over-compressed or small sized images. Also, Dumitrescu *et al.* [2,3] show that the test is valid under the following assumption:

Assumption: Let Π represent the set of all pixel intensity pairs (Θ, Δ) where either Δ is even and $\Theta < \Delta$, or Δ is odd and $\Theta > \Delta$. Also let Λ represent the set of all pixel intensity pairs (Θ, Δ) where either Δ is even and $\Theta > \Delta$, or Δ is odd and $\Theta < \Delta$. Statistically it should be $|\Pi| = |\Lambda|$.

In our method the above assumption is not considered. According to Dumitrescu *et al.* [2,3] mention Assumption 1 is valid only for natural images. Also our method works for all kinds of cover images. So theoretically, our method is not breakable by the dual statistics method.

StirMark Analysis: Using some standard tests we can test the strength and robustness of any steganographic algorithm. These tests can be performed using StirMark 4.0 [19] and our proposed algorithm is able to provide good results.

Table 2. StirMark analysis of Algorithm 1 on cover and stego version of sample cartoon image (size: 800 × 565) of Fig. 1

	Factor	Cover	Stego
Self Similarities	1	30.1953 dB	30.1954 dB
Self Similarities	2	47.6340 dB	47.6320 dB
Self Similarities	3	29.2784 dB	29.2779 dB
PSNR	10	38.5896 dB	38.5895 dB
AddNoise	20	9.07605 dB	9.0928 dB
AddNoise	40	7.6766 dB	7.6767 dB
AddNoise	100	6.9311 dB	6.9306 dB
SmallRandom Distortions	0.95	14.3199 dB	10.5026 dB
SmallRandom Distortions	1.00	14.1777 dB	10.4052 dB
SmallRandom Distortions	1.05	14.0369 dB	10.3138 dB
ConvFilter	1.00	10.7997 dB	10.7997 dB
ConvFilter	2.00	-6.6931 dB	-6.6931 dB
MedianCut	3.00	30.1001 dB	30.0997 dB

The minute differences between the values corresponding to the cover and the stego image (as shown in Table 2) prove that our proposed method is robust.

4 Conclusion

In this work a novel approach for multi-bit steganography has been proposed that is capable of embedding a secret message in a cover image without any visible quality degradation of the image. The method discussed here uses both the frequency domain and also the spatial domain of the image, the former for selecting the potential pixels and the latter for embedding the message bits. It has been seen that our embedding capacity is also high. However, since we are embedding the message bits in the pixels with negative DCT, the AEC is low. So our future plan is to enhance the embedding capacity.

References

1. Chang, C.C., Tseng, H.W.: A Steganographic Method for Digital Images Using Side Match. Pattern Recognition Letters 25, 1431–1437 (2004)
2. Dumitrescu, S., Wu, X., Memon, N.: In Steganalysis of Random LSB Embedding in Continuous-tone Images. In: IEEE International Conference on Image Processing, pp. 641–644 (2002)
3. Dumitrescu, S., Wu, X., Wang, Z.: Detection of LSB Steganography via Sample Pair Analysis. IEEE Transactions on Signal Processing 51(7) (2003)
4. Fridrich, J., Goljan, M., Dui, R.: Reliable Detection of LSB Steganography in Color and Grayscale Images. In: Proceedings of the ACM Workshop on Multimedia and Security, pp. 27–30 (2001)
5. Fridrich, J., Lisonek, P.: Grid Colorings in Steganography. IEEE Trans. on Information Theory 53(4), 1547–1549 (2007)
6. Kahn, D.: The History of Steganography. In: Anderson, R. (ed.) IH 1996. LNCS, vol. 1174, pp. 1–17. Springer, Heidelberg (1996)
7. Mandal, J.K., Das, D.: Colour Image Steganography Based on Pixel Value Differencing in Spatial Domain. International Journal of Information Sciences and Techniques 2(4) (2012)
8. Mukherjee, I., Paul, G.: Efficient Multi-bit Image Steganography in Spatial Domain. In: Bagchi, A., Ray, I. (eds.) ICISS 2013. LNCS, vol. 8303, pp. 270–284. Springer, Heidelberg (2013)
9. Park, Y.-R., Kang, H.-H., Shin, S.-U., Kwon, K.-R.: An Image Steganography Using Pixel Characteristics. In: Hao, Y., Liu, J., Wang, Y.-P., Cheung, Y.-M., Yin, H., Jiao, L., Ma, J., Jiao, Y.-C. (eds.) CIS 2005. LNCS (LNAI), vol. 3802, pp. 581–588. Springer, Heidelberg (2005)
10. Paul, G., Davidson, I., Mukherjee, I., Ravi, S.S.: Keyless Steganography in Spatial Domain Using *Energetic* Pixels. In: Venkatakrishnan, V., Goswami, D. (eds.) ICISS 2012. LNCS, vol. 7671, pp. 134–148. Springer, Heidelberg (2012)
11. Petitcolas, F.A.P., Anderson, R.J., Kuhn, M.G.: Information Hiding: A Survey. Proceedings of the IEEE 87, 1062–1078 (1999)
12. Provos, N.: Defending against Statistical Steganalysis. In: Tenth USENIX Security Symposium, pp. 325–335 (2001)

13. Reddy, V.L., Subramanyam, A., Reddy, P.C.: Implementation of LSB Steganography and its Evaluation for Various File Formats. International Journal of Advanced Networking and Applications 2(5), 868–872 (2011)
14. Westfeld, A., Pfitzmann, A.: Attacks on Steganographic Systems. In: Pfitzmann, A. (ed.) IH 1999. LNCS, vol. 1768, pp. 61–76. Springer, Heidelberg (2000)
15. Wu, D.C., Tsai, W.H.: A Steganographic Method for Images by Pixel-value Differencing. Pattern Recognition Letters 24, 1613–1626 (2003)
16. Wu, H.C., Wu, N.I., Tsai, C.S., Hwang, M.S.: Image Steganographic Scheme Based on pixel value differencing and LSB replacement method. IEEE Proceedings on Vision,Image and Signal Processing 152(5), 611–615 (2005)
17. `http://www.webshots.com`
18. `http://221bakerstreet.org`
19. `http://www.petitcolas.net/fabien/watermarking/stirmark`

Service Adaptive Broking Mechanism Using MROSP Algorithm

Ashish Tiwari, Vipin Sharma, and Mehul Mahrishi

Arya Institute of Engineering and Technology, Jaipur
SKITM Jaipur, SKITM Jaipur, Rajasthan India
{er.ashish.tiwari89,ervipin85,Mehul.Marshi}@gmail.com

Abstract. Cloud computing is an effort in delivering resources as a service. It represents a shift away from the era where products were purchased, to computing as a service that is delivered to consumers over the internet from large-scale data centers or clouds. As cloud computing is gaining popularity in the IT industry, academia appeared to be working in parallel for the rapid developments in this field. In a cloud computing environment now a days, the role of service provider is divided into two: Cloud Broker who manage cloud platforms and lease resources according to a usage-based pricing model, and service providers, who rent resources from one or many infrastructure providers to serve the end users. The aim of this research work is to deal with the scheduling of the requests on the basis of some parameters that we have identified to achieve the best optimal paths or cloud service provider allotment to the users. We have used rough set theory to generate the mathematical model. The algorithm is implemented in the cloud simulator CLOUDSIM in which cloudlets, datacenters, cloud brokers are created to perform the algorithms. Finally, we created a GUI for the user convenience so that both Cloud Service Provider and users can themselves analyze each other's performance. We have reused some inbuilt packages of Cloudsim net beans to simulate the process.

Keywords: Cloud Computing, Cloud Service Providers, Rough Set Theory, Datacenters, Cloudsim.

1 Introduction

In the day today world Computing is moving in its seven havens where it is making its place like a worm in a fruit. The most important thing to see here is its play a vital in businesses. Number of people wants to save their money, time, space and improve performance of their day to day IT activities. This increment done is due to the people requirements in the web data. Advent of Internet plays a very vital role in developing concept of running the Wireless Mobile Phones, Internet Connectivity by Dongle or other wireless Devices, T.V, Radio etc. So from all over analysis the basic definition of the cloud computing can be given as the A style of computing where massively scalable (and elastic) IT-related capabilities are provided as a service to external

M.K. Kundu et al. (eds.), *Advanced Computing, Networking and Informatics - Volume 2*,
Smart Innovation, Systems and Technologies 28,
DOI: 10.1007/978-3-319-07350-7_43, © Springer International Publishing Switzerland 2014

customers using Internet technologies (Fig. 1 shows an Example). A paradigm of computing which tells about both the applications providing services with the help of internet and available scalable hardware and the software running on the systems that may provide the services. A cloud service provider is responsible for delivering the user demands as a service which results in cost reduction and efficiently amalgamation which treat with ability, security and reliability. Some major cloud service providers are: Ad host Internet, Blue Fire, Cloud more etc. Every CSP (Cloud Service Providers) has their own mechanism of providing services. For example Adhost provides dedicated web hosting featuring Microsoft servers, including windows 2008 and IIS whereas Enter host has their expertise in disaster recover solutions redundant storage and backup services as shown in Fig. 1 [3], [5]. The study of the artifacts provided by various cloud standard generating organizations which are providing the various certification and protocols, shows that each incoming request can be categorized into one of the 12 standard parameters.

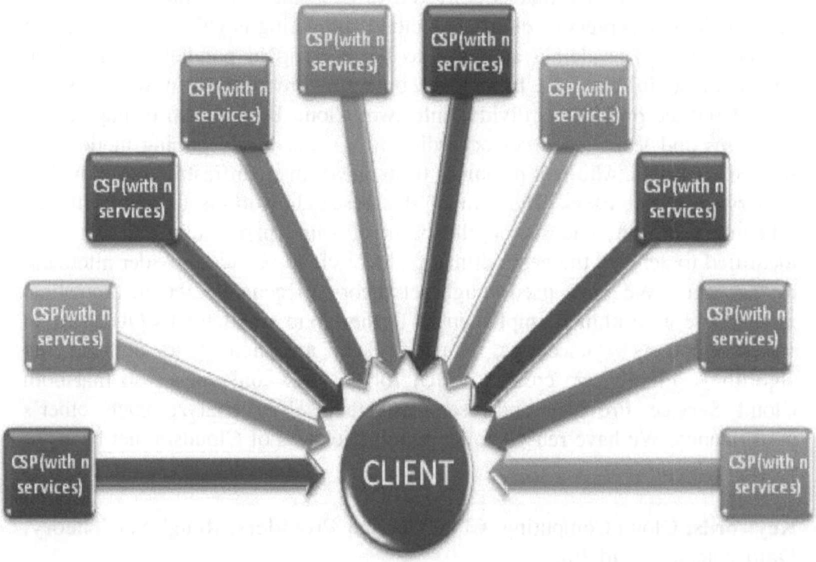

Fig. 1. Basic Structure of Cloud Computing which describes about the number of Clients and different types of CSPs services with different colors

2 Problem Statement

There is a need to standardize cloud. The technology is growing just like Internet use to grow in its time. It can be thought that some of the standards may be too early or too late and some may prove to be inadequate, duplicative, or inappropriate. There will be bumps on the roads, but cloud computing will be the major computing paradigm, and the development of standards will play a key role in facilitating the development of the marketplace. The method of selecting a Cloud Service Provider is evaluated on the basis of Which-Cloud Provider- Provides-What. The Basic Questions that comes in mind for the selection process are as follows. a) Is Data or information safe? b) What Security features the CSPs are providing? c) How should large enterprise create their own cloud? d) What is the role of opensource and proprietary software? e) How should one leverage existing data centers (cloud interconnections)?

3 Related Works

Few research works focus on Resource Management of Cloud requests. These researches mainly focus on the type of incoming requests. The user has to enter the type of request and on the basis of some algorithmic parameters like priority, reliability etc. the resource it allocated. The parameters also included are Security and Performance, Resource management, high System Performance, Quality of Services and Service level agreements measures. The main drawback of this algorithm was found that it does not support co-scheduling of resources and handling uncertainty. Another paper that we have cited is A Pragmatic Scheduling for Optimal priority. This research reflects the handling technical aspects of tasks. At Last author had calculated the total resultant priority based on demand divided incoming tasks. The conclusion that was found at the last is priority of each job that arrives and hybridization of Tp and Bp. An Ant Colony Optimization Scheduling Algorithm has the problems found under this research work are related to the quality of service requirements such as storage, calculations and evaluations of the cloud services, services with high reliability and low cost etc. The formulation uses DAG for minimization and scheduling algorithms based on ACS (Ant Colony System algorithm). At last author gets to know about the use of SLA monitoring module to monitor the running condition of cloud services. Efficient Scheduling Algorithms to maximize the utilization of the internal infrastructure and to minimize the cost of running tasks and computational and data transfer costs.

4 Methodology

We have seen that many researchers are working in the same direction but with different approaches. The goal is same i.e. to improve and optimize the service provider scheduling [2], [3]. The algorithms for task ordering on the basis of their arrival and task mapping are already proposed [2], [4]. We are proposing a similar kind of approach but to a level above that of job scheduling. It is a responsibility of the cloud computing service provider to provide the adequate service level satisfaction, so we are proposing to device an algorithm which made possible the cloud middleware to determine capability of CSP by using Rough Set analysis on the basis of level of satisfaction of service. Rough set model can be handled with objects and its characteristics. Here we are considering service providers as objects and its characteristics based on some defined standard parameters.

We have gone through number of artifacts and resource documents to search and summarize 12 major standard parameters. For the sake of simplicity and understanding of the algorithm we are just using 5 parameters out of 12. Data Operation, which deals with the different data operation such as searching of data either on the basis of content or location. Moreover some security and access mechanism issues are also discussed. Lastly, it also touches the data persistence methods. Risk Management, which we may deal about Organizations shall develop and maintain a cloud oriented risk management framework to manage risk as defined in the master agreement or industry best practices and standards .Legal Issues, which we have found that it deals with the legal issues that could be generated when data is moved to the cloud, issues regarding NDA and other agreements between CSP and the customer and other issues that are resolved under laws and litigations of a country. Compliance And Audit, which Customer and provider both must understand the importance and should follow the implications on existing compliances, audits and other best practices. Inter-portability and Portability, explained to interoperability features provided shall support security of the data exchanges and messages at protocol level, policy level and identity level. Rough Set Model Rough sets concept was developed by Zdzislaw Pawlak [10-12] in the early 1980s. It deals with classificatory analysis of data tables. The data can be acquired from measurements or from human experts. The main goal of the rough set analysis is to synthesize approximation concepts from the actual data. Here we perform classification of objects using mathematical tool rough set. A rough set Theory is based on the assumption that data and information is associated with every object of the universe. In general we may say sometimes an object neither belongs to positive nor negative then it is in the boundary. If the boundary is non-empty then we call it rough.
For each attribute value ρi

$$\begin{array}{ll} & \rho\gamma i \geq \text{or True} \\ 0 & \text{, Otherwise} \end{array} \tag{1}$$

We are applying the rough set modeling to the identified services to generate a kind of rating (On the basis of specialized Service Parameters) of the CSP's to improve the performance of Cloud scheduling. Rough set analysis will be on the basis of level of satisfaction of service Parameters (From Equation 1). We know that all service providers are not all rounders i.e. they cannot be counted non-stop utilization for all kind of services. By applying Rough set model we are able to categorize each CSP on the basis of its specialty in service providing.

$$\mu = \sum^{n}{}_{i=1} \delta i \tag{2}$$

The value 1 and 0 above represents that either a CSP is specialized in a particular service or not. If the resultant value is 1 that means the CSP can be considered for the request otherwise not (From Eq. 2). We also categorise the attributes on the basis of relevance to tenant and CSP's. The relevance generates a threshold value for each attribute out of a scale of 10.

Algorithm Part I

Input: Set of Parameters P, Set of Database d, Set of Service Providers S and Cloud Broker Criteria(threshold Value) C, this Algorithm deals with the Cloud Service Provider performances.

Output: Success full Assigning Set of Jobs J by Cloud Service Providers S.

```
Method: S' = S , P' = P, C'=C
For all S' the should be an Account A,
        If( A ≠ ϕ and P' ≠ ϕ )
        {
                S'= Select ( P', d, C')
                P' = ∑ (Pi – Pj ) where i>j
        }
        Select (P', d, S')
        {
                For each P', there is a Suitable value Entered by S'
                If ( p' = C')
                {
                        d = Value is follow the threshold Criteria C in Database.
                }
                Else
                        d = Value does not follow the threshold Criteria C in Database.
        }
}
```

Fig. 2. Algorithm first part deal with the modified ROSP for Cloud Service Providers

As per the Rough Set representation, we have represented the CSP and their attributes in a tabular form called Information System. The rows of the table contain the list of cloud service providers and the columns consist of the attributes of the respective cloud service provider. When we talk about attributes, that means the parameters we have identified in our study (Data Operations, Risk Management etc.) In our previous research paper we have generated the ROSP Algorithm. The main aim of the algorithm is to find the optimal fuzzy value for each cloud service provider and allot the tenant the CSP with the maximum fuzzy value. In this research paper we have enhanced our algorithm, removed some bugs and even provide a GUI environment so that the users feel it easy to operate (Fig. 2 shows an Algorithm Part 1). As per our Algorithm, there are number of Cloud Service Providers which may contain the number of Datacenters. The Part 1 of the algorithm elaborates how we are extracting the relevance values for Cloud Service Providers. (Fig. 2 shows an Algorithm Part 1). For each CSP, whenever it logs in to the middle ware, it needs to create an account which includes answering of certain relevant questions based on our identified parameters. Each question has some evaluation criterion, through which we calculate the average relevance factor which is placed in the table. As soon as the table is filled, the ROSP algorithm will execute at the CSP side and extract the optimal CSP on the basis of achieved fuzzy values. The Part 2 of the algorithm executes at the client side. After logged in and availing the services of cloud service provider, the tenant needs to answer some questions as similar to service provider (Fig. 3 shows an Algorithm Part 2). The feedback given by the tenant will be incorporated over the existing value of the CSP in table. Therefore we named it as user adaptive approach.

```
Algorithm Part 2

Input: Set of Parameters P, Set of Database d, Set of Service Providers S , Users U and Cloud
Broker Criteria (threshold Value) C, this Algorithm deals with the Cloud Service Provider
performances.

Output: Success full Assigning Set of Jobs J to Users U.

Method: U' = U, P' = P,A'=A, S'=S
For all U' the should be an Account A',
        If( A' ≠ ϕ )
        {
                A' = S'( By ROSP Algorithm)
        }
        else
                Create an Account A' then
                A'=S'(By ROSP)
        If (A'= Exit)
        {
                For each P', there is a Suitable value Entered by U'
                P' = ∑ ( Pi – Pj ) where i>j
                If ( P' = C')
                {
                        d = Value is follow the threshold Criteria C in Database.
                }
                Else
                        d = Value does not follow the threshold Criteria C in Database.
        }
        Then
                        S' = d
                        Updating the Same CSP provided to the User U'
        }
}
```

Fig. 3. Algorithm second part deal with the modified ROSP for Tenants

5 Result and Simulation

As proceeding with our previous ROSP Algorithm, in our Simulation we have used some existing packages of CloudSim to create Cloud Service Providers, Datacenters, Network etc. In addition to this we have created some packages in NetBeans. (Fig. 4 shows an Output Part 1) (Fig. 5 shows an Output Part 2).

5.1 Results between Cloud Users and CSPs

With an aim of checking the efficiency of the algorithms, we have taken two parameters time taken and CPU utilization of the algorithm for fixed values.

Result 1: The below graph represents the time taken in Millisecond (ms) to execute the algorithm by different CSPs with increasing number of Users . We can easily observe that the time taken is increasing exponentially as we increase the number of users. The increasing number of Datacenters also affects the resulting graph. We have calculated for 30 datacenters.6 The time taken by the CSP is directly proportional to the number of Cloud Parameters. We have initially taken 4 Parameters namely Virtualization, Application Security, Risk Management and Compliance and Audit, but the algorithm is subject to expand for any number of parameters. (Fig. 6 shows Time Taken between CSPs & Users).

Result 2: In the below graph we are representing the CPU Utilization Percentage to execute the Algorithm by different CSPs with increasing number of users. The number of users and CSPs are regularly increasing. The CPU Utilization also

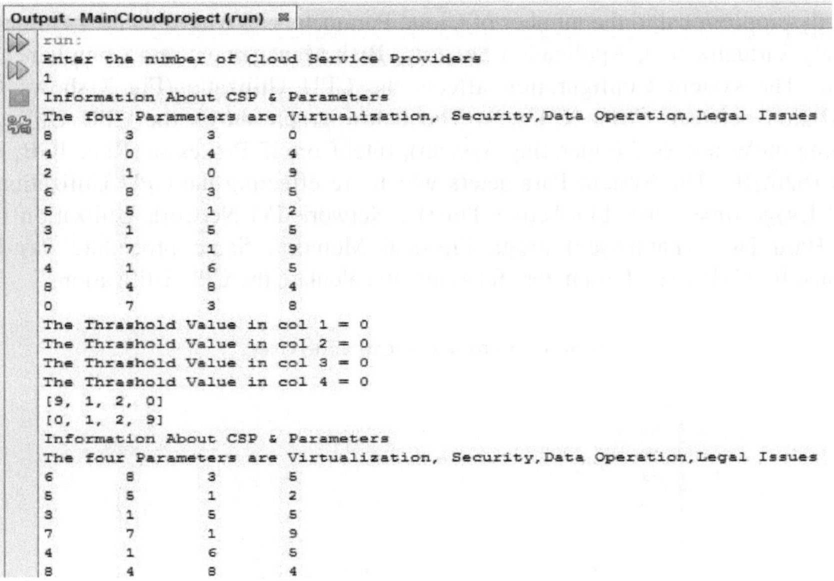

```
Output - MainCloudproject (run)
run:
Enter the number of Cloud Service Providers
1
Information About CSP & Parameters
The four Parameters are Virtualization, Security,Data Operation,Legal Issues
8       3       3       0
4       0       4       4
2       1       0       9
6       8       3       5
5       5       1       2
3       1       5       5
7       7       1       9
4       1       6       5
8       4       8       4
0       7       3       8
The Thrashold Value in col 1 = 0
The Thrashold Value in col 2 = 0
The Thrashold Value in col 3 = 0
The Thrashold Value in col 4 = 0
[9, 1, 2, 0]
[0, 1, 2, 9]
Information About CSP & Parameters
The four Parameters are Virtualization, Security,Data Operation,Legal Issues
6       8       3       5
5       5       1       2
3       1       5       5
7       7       1       9
4       1       6       5
8       4       8       4
```

Fig. 4. Output of the simulated result – Part 1

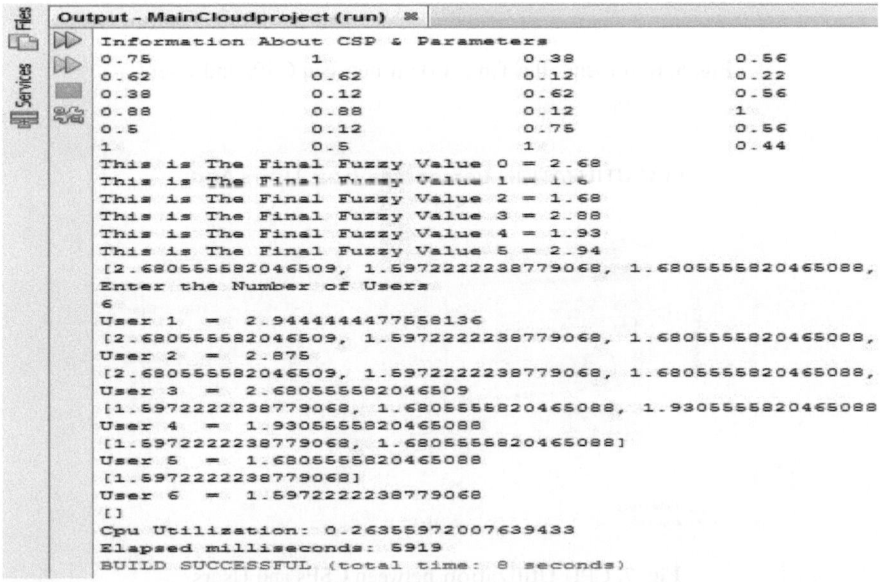

```
Output - MainCloudproject (run)
Information About CSP & Parameters
0.75          1             0.38          0.56
0.62          0.62          0.12          0.22
0.38          0.12          0.62          0.56
0.88          0.88          0.12          1
0.5           0.12          0.75          0.56
1             0.5           1             0.44
This is The Final Fuzzy Value 0 = 2.68
This is The Final Fuzzy Value 1 = 1.6
This is The Final Fuzzy Value 2 = 1.68
This is The Final Fuzzy Value 3 = 2.88
This is The Final Fuzzy Value 4 = 1.93
This is The Final Fuzzy Value 5 = 2.94
[2.680555582046509, 1.597222238779068, 1.680555820465088,
Enter the Number of Users
6
User 1  =  2.944444477558136
[2.680555582046509, 1.597222238779068, 1.680555820465088,
User 2  =  2.875
[2.680555582046509, 1.597222238779068, 1.680555820465088,
User 3  =  2.680555582046509
[1.597222238779068, 1.680555820465088, 1.930555820465088
User 4  =  1.930555820465088
[1.597222238779068, 1.680555820465088]
User 5  =  1.680555820465088
[1.597222238779068]
User 6  =  1.597222238779068
[]
Cpu Utilization: 0.2635597200763943З
Elapsed milliseconds: 5919
BUILD SUCCESSFUL (total time: 8 seconds)
```

Fig. 5. Output of the simulated result – Part 2

increases exponentially to the number of user but decreases with increasing number of CSPs. We can clearly conclude that if we increase the resources in a system, the CPU utilization decreases. The increasing number of Datacenters also affects the resulting graph. We have calculated for 30 datacenters. The CPU Utilization by the CSP is directly proportional to the number of Cloud Parameters. We have taken 4 Parameters namely Virtualization, Application Security, Risk Management and Compliance and Audit. The system Configuration affects the CPU Utilization(Fig. 7 shows CPU Utilization between CSPs & Users) The below graph shows the CPU Utilization running on Windows 7 (Operating System), Intel Core i5 Processor, Ram 4GB, Hard Disk (620GB). The System Parameters which are effecting the CPU Utilization are CPU Usage time, Disk I/O(Active Time) , Network I/O(Network Utilization time) and Hard Disk (Faults/sec) usage Physical Memory. Same procedure has been adopted for CSP's and Parameters for Users to calculate the CPU Utilization.

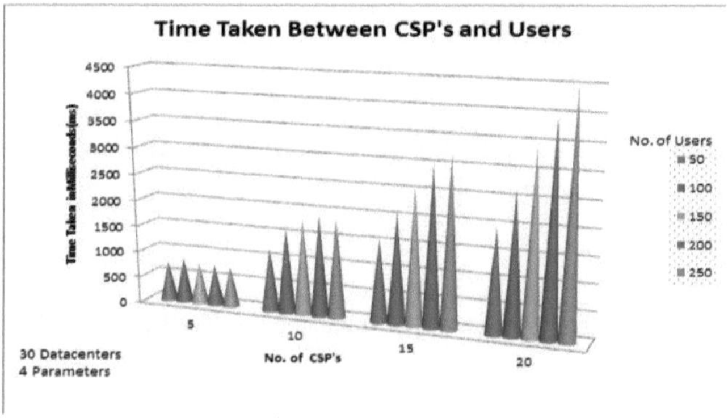

Fig. 6. Represents the Time Taken between CSPs and Users

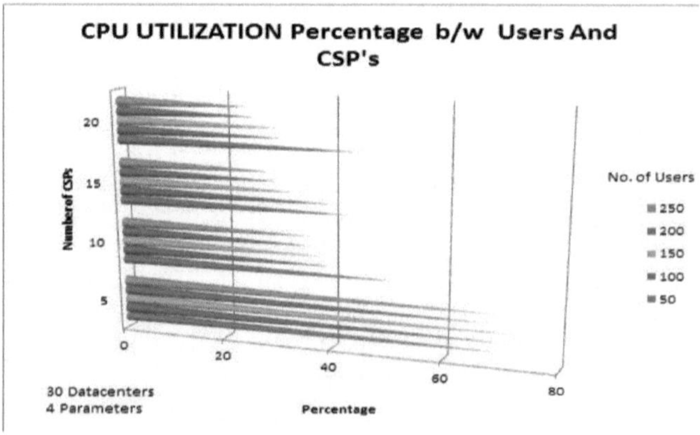

Fig. 7. CPU Utilization between CSPs and Users

6 Conclusion

In this research work we have proposed the scheduling algorithm for Cloud Broker which exists between the CSPs and Users. The user/tenant need not to identify the service requirements before submitting their job to the cloud. This is the job of middleware to implement the algorithm and rate CSPs on the basis of their capabilities. All the user/tenant needs to do is to fill the feedback from correctly as to improve the CSPs performance in the future. Any existing cloud simulators (CloudSim, Grid Sim etc.) can easily implement this model. The scope of this research can be extend further, in which apart from taking the cumulative cost function, we can also generate the cost on the basis of individual capability of CSP for an individual attribute. For example if a user needs a CSP whose cost function is maximum for security attribute, we can implement the algorithm at each node of the CSP rather than executing it at the broker.

References

1. Tiwari, A., Nagaraju, A., Mahrishi, M.: An Optimized Scheduling Algorithm for Cloud Broker Using Cost Adaptive Modeling. In: 3rd IEEE International Advanced Computing Conference (2013)
2. Ergu, D., Kou, G., Peng, T., Shi, Y., Shi, Y.: The analytic hierarchy process: task scheduling and resource allocation in cloud computing environment. The Journal of Supercomputing 64(3), 835–848 (2013)
3. Komorowski, J.: Rough Sets: A Tutorial. Department of Computer and Information Science Norwegian University of Science and Technology (NTNU) 7034 Trondheim, Norway
4. Calheiros, R.N.: CloudSim: A Novel Framework for Modeling and Simulation of Cloud Computing Infrastructures and Services. In: Grid Computing and Distributed Systems (GRIDS) Laboratory Department of Computer Science and Software Engineering. The University of Melbourne, Australia
5. Pawlak, Z.: Rough set theory and its applications. Journal of Telecommunication and Information Technology 3 (2002)
6. Sharma, A.K.: The Role and Use of Data Mining Techniques for Intrusion Detection Systems. International Journal of Research in IT and Management 2(2), 425–430 (2012)
7. Mahrishi, M., Sharma, D.K., Shrotriya, A.: Globally Recorded binary encoded Domain Compression algorithm in Column Oriented Databases. Global Journals of Computer Science and Technology 11(23), 27–30 (2011)
8. Tiwari, A., Tiwari, A.K., Saini, H.C., Sharma, A.K., Yadav, A.K.: A Cloud Computing using Rough set Theory for Cloud Service Parameters through Ontology in Cloud Simulator. In: ACITY (2013)
9. Nair, T.R.G., Sharma, V.: A Pragmatic Scheduling Approach for Creating Optimal Periority of jobs with Business Values in Cloud Computing. In: ACC (2012)
10. Tiwari, A.: Adaptive Cost Model for Cloud Broker application Rough Set and Fuzzy argumentation Technique. In: ACITE Sponsored National Seminar on Recent Trends in Embedded Systems (2014)
11. Mahrishi, M., Nagaraju, A.: Optimizing Cloud Service Provider Scheduling by Rough Set model. In: International Conference on Cloud Computing Techno and Management (2012)
12. Mahrishi, M., Nagaraju, A.: Rating Based Formulation for Scheduling of Cloud Service Providers. In: National Conference on Emerging Trends in IT

Performance Comparison of Hypervisors in the Private Cloud

P. Vijaya Vardhan Reddy[1] and Lakshmi Rajamani[2]

[1] GE Capital, Hyderabad, India
[2] Osmania University, Hyderabad, India
kavirajvj@gmail.com, dlakshmiraja@gmail.com

Abstract. To make cloud computing model Practical and to have essential characters like rapid elasticity, resource pooling, on demand access and measured service, two prominent technologies are required. One is internet and second important one is virtualization technology. Virtualization Technology plays major role in the success of cloud computing. A virtualization layer which provides an infrastructural support to multiple virtual machines above it by virtualizing hardware resources such as CPU, Memory, Disk and NIC is called Hypervisor. It is interesting to study how different Hypervisors perform in the Private Cloud. Hypervisors do come in Paravirtualized, Full Virtualized and Hybrid flavors. It is novel idea to compare them in the private cloud environment. This paper conducts different performance tests on three hypervisors XenServer, ESXi and KVM and explains the behavior and results of each hypervisor. In the experiment, CloudStack 4.0.2 (open source cloud computing software) is used to create a private cloud, in which management server is installed on Ubuntu 12.04 – 64 bit operating system. Hypervisors XenServer 6.0, ESXi 4.1 and KVM (Ubuntu 12.04) are installed as hosts in the respective clusters and their performances have been evaluated in detail.

Keywords: CloudStack, Fullvirtualization, Hypervisor, Management Server, Paravirtualization, Private Cloud, Virtualization Technology.

1 Introduction

Cloud computing is a model for enabling convenient, on-demand network access to a shared pool of configurable computing resources such as networks, servers, storage, applications, and services that can be rapidly provisioned and released with minimal management effort or service provider interaction [1].

Virtualization is the technology which increases the utilization of physical servers and enables portability of virtual servers between physical servers. Virtualization Technology gives the benefit of work load isolation, work load migration and work load consolidation. For being able to reduce hardware cost, cloud computing uses virtualization. Virtualization technology has evolved really quickly during past few years. Also it is particularly due to hardware progresses made by AMD [5] and Intel. Virtualization is a technology that combines or divides computing resources to present one or many operating environments using methodologies like hardware and

DOI: 10.1007/978-3-319-07350-7_44, © Springer International Publishing Switzerland 2014

software partitioning or aggregation, partial or complete machine simulation, emulation, timesharing, and many others [2]. A virtualization layer provides an infrastructural support using the lower-level resources to create multiple virtual machines that are independent and isolated from each other. Such a virtualization layer is also called Hypervisor [2].

Cloud computing allows customers to reduce the cost of the hardware by allowing resources on demand. Also customers of the service need to have guaranty of the good functioning of the service provided by the cloud. The Service Level Agreement brokered between the providers of cloud and the customers is the guarantees from the provider that the service will be delivered properly [3].

This paper provides a quantitative comparison of three virtualization hypervisors available for the x86 [4] architecture — XenServer 6.0, VMware ESXi Server 4.1 and KVM (Ubuntu 12.04) in the private cloud environment. A series of performance experiments were conducted on the three hypervisors using Microsoft Windows 2008 R2 server as the guest operating system. This technical paper discusses the results of these experiments. The discussion in this technical paper should help both IT decision makers and end users to choose the right virtualization hypervisor for their respective private cloud environments.

The experimental results show that XenServer and VMware ESXi Servers delivers almost equal, performance in all the tests except in CPU tests ESXi is performing marginally better than XenServer and in Memory and Network tests XenServer performing slightly better than that of ESXi Server. Furthermore, KVM performance is noticeably lower than that of XenServer and ESXi Server; hence it needs improvement in all the performance aspects.

2 Hypervisor Models

Paravirtualized Hypervisor: *XenServer* - Citrix XenServer 6.0 is an open-source, complete, managed server virtualization platform built on the powerful Xen Hypervisor. Xen [17] uses para-virtualization. Para-virtualization modifies the guest operating system so that it is aware of being virtualized on a single physical machine with less performance loss. XenServer is a complete virtual infrastructure solution that includes a 64-bit Hypervisor with live migration, full management console, and the tools needed to move applications, desktops, and servers from a physical to a virtual environment [8]. Based on the open source design of Xen, XenServer is a highly reliable, available, and secure virtualization platform that provides near native application performance [8]. Xen usually runs in higher privilege level than the kernels of guest operating systems. It is guaranteed by running Xen in ring 0 and migrating guest operating systems to ring 1. When a guest operating system tries to execute a sensitive privilege instruction (e.g., installing a new page table), the processor will stop and trap it into Xen [9]. In Xen, guest operating systems are responsible for allocating the hardware page table, but they only have the privilege of direct read, and Xen [9] must validate updating the hardware page table. Additionally, guest operating systems can access hardware memory with only non-continuous way because Xen occupies the top 64MB section of every address space to avoid a TLB flush when entering and leaving the Hypervisor [9]. XenServer is a complete virtual infrastructure solution that includes a 64-bit Hypervisor [8].

Fullvirtualized Hypervisor: *ESXi Server* - VMware ESXi is a Hypervisor aimed at server virtualization environments capable of live migration using VM motion and booting VMs from network attached devices. VMware ESXi supports full virtualization [7]. The Hypervisor handles all the I/O instructions, which necessitates the installation of all the hardware drivers and related software. It implements shadow versions of system structures such as page tables and maintains consistency with the virtual tables by trapping every instruction that attempts to update these structures. Hence, an extra level of mapping is in the page table. The virtual pages are mapped to physical pages throughout the guest operating system's page table [6]. The Hypervisor then translates the physical page (often-called frame) to the machine page, which eventually is the correct page in physical memory. This helps the ESXi server better manage the overall memory and improve the overall system performance [16]. VMware's proprietary ESXi Hypervisor, in the vSphere cloud-computing platform, provides a host of capabilities not currently available with any other Hypervisors. These capabilities include High Availability (the ability to recover virtual machines quickly in the event of a physical server failure), Distributed Resource Scheduling (automated load balancing across a cluster of ESXi servers), Distributed Power Management (automated decommissioning of unneeded servers during non-peak periods), Fault Tolerance (zero downtime services even in the event of hardware failure), and Site Recovery Manager (the ability to automatically recover virtual environments in a different physical location if an entire data center outage occurs) [7].

Hybrid Methods: *KVM* - KVM (Kernel-based Virtual Machine) is another open-source Hypervisor using full virtualization apart from VMware. And also as a kernel driver added into Linux, KVM enjoys all advantages of the standard Linux kernel and hardware-assisted virtualization thus depicting hybrid model. KVM introduces virtualization capability by augmenting the traditional kernel and user modes of Linux with a new process mode named guest, which has its own kernel and user modes and answers for code execution of guest operating systems [9]. KVM comprises two components: one is the kernel module and another one is user space. Kernel module (namely kvm.ko) is a device driver that presents the ability to manage virtual hardware and see the virtualization of memory through a character device /dev/kvm. With /dev/kvm, every virtual machine can have its own address space allocated by the Linux scheduler when being instantiated [9]. The memory mapped for a virtual machine is actually virtual memory mapped into the corresponding process. Translation of memory address from guest to host is supported by a set of page tables. KVM can easily manage guest Operating systems with kill command and /dev/kvm. User-space takes charge of I/O operation's virtualization. KVM also provides a mechanism for user-space to inject interrupts into guest operating systems. User-space is a lightly modified QEMU, which exposes a platform virtualization solution to an entire PC environment including disks, graphic adapters and network devices [9]. Any I/O requests of guest operating systems are intercepted and routed into user mode to be emulated by QEMU [9].

3 Related Work

The following papers were surveyed to know the related work which had happened in the selected research area. The virtualization overhead involves performances depreciation rather to native performances. Researches have been made to measure the overhead of the virtualization for different hypervisor such as XEN, KVM and VMware ESX [10-14]. For their researches Menon used a toolkit called Xenoprof which is a system wide statistical tool implemented specially for Xen [12]. Due to this toolkit they have managed to analyse the performances of the overhead of network I/O devices. Their study has been performed within uniprocessor as well as multiprocessor. A part of their research has been dedicated to performance debugging of Xen using Xenoprof. Those researches have permitted to correct bugs and improve by that the network performances significantly. After the debugging part it has been focused on the network performances. It has been observed that the performance seems to be almost the same between Xen Domain0 and native performances. However if the number of interfaces increase, the receive throughput of the domain0 is significantly smaller than the native performances. This degradation of network performances is cause by an increasing CPU utilisation. Because of the overhead caused by the virtualization there are more instructions that need to be managed by the CPU. This involves more information to treat and bufferization by the CPU which cause a degradation of receive throughput compared to native performances. More recent studies try to compare the differences between hypervisors and especially the performances of each one according to their overhead [11], [14]. They are using three different benchmark tools to measure the performances: LINPACK, LMbench and Iozone. Their experiment is divided in three parts according to the specific utilisation of each tool. With LINPACK Jianhua had tested the processing efficiency on floating point. Different pick value has been observed over the different systems tested which are native performance, Xen and KVM. The result of this show that the processing efficiency of Xen on floating point is better than KVM because Fedora 8 virtualized with Xen have performances which represent 97.28% of the native rather than Fedora 8 virtualized with KVM represent only 83.46% of the native performances. The virtualization of Windows XP comes up with better performances than with the virtualization of fedora 8 on Xen. This is explained by the authors by the fact that Xen own fewer enhancement packages for windows XP than for fedora 8because of that the performances of virtualized windows XP are slightly better than virtualized fedora 8.

After analysing the relevant study, we have chosen the below experimentation to compare the respective hypervisors which is a novel idea and had never tried before with CloudStack in the private cloud environment.

4 Test Methodology-Private Cloud: CloudStack with Hypervisors

In our experiment, the proposed test environment contains following infrastructure using open source cloud computing software. CloudStack is an Infrastructure as a service (IaaS) cloud based software which is able to rapidly build and provide private cloud environments or public cloud services. Supporting KVM, XenServer and

Vmware ESXi, CloudStack is able to build cloud environments with a mix of multiple different hypervisors. With rich web interface for users and administrators with operations of cloud use and operation being performed on a browser. Additionally, the architecture is made to be scalable for large-scale environments [18]. CloudStack is open source software written in java that is designed to deploy and manage large networks of virtual machines, as a highly available, scalable cloud computing platform. CloudStack offers three ways to manage cloud computing environments: an easy-to-use web interface, command line and a full-featured RESTful API [18]. Private clouds are deployed behind the firewall of a company where as public cloud is usually deployed over the internet. It is always ideal to use open source solutions to perform any experiment related to cloud computing.

In our test environment XenServer, ESXi and KVM are used as hypervisors (Hosts) in CloudStack to create a private cloud. One machine is Management Server, runs on a dedicated server. It controls allocation of virtual machines to hosts and assigns storage and IP addresses to the virtual machine instances. The Management Server runs in a Tomcat container and requires a MySQL database for persistence. In the experiment, Management Server is installed on Ubuntu (12.04 64-bit). On the host servers XenServer 6.0, ESXi 4.1 and KVM (Ubuntu 12.04 [19]) hypervisors are installed as depicted in Fig. 1. Front end will be any base machine to launch CloudStack UI using web interface (with any browser software IE, Firefox, Safari) to provision the cloud infrastructure by creating zone, pod, cluster and host in the sequential order. After respective hypervisors are in place, guest OS Windows 2008 R2 64-bit [20] installed on them to carry out all performance tests.

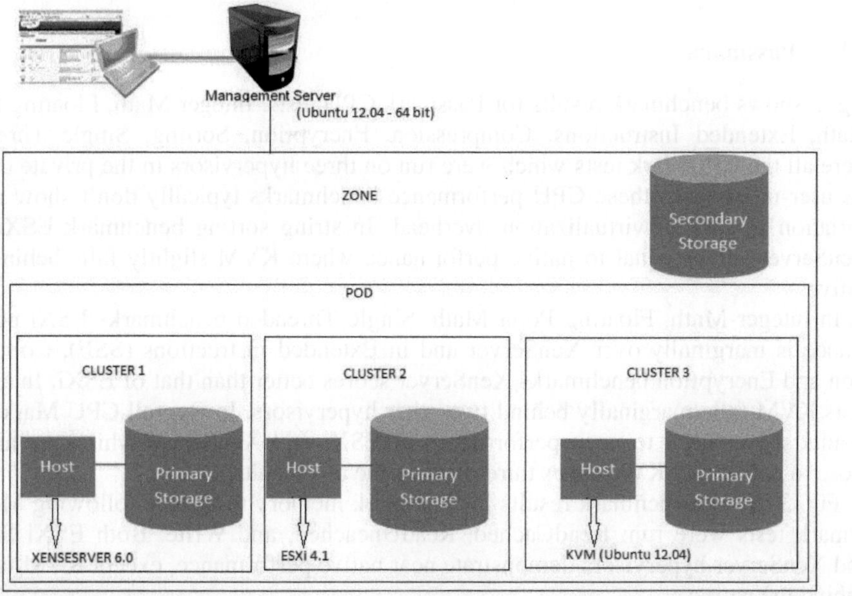

Fig. 1. Test Environment Architecture – Private Cloud (CloudStack with Multiple hypervisors)

A typical enterprise datacenter runs a mix of CPU, memory, and I/O-intensive applications. Hence the test workloads chosen for these experiments comprise several well-known standard benchmark tests, in which for CPU, Memory, Disk I/O Passmark is used. Passmark [15], a synthetic suite of benchmarks intended to isolate various aspects of workstation performance, was selected to represent desktop-oriented workloads. And for network performance Netperf is used in the tests. Netperf [21] was used to simulate the network usage in a datacenter. The objective of these experiments was to test the performance and scalability of the three virtualization hypervisors. The tests were performed using a configuration with a single virtual CPU and Windows 2008 R2 64-bit used as guest operating system for all the tests. The benchmark test suites are used in these experiments only to illustrate performance of the three virtualization hypervisors.

5 Results

This section provides the detailed results for each of the benchmarks run. All of the results have been normalized to native performance measures. Native performance is normalized at 1.0 and all other various benchmark results are shown relative to that number. Hence benchmark results of 90% of the native performance would be shown as 0.9 on the scale in the graph. Higher numbers indicate better performance of the particular virtualization platform, unless indicated otherwise. Near-native performance also indicates that more virtual machines can be deployed on a single physical server, resulting in higher consolidation ratios. This can help even if an enterprise plans to standardize on virtual infrastructure for server consolidation alone.

5.1 Passmark

Fig. 2 shows benchmark results for Passmark CPU tests. Integer Math, Floating Point Math, Extended Instructions, Compression, Encryption, Sorting, Single Threaded were all the CPUMark tests which were run on three hypervisors in the private cloud. As user-mode tasks these CPU performance benchmarks typically don't show much variation in case of virtualization overhead. In string sorting benchmark ESXi and XenServer shows equal to native performance where KVM slightly falls behind the native.

In integer Math, Floating Point Math, Single Threaded benchmarks ESXi performance is marginally over XenServer and in Extended instructions (SSE), Compression and Encryption benchmarks XenServer scores better than that of ESXi. In all the tests KVM falls marginally behind two other hypervisors. In Overall CPU Mark tests results shows neck to neck performance of ESXi and XenServer which are almost close to native and KVM takes third place in the test results.

Fig. 3 shows benchmark results for Passmark memory tests. The following Memorymark tests were run: ReadCached, ReadUncached, and Write. Both ESXi Server and XenServer hypervisors demonstrate near native performance, except KVM falling behind the native.

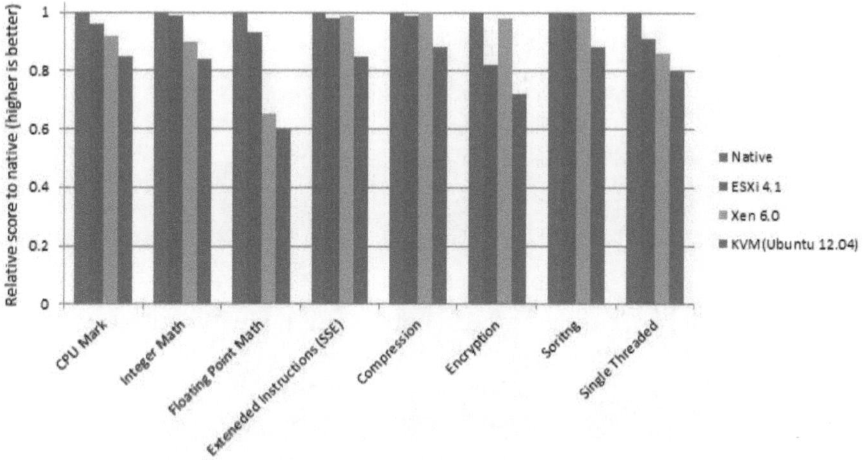

Fig. 2. Passmark – CPU results compared to native (Higher values are better)

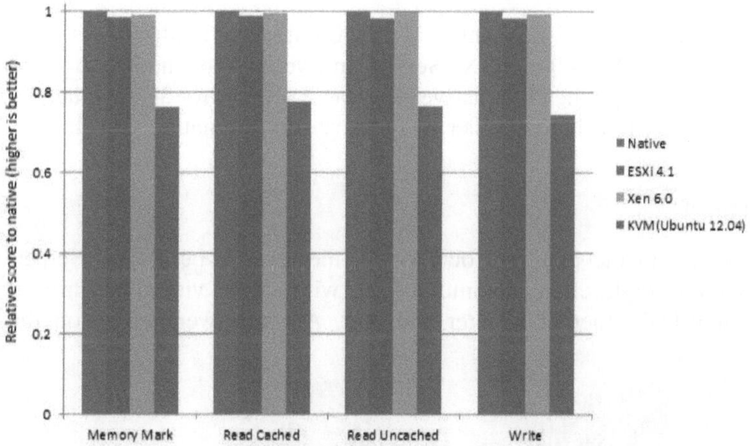

Fig. 3. Passmark – Memory results compared to native (Higher values are better)

XenServer exhibits almost native performance in the Read Uncached test, where ESXi shows 2% overhead vs native. In Read Cached and Write tests XenServer shows 1% overhead vs native while in the same tests ESXi shows close to 1.5 to 2% overhead vs native. In overall for memory mark XenServer demonstrates slightly better performance than that of ESXi. XenServer shows 1% overhead vs native and ESXi shows 1.5% overhead vs native. KVM performance in the memory benchmarks tests clearly indicates significantly lower than that of other two hypervisors and considerably falls behind the native performance.

Fig. 4 Shows benchmark results for Passmark Disk I/O read write tests. Sequential Read and Sequential Write were the disk mark tests which were conducted on the three hypervisors in the private cloud environment. Both XenServer and ESXi perform almost equal to native performance.

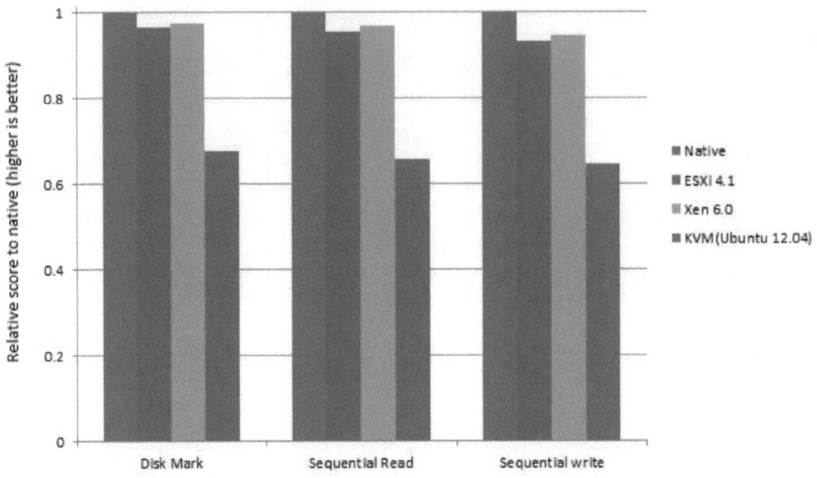

Fig. 4. Passmark – Disk I/O Read Write results compared to native (Higher values are better)

In Sequential Read and Sequential Write XenServer slightly shows better performance than that of VMWare ESXi Server. In overall disk mark performance XenServer shows 2.5% overhead vs native whereas ESXi shows 3% overhead vs native. KVM significantly falls behind other two hypervisors and native as well.

5.2 Netperf

For experiment, in the private cloud for all the three hypervisors, Netperf test involved running single client communicating with single virtual machine through a dedicated physical Ethernet adapter and port. All tests were based on the Netperf

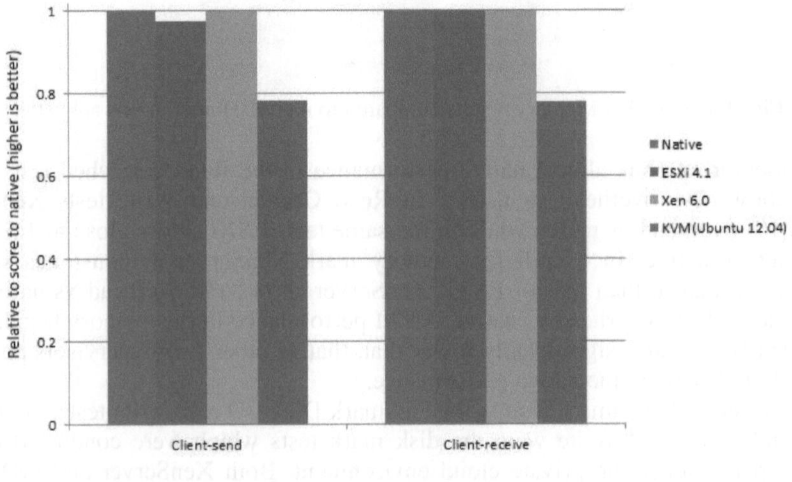

Fig. 5. Netperf results compared to native (higher values are better)

TCP_STREAM test. Fig. 5 shows the Netperf results for send and receive tests. XenServer and ESXi demonstrated near native performance in Netperf test, while KVM lags behind other hypervisors and native.

6 Conclusion

The objective of this experiment was to evaluate the performance of VMWare ESXi Server, XenServer and KVM Hypervisors in the private cloud environment. Performance results show convincingly that XenServer and ESXi Server both perform almost equally well and their performance is close to native performance in almost all tests without showing the signs any virtualization overhead. In CPU tests ESXi score over XenServer and in memory and I/O tests XenServer scores over ESXi. On overall two hypervisors are reliable, affordable and offer the windows or any other guest operating system IT professional a high performance platform for server consolidation for production workloads. KVM needs to improve up on almost all fronts if it has to become on par with other two hypervisors. ESXi and XenServer are matured hypervisors as compare to KVM and their Reliability, Availability and Serviceability (RAS) is significantly higher than that of KVM.

The tests were conducted in the private cloud with 64-bit Windows guest operating system. In the network test one client send and receive tests were performed on three hypervisors which are supported by CloudStack private cloud platform. The future work may include multiple client send and receive network tests for hypervisors. The future tests may include paravirtualized Linux guest operating system, as well as the scalability tests and can test with other hypervisors which are not covered in the experiment. And future work may also consider public cloud environment for experimentation. Virtualization infrastructure should offer certain enterprise readiness capabilities such as maturity, ease of deployment, performance, scalability and reliability. From the test results VMware ESXi Server and XenServer are better equipped to meet the demands of an enterprise datacenter than the KVM hypervisor. And KVM needs significant improvement to become an enterprise ready hypervisor. The series of tests conducted for this paper proves that VMware ESXi Server and XenServer delivers the production-ready performance needed to implement an efficient and responsive datacentre in the private cloud environment.

References

1. Mell, P., Grance, T.: The NIST Definition of Cloud Computing. National Institute of Standards and Technology. Information Technology Laboratory (2009)
2. Nanda, S., Chiueh, T.: A Survey on Virtualization Technologies. Technical report, Department of Computer Science, SUNY at Stony Brook, New York, 11794 – 4400 (2005)
3. Buyya, R., Yeo, C.S., Venugopal, S., Broberg, J., Brandic, I.: Cloud computing and emerging IT platforms: Vision, hype, and reality for delivering computing as the 5th utility. Future Generation Computer Systems 25(6), 599–616 (2009)
4. Adams, K., Agesen, O.: A Comparison of Software and Hardware Techniques for x86 Virtualization. In: Proceedings of the 12th International Conference on Architectural Support for Programming Languages and Operating Systems (2006)

5. AMD. AMD secure virtual machine architecture reference manual (2005)
6. Barham, P., Dragovic, B., Fraser, K., Hand, S., Harris, T., Ho, A., Neugebauer, R., Pratt, I., Warfield, A.: Xen and the art of virtualization. In: Proceedings of the Nineteenth ACM Symposium on Operating systems Principles, pp. 164–177 (2003)
7. Hostway UK VMware ESXi Cloud Simplified, http://www.hostway.co.uk/smallbusiness/dedicatedhosting/cloud/vmware-esxi.php
8. Fujitsu Technology Solutions, DataSheet Citrix XenServer, http://sp.ts.fujitsu.com/dmsp/Publications/public/ds-XenServer.pdf
9. Che, J., He, Q., Gao, Q., Huang, D.: Performance Measuring and Comparing of Virtual Machine Monitors. In: IEEE/IFIP International Conference on Embedded and Ubiquitous Computing (2008)
10. Apparao, P., Makineni, S., Newell, D.: Characterization of network processing overheads in Xen. In: First International Workshop on Virtualization Technology in Distributed Computing (2006)
11. Jianhua, C., Qinming, H., Qinghua, G., Dawei, H.: Performance Measuring and Comparing of Virtual Machine Monitors. In: IEEE/IFIP International Conference on Embedded and Ubiquitous Computing (2008)
12. Aravind Menon, A., Santos, J.R., Turner, Y., Janakiraman, G., Zwaenepoel, W.: Diagnosing Performance Overheads in the Xen Virtual Machine Environment. In: Conference on Virtual Execution Environments (2005)
13. Shan, Z., Qinfen, H.: Network I/O Path Analysis in the Kernel-based Virtual Machine Environment through Tracing. In: International Conference on Information Science and Engineering (2009)
14. VMware. A Performance Comparison of Hypervisors VMware. White paper (2007)
15. Passmark, http://www.passmark.com/products/pt.htm
16. VMware, The Architecture of VMware ESXi. White paper (2007)
17. Xen. How does Xen work (2009)
18. http://cloudstack.apache.org
19. http://www.ubuntu.com/download/cloud
20. http://www.microsoft.com/en-us/download/details.aspx?id=11093
21. Netperf, http://www.netperf.org/netperf/

An Ant Colony Based Load Balancing Strategy in Cloud Computing

Santanu Dam[1], Gopa Mandal[2], Kousik Dasgupta[2], and Paramartha Dutta[3]

[1] Future Institute of Engineering & Management, Kolkata-700 150, India
[2] Kalyani Government Engineering College, Kalyani-741 235, India
[3] Visva- Bharati University, Shantiniketan-731 235, India
sntndm@gmail.com

Abstract. Cloud computing thrives a new supplement of consumption and delivery model for internet based services and protocol. It provides large scale computing infrastructure defined on usage and also provides infrastructure services in a very flexible manner which may scales up and down according to user demand. To meet the QoS and satisfy the end users demands for resources in time is one of the main goals for cloud service provider. For this reason selecting a proper node that can complete end users task with QoS is really challenging job. Thus in Cloud distributing dynamic workload across multiple nodes in a distributed environment evenly, is called load balancing. Load balancing can be an optimization problem and should be adapting its strategy to the changing needs. This paper proposes a novel ant colony based algorithm to balance the load by searching under loaded node. Proposed load balancing strategy has been simulated using the CloudAnalyst. Experimental result for a typical sample application outperformed the traditional approaches like First Come First Serve (FCFS), local search algorithm like Stochastic Hill Climbing (SHC),another soft computing approach Genetic Algorithm (GA) and some existing Ant Colony Based strategy.

Keywords: Cloud Computing,CloudAnalyst, Ant Colony Optimization, Load Balancing.

1 Introduction

A new paradigm of large scale distributed computing is "Cloud". It utilizes the high speed of the internet to disperse the job from private PC to the remote computer clusters (Data Center owned by the cloud service providers). Cloud computing has become very popular for industry as well as academia for its sophisticated on demand services offered by its service providers like Google, Amazon [1]. Due to exponential growth of the internet in this decade computing infrastructure provided by its service providers may be used by industry or individuals from anywhere of the world. In future it has full potentiality to serve as computing as utility by the help of distributed virtualized elastic resource for end user [2]. Cloud service provider offers computing, software and storage as service. On demand provisioning and de-provisioning helps organization to reduce capital costs of software and hardware for this reason it has

M.K. Kundu et al. (eds.), *Advanced Computing, Networking and Informatics - Volume 2*,
Smart Innovation, Systems and Technologies 28,
DOI: 10.1007/978-3-319-07350-7_45, © Springer International Publishing Switzerland 2014

been adopted widely. As the size of the cloud may scale up and down the service providers have to provide computing power as lease to the users, in form of virtual machines (VM's)[3]. That makes Cloud computing a promising technology to provide resource on demand and to service the received request within time. Therefore high availability of resources is required and moreover management of resources is a big challenge to ensure QoS to end user and accelerates business performance of cloud service provider [4]. The primary challenges for the Cloud service provider is to scale up the performance or keep same. Cloud computing has a glorious future but many crucial problem still need to be realized. Load balancing is one of these problems where we have to distribute the local workload evenly to the whole cloud and ensures that at any instant of time all the processor or resources in the cloud does approximately the equal amount of work. This avoids the situation where some resources are heavily loaded while other are idle or doing very little amount of work (under loaded). To meet the criteria a good load balancing algorithm should be dynamic and adapt the environment [5].

In this paper a basic Ant Colony Optimization (ACO) has been proposed for load balancing of VMs in Cloud. ACO is a random search algorithm imitating the behavior of ant colonies. Ants are trailing from their nest to food and connect each other by pheromone which is volatile substance laid on paths traveled. CloudAnalyst a CloudSim based visual modeler used here for simulation and analysis of the proposed technique. The experimental result remarkably optimizes the entire system load.

The rest of paper is organized as follows. Section2 Introduces the CloudAnalyst toolkit. Section 3 Load Balancing of VM's using Ant Colony Algorithm in Cloud Computing. Section 4 details the proposed ACO algorithm. Section 5 presents the simulation results. Finally, Section 6 concludes this paper.

2 CloudAnalyst

Sometime it's very difficult and time consuming to measure the performance of the application or proposed policies in real world environment. In this conse-quence simulation is very much helpful to allow users or researchers with practical feedback without having real environment. This section portray the simulation in cloud to support application level infrastructure, services arises from cloud compu-ting paradigm such as modeling of on demand virtualized resources, which sup-ports cloud infrastructure. Different simulators are available today to adapt the real world situation like CloudSim [6] and CloudAnalyst [7]. CloudAnalyst has been used in this paper as simulation tool. CloudAnalyst is a GUI based visual model-ing and simulation tool based on the functionalities of CloudSim . Large scale applications that can be deployed on cloud infrastructures. CloudAnalyst enables developers to evaluate the large scale application in terms of geographic distribution of both computing servers and user's workload. A snapshot of the GUI of CloudAnalyst simulation toolkit is shown in figure 1(a) and its architecture in depicted figure 1(b). CloudAnalyst [7] developed as a visual modeler tool on CloudSim [8].

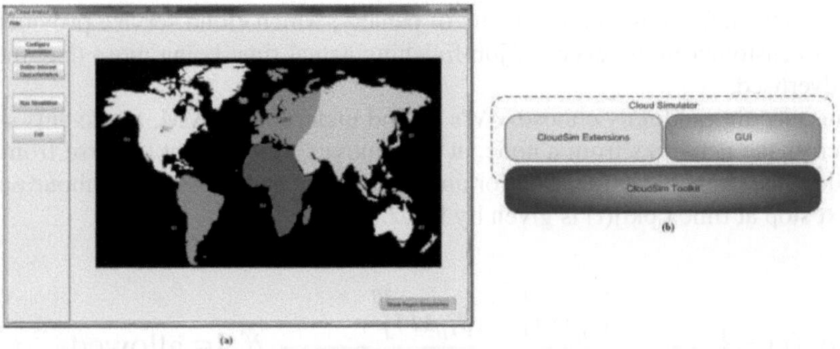

Fig. 1. Snapshot of CloudAnalyst (a) GUI of CloudAnalyst (b) Architecture of CloudAnalyst built on CloudSim

3 Load Balancing of VMs Using Ant Colony Algorithmin Cloud Computing

Ant Colony Optimization is basic foraging behavior of an ant that encouraged them to find the optimal shortest path from their nest to food introduced by Dorigo and Gambardella [8]. When ants are moving from their nest to food or vice versa they deposit a chemical substance called pheromone on their path. Paths are randomly chosen by ants initially. Chance of an isolated ant to follow a particular path among several possibilities always based on previously laid trail [10]. High concentrated pheromone helps an ant to choose a path and more ants are also attracted due to this high pheromone. By this way trail are reinforced with its own pheromone. Probability of an ant can separate the best optimal path from different set of paths is proportion to the concentration of a way's pheromone. As a result denser pheromone attracts more ants. It's a basically positive feedback mechanism that helps ants to find an optimal path finally.

3.1 The Proposed Method

As and when a job/request comes to the cloud service provider, they are allocated VMs in First Come First Serve manner and an index table is maintained to keep account about their current allocation. As the process continues a time will come due to vastness of Cloud when free available VMs are going to exhaust. In the situation artificial ants are created and dispersed to wander across the network to search under loaded VM's. Such an artificial ant is trying to choose a path from pheromone trail intensity that is initially assigned as given in Eq. 1.

$$\tau_{ij}(t=o) = f(MIPSJ, L, BWJ) \quad (1)$$

where, $\tau_{ij}(t=0)$ is the pheromone value in between two node i and j at turn t=0, MIPSJ (Million Instructions per Second) is the maximum capacity of each processor of VMJ and the parameter BWJ is related to the communication bandwidth ability of the

VMJ. L is the delay cost is an estimate of penalty, which cloud service provider needs to pay to customer in the event of job finishing actual time being more than the deadline advertised.

Thus any ant randomly choose VM's to find under loaded VM, as the ants starts its trip across the networks from a node, at each move of the kth ant traverse from node i to node j, the probability function for an ant at node i to choose a neighbour node j as its next stop at time t pkij(t) is given by Eq. 2.

$$p_{ij}^{k}(t) = \{ \frac{[\tau_{ij}(t)]^{\alpha} \ [\eta_{ij}(t)]^{\beta}}{\sum_{k \in \text{allowed}_k} [\tau_{ik}(t)]^{\alpha} \ [\eta_{ik}(t)]^{\beta}}, if \ j \in \text{allowed}_k \quad (2)$$

Where, allowed_k means the pheromone value updating due to the tour of the kth ant on its tabu (memory) list. The tabu list of the kth ant defined by tabu_k.α, β are two parameters for controlling the relative weight of the pheromone trail and heuristic value. τ_{ij} (t) is the pheromone value in between two node i and j, this value defined attractiveness. η_{ij} is the heuristic value given by Eq. 3.

$$\eta_{ij}(t) = 1/d_{ij} \quad (3)$$

where, d_{ij} is the hop distance between node i and node j.

Finally, the trip of an ant helps to identify the effectively underloaded VM within optimal distance. The information is updated in the index table globally. Correspondingly the pheromone values are updated as given in Eq. 4.

$$\tau_j(t+1)=(1-\rho)*\tau_j(t)+\Delta\tau_j \quad (4)$$

where, $\tau_j(t+1)$ is pheromone value of node j at time (t+1), ρ is the pheromone trail decay coefficient. If the value of ρ is greater, that shows less the impact of past solution. $\Delta\tau_j$ is local pheromone updating on the visited VMs when an ant completes its tour is given by Eq. 5.

$$\Delta\tau_j = 1/T_{ik} \quad (5)$$

where, T_{ik} be the optimal path distance that searched by k^{th} ant at the ith iteration.

4 Proposed Algorithm

Step 1: Maintain an index table which contains VmId and its corresponding requests (that are allocated for execution). Initially all VMs have current request 0.
Step 2: Schedule new request to VMs according to FCFS scheduling policy.
Step 3: Make corresponding change in the index table.
Step 4: If VMs are not available to allocate next job.

Step 4a: Create random number of ant with same pheromone value and place them randomly to traverse.

Step 4b: For m numbers of VMs and n numbers of random ants do

Step 4b-1: If an ant choose a VM then check whether the ant completes its tour or not.

Step 4b-2: If tour completed then update the pheromone value.

Step 4b-3: Check whether the solution is optimal and go to Step 5,

Step 4b-3: Else for non optimal solution, check whether all the ants have completed its tour. For non completion go to step 4b-2, else step-5.

Step 5: Store the current optimal solution and update pheromone value globally in the table.

Step 6: If all ants complete their tour then compare every local pheromone updates to output best possible solution.

5 Simulation Results and Analysis

The proposed algorithm is simulated in CloudAnalyst[8] by considering the scenario of "social networking site like Facebook". Suppositional configuration generated partitions the world into six "Regions" that is nothing but six continents as given in Table 1.

Table 1. Configuration of simulation environment

S.No	User Base	Region	Simultaneous Online Users During Peak Hrs	Simultaneous Online Users During Off-peak Hrs
1	UB1	0–N. America	4,70,000	80,000
2	UB2	1–S. America	6,00,000	1,10,000
3	UB3	2 – Europe	3,50,000	65,000
4	UB4	3 – Asia	8,00,000	1,25,000
5	UB5	4 – Africa	1,15,000	12,000
6	UB6	5 – Oceania	1,50,000	30,500

A single time zone is set for all user bases(UB) and for each UBs a sample online user during peak hour and off peak hour has been considered. Of the entire online user only one tenth approximately is available during off peak hours.

Each simulated data centre host has a particular amount of virtual machines (VMs) dedicated for the application. For simulation each of the Machines has been considered of 4GB of RAM, 100 GB storage and 1000MB of available bandwidth. Each Datacenter (DC) is assumed to be having 4 CPUs with a capacity of 10000 MIPS. Simulated hosts have x86 architecture, virtual machine monitor Xen and Linux operating system. Each user request (jobs) has been considered to be requiring 100 instructions to be executed.

The proposed algorithm is executed in several setups as tabulated in Table 2, where one DC is considered having initially 25, 50 and 75 VMs in each Cloud Configurations (CCs). Simulation scenario of Table 3 consists of two DCs with a variation of 25, 50 and 75 VMs. In Table 4, 5, 6 and 7 considers three, four, five and six DCs respectively with a mixture of 25, 50 and 75 VMs for all DCs. Average response time of the jobs are calculated for the proposed algorithm and tabulated. The performance of proposed algorithm is compared with some existing load balancing algorithm like .Genetic Algorithm (GA)[10], Stochastic Hill Climbing Algorithm (SHC)[11], Existing ACO[12] strategy and First Come First Serve (FCFS)[8]. Fig. 2, 3,4,5,6, and 7 make a comparative analysis of the proposed technique for the different scenarios and techniques. The comparative analysis confirms the novelty of the work.

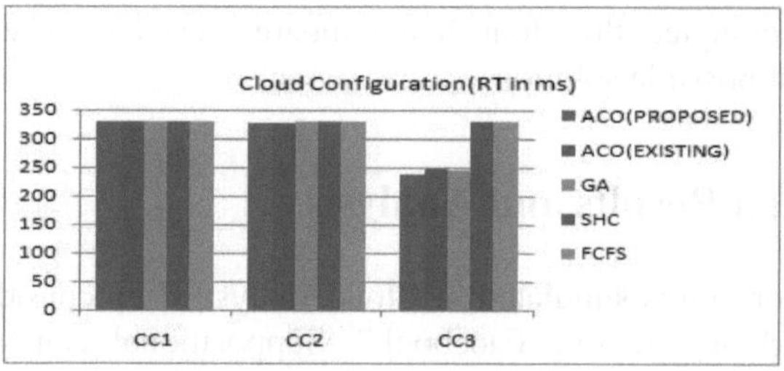

Fig. 2. Performance analysis of proposed ACO with GA, SHC and FCFS Result using one Datacenter

Table 2. Simulation scenario and calculated overall average response time (RT) in (ms) using one DC

Sl. No.	Cloud Configuration	Data Center specification	RT in ms for proposed ACO	RT in msfor existing ACO	RT in ms for GA	RT in ms for SHC	RT in ms for FCFS
1	CC1	One DC with 25 VMs	328.98	329.01	329.01	329.02	330.11
2	CC2	One DC with 50 VMs	327.63	328.63	328.97	329.01	329.65
3	CC3	One DC with 75 VMs	238.12	248.43	244	329.34	329.44

Table 3. Simulation scenario and calculated overall average response time (RT) in (ms) using Two DC

S.No	Cloud Cofigu-ration	Data Center specification	RT in ms for pro-posed ACO	RT in ms for existing ACO	RT in ms for GA	RT in ms for SHC	RT in ms for FCFS
1	CC1	Two DCs with 25 VMs each.	354.72	358.97	360.77	365.44	376.34
2	CC2	Two DCs with 50 VMs each.	349.89	354.21	355.72	360.15	372.52
3	CC3	Two DCs with 75 VMs each.	348.68	352.66	355.32	359.73	370.56
4	CC4	Two DCs with 25, 50 VMs.	346.57	348.64	350.58	356.72	368.87
5	CC5	Two DCs with 25, 75 VMs.	347.86	348.12	351.56	357.23	367.23
6	CC6	Two DCs with 75, 50 VMs.	350.47	351.45	352.01	357.04	361.01

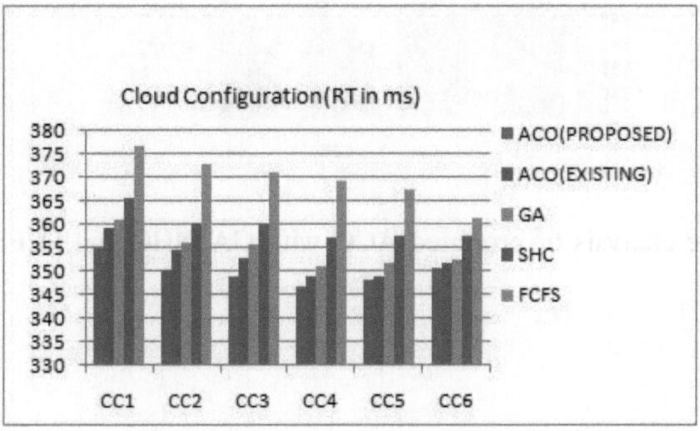

Fig. 3. Performance analysis of proposed ACO with GA, SHC and FCFS Result using Two Datacenter

Table 4. Simulation scenario and calculated overall average response time (RT) in (ms) result using Three Data Centers

Sl No.	Cloud Configuration	Data Center specification	RT in ms for proposed ACO	RT in ms for existing ACO	RT in ms for GA	RT in ms for SHC	RT in ms for FCFS
1	CC1	Each with 25 VMs .	345.68	348.57	350.32	356.82	363.34
2	CC2	Each with 50 VMs .	344.86	346.83	350.19	355.25	363.52
3	CC3	Each with 75 VMs	340.62	343.89	346.01	350.73	361.56
4	CC4	Each with 25, 50 ,75VMs.	340.86	343.53	345.98	350.01	360.87

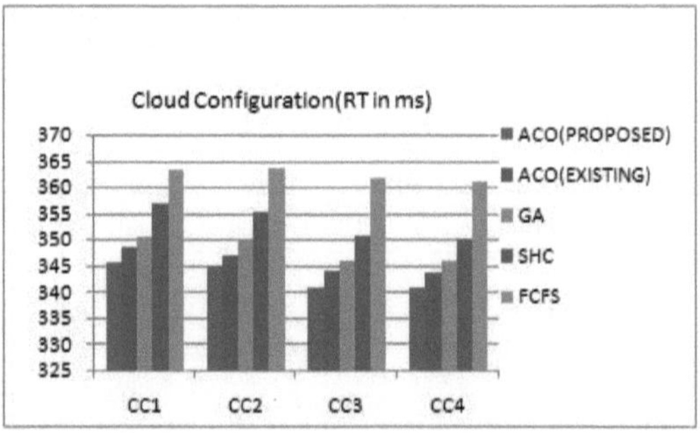

Fig. 4. Performance analysis of proposed ACO with GA, SHC and FCFS Result using Three Datacenter

Table 5. Simulation scenario and calculated overall average response time (RT) in (ms) result using Four Data Centers

Sl No.	Cloud Configuration	Data Center specification	RT in ms for proposed ACO	RT in ms for existing ACO	RT in ms for GA	RT in ms for SHC	RT in ms for FCFS
1	CC1	Each with 25 VMs.	341.46	346.57	348.85	354.35	360.95
2	CC2	Each with,50 VMs .	339.78	343.84	345.54	350.71	359.97
3	CC3	Each with 75 VMs	336.56	339.78	340.65	346.46	358.44
4	CC4	Each with 25, 50 ,75VMs.	334.32	335.43	337.88	344.31	355.94

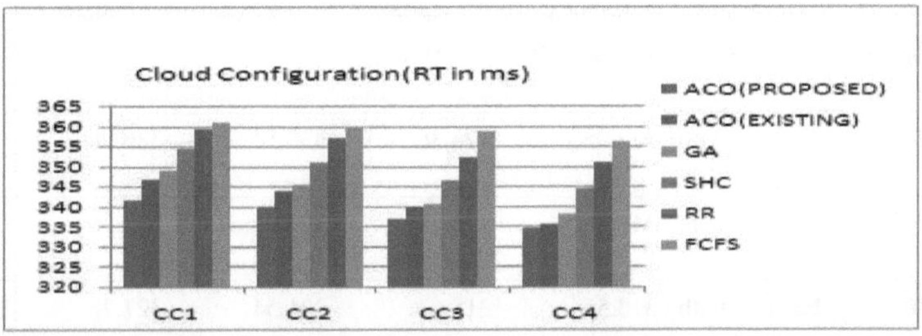

Fig. 5. Performance analysis of proposed ACO with GA, SHC and FCFS Result using Four Datacenter

Table 6. Simulation scenario and calculated overall average response time (RT) in (ms) result using Five Data Center

Sl. No	Cloud Configuration	Data Center Specification	RT in ms proposed ACO	RT in ms for existing ACO	RT in ms for GA	RT in ms for SHC	RT in ms for FCFS
1	CC1	Each with 25 VMs.	331.45	334.80	335.64	342.86	352.05
2	CC2	Each with 50 VMs.	321.12	325.59	326.02	332.84	345.44
3	CC3	Each with 75 VMs	319.89	321.48	322.93	329.46	342.79
4	CC4	Each with 25, 50 , 75 VMs.	317.65	319.04	319.98	326.64	338.01

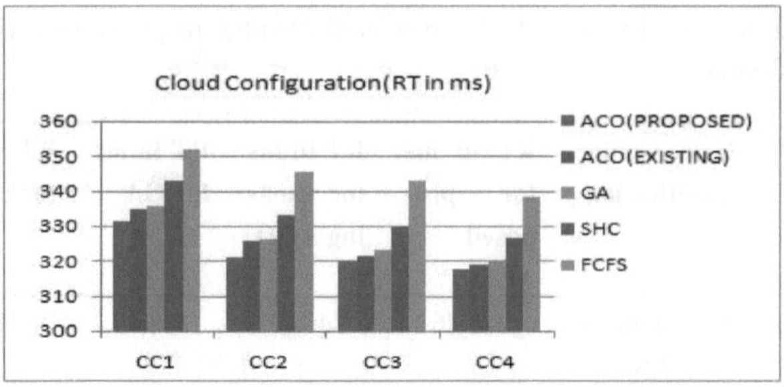

Fig. 6. Performance analysis of proposed ACO with GA, SHC and FCFS Result using Five Datacenter

Table 7. Simulation scenario and calculated overall average response time (RT) in (ms) result using Six Data Center

Sl. No.	Cloud Configuration	Data Center specification	RT in ms for proposed ACO	RT in ms existing ACO	RT in ms GA	RT in ms SHC	RT in ms FCFS
1	CC1	Each with 25 VMs .	323.98	326.36	330.54	336.96	349.26
2	CC2	Each with 50 VMs .	316.48	321.73	323.01	331.56	344.04
3	CC3	Each with 75 VMs.	313.56	318.64	321.54	327.78	339.87
4	CC4	Each with 25,50,75	309.66	312.32	315.33	323.56	338.29

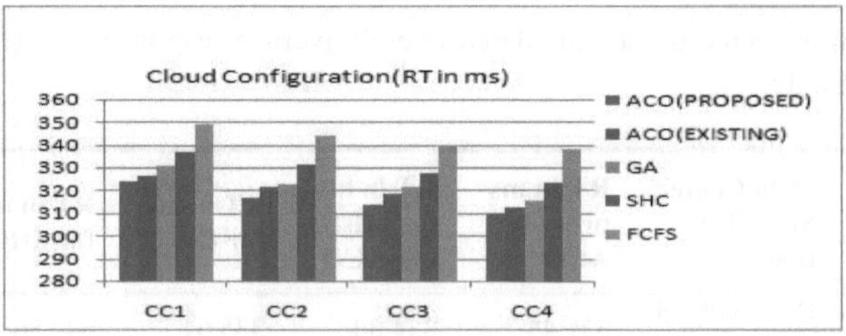

Fig. 7. Performance analysis of proposed ACO with GA, SHC and FCFS Result using Six Datacenter

6 Conclusion

In this paper, soft computing based algorithm on ant colony optimization has been proposed to initiate the load balancing under cloud computing architecture. Detail analysis of the results, indicates that the proposed strategy for load balancing not only outperforms a few existing techniques but also guarantees the QoS requirement of customer job. Though fault tolerance issues does not consider and all jobs are predicted with same priority here, which may not be the actual scenario. Researchers can proceed to include the fault tolerance and different function variation to calculate the pheromone value can be used for further research work.

References

1. Buyya, R., Broberg, J., Goscinski, A.: Cloud Computing: Principles and Paradigms. John Wiley & Sons (2011)
2. Li, K., Xu, G., Zhao, G., Dong, Y., Wang, D.: Cloud Task scheduling based on Load Balancing Ant Colony Optimization. In: 2011 Sixth Annual ChinaGrid Conference (2011)
3. Dikaiakos, M.D., Pallis, G., Katsa, D., Mehra, P., Vakali, A.: Cloud Computing: Distributed Internet Computing for IT and Scientific Research. Proceedings of IEEE Journal of Internet Computing 13(5), 10–13 (2009)
4. Srinivasan, R.K., Suma, V., Nedu, V.: An Enhanced Load Balancing Technique for Efficient Load Distribution in Cloud-Based IT Industries. In: Abraham, A., Thampi, S.M. (eds.) Intelligent Informatics. AISC, vol. 182, pp. 479–485. Springer, Heidelberg (2013)
5. Li, K., Xu, G., Zhao, G., Dong, Y., Wang, D.: Cloud Task scheduling based on Load Balancing Ant Colony Optimization. In: 2011 Sixth Annual ChinaGrid Conference (2011)
6. Calheiros, R.N., Ranjan, R., Beloglazov, A., Rose, C., Buyya, R.: Cloudsim: A toolkit for modeling and simulation of cloud computing environments and evaluation of resource provisioning algorithms. In: Software: Practice and Experience, vol. 41(1), Wiley (2011)
7. Wickremasinghe, B., Calheiros, R.N., Buyya, R.: Cloudanalyst: A cloudsim-based visual modeller for analysing cloud computing environments and applications. In: Proceedings of Proceedings of the 24th International Conference on Advanced Information Networking and Applications (AINA 2010), pp. 446–452 (2010)
8. Dorigo, M., Gambardella, L.M.: Ant colony system: A cooperative learning approach to the traveling salesman problem. IEEE Transactions on Evolutionary Computation, 53–66 (1997)
9. Suryadevera, S., Chourasia, J., Rathore, S., Jhummarwala, A.: Load Balancing in Computational Grids Using Ant Colony Optimization. International Journal of Computer & Communication Technology 3(3), 20–23 (2012)
10. Dasgupta, K., Mondal, B., Dutta, P., Mondal, J.K., Dam, S.: A Genetic Algorithm (GA) based Load Balancing Strategy for Cloud Computing. In: Proceedings of Computational Intelligence: Modeling, Techniques and Applications, pp. 340–347 (2013)
11. Mondal, B., Dasgupta, K., Dutta, P.: Load Balancing in Cloud Computing using Stochastic Hill Climbing-A Soft Computing Approach. In: Proceedings of 2nd International Conference on Computer, Communication, Control and Information Technology, pp. 783–789 (2012)
12. Nishant, K., Sharma, P., Krishna, V., Rastogi, N., Rastogi, R.: Load Balancing of Nodes in Cloud Using Ant Colony Optimization. In: Proceedings of the 14th International Conference on Modelling and Simulation, pp. 1–9 (2012)

Scalable Cloud Deployment on Commodity Hardware Using OpenStack

Diplav Dongre, Gourav Sharma, M.P. Kurhekar, U.A. Deshpande, R.B. Keskar, and M.A. Radke

Department of Computer Science and Engineering,VNIT, Nagpur, India
{diplav.tkiet,gouravsharma029}@gmail.com,
{uadeshpande,manishkurhekar,rbkeskar,mansi.radke}@cse.vnit.ac.in

Abstract. OpenStack is a cloud computing project aimed at providing infrastructure as a service (IaaS). In this paper we describe our experience in deploying OpenStack cloud over commodity hardware. We have made an effort to build a large computational facility by sharing the computational resources of our institute through the use of the OpenStack cloud platform. In this paper, we give an overview ofthe OpenStack cloud platform and various services offered by it. We describe two multi-node cloud architectures that we have implemented. In the first architecture, we have deployed the cloud over few machines connected by a closed network. The second architecture allowed us to use geographically separated nodes. We describe the steps required for installation of the cloud for eachof these architectures and provide automated scripts for the same. These automated scripts are available at the following website:http://vnit.ac.in/images/openstack/openstack_grizzly.rar.

Keywords: OpenStack Grizzly, Private Cloud, Commodity Hardware.

1 Introduction

Typically, a desktop workstation is unused 50-70% of time. Since the computing capacity of desktops steadily increases, these numbers are likely to grow [8]. In our educational institute, we have several machines connected to each other over LAN. Most of these machines are used for executing programs written in different programming languages and for deploying the web server and database servers. We have observed that the computational power of these machines is not completely utilized. On the other hand, there are several research projects having a very high demand for computing resources, such as those running time consuming simulations. It is also not always feasible to buy new hardware dedicated for each of these projects. A user is constrained to use the computational power available with his/her machine. These observations imply that there is a huge idle computational power available which, even though accessible through the network, remains unused. In view of this, we started with an objective to share the computational resources of the machines in our institute and channelize the resources wherever required. In order to achieve this, we have made an effort to implement a private cloud for our institute.

M.K. Kundu et al. (eds.), *Advanced Computing, Networking and Informatics - Volume 2*, Smart Innovation, Systems and Technologies 28,
DOI: 10.1007/978-3-319-07350-7_46, © Springer International Publishing Switzerland 2014

During the last few years, several highly scalable systems have been built using the cloud software stack by various organizations in different domains. A cloud is built on existing technologies to virtualize hardware, software and storage, into flexible units that can be quickly allocated to meet the demand. Cloud computing is a distributed computing model, where computer resources such as computing power, software and storage are provided as network based services.

The existence of free and open source cloud platforms is essential to further drive the propagation of cloud computing environments. OpenStack is one such software, originally released by Rackspace and NASA into the open source domain. OpenStack cloud is a collection of open source projects that canbe used to setup and run computational infrastructures by enterprises/service providers. It is designed to run on commodity hardware e.g. ARM and x86.

Some of the characteristics of OpenStack that attract the cloud community are scalability, flexibility and openness [9].

In this work, we discuss the steps required for a fully operational multi-node OpenStack installation using commodity hardware. Every major step discussed in this paper has been implemented modularly as a separate shell script.

2 Openstack Cloud Platform

The OpenStack software [10] is divided into several services shown in Fig.1. The "compute", "network" and "storage" services collectively provide the OpenStack cloud management capabilities.

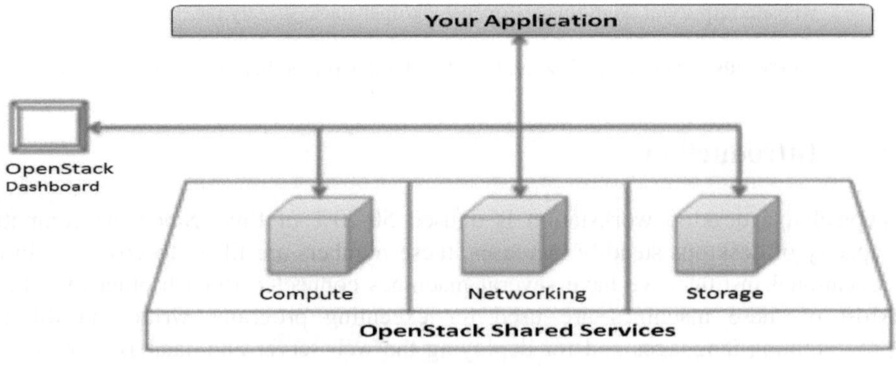

Fig. 1. OpenStackservices [5]

The main OpenStack services are [3]:

1. *OpenStack Compute (Nova)*: Nova is the Computing controller for the OpenStack cloud; it handles the scheduling ofVirtual Machine (VM) instances.
2. *OpenStack Storage*: It provides block and object storage used by the VM instances. The block storage system allows the users to create block storage devices and dynamically attach and detach them from the VM instances using the dashboard or API.

3. *OpenStack Networking*: It provides software driven network and IP address management capabilities. This service allows the users to create their own software driven networks with dynamic or static IP addresses to VM instances.
4. *OpenStack Dashboard (Horizon):*Horizon is the python based dashboard, used to administerOpenStack services.
5. *OpenStack Identity (Keystone):*Keystone provides authentication and authorizationservices for all OpenStack activities. It works like a KDC [11] (Key Distribution Center).
6. *OpenStack Image (Glance):* Glance is used to store and retrieve VM images in the cloud. It can be configured to use different storage backend. The service supports multiple VM image formats likeISO, VHD, VDI, and OVF.

3 Basic Deployment Architecture

While deploying the cloud software stack we have to first select a machine which will act as Controller and manage most of the cloud by running the OpenStack services. Other machines, called Compute nodes, are used as the resources for it. As shown in Fig.2, this architecture has SERVER1 as a Controller and has services Glance, Keystone, Nova and Horizon installed on it. Machine named COMPUTE1 acts as a Compute node and has services Nova-Compute and Nova-Network installed on it. The controller node SERVER1 manages compute node COMPUTE1.

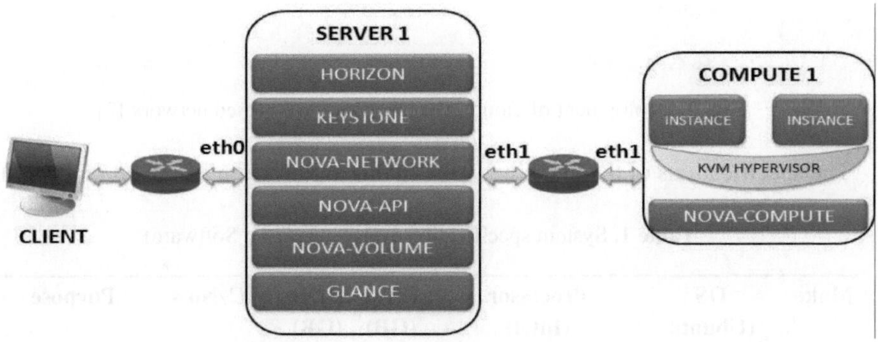

Fig. 2. Basic deployment architecture

3.1 Architecture of Our Cloud Installation

While designing our system, we started with a singlenode and progressively scaled the system to include other machines as compute nodes. On the first node, we installed the Controller, the Network and the Compute services. After successfully deploying it, we moved to multi-node architecture [12]. We designed two multi-node architectures, deployed them and tested them successfully. The details follow in the subsequent sections.

3.1.1 Multi-node Private Cloud Architecture

Our cloud consists of four machines, as shown in Fig.3. A private network was created among them using a network switch. We chose a node with two NICs as the Controller. The Controller connects to our private network through eth0 and to the public network through eth1. Thus, the Controller node acts as a gateway for all the Compute nodes. Compute nodes are equipped with only one NIC. Compute nodes are required to have OpenStack services- Nova-Compute and Nova-Network. This model gives us a large computing and storage power, which actually is the collective computing and storage power of all the nodes in this model.

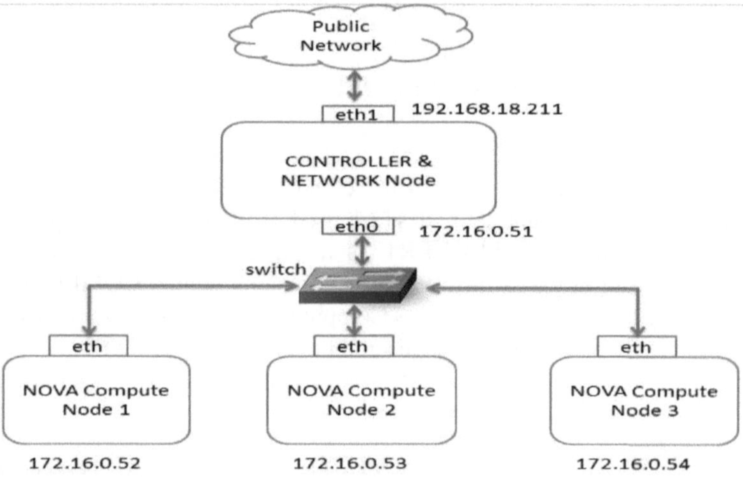

Fig. 3. Deployment of cloud over machines in a closed network [2]

The specifications of the machines used in this architecture are given in Table 1.

Table 1. System specification 1(Hardware and Software)

Make	OS (Ubuntu 12.04 LTS)	Processor (Intel)	RAM (GB)	HDD (GB)	NIC	No's	Purpose
HP	64 bit	Core 2Duo 2.93 GHz	2	310	2	1	Controller & Network
Lenovo	64 bit	Core i3 3.40 GHz	4	500	1	2	Compute
Lenovo	32 bit	Core i3 3.40 GHz	2	149	1	1	Compute

3.1.2 Multi-Node Installation in Public Network

The problem with our previous architecture is that it can only be accessed from machines that are part of the model, i.e. from either Compute or Controller nodes. Machines outside the private network cannot access the virtual machine instances. However, clients can access the dashboard and can do some basic tasks (adding/deleting image, creating/terminating instance, etc.). External machines cannot access the virtual machines because there is no direct connection between the client machines and the compute nodes; client machines can only access the Controller node through the NIC connected to the public network. To overcome this problem, we used our institute's network to carry out communication to and from the cloud. Every machine inside the campus is accessible to every other machine. Due to this, compute machines are directly accessible to any client machine inside the campus.

The machines we have chosen can be geographically distributed, as shown in Fig. 4.We have placed some of the Compute machines with the Controller machine in Computer Lab 1 and rest of the Compute machines in Computer Lab 2. We have also used a mix of 64 and 32 bit operating systems while deploying our multi-node architecture. Table 2 gives the specification of the nodes.

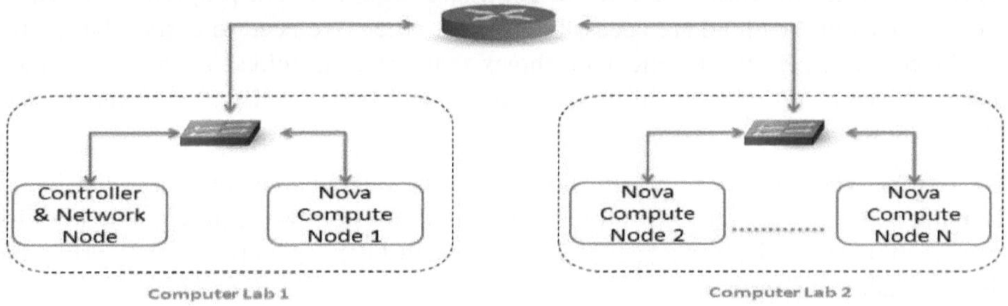

Fig. 4. Deployment of cloud over machines connected by our institute's LAN

Table 2. System specification 2 (Hardware and Software)

Make	OS(Ubuntu 12.04 LTS)	Processor (Intel)	RAM (GB)	HDD (GB)	NIC	No's	Purpose
Lenovo	64 bit	Core i3 3.40 GHz	4	500	1	1	Controller
Lenovo	64 bit	Core i3 3.40 GHz	4	500	1	4	Compute
Lenovo	32 bit	Core 2Duo 3 GHz	2	149	1	1	Compute

Our public network architecture has the following advantages:

1. Only Single NIC machines beused as a Controller/Compute node.
2. Fully Scalable architecture: any machine within campus can be used as a Compute node dynamically.

For the above two multi-node architectures, we followed similar installation steps except in the case of network configuration which are described in the next section.

4 Installation Procedure for Multi-Node Architectures

We have created shell scripts, which can be used for deploying the OpenStack cloud services with minimal effort. The scripts and the instructions for their use are available at [7]. These have been tested on Ubuntu 12.04 LTS platform. A brief description of the scripts follows.

4.1 Common Configurations

Operating System: We have installed Ubuntu Desktop 12.04 LTS [14] 64 bit OS on most of the machines (Controller and some Compute nodes) and Ubuntu Desktop 12.04 LTS 32 bit OS on one of the Compute nodes.

Network Configuration: We have to configure basic network properties so that all the machines in the cloud are accessible to each other. We need to ensure that all the machines are accessible to client machines that are using cloud services. We have assigned static IP address in the range of 192.168.18.XXX for Compute and Controller machines.

Grizzly Sources: We have added the Grizzly package sources in source list file of Ubuntu, and upgraded the OS.Before upgrading the system we require public keys for downloading the Grizzly packages. So, we first installed Ubuntu-cloud-Keyring and then upgraded our system.

Network Time Protocol [15] (NTP): We have installed NTP package on the Controller and Compute nodes. In our configuration, Controller is working as NTP server and all Compute nodes act as NTP clients. This package is necessary to synchronize time between all the nodes.

4.2 Packages for Controller Node

MySQL & RabbitMQ: We have installed MySQL database server [16] and created databases for each of the cloud service. These databases should be accessible to the installed cloud services. We have created separate databases for each service with the same name as the service name, for example Keystone service will have database named keystone. For each of these databases, we have added users with the same name as the database name and provided all privileges to them over that database, for example, keystone database will have keystone user with all the privileges on keystone database. The layout of the database is shown in the Fig. 5.

In a cloud, Advanced Message Queuing Protocol [17] (AMQP) is used for message communication. There are several packages that implement the AMQP architecture. However, RabbitMQ is a preferred package for OpenStack.

For simplicity we have configured all the cloud services to use RabbitMQ with default credentials. However for security reasons we suggest creating new user with proper privileges.

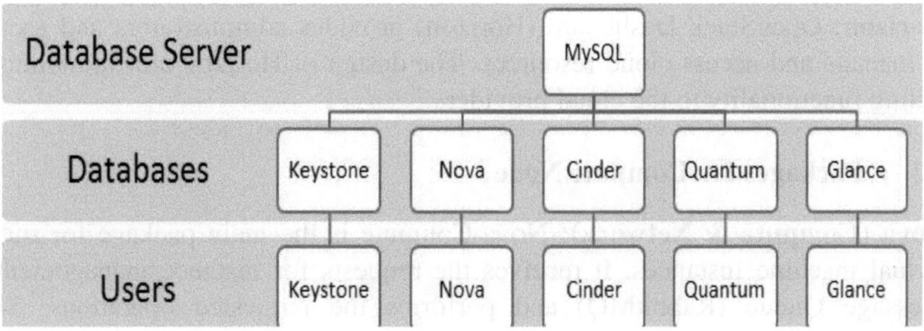

Fig. 5. Database layout

Keystone: OpenStack Keystone provides a central repository of users who are registered to the OpenStack. It is a centralized authentication system across the cloud. It mainly supports two forms of authentication, tokenbased and credentials based (username and password). Additionally, the keystone directory provides a list of all of the deployed services in an OpenStack cloud.

We have installed and configured Keystone on the Controller node. Initial configuration of tenants, users, roles and endpoints makes OpenStack deployment difficult for novice users. To make it easier, we have provided a script.

Glance: OpenStack Image Service (Glance) provides finding, and delivery services for VM images. Stored images can be used as a template. The Image Service can store VM images in a variety of storage back-ends like OpenStack Object Storage and block storage. The Image Service API offers Representational state transfer (REST) interface for inquiring information about disk images [1,6].

We have installed the Glance package and configured it as per our requirement. We have also uploaded image of Ubuntu 10 and CirrOS [4] operating system for testing purpose. The provided script performs this task. We have uploaded several images through OpenStack dashboard and used those images for creating instances.

Cinder: OpenStack Block Storage (Cinder) provides persistent block-level storage devices for use with OpenStackVM instances. The cinder service manages the launching, attaching and detaching of the block devices to the VM instances. Block storage volumes are unified into OpenStackto allow cloud users to accomplish their own storage requirements [1,6].

Quantum: As cloudentrepreneurs began to provide their cloud services upon resource virtualization, they found that the complexity of networks required more automation. They came up with the first functioning software-defined networking (SDN) which is called as Quantum.

Nova (without Nova-Compute): Another crucial part of the OpenStack cloud deployment is Nova. This is because, using of the computing power of all machines is done using the Nova service. There are different packages of Nova service for Controller and Compute nodes. All the packages (e.g. NOVA-API, NOVA-Scheduler, etc.)required for handling the distributed Compute nodes are installed in the Controller node.

Horizon: OpenStack Dashboard (Horizon) provides administrators and users a GUI to manage and access cloud resources. The design of Horizon allows monitoring and billing functionality to the cloud providers.

4.3 Packages for Compute Node

Nova (Compute & Network): Nova-Compute is the main package for running the virtual machine instances. It receives the requests for instance management via the Message Queue (RabbitMQ) and performs the requested operations. There are numerous Compute nodes in a typical cloud production environment. In our architecture, we have five Compute nodes. Nova-network is a Network service that controls DHCP, DNS, and routing. It performs network management tasks like IP addresses allocation, VLANconfiguration [13]and security group implementation.

5 Results

The observed network utilization on Controller node while creating instance was 6.5-12 MiB/s as shown in Fig. 6.

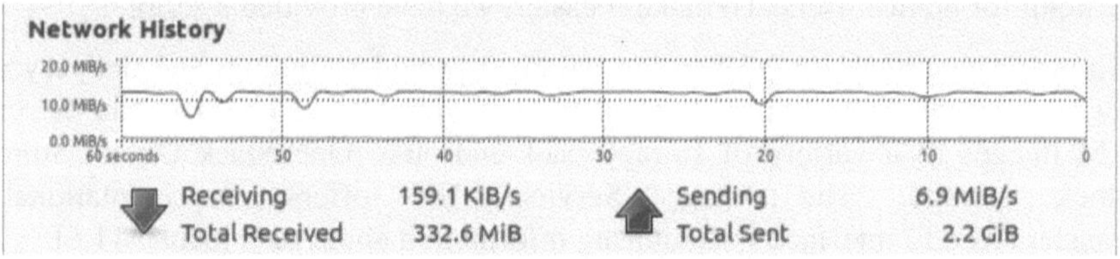

Fig. 6. Network statistics on the Controller

Similar network statistics are observed on the Compute node where the actual instances were created.

We were successful in creating the compute nodes using NOVA-COMPUTE package and add these compute nodes to the cloud. We were able to create virtual machine instances over the cloud. The controller created the VM instances and assigned them to one of the compute nodes. Fig. 7 shows that the CPU utilization at a compute node before assignment of a VM instance is below 20%.

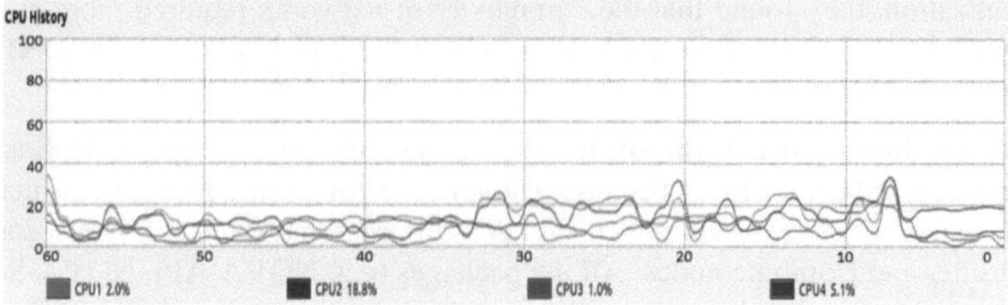

Fig. 7. CPU utilization at compute-node when no VM instance has been assigned to it

We have observed that the CPU utilization of a compute node increased when VM instances were assigned to it and accessed by users, as shown in Fig. 8. Here average CPU utilization is above 65%.

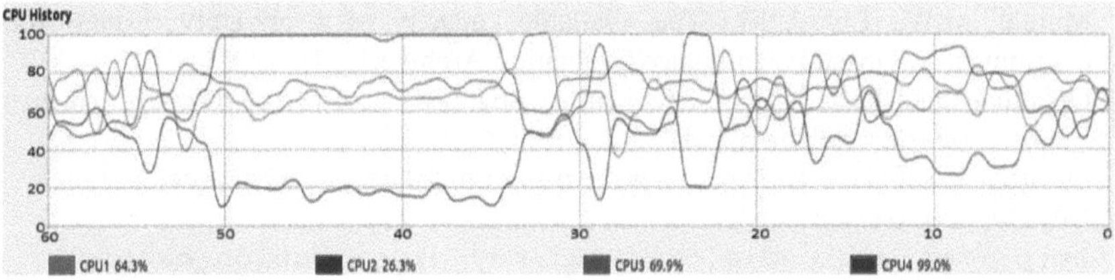

Fig. 8. CPU utilization at compute-node when VM instances have been assigned to it

6 Conclusion

In this work, we have discussed various steps in OpenStack installation on commodity hardware. We started with installing Ubuntu 12.04 LTS on all nodes and progressed to installation of various packages on Controller and Compute nodes. We have also described the required database structure for OpenStack installation. We have provided scripts automating this process. The cloud has been deployed on readily available heterogeneous commodity hardware, hence, incurring no additional cost. We have successfully deployed the cloud in two scenarios. First, we have deployed cloud over machines connected by a closed network of few machines. Then we have used the existing network infrastructure, which allows us to potentially use geographically separated nodes.

Since OpenStack is highly scalable, we hope to utilize the existing infrastructure to its fullest potential. Research projects where the demand for resources is very high will be able to meet their demand without requiring additional specialized hardware. Users can put forwardtheir demands for the cloud resources and the cloud identifies idle resources and fulfills those demands. We believe that this work will be helpful for coming up with a fully functional multi-node OpenStack installation using commodity hardware.

Acknowledgement. This work was carried out with a funding received from IBM under the Shared University Research (SUR) grant.

References

1. OpenStack Documentation, http://docs.openstack.org/grizzly/
2. IBM Developer works, `http://www.ibm.com/developerworks/cn/cloud/library/1209_zhanghua_openstacknetwork/`
3. Jackson, K.: OpenStack Cloud Computing Cookbook. Packt Publishing, Birmingham (2012)

4. CirrOS, https://launchpad.net/cirros
5. OpenStack wiki., https://wiki.openstack.org/wiki/Main_Page
6. OpenStack LLC. OpenStack Install and Deploy Manual (2012)
7. OpenStack installation script, http://vnit.ac.in/images/openstack/openstack_grizzly.rar
8. Mutual, M.W., Livny, M.: The available capacity of a privately owned workstation environment. Journal Performance Evaluation Archive 12(4) (1991)
9. OpenStackdatasheet, http://www.openstack.org/downloads/openstack-overview-datasheet.pdf
10. OpenStack Software, http://www.openstack.org/software/
11. Key Distribution Center, http://en.wikipedia.org/wiki/Key_distribution_center
12. OpenStackMultinode Architecture, http://docs.openstack.org/folsom/openstack-ops/content/example_architecture.html
13. Virtual LAN, http://en.wikipedia.org/wiki/Virtual_LAN
14. Ubuntu 12.04 LTS, http://www.ubuntu.com/download/desktop
15. NTP, http://www.ntp.org/documentation.html
16. MySQL, http://www.mysql.com
17. AMQP, http://www.amqp.org/

A Demand Based Resource Provisioner for Cloud Infrastructure

Satendra Sahu[1], Harshit Gupta[2],
Sukhminder Singh[1], and Soumya Kanti Ghosh[1]

[1] School of Information Technology
[2] Department of Computer Science and Engineering
Indian Institute of Technology Kharagpur, India
{harshitgupta1337,sahu.satendra1988,ss.comp.engg}@gmail.com,
skg@iitkgp.ac.in

Abstract. Resource management in cloud environment poses unique challenges. Resources, in the form of *virtual machines* (VM) are to be provisioned on the fly while using the underlying infrastructure efficiently and still meeting the performance parameters. This involves collecting system resource statistics for decision making by other components of cloud environment. In this paper, the process of resource (VM) management in the cloud is mapped to *demand based* system wherein the VMs that require additional resources or need to relinquish their resources send requests to a centralized controller. Further, since resources are limited, dynamic resource allocation forms a classical optimization problem. This paper proposes a one-dimensional knapsack optimization solved using dynamic programming, to achieve efficient resource allocation. The performance of the proposed algorithm has been compared with brute force algorithm.

Keywords: Cloud computing, Operations research, Dynamic programming, Just in time, Demand based model.

1 Introduction

The advent of cloud computing has led to a paradigm shift in the way computing facilities are managed. There has been an accelerated growth in the demand for more and more flexible computing resources. In response to this, cloud computing has emerged as a boon for the industry over the past few years. IaaS services over cloud [1] have enabled convenient and on-demand network accessed shared pool of configurable computing resources like servers, storage, applications, and services that can be rapidly purveyed and released with minimal management effort or service-provider interaction. The accompanying flexibility gives rise to a vast array of resource management decisions [2]. From a cost perspective, it is important to optimize cloud computing and ensure maximum utilization of the available resources. Hence it can be argued that the cloud stakeholders could benefit from an operations research (OR) perspective due to similar nature of the problems faced.

M.K. Kundu et al. (eds.), *Advanced Computing, Networking and Informatics - Volume 2,* Smart Innovation, Systems and Technologies 28,
DOI: 10.1007/978-3-319-07350-7_47,

Demand based approach is a classical model for producer-consumer interaction where-in inflow of infrastructure (resources) is signaled by the consumer (VM) of the service in a continuous flow mode. Whenever, a trigger is raised by the consumer (VM), resource allocation process is triggered. This approach ensures that unnecessary network load is not generated and the consumers correctly predict their resource needs as compared to the supplier. This approach also eliminates the need for regular monitoring of resource status.

There are a number of works for analyzing resource allocation in grid platforms based on various OR models [3], [4], [5], [6].

The rest of the paper is organized as follows. Section 2 describes the proposed resource allocation model for cloud. In this section, the resource allocation model and details of allocation and de-allocation algorithm are presented. Section 3 and Section 4 show the implementation and experimental results of the proposed allocation model respectively.

2 Demand Based Model for Resource Allocation in IaaS Cloud

Drawing parallels between the process of resource management in cloud and *Just In Time*(JIT) system one can see that resource users in cloud are the applications running on virtual machines. Supplier of resources is the cloud infrastructure (resource manager) and the resources are allocated to the requesting process in the form of VMs. The proposed resource management model can be divided into three sub-processes: Demand Generation Process, Demand Collection Process, Demand Service Process.

Demand Generation Process: The primary objective of this module is to monitor the resource utilization status of every VM. It runs on every VM and periodically checks the performance of the VM. If there is a degradation in performance of the running service then this process issues a demand card to the Demand Collection Process requesting allocation of additional resources. On the other hand, if an under-utilization of resources is observed, then this process issues a demand card to the Demand Collection Process to de-allocate surplus resources.

Demand Collection Process: This process runs on the cloud controller. The primary objective of this module is to collect all the demand cards sent by the Demand Generation processes and en-queue them.

Demand Service Process: It runs on the cloud controller and services the resource allocation/de-allocation requests based on the dynamic resource allocation/de-allocation algorithms discussed in the subsequent sections. It checks the resource availability in the cloud to assign new VMs and stores information about running instances of applications.

The demand card comprises of the following information:

- ID of the application running on requesting VM (appID)
- The instance ID of the requesting VM (instanceID)
- Type of the VM being requested by the demand card (vM)
- Request Type [allocation(1),de-allocation(0)] (dT)
- Priority of the request (ρ)
- Time-stamp of the instant when the request is received by the *Demand Collection Process* (t_0)
- The performance metrics that triggered the generation of the *demand card*
- Service class of the application running on the VM [$Premium(5)$, $Economy(3), General(1)$] (sC)

The proposed model is based on the following assumptions:

- At least one suitable VM type (in terms of resource supply) exists for each request. In the worst case, one specific VM type is the sole suitable type for all requests.
- Each resource request is fulfilled by one VM instance.
- Only one application can run on a VM.
- Resources can only be allocated in the form of VM and not independently

At each instant a certain number of resource demands enter into the system in the form of demand cards. All the demands present in the system at any time instant may or may not be served completely. The objective is to decide which demands to service first and which type of resource (i.e. VM) to allocate, or to degrade. Serving a resource allocation request will generate a contribution factor and will change the state of the allocated resources (available, pending, running, terminated). The objective is to find an optimal solution in terms of total contribution in reasonable time.

The following notations are used in the proposed model:
λ = set of time instants at which the *Demand Service Process* is run
= { nT | n $\in \mathbb{N}$ and T is the periodicity of execution of the *Demand Service Process* }
ν = set of possible states of the VMs
μ = set of resource Types [1]
W_t = Set of all resource demands present in the system at time t where $t \in \lambda$
$W_t^{'} \subseteq W_t$ such that $W_t^{'}$ is the set of all resource demands serviced at time t such that servicing $W_t^{'}$ generates highest total contribution
E = Set of SLA based Service Classes of the applications running on the VMs
ρ_{kt} = Priority of the request $k \in W_t$ at time $t \in \lambda$

$$\rho_{kt} = \begin{cases} sC_k & \text{if t - } t_0 \leq \text{T} \\ \rho_{k(t-T)} + \chi_{kt} & \text{otherwise} \end{cases} \tag{1}$$

[1] The value associated with each instance type has been decided based on the number of cores associated with the instance type in a typical Eucalyptus [7] cloud setup

χ_{kt} = Penalty factor in resource demand $k \in W_t$ at time $t \in \lambda$.

$$\chi_{kt} = \rho_{kt} * \lfloor (t - t_0)/T \rfloor * \left(1 - \frac{|\{j \in W_t' \mid sC_j = sC_k\}|}{|\{l \in W_t \mid sC_l = sC_k\}|}\right) \quad (2)$$

where t_0 is the instant at which the Demand Collection Process received the request k
C_k = Contribution obtained by servicing a resource demand k

$$C_k = \rho_k * vM_k \quad (3)$$

R_{it}^v = number of resources of type $v \in \mu$ in the state $i \in \nu$ at time $t \in \lambda$

$$R_t = [R_{it}^v]_{v \in \mu, i = Available} \quad (4)$$

Given R_t and W_t, the feasible solution space consisting of all the possible combinations is defined by

$$X_t(R_t, W_t) \quad (5)$$

where X_t is defined as the set of all the possible solutions. Each possible solution $x \in X_t$ has a value C_x (C_x is the sum of the contributions C_k obtained by servicing demands request $k \in x$). So the optimal solution at time $t \in \lambda$ can be found by solving the following optimality Eq-7.

$$C_x = \sum_{\forall k \in x} C_k \quad (6)$$

C_t = Contribution of the system obtained by servicing the optimal solution set at time t, $t \in \lambda$

$$C_t = (max[C_x : x \in X_t]) \quad (7)$$

3 Implementation of Demand Based Resource Provisioning Using Dynamic Programming

To resolve the problem of resource allocation in a cloud we implement the proposed demand based resource allocation framework using *Dynamic Programming* approach.

3.1 Resource Allocation Algorithm

Let C_t be the highest contribution that can be obtained by servicing a subset of demands $W_t' \subseteq W_t$. Let R_t (refer equation 4) be the number of resources of type $v = m1.small$ in the state $i = available$ at time $t \in T$. Let $S_{i,b}$ denote a subset of resource requests in the form of demand cards $\{k_1, k_2, \cdots, k_i\}$ that if serviced generates maximum contribution, given the available resources are b, and $1 \leq i \leq |W_t|$. Let $A(i, b)$ denote the contribution that can be obtained by servicing the set $S_{i,b}$. For A(i,b) with $b == 0$.

```
Input: kCard[1, ···, W]: array of resource requests in the form of
demand cards
R: number of available resources in cloud
Parameters:
C: contribution obtained after servicing the requests
A[W+1][B+1]: A[i][j] stores A_ij
soln[W+1][B+1]: 2D boolean array
T: size of serviceList
if verify(B) is true then
 | service all requests from kCard
end
else
    for each kc in kCard do
        if t - kc.t0 > T then
            updatePriority(kc)
        end
    end
    for n = 1 → W do
        for w = 1 → B do
            leave ← A[n-1][w]
            service ← 0
            if kCard[n].vM ≤ w then
                service ← kCard[n].priority + A[n-1][w-kCard[n].vM]
                A[n][w] ← max(leave,service)
                soln[n][w] ← (service > leave)
            end
        end
    end
    initialize w ← B
    for n = W → 0 do
        if soln[n][w] then
            add kCard[n] to serviceList
            w ← w-kCard[n].vM
        end
    end
    for each kc in serviceList do
        C ← ∑ (kc.sC*kc.vM)
    end
    service all requests in serviceList
end
```

Algorithm 1. Resource Allocation Policy (Dynamic Programming)

The base case $A(i,0)$=0 $\forall\ 1 \leq i \leq n$, denotes no contribution when resources are zero. The following equation can be used to calculate all the values of $A(i+1,b)$.

$$A(i+1,b) = \begin{cases} max\{A(i,b), C_{k_{i+1}} + A(i,b-vM_{k_{i+1}})\} \\ \qquad\qquad\qquad \text{if } vM_{k_{i+1}} \leq b \\ A(i,b) \qquad\quad \text{otherwise} \end{cases} \tag{8}$$

The optimal solution then corresponds to the set $S_{|W_t|,B_t}$ for which the benefit is maximized. Algorithm 1 presents the dynamic programming solution of resource allocation problem.

3.2 De-allocation Algorithm

Algorithm 2 shows the de-allocation policies of proposed architecture. De-allocation algorithm services the de-allocation and degradation demand from the requesting instances.

```
Input: kCard (emi, instanceID, dT, priority, previous, metrics, sC):
list of resource requests
Output: degradeList: list of resource requests for degradation
Parameters:
listTotal: list of running instances (i.e VMs) of applications
running in cloud
removeList: list containing false requests
currentVM: resource type of running instance
serviceVM: service type of running instance
state: current state of instance [ pending, running, terminated ]
for each kc in kCard do
    if state == running and canDegrade ==1 then
        degradeVM
    end
    else if currentVM == m1.small then
        if listTotal.appID==1 then
            add kc to removeList
        end
        else
            shut-down the Running instance
        end
    end
end
remove all requests from removeList
service all requests from degradeVM
```

Algorithm 2. Resource De-allocation/Degradation Policy

4 Experiments and Simulation

A third party API typica [8] was used to communicate with Eucalyptus cloud [7].

To compare the performance of brute force algorithm and the proposed approach two experiments were carried out. Request arising as well as resource relinquishing both have been modeled based on in-homogeneous Poisson distribution [9] with their intensity following a Gaussian distribution.

Comparison of execution times: The response times of the two algorithms (i.e. brute force algorithm and the proposed algorithm) were compared. The nature of the execution time of the brute force algorithm was exponential with respect to the number of requests while that of the proposed algorithm was a linear one.

Comparison of service percentage: This experiment aims at the evaluation and comparison the service percentages of the service classes for the two algorithms (for Brute Force Algorithm as shown in Fig. 1 and for Dynamic Programming Algorithm as shown in Fig. 2).

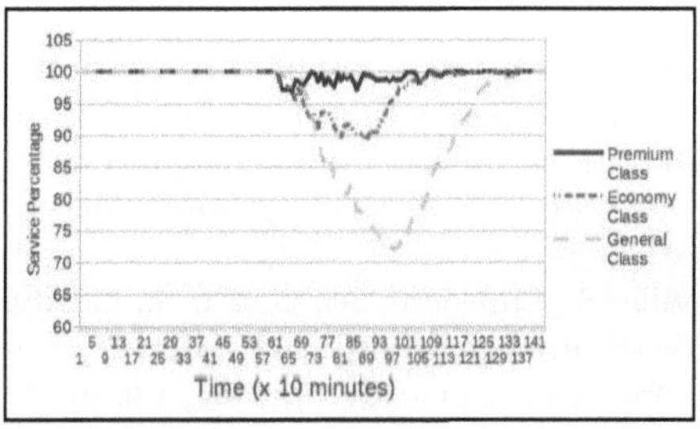

Fig. 1. Service percentage of *service_class*(E) for Brute Force algorithm

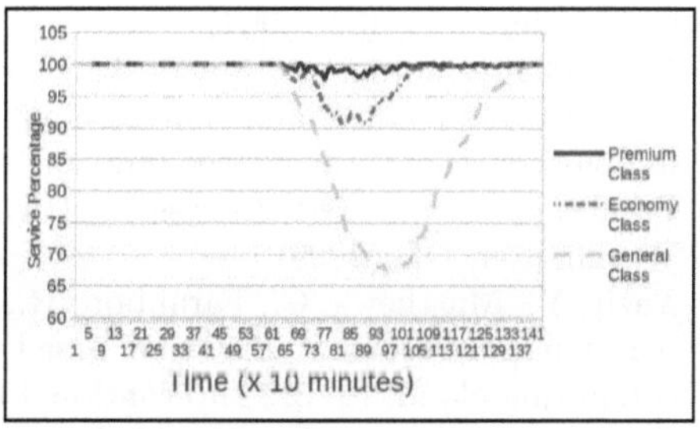

Fig. 2. Service percentage of *service_class*(E) for Dynamic Programming algorithm

Both the approaches show similar service percentages for every service class. As is evident from the graphs, the satisfaction rate of the various service classes is comparable between the two approaches.

From the experimentation, it can be inferred that the overall request response time (in case of both allocation and deallocation) will be enhanced significantly.

5 Conclusion

Cloud computing promises a computing environment that provides simple, intuitive, on-demand provisioning of resources to meet their application needs. The demand based resource manager addresses the scalable collection and analysis of resource metrics from resource utilization perspectives. Allocation algorithm in demand based resource manager has been formulated on a basic optimization model that provides dynamic scaling of resources. The experimental results show that, compared with the brute force algorithm, the proposed allocation algorithm effectively reduces the execution time and at the same time provides almost the same satisfaction rate to the requests.

References

1. Mell, P., Grance, T.: Perspectives on cloud computing and standards. Technical report, National Institute of Standards and Technology (NIST), Information Technology Laboratory (2009)
2. Buyya, R., Yeo, C.S., Venugopal, S., Broberg, J., Brandic, I.: Cloud computing and emerging it platforms: Vision, hype, and reality for delivering computing as the 5th utility. Future Generation Computer Systems 25(6), 599–616 (2009)
3. Tierney, B., Aydt, R., Gunter, D., Smith, W., Swany, M., Taylor, V., Wolski, R.: A grid monitoring architecture. The Global Grid Forum Draft Recommendation, GWD-Perf-16-3 (2002)
4. Cooke, A., Gray, A., Nutt, W., Magowan, J., Oevers, M., Taylor, P.: The relational grid monitoring architecture: Mediating information about the grid. Journal of Grid Computing, 323–339 (2004)
5. Brandt, J., Gentile, A., Mayo, J., Pebay, P., Roe, D., Thompson, D., Wong, M.: Resource monitoring and management with ovis to enable hpc in cloud computing environments. In: Proceedings of the 2009 IEEE International Symposium on Parallel & Distributed Processing, IPDPS 2009, pp. 1–8. IEEE Computer Society, Washington, DC (2009)
6. Yazir, Y., Matthews, C., Farahbod, R., Neville, S., Guitouni, A., Ganti, S., Coady, Y.: Dynamic resource allocation based on distributed multiple criteria decisions in computing cloud. In: 3rd International Conference on Cloud Computing, pp. 91–98 (2010)
7. Nurmi, D., Wolski, R., Grzegorczyk, C., Obertelli, G., Soman, S., Youseff, L., Zagorodnov, D.: The eucalyptus open-source cloud-computing system. In: 9th IEEE/ACM International Symposium on Cluster Computing and the Grid, CC-GRID 2009, pp. 124–131. IEEE (2009)
8. Kavanagh, D.: Typica: A java client library for a variety of amazon web services (2008), `https://code.google.com/p/typica/`
9. Arlitt, M.F., Williamson, C.L.: Internet web servers: Workload characterization and performance implications. IEEE/ACM Transactions on Networking (ToN) 5(5), 631–645 (1997)

Large Scale Cloud for Biometric Identification

Sambit Bakshi and Rahul Raman

Department of Computer Science and Engineering,
National Institute of Technology Rourkela,
Odisha - 769008, India
{sambitbaksi,rahulraman2}@gmail.com

Abstract. This article aims to propose a large-scale cloud architecture to serve for biometric system that enrols large population. In identification mode of biometric system, a query template is matched with all stored templates in the database and a match is said to occur with the one with which match-value becomes highest. Hence the identification time = $n \times t$ where n = database size and t = 1:1 matching time. As the database size n becomes sufficiently large, the identification time increases significantly. This leads to long response time of the system. However, achieving the n matching processes in parallel can bring down the total identification system from nt to t. This speeds up the proposed system n times than its sequential counterpart with the trade-off of the cost of resources for cloud and extra communication. The proposed architecture also takes care of threat to compromise secured data as they are passed to different nodes. This architecture passes inputs to cloud nodes hiding the identity-holders information so that stealing the identity data of an individual will not compromise the security of the system.

Keywords: Cloud architecture, biometric authentication, large database.

1 Introduction

Biometric security is the science of authenticating a person through his behavioral or physical trait which he bears. The biological traits considered for authentication are unique and cannot be easily spoofed. A typical biometric system consists of three modules: sensor and preprocessing module that senses and extracts the region of interest of a trait of a subject while a subject enrols for the first time or when a returning subject attempts to authenticate him. Local feature extraction module extracts features from the live query and previously stored data fetched from the database. The matching module generates a score indicating the degree of match between two sets of features and a decision, based on the score, is taken whether the live query matches the template fetched from database or not. Fig. 1 shows a typical block diagram of a biometric system. The system consists of three primary modules:

– Sensor and preprocessing module, which senses the trait from a subject and selects the region of interest from the sensed data.

M.K. Kundu et al. (eds.), *Advanced Computing, Networking and Informatics - Volume 2*, Smart Innovation, Systems and Technologies 28,
DOI: 10.1007/978-3-319-07350-7_48, © Springer International Publishing Switzerland 2014

– Local feature extraction module, which extracts local features from the preprocessed template. The features are responsible for matching between live query and stored templates.
– Matching module, that takes input from the stored database and the live query, computes a match score between the two, and depending upon a threshold, concludes the live query to be either imposter or genuine.

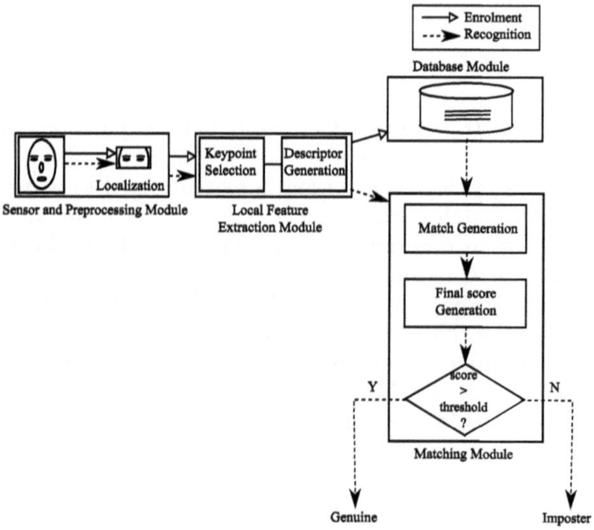

Fig. 1. Working model of biometric authentication

The challenge arises when biometric authentication has to be done in real time for a large database. There are few factors for a biometric security that require to be modelled in cloud architecture:

– To attain a high accuracy from biometric system, it is necessary that a biometric template is sufficiently large. Compared to simple features, multiscale features of the templates tend to reveal high accuracy. However large biometric template and multiscale feature analysis requires more matching time and computation.
– Practical biometric systems require real time response to clients. Furthermore there can be concurrent queries to the biometric server from different clients.
– In practical cases the biometric database is very large, may be as large as the number of citizens of a nation. Further, multimodal biometric systems increase the database size in several folds by storing more than one trait. Accessing the database and reconstructing the database with new enrolments (for indexed databases) needs high response time if the whole data is stored in a single central database.

– For identification biometric systems, a live query is matched to every stored data in the database and a match is said to be found when the live query yields highest matching score with an individual.

This article proposes a secure cloud architecture for biometric identification to resolve the aforementioned issues.

2 Related Works

A computation between all existing pairs present in a set has been marked as time consuming as it needs $\binom{n}{2} = O(n^2)$ computations. As a trade-off to save computation time, parallelism is exploited employing multiple processors in execution. Sometimes there is a need to abstract these processing operations from user and to deliver only the end result. Moretti *et al.* [3] illustrates all-pair computation abstraction in cloud. Cloud is marked as a biomedical information sharing and processing infrastructure due to its speed-up and abstraction as indicated in [6]. Following the initial experiments on cloud, there have been researches indicating the deployment of biometric system on cloud to achieve parallelism and thus real-time response. Be it traditional physical trait like fingerprint or challenging behavioral trait like gait, cloud-computation has been adopted by a wide span of biometric systems. Yang *et al.* [9] describes a new fingerprint recognition scheme based on a set of assembled geometric moment and Zernike moment features in cloud computing communications tested on SIFT-enhanced images obtained from four datasets of FVC2002 and claims to achieve satisfactory Equal Error Rate (EER). Panchumarthy *et al.* [5] records cloud computation for gait biometric recognition performed on Amazon EC2 instances using up to 26 CPU cores, which marks significant decrease in execution time compared to recognition in local machine. Shelly and Raghava[7] describes iris recognition system on Hadoop, which is an open source cloud computing environment. To process large amount of data, Hadoop uses Map/Reduce framework in Java. The biometric system thus implemented uses 512 byte biometric templates and 8 processing nodes on private network to perform the feature matching. As biometric dataset size deployed on this system is varied from 1.1 GB to 13.2 GB, speed-up varies from 1.36 to 6.645. A large dataset requires long time for sequential approach while cloud computing reduces the time significantly. Kohlwey *et al.* [2] presents a prototype system for generalized searching of cloud-scale biometric data. Instead achieving high speed-up, there have been scope to improve over this work by using more nodes connected to cloud and exploiting more parallelism. However this could not be done without proving secure computation since computation on cloud-nodes connected to internet may lead to data excavation and other threats. In [1], the authors have addressed the problem of secure outsourcing of large-scale biometric experiments to a cloud, where privacy of the data has been preserved and the client can verify with very high probability that the task had been computed correctly. Considering the plausibility of secured verification, the computation load can be deployed on cloud-nodes connected through internet with high confidence.

For practically utilizing cloud-based biometric recognition has faced challenges like mobile connectivity, and load balancing of big-biometric databases. Omri *et al.* [4] and Stojmenovic [8] have approached to attain biometric recognition from mobile devices. The mobile device acts as the client to fetch the live query template to the cloud. The mobile device hence does not need any computation for recognition. The only limiting factors in this direction have been found to be the battery power of the device and throughput of the communication channel, which have been resolved through use of sophisticated devices and high-bandwidth wireless communication techniques like 3.5G.

3 Large Scale Cloud for Biometric Authentication

In contrast to centralized biometric systems running on a single processor, the article proposes a cloud infrastructure tailored for large biometric authentication system. The proposed architecture which composes of a client and the database and application server modules together with a computer network to support the computation requested by database module and control node. The control node, database, application servers and networked computers are secured from client side and belong to the cloud which the client can access only through a live query. Hence the cloud remains secured from attacks from the client side. Furthermore, any node in the computer network does not receive the owner information of the template it is fetched to process. Hence any compromised node cannot extract any useful information from the template fetched to it.

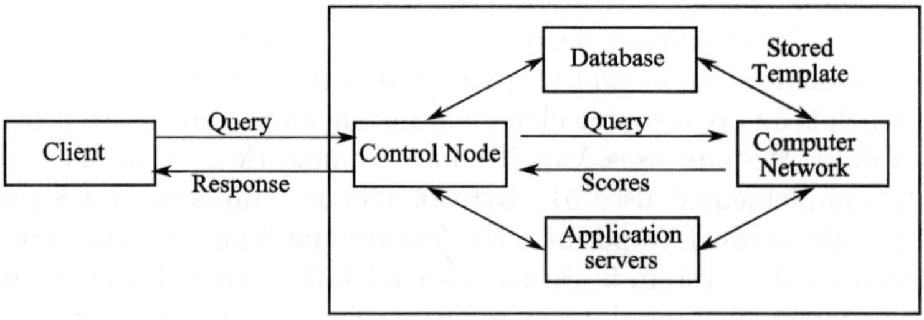

Fig. 2. Working model of biometric authentication through secured cloud

4 Schematic Representation of the Biometric Cloud

A schematic block diagram of the proposed biometric cloud is shown in Fig. 2 It has a thin client that only triggers the query and accepts the response. Within the cloud, there are three modules: database, application server, and computer network. A control node comes into play to synchronize these modules. When the query is fired by the client, the control node activates the database and the application servers. For concurrent queries, each query is processed by a

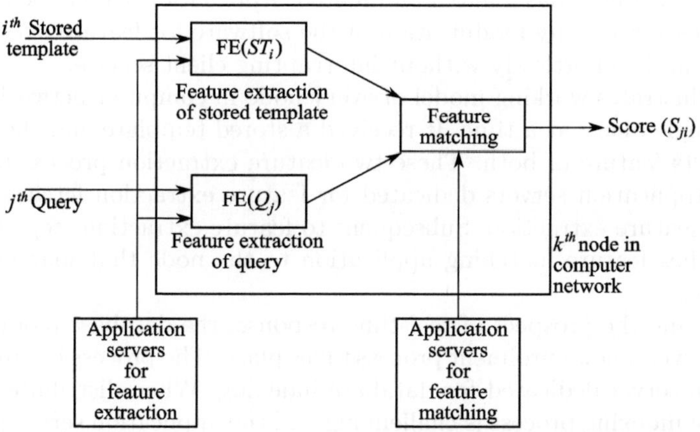

Fig. 3. Working model of every node in the computer network

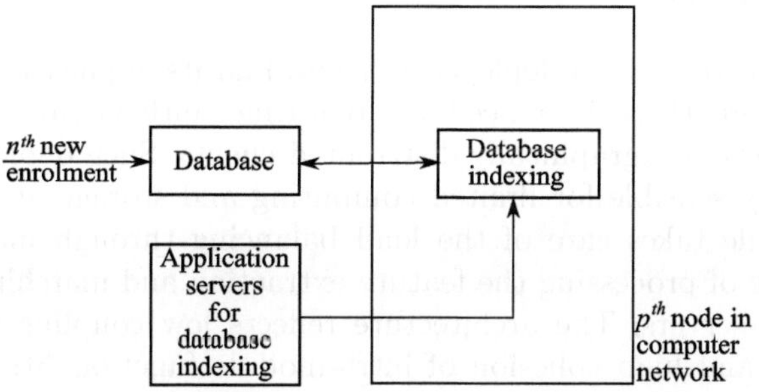

Fig. 4. Working model of database updating and indexing when a new enrolment takes place

single node in computer network. Hence the maximum number of nodes in the computer network required for real-time response should essentially be more than or equal to the number of concurrent queries. If highest number of concurrent queries is p, then the architecture needs only p nodes in the network to process the queries in real-time without waiting delay.

Every node receives a stored template from database and the query. Employing the application servers for feature extractor, the features from the stored template and the query are extracted in the form of vectors. Subsequently, both the feature vectors are subjected to feature matching through application servers dedicated for feature matching. Finally a score is generated which is returned to the control node by every node in the computer network. Based on all scores received from the nodes of the network, the control node responds to the client generating a single response pointing to the individual stored in the database with which the query has a maximum match. Application servers being separate from the databases, any modification of the software for feature extraction and matching can be effortlessly without interrupting client service.

Fig. 3 illustrates working model of every node in computer network that processes a single query at a time. It received a stored template and the live query and extracts feature of both. These two feature extraction process proceeds in parallel. Application servers dedicated for feature extraction fetches the application for feature extraction. Subsequent to feature extraction step, application server fetches feature matching application to the node that matches the two feature sets.

Considering the prospect of real-time response, the database module indexes itself whenever a new enrolment process takes place. The process is handled by an application server dedicated for database indexing. When distributed database is used, the indexing process is challenging and the application server performing this job should handle distributed environment. Fig. 4 illustrates the activity of a database module in biometric cloud.

5 Conclusions

Proposed biometric system deployed in cloud find its application for large scale population where there is a need for real-time authentication of concurrent queries made from geographically distributed clients. Such deployment will also be economically sensible for limited computing and storage at client side. The database module takes care of the load balancing through indexing, and the nodes take care of processing the feature extraction and matching as instructed by application servers. The architecture reflects low coupling of inter-module functionalities and high cohesion of intra-module functionalities, which makes the architecture easily upgradable, and hence scalable.

References

1. Blanton, M., Zhang, Y., Frikken, K.B.: Secure and Verifiable Outsourcing of Large-Scale Biometric Computations. In: IEEE Third International Conference on Social Computing (SOCIALCOM), pp. 1185–1191 (2011)
2. Kohlwey, E., Sussman, A., Trost, J., Maurer, A.: Leveraging the Cloud for Big Data Biometrics: Meeting the Performance Requirements of the Next Generation Biometric Systems. In: IEEE World Congress on Services (SERVICES), pp. 597–601 (2011)
3. Moretti, C., Bui, H., Hollingsworth, K., Rich, B., Flynn, P., Thain, D.: All-Pairs: An Abstraction for Data-Intensive Computing on Campus Grids. IEEE Transactions on Parallel and Distributed Systems 21(1), 33–46 (2010)
4. Omri, F., Hamila, R., Foufou, S., Jarraya, M.: Cloud-Ready Biometric System for Mobile Security Access. In: Benlamri, R. (ed.) NDT 2012, Part II. CCIS, vol. 294, pp. 192–200. Springer, Heidelberg (2012)
5. Panchumarthy, R., Subramanian, R., Sarkar, S.: Biometric Evaluation on the Cloud: A Case Study with HumanID Gait Challenge. In: IEEE/ACM Fifth International Conference on Utility and Cloud Computing, pp. 219–222 (2012)

6. Rosenthal, A., Mork, P., Li, M.H., Stanford, J., Koester, D., Reynolds, P.: Cloud computing: A new business paradigm for biomedical information sharing. Journal of Biomedical Informatics 43(2), 342–353 (2010)
7. Shelly, Raghava, N.S.: Iris recognition on Hadoop: A biometrics system implementation on cloud computing. In: IEEE International Conference on Cloud Computing and Intelligence Systems (CCIS), pp. 482–485 (2011)
8. Stojmenovic, M.: Mobile Cloud Computing for Biometric Applications. In: 15th International Conference on Network-Based Information Systems (NBiS), pp. 654–659 (2012)
9. Yang, J., Xiong, N., Vasilakos, A.V., Fang, Z., Park, D., Xu, X., Yoon, S., Xie, S., Yang, Y.: A Fingerprint Recognition Scheme Based on Assembling Invariant Moments for Cloud Computing Communications. IEEE Systems Journal 5(4), 574–583 (2011)

AFMEACI: A Framework for Mobile Execution Augmentation Using Cloud Infrastructure

Karishma Pawar[1], Vandana Jagtap[1], Mangesh Bedekar[1], and Debajyoti Mukhopadhyay[2]

[1] Maharashtra Institute of Technology, Department of Computer Engineering, Pune 411038
kvppawar@gmail.com,
{vandana.jagtap,mangesh.bedekar}@mitpune.edu.in
[2] Maharashtra Institute of Technology, Department of Information Technology, Pune 411038, Maharashtra, India
debajyoti.mukhopadhyay@gmail.com

Abstract. Due to advancement in mobile device technology, mobile devices are becoming an inevitable part of human lives. In context of running massive applications on mobile devices, users can't utilize the potential of mobile devices in an efficient manner since mobile devices are constrained by processing power, memory requirements and battery capacity. To alleviate this resource scarcity problem of the mobile devices, mobile cloud computing is the most promising solution which combines the technologies from both the mobile computing and the cloud computing. The execution of heavy application on mobile devices is augmented by powerful and resource-abundant cloud servers. This is achieved by partitioning an application into tasks such that the computational intensive tasks are offloaded to cloud and after executing task on cloud, results are sent back to mobile device, referred to as computation offloading. In this paper, we have put forth a scalable, fault-tolerant framework for dynamically and optimally partition the application using our proposed genetic algorithm.

Keywords: Augmented Execution, Computation Offloading, Mobile Cloud Computing, Partitioning.

1 Introduction

The mobile devices like smartphones are playing significant roles in human lives due to seamless features and productivity. The ease of user friendly interface, high definition graphics, customized application installment, compact size, portability, multi-card Subscriber Identity Modulefacility, multi-mode service operation, multi-band connectivity, high responsiveness, multimedia application support, etc features of mobile devices have attracted attentions of users worldwide. The demand of mobile phones is increasing in a tremendous way. A survey states that, in 2014more than1.1 billion phones will be shipped worldwide and is expected to rise over 1.5 billion in 2017 as predicted by International Data Corporation Market Research Company [1].

Smart Innovation, Systems and Technologies 28,
DOI: 10.1007/978-3-319-07350-7_49, © Springer International Publishing Switzerland 2014

In-spite of advancement in computer and communication technology, mobile devices can't reach full potential for the execution of high-end applications. These include computer vision applications, augmented reality applications, face and object recognition applications, natural language processing applications, file indexing in mobile system applications, virus scanning applications, optical character recognition applications, image and video processing applications, health monitoring applications, 3D gaming applications, etc. These applications require high processing speed, large memory and battery backup to execute efficiently and accurately. However, these application requirements are not fulfilled by mobile devices. Therefore, to ameliorate resource scarcity of mobile phones, computation offloading provides the best solution in mobile cloud computing. According to Wikipedia [2], mobile cloud computing is defined as *"The state-of-the-art mobile distributed computing paradigm comprising of three heterogeneous domains of mobile computing, cloud computing, and wireless networks that aims to enhance computational capabilities of resource-constrained mobile devices towards rich user experience."*

The computation offloading is the method of migrating entire or partial application tasks from mobile device to the resource-abundant and powerful cloud servers for execution bringing optimization in objective function.

From contribution point of view, we have proposed concurrent, scalable, fault-tolerant, distributed application processing framework supporting the dynamic partitioning approach for partitioning of massive application applied in mobile cloud computing paradigm. A genetic algorithm is devised to optimally partition the application providing run-time support.

The contents of the paper are portrayed as follows: The Section 2 deals with the literature survey of partitioning techniques. The system architecture and implementation details of the proposed method are discussed in Section 3. The design of optimal partitioning genetic algorithm is elaborated in Section 4. The paper is concluded in Section 5.

2 Related Work

This section deals with the types of partitioning methods adopted in various offloading environments. How the partitioning can be achieved is discussed along with pros and cons of each partitioning approach in our survey paper augmented execution in mobile cloud computing [3]. According to us, application partitioning is meant as the methodology of splitting the resource-consuming application into computationally intensive tasks (parts) with intention of execution on cloudservers having powerful resources. Generally, two types of partitioning are considered viz. static partitioning and dynamic partitioning. The static partitioning involves separation of computational intensive components at compile time statically [4], [5]. It results into fixed number of partitions. They are only suitable in conventional Internet computing. They don't consider varying load of CPU (Central Processing Unit), Network parameters and can't utilize elastic property of cloud resources. So, dynamic partitioning seems to be the best option for partitioning purpose.

Dynamic partitioning is a robust technique which partitions application at run time taking into account varying CPU load on mobile device and network parameters and utilizes elastic resources of cloud server [6], [7], [8], [9], [10].MAUI (Mobile Assistance Using Infrastructure) [6] uses managed code environment feature of .NET Common Language Runtime by Microsoft. The code portability feature allows MAUI to write the separate application code for both the smartphone and cloud server's CPUs exhibiting different instruction sets. MAUI suffers from scalability problem. Also programmer has to write the code for cloud as well as smartphone for running the framework, hence, increases the burden on programmer. CloneCloud [7] supports the thread level migration. When the massive application is launched, then depending upon the current available cloud resources and network conditions, partition file is generated which contains migration and integration points for methods. CloneCloud does not support on-demand allocation of resources. Agent-based application partition is implemented in [8]. The mobile agent encapsulates the application code and gets executed on any server platform on behalf of the owner. This approach lacks mobile agent authentication and so fake mobile agents can access the cloud service. In [9], the data stream application is represented as a dataflow graph representing the components. Adaptive execution of partitioning is supported which checks the CPU load, bandwidth at run time. Maximizing the throughput of data stream applications is the main objective function. ThinkAir [10] provides scalability in virtual machines at cloud side and supports parallel execution of tasks by assigning more virtual machines. But it does not give mechanism for efficient data transfer between mobile device and cloud. Like MAUI, it too increases coding overhead on programmer.

Calling the cloud [11] uses R-OSGi (Remote Services for OSGi-Open Source Gateway initiative) for module management and Alfred-O model [12] as a deployment tool. Bundles (Software modules) are offloaded to cloud for execution. Due to coarse level of migration granularity, bundle can expose to unauthorized access and security of data might get hampered. Dynamic partitioning best suits in offloading context which exploits the elastic resources of cloud and execute applications adaptive to changing mobile's and network parameters.

3 System Model

This section deals with the system architecture of the proposed approach and detailed discussion of implementation of the system.

a. System Architecture

The computation offloading scenario is depicted in Fig.1. Broadly speaking, the offloading scenario consists of 3 crucial components viz. mobile device, Internet wireless network and cloud server. The mobile device is connected to cloud using wireless technology like 3G, LTE (Long Term Evolution), or Wi-Fi (Wireless Fidelity) for accessing services from cloud. Due to mobility requirements of mobile device, mobile device runs location based services that consumes a lot of resources on it.

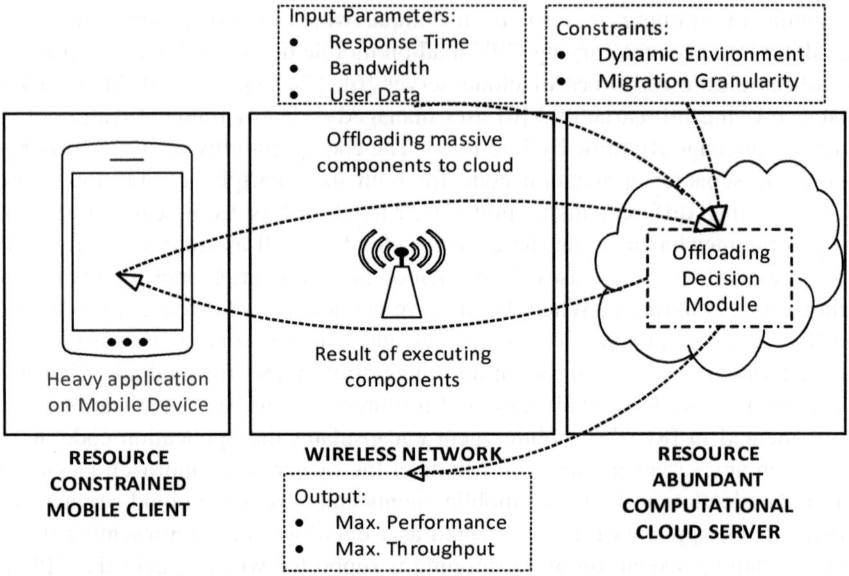

Fig. 1. Proposed System Architecture

The Internet bandwidth goes on fluctuating continuously. Therefore, mobile device is said to be resource-constrained by processor speed, memory, battery backup and bandwidth. On the contrary, cloud server has ample of processing speed, storage and power. So it is resource-abundant in nature.

When any massive application gets launched on mobile device, its execution speed can be accelerated by offloading the heavy components onto cloud. Due to this, the computation time is saved as the application's components are executed by powerful processors on cloud and hence performance of mobile device gets enhanced. But, the question arises which components should be offloaded to cloud so that the execution time of the application can be enhanced. This decision is taken by the offloading decision module deployed on cloud which runs the genetic algorithm put forth by us. This decision is taken on-the-fly considering continuously changing parameters like bandwidth and response time of executing the components under varying CPU load conditions.In this way, the proposed approach works on dynamic partitioning with the objective function of enhancing the throughput of application considering response time of component to execute the task.

The two costs are considered for partitioning viz. computation cost and communication cost. The computation cost constitutes the time required by processing the components on both the mobile device and the cloud servers. The communication cost is the cost required for transferring the data and control information between the mobile device and the cloud. The data required for executing the application is sent to the cloud. The data in the form of results is sent back to mobile device from cloud.

The control information constitutes the mobile phone number, name of component, etc is sent to cloud. This is explained in more detailed manner in section 3.2.

b. Implementation Details

The Fig.2. shows the application framework for augmented execution in mobile cloud computing paradigm. The mobile client side monitors the response time of the execution of components under varying CPU load and bandwidth.

When an application is launched on mobile device, then request from mobile device is sent to the *cloud controller* in cloud. The *cloud controller* assigns *node* for the mobile device and allocates the resources for *node*. The *node controller* acts as an interface for accessing the services offered by *node*and maintains the status of *node*. Initially, *node* asks the device characteristics (Response time) from mobile client.

Taking this information as an input, *node* generates partitioning result using our proposed genetic algorithm. Partitioning result states which components should be executed on cloud so that processing of application execution can be enhanced.

To achieve concurrent, fault tolerant, distributed and scalable framework, we have implemented the actor-based programming model provided by akka toolkit (version 2.2.3) [11] which works on the message-passing paradigm. Whenever message is sent to the actor-based model, the message is first accepted by actor's reference. All messages are queued up in the actor's *mailbox*. *Dispatcher* sends these messages to *Actor* for execution when *mailbox* dequeue message. After execution, *Actor* sends task execution notification to *dispatcher* to dequeue next message from *mailbox* for execution. After completing the whole procedure of execution, *actor* sends the task execution status to cloud controller. Akka toolkit frees the programmer from issues of resource management and managing scalability.

The communication between all components deployed on both the mobile and cloud is governed by HTTP (Hypertext Transfer Protocol). The node is responsible for managing the data and control flow from the mobile client. To uniquely identify which mobile is requesting for execution, *profiler* on the mobile device maintains and communicates following details to the *profiler* on node of cloud using a reliable Redis message queue [12].

- Mobile phone number to uniquely identify the mobile device onto cloud server
- Time required by component to execute task on the mobile device i.e. response time of component and Component name
- Name of application currently being executed by mobile device or cloud
- Current status of executing location of component i.e. where the component currently get executed (either mobile device or cloud)

When the mobile device connects with server, above information along with the task list is pushed in message queue. The *controller* on mobile device is responsible for maintaining the thresholds on performance measuring parameters mentioned.

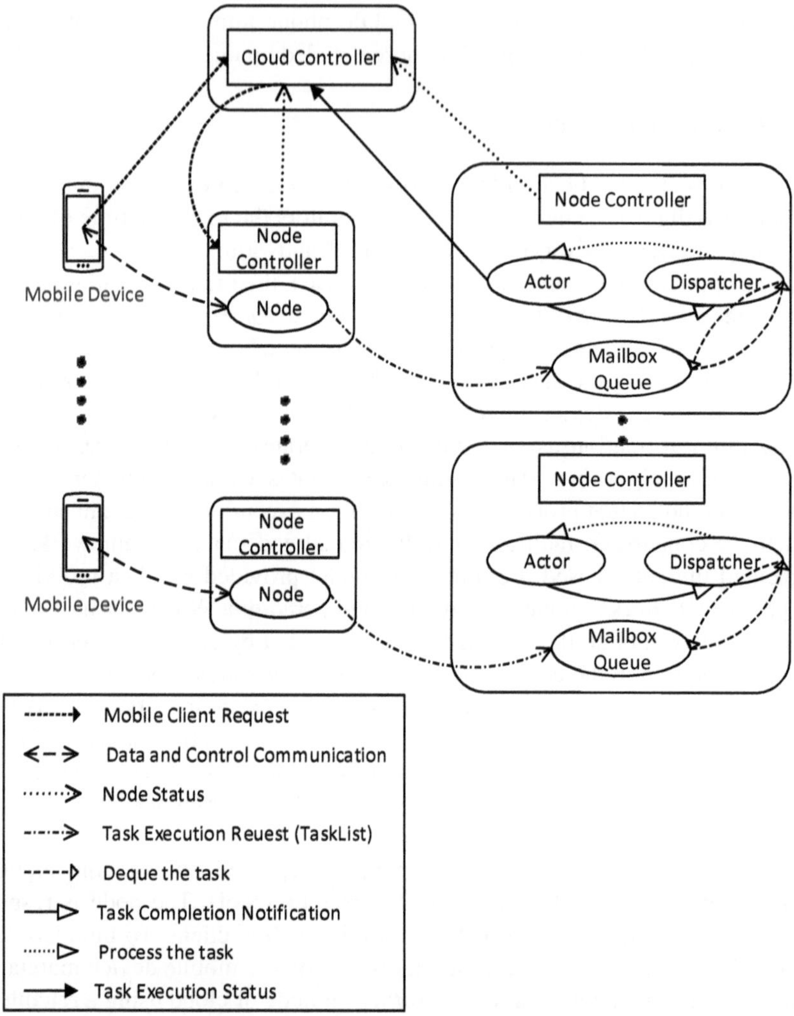

Fig. 2. Application Framework Overview

We have considered threshold as *"When 80% resources of performance measuring parameters are consumed then the request for modifying the partitioning result is sent to the solver present on node of cloud"*. The *solver* then runs genetic algorithm in an iterative manner to get updated partitioning so that the processing speed of an application can be maximized. The genetic algorithm takes response time of the components as an input parameter for calculating the throughput. The algorithm chooses the partitions giving maximized throughput. It is a robust algorithm and gives the best results from huge solution search space even though the input parameters are changing continuously.

We have deployed whole program (code) of the framework on cloud. So, for heavy components to execute on cloud, data required for running those components is offloaded to *executer* on cloud from mobile device. We have developed a framework using java. For achieving the partitioning, there should be loose coupling between the Java components/classes. To achieve the class level granularity, we have used Dependency Injection feature of spring framework (version 3.2.5) [13-15].

4 Design of Genetic Algorithm for Partitioning

For optimally partitioning the application we have designed genetic algorithm shown in Algorithm 1.The parameters used are denoted as:

- Components running on mobile device = $C_1, C_2,.,C_n$
- Total number of components running on mobile = n
- Virtual Machines available on cloud = $V_1, V_2,.,V_m$
- Virtual Machines consumed by cloud = $V_1', V_2',.,V_p'$
- Total number of virtual machines available on cloud = m
- Total number of virtual machines consumed by cloud= p
- The condition $p \bullet m$ must be followed
- Total mobile CPU usage = $Usage_{CPU}$
- Total mobile memory usage = $Usage_{MEM}$
- Threshold CPU = Th_{CPU}
- Threshold memory = Th_{MEM}
- Time at which input data is given to component $i$for processing = INC_i
- Time at which component i gives the output after processing data = $OUTC_i$
- Time in milliseconds required by component i to execute on mobile device = RET_i

The optimization problem is given as *"Select the massive components of a partitioned application such that their response time is maximum and offload them to cloud"*. The Performance in terms of maximum Response time is calculated as eq. 1.

$$Max.\ Response\ Time\ = RET_i = [OUTC_i - INC_i]_{max} \quad \text{for } i = 1,2,..,n \qquad (1)$$

We consider the partitions of an application as the population of individuals. The individuals which higher response time will be selected by the fitness function and offloaded to cloud for execution. Let us say, application is defined by the total set of components running as $\{C_1, C_2, C_3,.., C_n\}$.

From Genetic Algorithm point of view using binary encoding, we will denote these components as the bits 0 and 1. 0 is used for denoting the components executing on cloud and 1 is used for the components executing on mobile device. For example, if an application has 5 components. If the components C_1, C_2, C_3 executes on cloud while C_4 and C_5executes on mobile device, then the partition is represented as $\{0,0,0,1,1\}$ according to binary encoding.

Initially, we generate random population from the input population size *PS* (Step 1). We have applied elitism in our algorithm which facilitates to keep the individual with the best fitness in the population without undergoing mutation. *elitism* – boolean variable is used for checking the elitism. If elitism is to be applied (Step 3), then the best individual is obtained from the population (Step 4) and saved (Step 5). The flag *elitismOffset* is set to 1 to specify that the selected Best individual won't undergo mutation (Step 6); otherwise it is set to 0 (Step 9).

```
Algorithm: Optimally Partitioning Genetic Algorithm
Input: PS, TS, CR, MR
    PS: Population Size
    TS: Tournament Size
    CR: Crossover Rate
    MR: Mutation Rate
Output: Optimal partitioning result

1.  newPopulation• GenerateRandomPopulation(PS)
2.  while (Request_To_Stop ! = true) do
3.  if (elitism)
4.          Best_Individual•GetMaxFitness(newPopulation)
5.          newPopulation • SaveIndividual(Best_Individual)
6.          elitismOffset•1
7.      if end
8.  else
9.          elitismOffset• 0
10. else end
11. fori •elitismOffsettoi<PSdo
12.         indiv1   • TournamentSelection(newPopulation, TS);
13.         indiv2 • TournamentSelection(newPopulation, TS);
14.         newIndiv• UniformCrossover(indiv1, indiv2, CR);
15.         newPopulation • SaveIndividual(newIndiv);
16. end for
17. for i •elitismOffsettoi< Size(newPopulation) do
18.         newIndiv •Mutate(getIndividual[i], MR);
19. end for
20. Request_To_Stop •   CheckRequestStatus();
21. end while
22. RET• GetFitness(newIndiv)
23. k • GetMaxFitnessIndex(RET)
24. return newIndiv[k]
```

Algorithm 1. Genetic Algorithm for optimal partitioning

We have used Tournament Selection method for selecting the individuals from a population. It selects the best individual from the given population for crossover. For adjusting the selection pressure, the variable *Tournament Size (TS)* is chosen. The two individuals as *indiv1* and *indiv2* are selected (Step 12 and 13).

The Uniform crossover is applied on these individuals to get the *newIndiv*. We uniform Crossover rate *CR* is set as 0.5 which states that half of the genes in the new offspring come from the first individual *indiv1* and the remaining half genes from the second individual *indiv2*. The population is updated by adding the newly created individual *newIndiv*.

The mutation is applied on the individuals for maintaining the genetic diversity (Step 18). The individuals are again inserted into population for next iteration. The user request for running the Genetic Algorithm is checked (Step 20). As soon as the user experiences the fine tuned execution of application, the user stops sending the request for iterating the algorithm. The (partition) Individual having the maximum fitness is retuned as the optimal partitioning result.

5 Conclusion

The augmented execution in mobile cloud computing paradigm is a nascent technology to augment the capabilities of weaker mobile devices by utilizing the services of resource-rich and powerful cloud servers. A concurrent, scalable, fault-tolerant, distributed application processing framework supporting the dynamic partitioning approach for partitioning of massive application applied in mobile cloud computing paradigm is put forth in this paper. A genetic algorithm is proposed to partition the application in an optimal manner considering the response time as the performance metric.

The emphasis can be given on framework in which offloading decision will be taken by the operating system of mobile device will take offloading decision. Research is required to establish a single access platform for mobile cloud computing on the top of various operating systems platforms.

References

1. Global Smartphone Shipments, `http://www.statista.com/statistics/263441/global-smartphone-shipments-forecast/`
2. Mobile Cloud Computing, `http://en.wikipedia.org/wiki/Mobile_cloud_computing`
3. Jagtap, V.S., Pawar, K.V., Pathak, A.R.: Augmented Execution in Mobile Cloud Computing: A Survey. In: Proceedings of International Conference on Electronic Systems, Signal Processing and Computing Technologies, pp. 237–244 (2014)
4. Chun, B., Maniatis, P.: Augmented smartphone applications through clone cloud execution. In: IEEE 8th Workshop onHotOS (2009)
5. Dou, A., Kalogeraki, V., Gunopulos, D., Mielikainen, T., Tuulos, V.: Misco: a mapreduce framework for mobile systems. In: PETRA (2010)
6. Cuervoy, E., Balasubramanianz, A., Cho, D.: MAUI: Making smartphones Last Longer with Code Offload. In: MobiSys 2010 (2010)
7. Chun, B., Ihm, S., Maniatis, P., Naik, M., Patti, A.: CloneCloud: Elastic Execution between Mobile Device and Cloud. In: ACM Workshop EuroSys, pp. 301–314 (2011)

8. Angin, P., Bhargava, B.: An Agent-based Optimization Framework for Mobile-Cloud Computing. Journal of Wireless Mobile Networks, Ubiquitous Computing, and Dependable Applications 4(2), 1–17
9. Yang, L., Cao, J., Tang, S., Li, T., Chan, A.T.S.: A Framework for Partitioning and Execution of Data Stream Applications in Mobile Cloud Computing. In: IEEE Fifth International Conference on Cloud Computing, pp. 794–802 (2012)
10. Kosta, S., Aucinas, A., Hui, P., Mortier, R., Zhang, X.: ThinkAir: Dynamic resource allocation and parallel execution in the cloud for mobile code offloading. In: IEEE INFOCOM, pp. 945–953 (2012)
11. Giurgiu, I., Riva, O., Juric, D., Krivulev, I., Alonso, G.: Calling the cloud: Enabling mobile phones as interfaces tocloud applications. In: Proceedings of the ACM 10th International Conference on Middleware, pp. 83–102 (2009)
12. Rellermeyer, J.S., Riva, O., Alonso, G.: AlfredO: An Architecture for Flexible Interaction with Electronic Devices. In: Issarny, V., Schantz, R. (eds.) Middleware 2008. LNCS, vol. 5346, pp. 22–41. Springer, Heidelberg (2008)
13. AkkaToolkit, http://akka.io/
14. Redis, http://redis.io/
15. Spring Framework, http://projects.spring.io/spring-framework/

Implementation of Symmetric Functions Using Quantum Dot Cellular Automata

Subhashree Basu[1], Debesh K. Das[2], and Subarna Bhattacharjee[3]

[1,3] Computer Science and Engineering Department,
St. Thomas College of Engineering and Technology, Kolkata - 700023, India
subhashreebasu1984@yahoo.co.in, neel bee@yahoo.com
[2] Computer Science and Engineering Department, Jadavpur University,
Kolkata - 700032, India
debeshd@hotmail.com

Abstract. VLSI technology has made possible the integration of massive number of components into a single chip with the minimum power dissipation. But concerned by the wall that Moore's law is expected to hit in the next decade, the integrated circuit community is turning to emerging nano-technologies for continued device improvements. Quantum dot cellular automata(QCA) is a technology which has the potential of faster speed, smaller size and minimum power consumption compared to transistor –based technology. In quantum dot cellular automata, the basic elements are simple cells. Each quantum cell contains two electrons which interact via Coulomb forces with neighboring cells. The charge distribution in each cell tends to align along one of two perpendicular axes, which allows the encoding of binary information using the state of the cell. These cells are used as building blocks to construct gates and wires. This paper utilizes these unique features of QCA to simulate symmetric functions. A general equation for the minimum number of gates required to an arbitrary number of input variables causing synthesis of symmetric function is achieved. Finally a general expression for the number of gates in benchmark circuits is also deduced. It provides significant reduction in hardware cost and switching delay compared to other existing techniques.

Keywords: Quantum dot cellular automata, symmetric functions, Coulomb force.

1 Introduction

Silicon technology has experienced an exponential improvement in virtually any figure of merit, following Gordon Moore's famous dictum remarkably closely for more than three decades. However, there are indications now that these progresses may slowdown in future. Among the chief technological limitations responsible for this expected slowdown of silicon technology are the interconnect problem and power dissipation. This slow-down of silicon ULSI technology is providing an opportunity for alternative device technologies. QCA [2, 4-6, 10-12, 14-17, 19, 22-24] can be an

M.K. Kundu et al. (eds.), *Advanced Computing, Networking and Informatics - Volume 2*, Smart Innovation, Systems and Technologies 28,
DOI: 10.1007/978-3-319-07350-7_50, © Springer International Publishing Switzerland 2014

interesting alternative. Since QCAs were introduced in 1993, several experimental devices have been developed. Although they may not be ready for commercial uses, recent papers show that QCAs may eventually achieve high density, fast switching speed and room temperature operation.

Synthesis of Boolean functions by using QCA has drawn attention of several researchers. This paper proposes a design to synthesize the symmetric Boolean functions using QCA. Many computing, control and communication circuits are described by symmetric functions. Symmetric functions have huge applications in cryptography, VLSI and nanotechnology designs. Symmetric functions have been studied extensively in digital logic. Realization of symmetric functions by quantum reversible gates was described in [1, 3, 7-9, 13, 16, 18, 20-21, 25]. Although the realization of symmetric functions using other approaches have already been described in many works, the domain of synthesizing symmetric functions using QCA gates have been explored only in few instances [16,24].In these cases, the synthesis of symmetric function is done by digital logic gates (AND, OR, NAND etc) and these gates are implemented by QCA. It is generally a common approach in synthesizing Boolean functions by QCA gates to represent the function by basic digital logic gates while these gates are realized by QCA gates. We believe that these approaches fail to take the inherent advantages of the basic building block of QCA majority gate. In this paper, our synthesis approach starts with the building blocks of QCA majority gate. We first produce a design for unate symmetric functions. Then this design is extended to produce the designs for all symmetric functions using basic QCA majority gates.

This paper is organized as follows. In Section 2, the background of QCA technology is presented. Section 3 gives a brief overview of symmetric functions and the design and realization of symmetric function. Simulation results are shown in section 4 followed by conclusion in Section 5.

2 QCA Design Schemes

The quantum dot cellular automata [2,4-6,10-12,14-19,22-24] involves using a binary representation of information, by replacing the current switch with a cell having a bistable charge configuration. One configuration of charge represents a binary “1”, the other a "0," but no current flows into or out of the cell. The field from the charge configuration of one cell alters the charge configuration of the next cell. Remarkably, this basic device-device interaction is sufficient to support general purpose computing with very low power dissipation.

2.1 QCA Cell

The essential feature of a QCA cell is that it possesses an electric quadrupole which has two stable orientations as seen in Fig. 1. These two orientations are used to represent the two binary digits, "1" and "0." The simplest implementation of this is a 4-dot cell composed of dots at the corners of a square, with two mobile charges. The two quadrupole states then correspond simply to the two ways of occupying antipodal dots. The bit information is contained in the sign of the in-plane quadrupole moment. The polarization *P* of the cell is the normalized quadrupole moment. If there are no other fields, either from the environment or from the neighboring cells, then the two orientations of the quadrupole moment have the same electrostatic energy. The presence of nearby cells causes one arrangement to be the favored low-energy configuration.

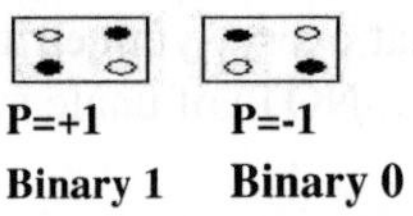

Fig. 1. Basic QCA cell and Two Possible Polarizations

2.2 QCA Wires

A series of QCA cells acts like a wire as shown in Fig. 2. During each clock cycle, half of the wire is active for signal propagation, while the other half is unpolarized. During the next clock cycle, half of the previous active clock zone is deactivated and the remaining active zone cells trigger the newly activated cells to be polarized. Thus, signals propagate from one clock zone to the next.

The circuit area is divided into four sections and they are driven by four phase clock signals. In each zone, the clock signal has four states: high-to-low, low, low-to-high, and high. The cell begins computing during the high-to-low state and holds the value during the low state. The cell is released when the clock is in the low-to-high state and inactive during the high state.

Fig. 2. QCA wires

2.3 QCA Gates

Logic gates are required to implement arithmetic and digital circuits. In QCA, inverter and Majority voter gates act as the basic gates. Fig. 3 and Fig. 4 show the gate symbols and their layouts. The governing equation of the Majority voter gate is

$$M(A,B,C,D,E)=ABC+ABD+ABE+ACD+ACE+ADE+BCD+BCE+BDE+CDE.$$

Three input AND and OR gates can be implemented using three input majority voter gate by making one input constant.

$M(A, B, C, 0, 0) = ABC$
$M(A, B, C, 1, 1) = A+B+C$

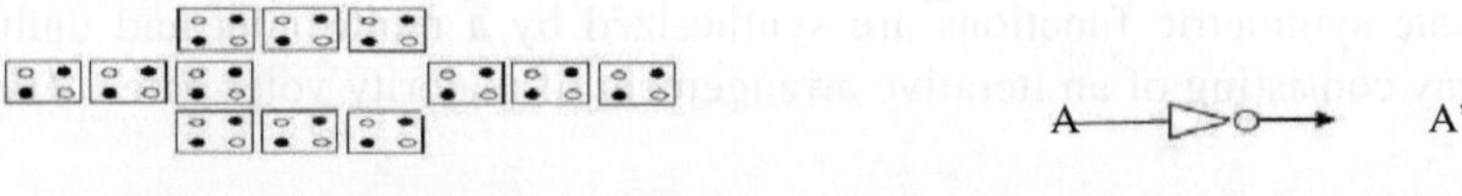

Fig. 3. Implementation of NOT Gate with QCA gates

Fig. 4. QCA Majority voter gate

In this paper we show that every symmetric function can be expressed as a logical composition (e.g., AND, OR, NOT) of unate symmetric functions.

3 Symmetric Functions

3.1 Preliminaries

Let $f(x_1,x_2,\ldots,x_n)$ denote a switching function of n Boolean variables. A *vertex (minterm)* is a product of variables in which every variable appear once. The *weight w* of a vertex *v* is the number of uncomplemented variables that appear in *v*. A Boolean function is called *negative (positive) unate*, if each variables appear in *complemented (uncomplemented)* form (but not both) in its minimum sum-of-products(s-o-p) expressions.

A switching function $f(x_1, x_2,\ldots,x_n)$[13] is called *totally symmetric* with respect to the variables $(x_1,x_2,\ldots x_n)$, if it is invariant under any permutation of the variables. Henceforth, by a symmetric function, we would mean a function with total symmetry. An n-variable symmetric function is denoted as $S^n_{(ai,\ldots,aj,\ldots ak)}$..

A symmetric function is called consecutive, if the set A consists of only consecutive integers $(a_l, a_{l+1},\ldots a_r)$.such a consecutive symmetric function is expressed by $S^n(a_l \quad a_r)$ where $l<r$. For n variables, we can construct 2^{n+1}-2 different symmetric functions (excluding constant functions 0 and 1).A totally symmetric function $S^n(A)$ can be expressed uniquely as a union of maximal consecutive symmetric functions such that $S^n(A)=S^n(A_1)+S^n(A_2)+\ldots+S^n(A_m)$, such that m is minimum and $i,j,1\leq i,j\leq m$, $A_i \cap A_j=0$ whenever $i\neq j$.

A function is called unate symmetric [18] if it is both unate and symmetric. A unate symmetric function is always consecutive and can be expressed as $S^n(a_l \quad a_r)$ where either $a_l=0$ or $a_r=n$.

3.2 Design Approach for Unate Symmetric Functions

Unate symmetric functions are synthesized by a multi-input and multi-output logic array consisting of an iterative arrangement of majority voter gates. There are n input lines $x_1,x_2,\ldots x_i, 1\leq i\leq n$ and n output lines $u_1,u_2,\ldots u_j$,$i \leq j \leq n$. The network is so designed that the output lines u_j assumes value 1 if there are 1s in at least j inputs. The output functions are as follows.

$u_1(n)=S^n_{(1,2,3,\ldots,n)}=\sum x_i$ for i =1 to n;

$u_2(n) = S^n_{(2,3,\ldots,n)}= \sum x_i x_j$ for i, j =1 to n;

……………………………………………………

$u_n(n)= x_i x_j x_k \ldots x_n$

For example if $n=5$, there will be 5 output functions i.e. $u_1(5),u_2(5),u_3(5),u_4(5)$ and $u_5(5)$.According to our design it is seen that n majority voter gates are required to get these n outputs.

Example 1

If n=5 then a 9 input lined majority gate with 4 input lines equal to 0 will give $u_5(5)$ and 4 input lines equal to 1 will give $u_1(5)$. Hence to get $u_1(5)$ and $u_5(5)$ we need two 9 input lined majority gates. A 7 input lined majority gate with 2 input lines equal to 0 will give $u_4(5)$ and 2 input lines equal to 1 will give $u_2(5)$. Hence to get $u_2(5)$ and $u_4(5)$,we need two 7 input lined majority gates. $u_3(5)$ can be derived by one 5 input lined majority gate.

Example 2

If n=6 then a 11 input lined majority gate with 5 input lines equal to 0 will give $u_6(6)$ and 5 input lines equal to 1 will give $u_1(6)$. Hence to get $u_1(6)$ and $u_6(6)$ we need two 11 input lined majority gates. A 9 input lined majority gate with 3 input lines equal to 0 will give $u_5(6)$ and 3 input lines equal to 1 will give $u_2(6)$. Hence to get $u_2(6)$ and $u_5(6)$,we need two 9 input lined majority gates. A 7 input lined majority gate with 1 input line equal to 0 will give $u_4(6)$ and 1 input line equal to 1 will give $u_3(6)$.Hence to get $u_3(6)$ and $u_4(6)$,we need two 7 input lined majority gates.

Lemma1. If there are n input lines $x_1,x_2,\ldots x_n$,realization of the output functions $u_1,u_2,\ldots u_n$ need n number of majority gates which are arranged in the following manner.

If n is odd then one is *n input lined* majority gate, two are *(2n-1) input lined* majority gates, two are *(2n-3) input lined* majority gates, two are *(2n-5) input lined* majority gates up to *(n-1)* number of gates.

If n is even then two are *(2n-1) input lined* majority gates, two are *(2n-3) input lined* majority gates, two are (*2n-5) input lined* majority gates up to *(n+1) input lined* majority gates.

Fig. 5(a) and Fig. 5 (b) shows the schematic and layout diagram of the circuit to synthesize unate symmetric functions for 5 inputs.

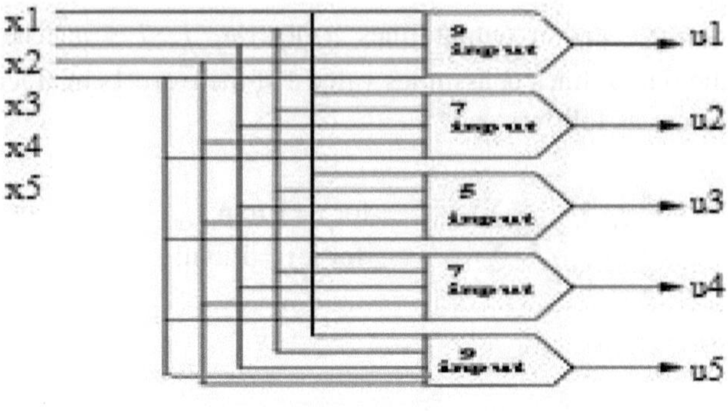

(a)Schematic

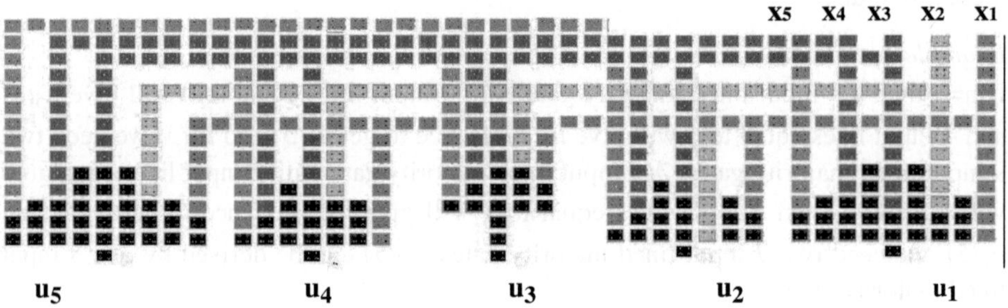

(b)Layout

Fig. 5. Synthesis of Unate Symmetric Functions for 5 Inputs Using Majority Voter Gates: (a) Schematic (b)Layout

3.3 Design Approach for General Symmetric Functions

Consecutive Symmetric Functions

A consecutive symmetric function consists of only consecutive integers (a_l , a_{l+1},.. . , a_r). Such a consecutive symmetric function is expressed by $S^n(a_l - a_r)$ where $l < r$. A consecutive symmetric function $S^n(a_l \;\; a_r)$, $a_l \neq a_r$, $l < r$, can be expressed as a composition of two unate symmetric functions. For example $S^5_{(2,3)}$ is realized as $S^5_{(2\text{-}5)}$. $S^5_{(4-5)} = u_2\,(5).\ u_4\,(5)$.Fig. 6 shows the schematic of the circuit to synthesize consecutive symmetric functions for 5 inputs. To synthesize a consecutive symmetric function that is not unate, we use the result in the previous module.

Nonconsecutive Symmetric Functions

To synthesize a non consecutive symmetric function, it is first expressed as a union of several maximal consecutive symmetric functions, and then each of the constituent consecutive symmetric functions is realized by combining the appropriate outputs of unate decomposition. Finally they are OR-ed together. For example $S^5_{(1,3,5)}$ is realized as

$$S^5_{(1)} + S^5_{(3)} + S^5_{(5)} = u_1(5).\ u_2(5) + u_3(5).\ u_4(5) + u_5(5)$$ as shown in Fig. 7.

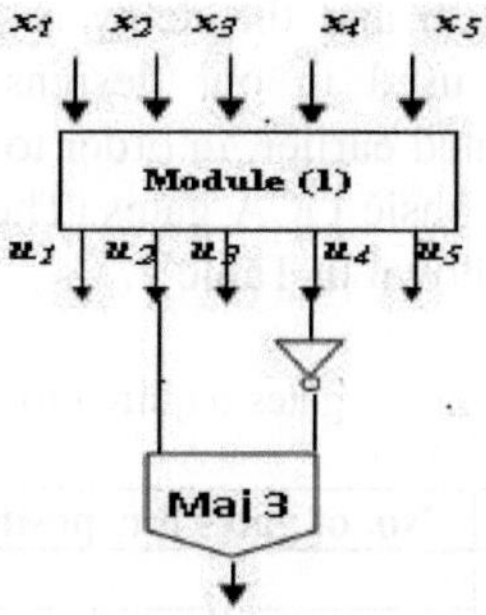

Fig. 6. Synthesis of Consecutive Symmetric functions

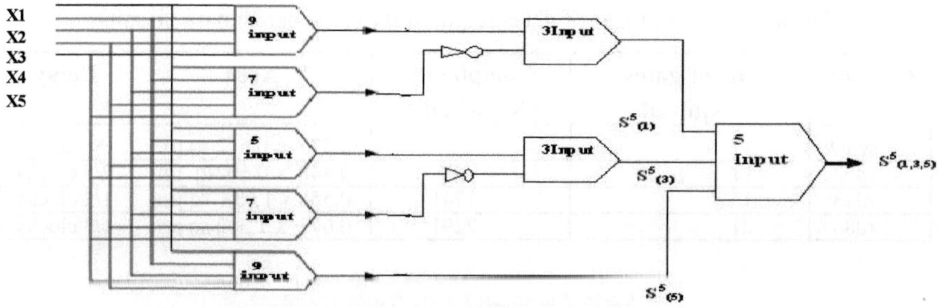

Fig. 7. Synthesis of Non-Consecutive Symmetric functions

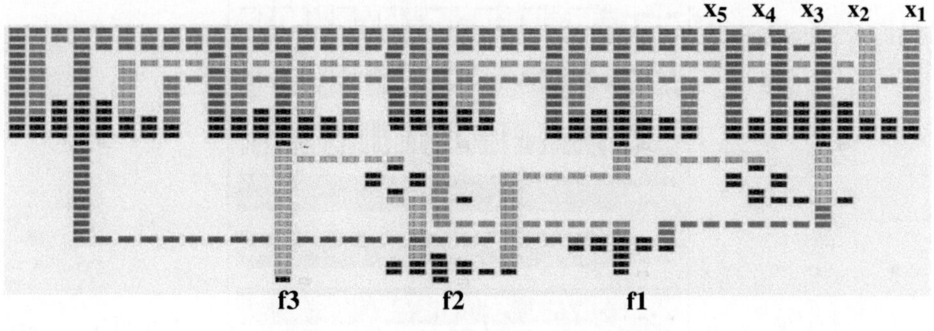

Fig. 8. Synthesis of Benchmark Circuit rd53

4 Results

Simulation Result

The layout of an rd53 circuit is demonstrated in Fig. 8.Simulation is performed using QCADesigner. The simulation result of rd53 is shown in Fig. 9. The circuit requires an area of 0.448 μm x 0.602 μm and the delay is *2/5 clocks.* Table 1 shows the numbers of basic QCA gates used in our designs to realize all positive unate symmetric functions. As mentioned earlier, in order to realize a circuit with *n* number of inputs, we need *n* number of basic QCA gates. The complexity, size and delay of various benchmark circuits are shown in Table 2.

Table 1. Comparative study of the no. of gates required for synthesis of symmetric functions

No. of inputs	No. of outputs	No. of gates for positive unate symmetric functions
2	6	2
3	14	3
n	2^{n+1}-2	n

Table 2. Total number of gates required for some benchmark circuits

Circuits	No. of gates required	Complexity (No. ofcells)	Area	Delay
sym9	11	523	0.576 x 0.612 sq μm	¾ clocks
rd53	11	595	0.448 x 0.602sq μm	2/5 clocks
rd73	15	1741	0.558 x 1.638 sq μm	1/2 clocks
rd84	21	2291	0.672 x 1.302 sq μm	1/3 clocks

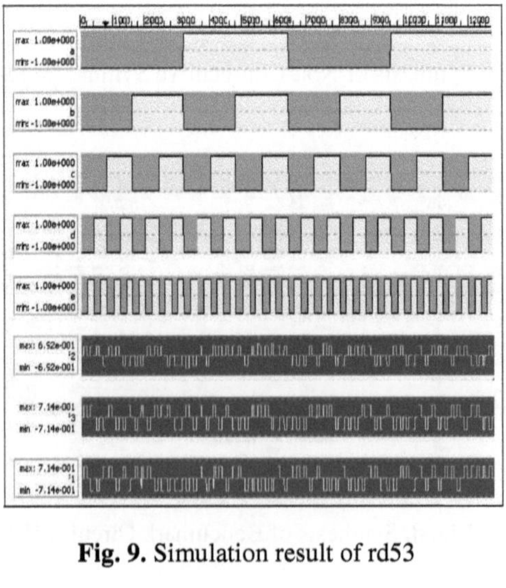

Fig. 9. Simulation result of rd53

5 Conclusion

This paper proposes the implementation of symmetric functions using QCA designs. The proposed procedure for implementing symmetric functions exploits the strength of the inherent characteristics of QCA majority gates. We first design the all positive unate symmetric functions using our technique. To synthesize any non-unate symmetric functions, we use the unate decomposition approach. Although QCA solves most of the limitations of silicon VLSI technology, the actual speed may be in the range of megahertz for solid state QCA instead of in the range of terahertz as suggested by many researches. Additionally, solid-state QCA devices cannot operate in room temperature. The only alternative to this temperature limitation is the recently proposed "Molecular QCA". Molecular QCA is also considered to be the only feasible implementation method for mass production of QCA devices.

References

1. JaJa, Wu, S.M.: A new approach to realize partially symmetric functions. Tech. Rep. SRCTR86-54, Dept. EE, University of Maryland (1986)
2. Lent, C.S., Taugaw, P.D., Porod, W., Berstein, G.H.: Quantum Cellular Automata. Nanotechnology 4(1), 49–57 (1993)
3. Picton, P.: Modified Fredkin gates in logic design. Microelectronics Journal 25, 437–441 (1994)
4. Orlov, A.O., Amlani, I., Bernstein, G.H., Lent, C.S., Sinder, G.L.: Realization of a Functional Cell for Quantum Dot Cellular Automata. Science 277(5328), 928–930 (1997)
5. Lent, C.S., Taugaw, P.D.: A Device Architecture for Computing with Quantum Dots. Proceedings IEEE 85(4), 541–557 (1997)
6. Amlani, I., Orlov, A.O., Toth, G., Lent, C.S., Bernstein, G.H., Sinder, G.L.: Digital Logic Gate using Quantum Dot Cellular Automata. Science 284(5412), 289–291 (1999)
7. Yanushekvich, S.N., Butler, J.T., Dueck, G.W., Shmerko, V.P.: Experiments on FPRM expressions for partially symmetric functions. In: Proc. of the 30th IEEE International Symposium on Multiple Valued logic, pp. 141–146 (2000)
8. Perkowski, M., Kerntopf, P., Buller, A., Chrzanowska Jeske, M., Mishchenko, A., Song, X., AlRabadi, A., Jo Zwiak, L., Coppola, A., Massey, B.: Regularity and symmetry as a base for efficient realization of reversible logic circuits. In: IWLS, pp. 245–252 (2001)
9. Perkowski, M., Kerntopf, P., Buller, A., ChrzanowskaJeske, M., Mishchenko, A., Song, X., AlRabadi, A., Jo Zwiak, L., Coppola, A., Massey, B.: Regular realization of symmetric functions using reversible logic. In: EUROMICRO Symp. on Digital Systems Design, pp. 245–252 (2001)
10. Lieberman, M., Chellamma, S., Varughese, B., Wang, Y., Lent, C.S., Bernstein, G.H., Snider, G.L., Peiris, F.: Quantum Dot Cellular Automata at a Molecular Scale. Annals of the New York Academy of Sciences 960, 225–239 (2002)
11. Zhang, R., Walus, K., Wang, W., Jullien, G.A.: A Method of Majority Logic Reduction for Quantum Cellular Automata. IEEE Trans. on Nanotechnology 3(4), 443–450 (2004)
12. Walus, K., Schulhof, G., Jullien, G.A., Zhang, R., Wang, W.: Circuit Design Based on Majority Gates for Application with Quantum Dot Cellular Automata. IEEE Trans. Signals, Systems and Computers 2, 1354–1357 (2004)

13. Rahaman, H., Das, D.K., Bhattacharya, B.B.: Implementing Symmetric Functions with Hierarchical Modules for Stuck-At and Path-Delay Fault Testability. Journal of Electronic Testing 22(2), 125–142 (2006)
14. Momenzadeh, M., Tahoori, M.B., Huang, J., Lombardi, F.: Characterization, Test and Logic Synthesis of AND-ORINVERTER (AOI) Gate Design for QCA Implementation. IEEE Trans. on Computer Aided Design of Integrated Circuits and Systems 24, 1881–1893 (2005)
15. Townsend, W.J., Abraham, J.A.: Complex Gate Implementations for Quantum Dot Cellular Automata. In: 4th IEEE Conference on Nanotechnology, pp. 625–627 (2004)
16. Rahaman, H., Sikdar, B.K., Das, D.K.: Synthesis of Symmetric Boolean Functions Using Quantum Cellular Automata. In: International Conference on Design & Test of Integrated Systems in Nanoscale Technology (DTIS 2006), Tunis, Tunisia, pp. 119–124 (2006)
17. Xu, Z.Y., Fenga, M.W., Zhang, M.: Universal Quantum Computation With Quantum-Dot Cellular Automata in Decoherence-Free Subspace. Quantum Information and Computation, pp. 0000–000c. Rinton Press (2008)
18. Azghadi, M.R., Kavehei, O., Navi, K.: A Novel Design for Quantum-dot Cellular Automata Cells and Full Adders
19. Maslov, D.: Efficient reversible and quantum implementations of symmetric Boolean functions. Circuits, Devices and Systems, IEEE Proc. 153(5), 467–472 (2006)
20. Cho, H.: Adder Designs and Analyses for Quantum-Dot Cellular Automata. IEEE Transactions on Nanotechnology 6(3) (2007)
21. Keren, O., Levin, I., Stankovic, S.R.: Use of gray decoding for implementation of symmetric functions. In: International Conference on VLSI, pp. 25–30 (2007)
22. Lauradoux, C., Videau, M.: Matriochka symmetric Boolean functions. In: IEEE ISIT, pp. 1631–1635 (2008)
23. Géza, T., Lent, C.S.: Quantum computing with quantum-dot cellular automata. Physical Review A 63, 052315
24. Bhattacharjee, P.K.: Use of Symmetric Functions Designed by QCA Gates for Next Generation IC. International Journal of Computer Theory and Engineering 2, 1793–8201 (2010)
25. Jagarlamudi, H.S., Saha, M., Jagarlamudi, P.K.: Quantum Dot Cellular Automata Based Effective Design of Combinational and Sequential Logical Structures. World Academy of Science, Engineering and Technology 60 (2011)

Dynamic Page Replacement at the Cache Memory for the Video on Demand Server

Soumen Kanrar[1,2] and Niranjan Kumar Mandal[2]

[1] Vehere Interactive, Calcutta-53, WB, India

[2] Vidyasagar University, Computer Science, Midnapore-02, WB, India

soumen.kanrar@veheretech.com, niranganmandal54@gmail.com

Abstract. The audio/video stream retrieves from the storage server depends upon the cache refreshment polices. The replacement policy depends upon the efficiency of handle the cache hit ratio and cache miss ratio at real time. The cache size is limited with compare to the auxiliary memory size, and it is only the fraction of the auxiliary memory. The cache memory maintains two blocks one for the Least Recently Used (LRU) and other for the Least Frequency Used (LFU). The Least Recently Frequency used (LRFU) pages store into the cache memory. Since the size of the cache is limited by using the exponential smoothing parameter, dynamically the cache replaces the page with the smallest hit count from the LRU. The request page from the submitted request stream increment the hit counts for the already listed pages. In this paper, we present the LRFU polices and the impact of that polices for the limited cache size with a huge submitted stream of requests in a very small interval of time.

Keywords: Cache Memory, LRFU, LRU, LFU, Replacement policies.

1 Introduction

The rapid modification in the area of computer hardware brings new demand for adaptive polices to handling massive request stream (Audio/video) in a very short interval of time. The huge amount of submitted requests to the (Audio/Video) server is approximated by Zipf like distribution [7], [11] in a better way. The performance of the system highly depends upon the response from the server [10], [8]. The enhancement of the storage portal depends on the handling of the cache memory [1], [4], [5], which itself is a fraction of the main storage capacity of the portal. The researcher world gave various types of concepts over the last decreed to handle the cache memory for distributed multi-tier computing. Due to the growth of the Internet, smart mobile phone, on demand audio/ video user grows exponentially. It has been observed that the Major traffic load experience at the 'audio/video' portal for a particular session. Multi queues' replacement polices at second level buffer cache [4] handle the stream of requests. Due to proper handling polices, server cache experiences poor hit ratio [2-3]. Recent trend to indexing the cache pages in the real time. The improvement over the cache performance focuses on changing afresh a reference pattern [6-7]. The page conflicts arise in the mapping, but the main problem

M.K. Kundu et al. (eds.), *Advanced Computing, Networking and Informatics - Volume 2*, Smart Innovation, Systems and Technologies 28,
DOI: 10.1007/978-3-319-07350-7_51, © Springer International Publishing Switzerland 2014

on the static mapping to calculate the average page conflicts by binomial distribution of the 'a' of 'A' pages fall in one of the B bins for the N size cache memory . B=N/A where A is the associative. Multiprocessor memory hierarchy used to optimize the page replacement in real time [7]. The paper is organized in this way. Section 1 describes the introduction and paper survey. Section 2 is the explanation about the problem. Section 3 presents the algorithm. Section 4 is an explanation for the simulation results and discussion. Section 5 discusses about the improvement and reference.

2 Problem Description

In this work, we study the real time page replacement by LRFU (Least Recently Frequency Used). It initially used the statistical decay with the count of page hits. Here we maintain LRU and LFU. The Least Recently Used (LRU) concepts have been widely employed for buffer cache management [9], [6]. LRU concepts used to replace the unused block for longer time. The longer time is depended upon the system hardware.

The identified blocks which have been referenced in the recent past will likely be referenced again soon. However, for the multilevel buffer caches, this observation is no longer present, or it existence is too less. Due to that reason, only LRU does not perform well for file server caches. These are also true or other workloads such as database back-end storage servers. The time complexity of this procedure is constant. The Least Frequently Used (LFU) concept is the replacement policy. It replaces the block that is least frequently used. This concepts present that some blocks are accessed more frequently compare to other blocks, so that the frequency of reference can be used as an estimate of the probability of a block being referenced within a time span. The LFU usually performs better than the original LRU because the former gives different weight to the recent references and very old references. The time complexity of this procedure is O ($log(n)$).The Least Frequently Recently Used (LFRU) concept was proposed by Lee *et al.* in 1999-2000 to Cover a spectrum of replacement algorithms that include LRU at one end and LFU at the other [8]. It enforced to replace those particular blocks that are the least frequently referred and not recently referred within a specified time span. It associates a combined value of the occurring immediately before the present and frequency of referred of that memory block.

The goal is to replace the block with the minimum value. Each reference to a block contributes to its value. A reference's contribution is determined by a weighting function within a particular the time span from the reference to the current time. By changing the parameters for the weighting function, LFRU can implement either LRU or LFU. The time complexity of this algorithm is between O(1) and O(log (n)). LRFU policy combines an effect of LRU and LFU. Initially, LRFU policy assigns a value Hit_count(x) = 0 to every page x and that value be updated every instance of time. The analytic form is

Hit_count (x) = $1 + 2^{-\lambda} \times$(Hit_count) if page (x) is referenced at time Δt

= $2^{-\lambda} \times$ (Hit_count) else (x) is not referenced at time Δt

λ is an integer variable that controls the behavior of the process flow.

LRFU policy replaces the page with the smallest Hit_count. If $\lambda \rightarrow 0$ then the Hit-count simply incremented with 1 i.e. the result is the frequency of appearance of that page during a specified interval of time. Then clearly LRFU converges to LFU. Otherwise, for $\lambda \rightarrow 1$, here λ works as an exponential decay. Hit_count occurs at a time immediately before the present. LRFU works as LRU. The performance of page replacement depends upon the value of $\lambda \in (0,1)$.

3 Algorithm Development

3.1 Initial Assumption

The processing of jobs depends on the system hardware configuration.

Buffer 1 : Listed up of all the Video /Audio file names that currently exists in the cache memory.
Buffer 2 : Start at t, storing incoming submitted request. At $t + \Delta t$ buffer is flash out.
Hit Count: Store the frequency of referenced 'Audio/Video pages' with respect to the absolute index position in the buffer1.
L1: store the least recently used pages "with reference high hit count" **LRU**
L2: store the least frequency used pages **LFU**

Size of L $\leftarrow$ C

$$L = L_1 \cup L_2$$

$$0 \leq L_1 \leq C$$

$$0 \leq L_2 \leq C$$

$$0 \leq L \leq C$$

```
Procedure: Eliminate repeated pages in the availability
list
```

```
Initialize the name of the Audio/Video file to the
Biffer1 (manually insert)
// check the listed file name are without repetition
  for i, j∈I≥0
  if (i≠j)
            Buff1 [i]≠ Buff1[j]
End
```

```
Procedure: Store all submitted requests to the proxy
server.

Start at time t
// C is the size of the Cache
// request signal for submitted request string
// Buffer2 is the Intermediate Storage
Var
Buffer2:  array [1, 2, 3,…, n] of string
// Buffer2[1][i] ← String, Buffer2[2][i]← Boolean
// size of buffer2 is A≻≻C
Int: counter, trial, Hit_count
Request: Boolean
While (request == true) do
      If (counter!= A)
      {
         Buffer2 [1][counter] <- load the request
         Buffer2 [2][counter] =  False
         Counter ++
      }
      Else
      // No more request be accepted, request discarded
End
```

```
Procedure:  Hit count
Var
trial, index, counter : integer
hit_count  :  array[1,….,C]  of integer
Buffer2 : array [1,2,1,…n] of string
// Buffer2[1][i]←String, Buffer2[2][i]←Boolean
   Index = 0
   While(index<= A)
    for     trial   = 1 to C do
     counter = 0;
     if (Buff1[trial==Buffer2[1][index])
     {
      counter++
      Buffer2 [2][index] = True
      Hit_count[trial]=counter+2^-λ× Hit_count[trial]
      L1← L1 ⋃ {Buff1[trial]}
     }
     //λ→0 implies LRFU→LFU and λ→1 implies LRFU→ LRU
     else
     {
```

```
      Hit_count [trial] = 2^-λ × Hit_count [trial]
      L2← L2 ∪ {Buff1[trial]}
      }
    End for
   End  while
At   time   t+Δt   flush the buffer2
// Ready for the next session to accept the submitted
request
End
```

```
Procedure:   New page request to be loaded to L2 by
replacing the existing page from LFU (Least Frequency
Used)
```

```
 At time t+2Δt
 Index=0;
 While (index <= A)
  if (Buffer2 [2][index] == True ) then
  {
   for i = 1  to A  do  // check the repetition
   if (hash(Buffer2[2][index]==hash(Buffer2[2][index+ i])
    Buffer2[2][index] = False
   End for
  }
  else continue
 End While
 While (index <= A)
  if  (Buffer2 [2][index] == True )
   Minimum {Hit_count (L2)} ← Buffer2[2][index]
   Hit_count (of the page in L2)++
 End While
End
```

4 Result and Discussion

The simulation environment is built up with the limited cache memory size 20 Mb The L1 and L2 portion block size varies within the limit zero Mb to 20 Mb, but the total value is 20 Mb.

The value of λ varies within the limit (0, 1). We consider the size of the 'stream of heterogeneous request pattern' varies in the range [10 to 1000]. We run the simulation by considering the variable of size of the requested sample.

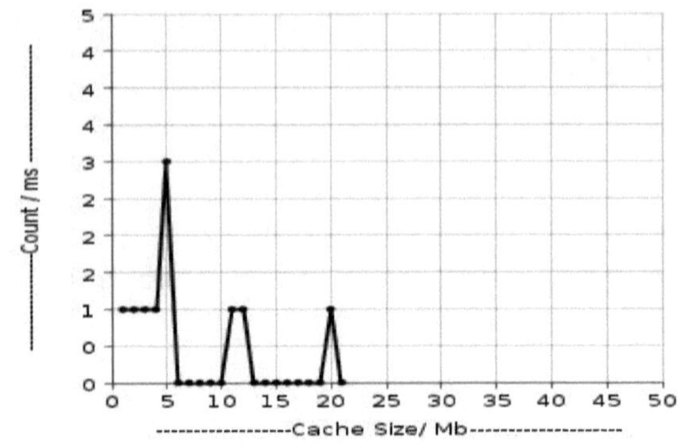

Fig. 1. Hit count vs cache size 10

In Fig. 1 we present the simulation for submitted request size of 10. The cache processes that stream within one millisecond. Initially, Number of request size is less than the size of the cache. Fig. 2 Cache sizes 20; number of the submitted requests is 15 in one millisecond.

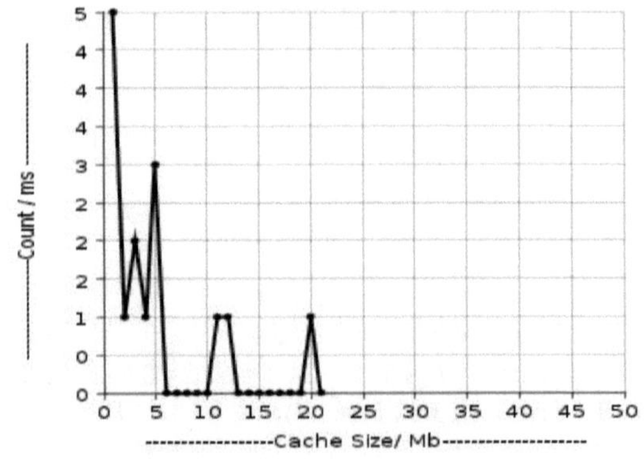

Fig. 2. Hit count vs cache size 20

Fig. 3 presents the graphical view of the Cache of fixed size 20, the number of the submitted requests is 20. The algorithm runs in 1 millisecond. Fig. 1, Fig. 2, Fig. 3 show the listed page hit in a millisecond. Starting from location Zero Mb to five Mb used for L1 (Logical Block) and (five Mb to 20Mb for L2) for the other logical block when the system has very low traffic loads.

The simulation result shows that when the number of request in the system is less than the total available cache size including L1 and L2 both, no request loss. As the sum of the Hit count to the memory block is all most equal to 20.

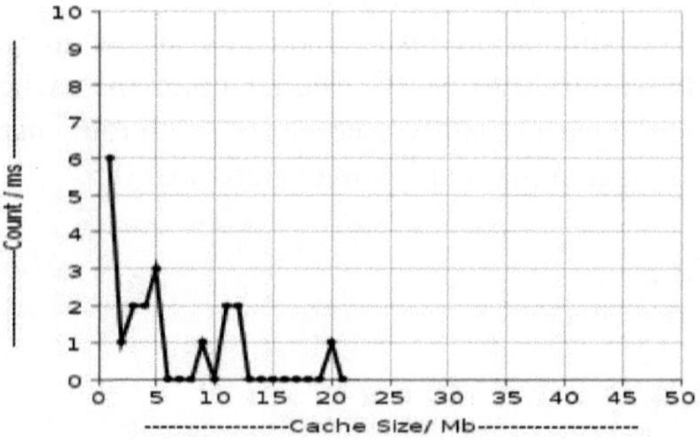

Fig. 3. Cache size 20, number of the submitted requests is 20 in one millisecond

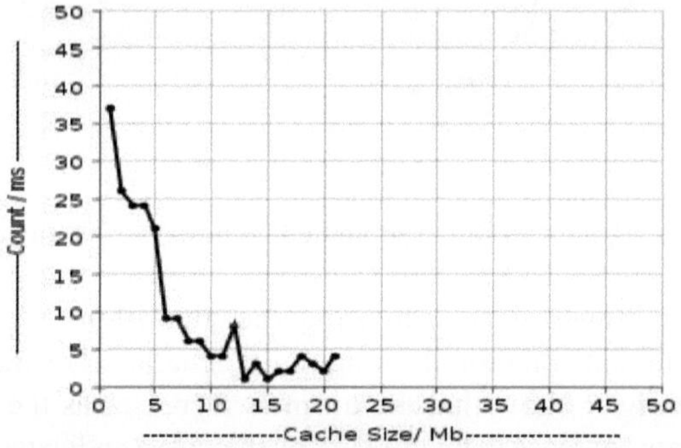

Fig. 4. Cache size 20, numbers of submitted requests is 200 in one millisecond

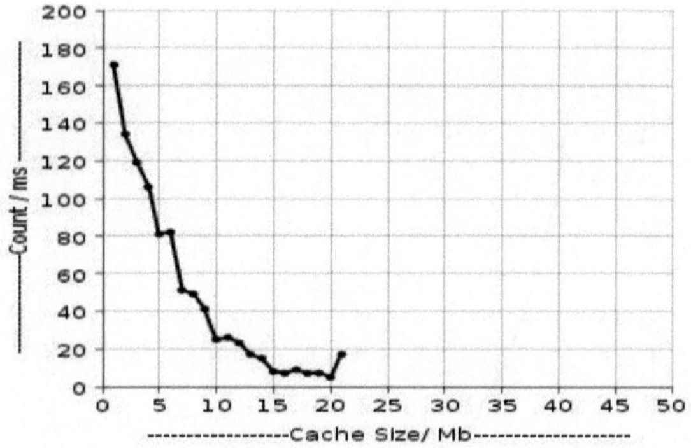

Fig. 5. Cache size 20, number of the submitted requests is 1000 in one millisecond

The Simulation results for the huge traffic rate are plotted in the Fig. 4, Fig. 5, Fig. 6 and Fig. 7. The simulation result indicates at the initial stage when the submitted traffic loads within 20 to $2x10^3$. The L1 holds the cache memory block of size 0 to 10 as L1 store the least recently used pages "with reference high hit count for the Fig. 4 and 5. In this stage λ maintains somewhere near 0.5.

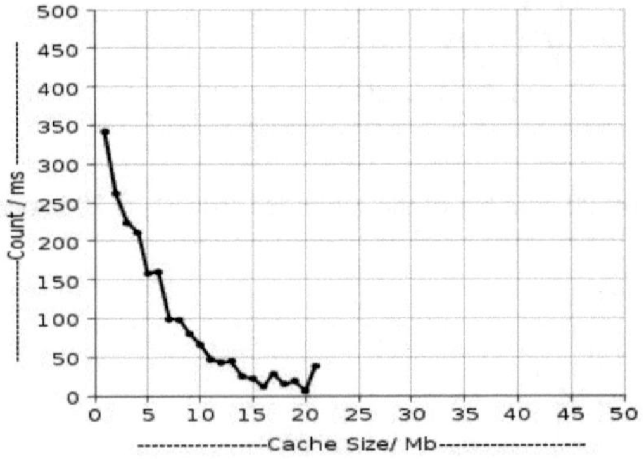

Fig. 6. Cache size 20, number of the submitted requests is 2000 in one millisecond

When the number request size increasing the algorithms try to expand the L_1 portion dynamically in that situation the value of λ generally lies like$\lambda \rightarrow (0.5, 1)$.

Fig. 5, Fig. 6 and Fig. 7 show that the hit miss increases as the size of the heterogeneous request stream increase. The tail side of the curve indicates the Hit miss. The simulation result shows for the huge traffic load L_1 portion of the cache memory block stores the reference of least recently used.

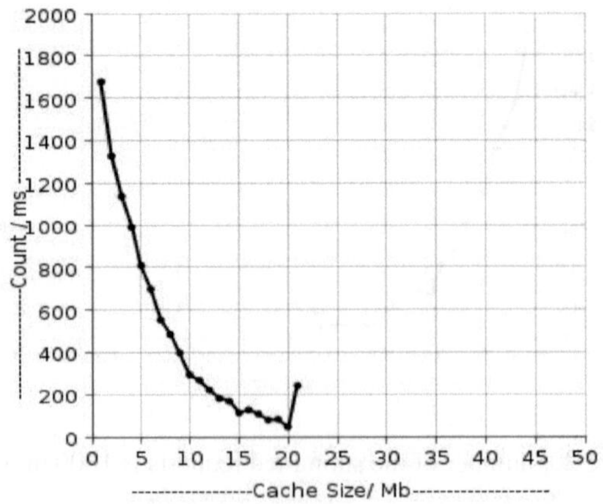

Fig. 7. Cache size 20, numbers of submitted request is 10000 in 1 millisecond

5 Remark and Conclusion

In this work, we present the Least Recently Frequency Used (LRFU) policy to the cache memory for handling the stream of heterogeneous requests. For the burst request, the LRFU is one of the optimize policy to minimize the request loss. Cache memory only stores the link to the video file that link to the distributed data bases. In the Video on the demand system, the total system performance depends on the cost of the database query execution and the load balancing for burst requests.

References

1. Muntzand, D.D., Honeyman, D.D.: Multi-level Caching in Distributed File Systems-oryour cache ain't nuthin' but trash. In: Proceedings of the Usenix Winter Technical Conference (1992)
2. Zhou, Y., Philbin, J.F.: The multi-queue replacement algorithm for second level buffer caches. In: Proceedings of USENIX Annual Technical Conference, pp. 91–104 (2001)
3. Gramacy, R.B., Warmuth, M.K., Brandt, S.A., Ari, I.: Adaptive caching by refetching. In: NIPS (2002)
4. Ari, I., Amer, A., Gramarcy, R., Miller, E., Brandt, S., Long, D.: ACME: Adaptive caching using multiple experts. In: Proceedings of the Work shop on Distributed Data and Structures (WDAS), Carleton Scientific (2002)
5. Wong, T.M., Wilkes, J.: My cache or yours? Makings to storage more exclusive. In: Proceedings of USENIX Annual Technical Conference, pp. 161–175 (2002)
6. Hsu, W.W., Smith, A.J., Young, H.C.: The automatic improvement of locality in storage systems. Tech. Rep., Computer Science Division, Univ. California, Berkeley (2001)
7. Lee, W., Park, S., Sung, B., Park, C.: Improving Adaptive Replacement Cache (ARC) by Reuse Distance. In: 9th USENIX Conference on File and Storage Technologies (FAST 2011) (2011)
8. Sha, E.H.-M., Chantrapornchai, C.: Optimizing Page Replacement for Multiple – Level Memory Hierarchy
9. Balamash, A., Krunz, M.: An Overview of Web Caching Replacement Algorithms. IEEE Commun. Surveys & Tutorials 6(2) (2004)
10. Kanrar, S.: Performance of distributed video on demand system for multirate traffic. Retis, pp. 52–56 (2011)
11. Kanrar, S.: Analysis and Implementation of the Large Scale Video-on-Demand System. IJAIS 1(2), 41–49 (2012)

Hybrid Function Based Analysis of Simple Networks

Sangramjit Sarkar[1], Aritra Purkait[2], Anindita Ganguly[3], and Krishnendu Saha[4]

[1] Development Consultants Pvt. Ltd, Kolkata 700091, India
[2,4] Cognizant Technology Solutions
[3] St. Thomas' College of Engineering & Technology, Kolkata-700023, India
{sangramjitsarkar,aritrap06,aninditaganguly80, krishnendurocks07}@gmail.com

Abstract. This paper represents a method of analysis of simple networks with the help of MATLAB-SIMULINK models and microcontroller based hardware. The mathematical foundation of this method is an algorithm based on orthogonal Hybrid Function.

Keywords: Hybrid function, microcontroller, MATLAB, analysis.

1 Introduction

Marriage of sample-and-hold functions (SHF) [1] and triangular functions (TF) [1] has given birth to a set of orthogonal hybrid functions (HF) [2-3]. Among various utilities and applications of HF, this paper is focused on System Analysis.

2 Hybrid Function – A Combination of Shf and Tf

We can use a set of sample-and-hold functions and the set of triangular functions to form a function set, which we call a 'Hybrid Function' set. To define this set, we express the (i+1)-th member $H_i(t)$of the m-set hybrid function $\mathbf{H}_m(t)$

$$H_i(t)= a_iS_i(t)+ b_iT2_i(t) \text{in } 0 \leq t < T, \tag{1}$$

where, $i = 0, 1, 2, \ldots, (m-1)$, a_i and b_i are scaling constants.

We can approximate a square integrable function f(t) of Lebesgue measure by Hybrid Function set. Fig. 1 illustrates this idea. f(t) is sampled at three equidistant points A,C,E. c_0, c_1, $c_2$are the magnitude of these samples. Linearly joining A, C& C,E we get a SHF set composed of □ABFO &□CDGF and TF set constituted with ΔACB & ΔCED. Hence, piecewise linear wave which we get in this manner, is nothing but HFapproximation of f(t).

M.K. Kundu et al. (eds.), *Advanced Computing, Networking and Informatics - Volume 2*, Smart Innovation, Systems and Technologies 28,
DOI: 10.1007/978-3-319-07350-7_52, © Springer International Publishing Switzerland 2014

That is,

$$\begin{aligned} f(t) &= H_0(t)+H_1(t) \\ &= \{c_0S_0(t)+ c_1S_1(t)\} + \{(c_1-c_0)T_0(t) + (c_2-c1)T_1(t)\} \\ &\triangleq C^T S_{(2)}(t) + D^T T_{(2)}(t) \end{aligned} \tag{2}$$

where, $[c_0 \;\; c_1] = C^T$ and $[(c_1-c_0) \;\; (c_2-c1)] = D^T$ and $[\ldots]^T$ denotes transpose.
Equation (2) represents f(t) in hybrid function domain.

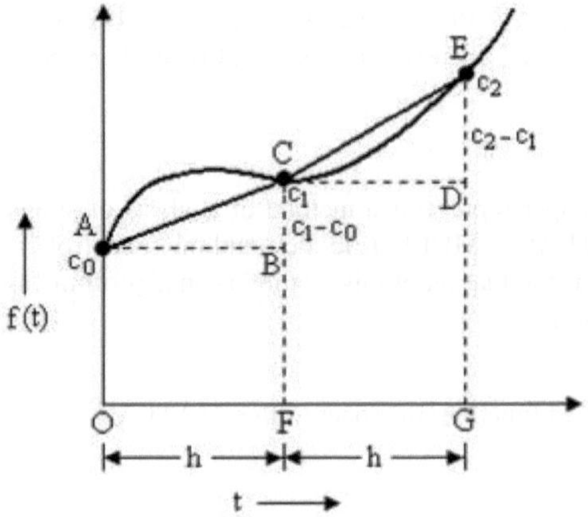

Fig. 1. A function f(t) represented via hybrid functions

3 Methodology

The continuous time input signal is sampled at a constant rate. From these samples the output samples are obtained in the following two ways-

1. Plant (1st or 2nd order simple Network) is physically excited by piecewise linear (HF approximation) waveform obtained from continuousi/p signal.
2. The output samples are directly computed from the input samples using the HF based algorithm [4].

Finally, the accuracy of the HF based algorithm as a tool of system analysis is checked by comparing the plots of the output samples obtained mathematically and physically as described above.

We have taken1 Hz sine wave as input signal.

Since the input is of a known (sinusoidal) nature, instead of actually sampling the wave, at the initial stage, the samples are pre-calculated for a 1Hz sine wave and directly used to construct the SHF (Sample and Hold Function) or ZOH (Zero order wave) as explained later.

System or Plant is initially taken as 1st order RC circuit for the sake of simplicity and this method of analysis is further extended to 2nd order RC circuit.

4 Schematic Diagram of the Overall System

A simplified schematic diagram of the overall system is shown in Fig. 2.

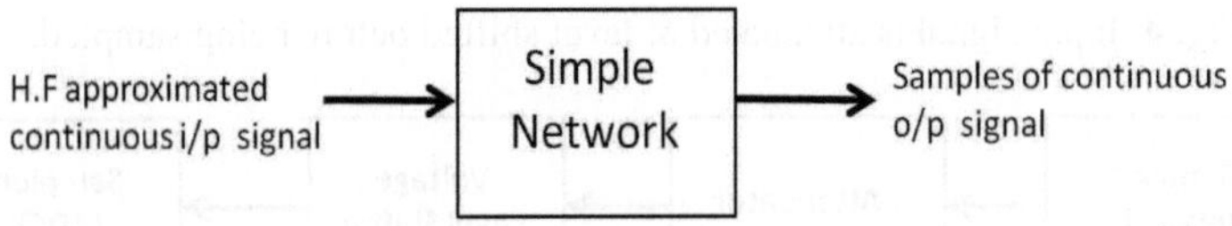

Fig. 2. Simplified Block diagrammatic representation of the overall system

Now the question arises how to physically approximate a continuous time signal into HF domain. To describe this, detailed schematic diagram is introduced in Fig. 3.

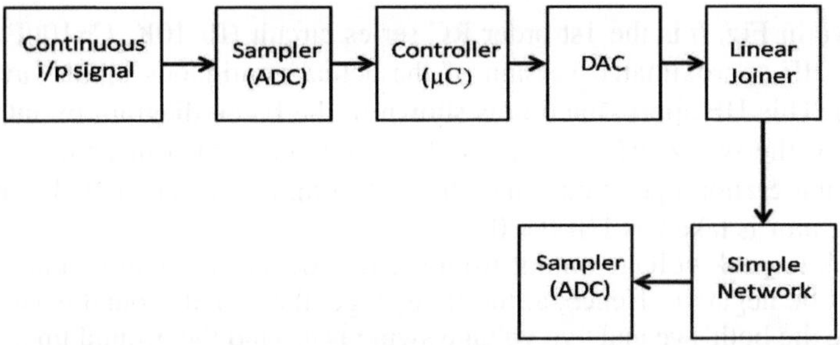

Fig. 3. Detailed Block diagrammatic representation of the overall system

It is mentioned earlier that a HF set is composed of a SHF set and a TF set. SHF approximation is realized by **Sampler (ADC), Controller (μC)& DAC** whereas TF approximation is realized by **Linear Joiner.**

The continuous input waveform is sampled by an ADC present inside the microcontroller [5]. Microcontroller controls the sampling rate of ADC and stores the samples in memory. It processes the signal and gives digital o/p which is converted into analog o/p with the help of a DAC (we have used 8 bit R-2R ladder). This o/pis nothing but the SHF waveform of the i/p signal. In the next step, 'Linear joiner' is used to accomplish TF realization of the i/p signal. It actually integrates step voltages by differencing two consecutive sample values of i/p signal. Hence, the o/p of the 'Linear Joiner' is a piecewise linear wave (HF approximation of i/p signal).

The piecewise linearly joined wave is fed to simple network (1st order and 2nd order RC circuits) and the response is viewed in a DSO screen.

It is important to mention that ADC present in the microcontroller can only measure voltages between 0 – 5 V. Thus, the input signal has to be attenuated to restrict its peak to peak value within 5 V. Some amount of positive voltage is injected into the signal to make its amplitude entirely positive as shown in Fig.4. Similar steps are followed while sampling the response of the simple networks as shown in Fig. 5.

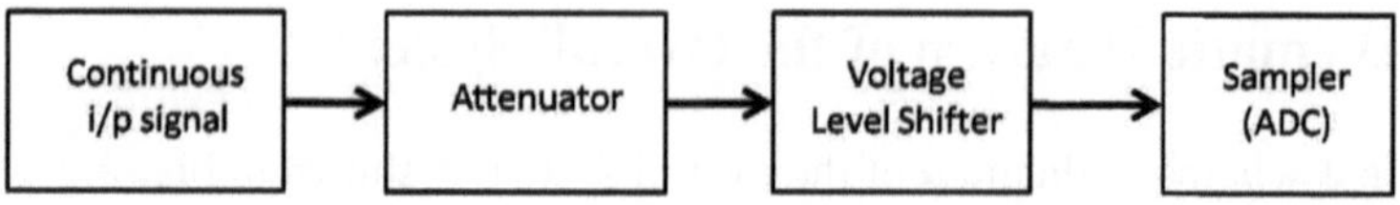

Fig. 4. Input signal is attenuated & level shifted before being sampled.

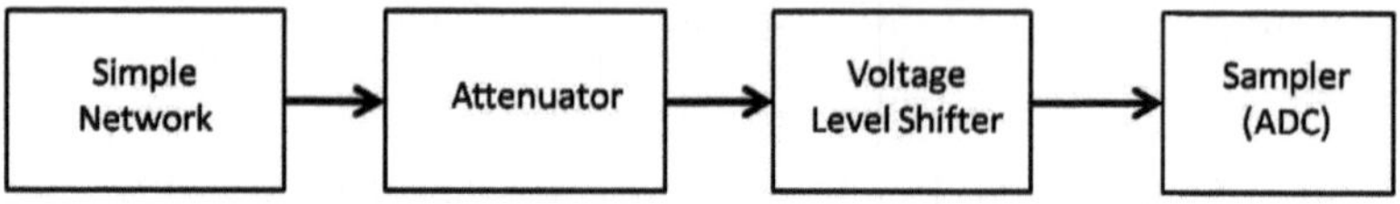

Fig. 5. Response is attenuated & level shifted before sampled

5 SIMULINK Model for 1st Order System

Plant shown in Fig. 6 is the 1st order RC series circuit (R=10K, C=10μF). It is to be excited by HF approximated version of the actual continuous input wave (taken as sine wave). This HF approximation is shown in the block diagram by integrating the difference of the two ZOH sine waves. The block entitled 'Linear Interpolator' performs this integration operation. Since the sampling rate is taken 10Hz time constant of the integrator is taken as 1/100 = 0.1.

The peak to peak voltage output from the microcontroller can be maximum 5 volt and it can't be negative. Hence, at the final stage, the signal ground is shifted by -2.5 V to ensure the both +ve and –ve voltage swing is around the ground line.

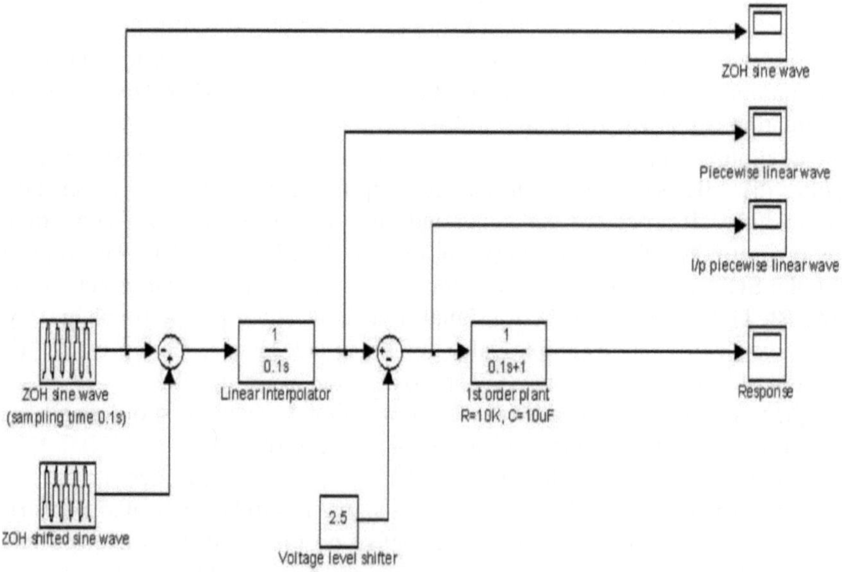

Fig. 6. SIMULINK model for 1st order plant

6 SIMULINK Model for 2nd Order System

The 2nd order system shown in Fig.7 is chosen to be an under damped system with damping ratio (zeta) = 0.44.

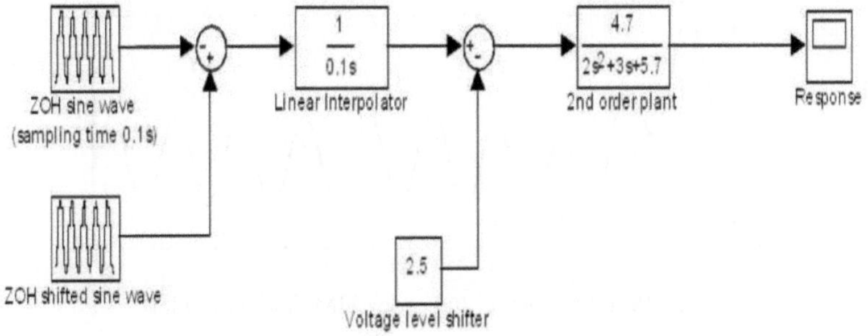

Fig. 7. SIMULINK model for 2nd order plant

7 Response of 1st Order Plant

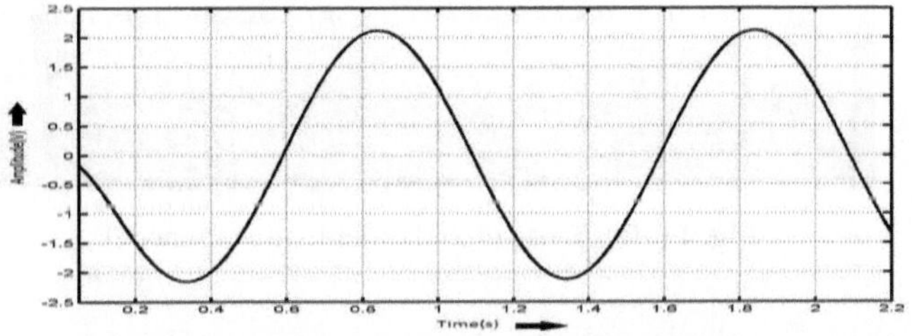

Fig. 8. Response of 1st order SIMULINK model

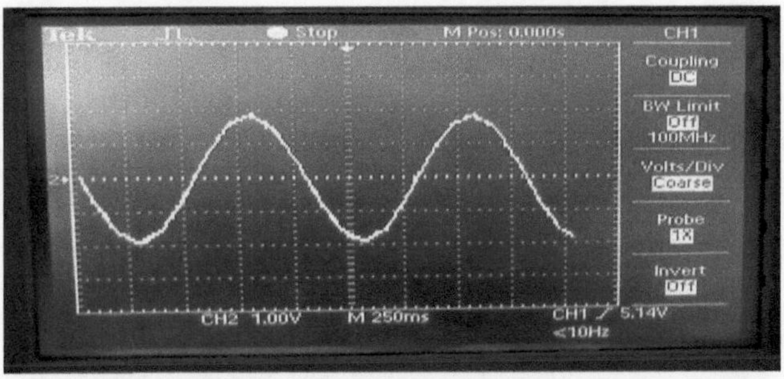

Fig. 9. Response of 1st order hardware model

8 Response of 2ndOrder Plant

Fig. 8 and Fig. 9 show responses of the SIMULINK model and respective hardware model for 1st order system where as Fig. 10 and Fig. 11show responses of the SIMULINK model and respective hardware model for 2nd order system.

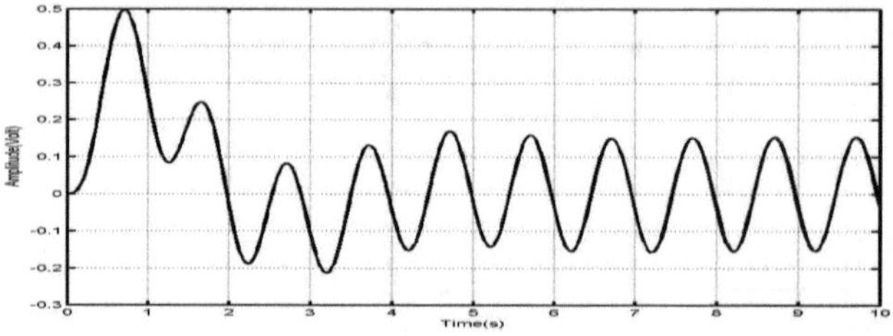

Fig. 10. Response of 2nd order SIMULINK model

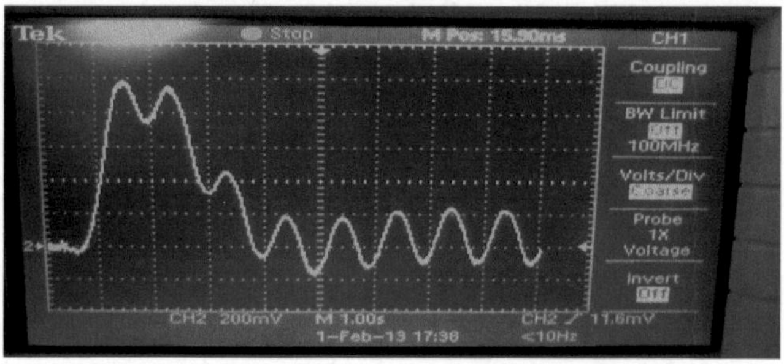

Fig. 11. Response of 2nd order hardware model

9 Result Analysis

The plots shown in the Fig. 12 and Fig. 13illustrate the comparative study between SIMULINK response and response obtained by HF based algorithm for 1st and 2nd order system respectively. Two different coloured (**red** and **blue**) plots are almost indistinguishable except at some regions which give a qualitative idea about the error.

This can also be reduced to a greater extent by enhancing the sampling rate.

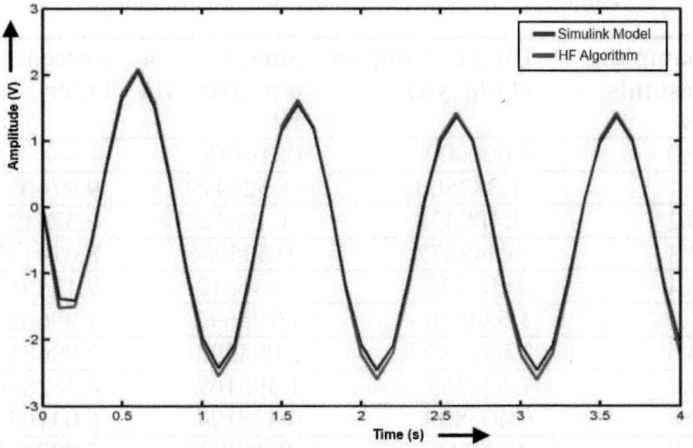

Fig. 12. Response plot of 1st order plant using samples of SIMULINK model and samples computed by HF algorithm

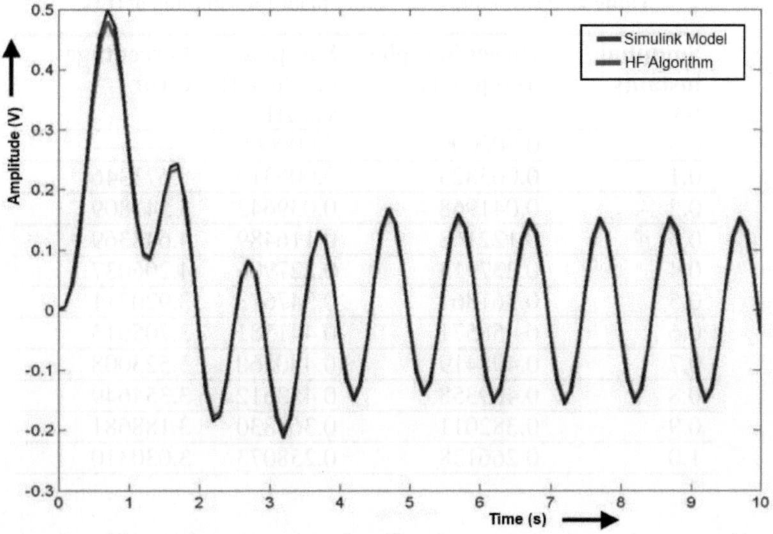

Fig. 13. Response plot of 2nd order plant using samples of SIMULINK model and samples computed by HF algorithm

10 Percentage Error

Quantitatively percentage error is computed and tabulated in Table 1 (for 1st order plant) and Table 2 (for 2nd order plant). This error analysis shows that the absolute error is within 10%.

Table 1. Error table of 1st order plant for sampling at 0.1s

Sampling instants (s)	Direct Samples of o/p y(t)	Samples of o/p y(t) via HF	Percentage error
0.0	0.000000	0.000000	----------
0.1	-1.382500	-1.520419	-9.976054
0.2	-1.408437	-1.49822	-6.375129
0.3	-0.548372	-0.595966	-8.679117
0.4	0.603735	0.646812	-7.134959
0.5	1.624526	1.705697	-4.996622
0.6	2.051280	2.094293	-2.096853
0.7	1.522355	1.595167	-4.782816
0.8	0.403800	0.428194	-6.041072
0.9	-0.905658	-0.959006	-5.890467
1.0	-1.979414	-2.085004	-5.334416

Table 2. Error table of 2nd order plant for sampling at 0.1s.

Sampling instants (s)	Direct Samples of o/p y(t)	Samples of o/p y(t) via HF	Percentage error
0.0	0.000000	0.000000	------------
0.1	0.005825	0.005319	8.673346
0.2	0.041968	0.039642	5.541809
0.3	0.122168	0.116489	4.648369
0.4	0.237913	0.227906	4.206037
0.5	0.361861	0.347673	3.920711
0.6	0.458571	0.441581	3.705013
0.7	0.498419	0.480860	3.523008
0.8	0.469358	0.453612	3.354649
0.9	0.382011	0.369830	3.188681
1.0	0.266138	0.258073	3.030310

11 Comparison between the Performance of the Hardware Model and the Hf Algorithm Based Computation Method

Fig. 14 shows two response plots of 1st order system. Solid one is plotted by linearly joining the samples of the hardware response tabulated in Table 3 and dotted one is plotted by linearly joining the samples computed directly using HF based algorithm.

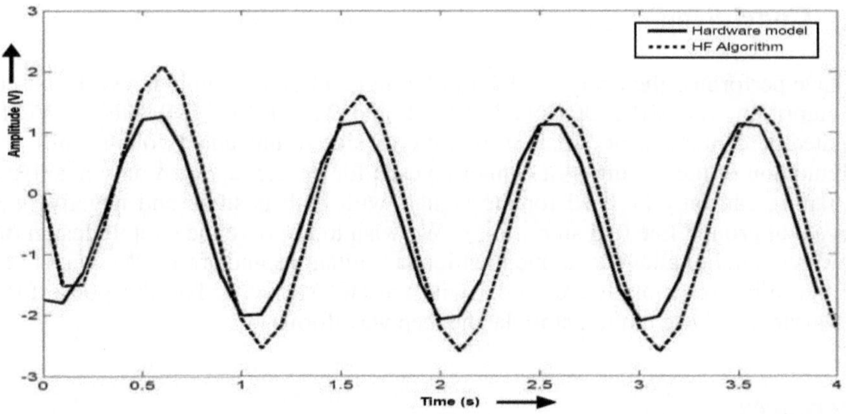

Fig. 14. Response plots of 1st order plant using output samples of HARDWARE model and samples computed by HF algorithm

Table 3. Samples of the response collected using ADC

ADC reading	ADC (V)	Offset(V)	Actual(V)
145	0.71	2.40	-1.70
133	0.65	2.40	-1.75
232	1.13	2.40	-1.27
417	2.04	2.40	-0.37
616	3.01	2.40	0.61
748	3.66	2.40	1.25
759	3.71	2.40	1.30
642	3.14	2.40	0.73
437	2.14	2.40	-0.27
218	1.07	2.40	-1.34
90	0.44	2.40	-1.96
92	0.45	2.40	-1.96
201	0.98	2.40	-1.42
393	1.92	2.40	-0.48
597	2.92	2.40	0.51
728	3.56	2.40	1.15

12 Problems Encountered

Significant deviation between solid and dotted traces in Fig. 14 shows appreciable error. Probable cause of the error is improper voltage shift by voltage level shifter circuits.

13 Conclusion

We have performed the analysis of 1st order and 2nd order simple networks based on HF algorithm via MICROCONTROLLER and MATLAB-SIMULINK. We have tabulated percentage error. Apart from the error due to unwanted voltage shift, a major limitation of the circuit is- it cannot be used for the signals with only positive part (e.g ramp), can only be used for the signals with both positive and negative excursions about ground line (e.g sine wave). We wish to improve the overall design of the hardware model to alleviate above mentioned limitations and errors. We wish to compare the response of hardware model with the HF Algorithm basedresponse for 2nd order system and use ramp, parabola and step waveforms.

References

1. Deb, A., Sarkar, G., Sengupta, A.: Triangular orthogonal functions for the analysis of continuous time systems. Anthem Press (2011)
2. Deb, A., Sarkar, G., Mandal, P., Biswas, A., Ganguly, A., Biswas, D.: Transfer function identification from impulse response via a new set of orthogonal hybrid functions (HF). Applied Mathematics and Computation 218(9), 4760–4787 (2012)
3. Deb, A., Sarkar, G., Ganguly, A., Biswas, A.: Approximation, integration and differentiation of time functions using a set of orthogonal hybrid functions (HF) and their application to so-lution of first order differential equations. Applied Mathematics and Computation 218(9), 4731–4759 (2012)
4. Deb, A., Ganguly, A., Sarkar, G., Biswas, A.: Computation of Convolution via ANew Set of Orthogonal Hybrid Functions (HF) for Linear Control System Analysis and Identification. In: IEEE India Conference (INDICON), pp. 682–688 (2012)
5. Atmel Atmega8 datasheet. Atmel Corporation, USA (2011)

Enhancing Memory Deduplication Using Temporal Page Sharing Behaviors in Virtual Environments

T. Veni and S. Mary Saira Bhanu

Department of Computer Science and Engineering,
National Institute of Technology,
Tiruchirappalli-620015, Tamil Nadu, India
{406111001,msb}@nitt.edu

Abstract. The performance and scalability of the virtualized systems are affected by the size and speed of main memory. The memory deduplication is a prominent approach which increases memory savings by sharing of duplicate memory pages across virtual machines. Stability of shared pages is the important factor for page sharing mechanism. If sharing is short lived, it triggers CoW exception handler instantly which significantly impact the performance of memory deduplication process. The proposed approach uses *hinting* mechanism to evade instability pages from scanning and merging process, thereby enhancing the effectiveness and efficiency of memory deduplication process. The main advantage of proposed approach is that it does not need any guest OS modification and it is implemented using Kernel Samepage Merging (KSM) - a memory deduplication daemon in the linux kernel. The evaluation with several benchmark workloads shows that the proposed approach achieves a significant improvement over vanilla KSM memory deduplication process.

Keywords: Memory Deduplication, Virtualization, Server Consolidation, KSM, Page Sharing.

1 Introduction

The virtualization technology is a key technology which facilitates on-demand and flexible resource provisioning, strong service isolation as well as live migration of services in cloud computing scenarios [1]. Moreover, this technology allows packing of several Virtual Machines (VMs) into fewer physical servers (called server consolidation) to increase the resource utilization and power savings. However, memory is a bottleneck resource which limits the degree of server consolidation and performance of virtualized systems. It is evident from existing literature that VMs contain substantial amount of redundant memory ranging from 11% to 79% depending upon OS and workloads running on them [2], [3].

Memory deduplication is a promising approach that harnesses these redundancies through sharing of the duplicate memory, i.e., this method shares pages with same content on page level granularity and remapping them into a single

M.K. Kundu et al. (eds.), *Advanced Computing, Networking and Informatics - Volume 2*, Smart Innovation, Systems and Technologies 28,
DOI: 10.1007/978-3-319-07350-7_53,

Copy-on-Write (CoW) page [4]. Many techniques have been proposed to leverage the benefits of memory deduplication process which can be either by using para virtualization based approach or periodical memory scanning based approach. The para-virtualization based approach [5–7] enhances deduplication efficiency through guest OS modification which is not viable to closed-source OSs due to the limitations of source code or copy-right issues.On the other hand, the memory scanning based deduplication approach [4], [8–11] should maintain a tradeoff between memory savings and CPU overhead. High page scan rate increases the memory savings, but leads to more computational overhead.

In addition, memory deduplication process incurs a significant overhead when shared pages are subject to frequent modification resulting in early breaking of sharing mechanism. The proposed approach uses *hinting* mechanism, where hints can be developed through temporal analysis of page sharing behaviors. With these hints, the proposed approach evades the instability pages from scanning and merging process and thereby speed-up the memory deduplication process.

The main advantage of the proposed approach is that it does not need any guest OS modification. The evaluation with several benchmark workloads shows that the proposed approach achieves a significant improvement in terms of memory savings and CPU overhead in comparison to vanilla KSM memory deduplication process.

The remainder of this paper is organized as follows: Section 2 surveys existing literature work related to memory deduplication approach. The proposed approach is presented in Section 4. Section 5 discusses the evaluation of the proposed approach with various benchmark workloads and finally Section 6 concludes the paper.

2 Related Work

The existing literature can be broadly classified into para-virtualization based memory deduplication and memory scanning based memory deduplication.

2.1 Para-virtualization Based Deduplication Approaches

Disco [5] system introduced the concept of transparent page sharing among VMs. The special DMA-based virtual disk interface is developed to explicitly track each VM's I/O request. If the requested I/O pages are already in main memory, the VM obtains the same pages mapped in their address space. Hence, Disco reduces the memory foot print of VMs and increases the performance by avoiding further disk access. Satori [6] further extends the concept used by Disco. In Satori, the identical memory pages that stem from background storage (multiple VDI images) can be shared via sharing-aware block devices. However, Satori and Disco exploit sharing opportunities only on the pages that stem from background storage. The IBMs Collaborative Memory Management (CMM) [7] exploits para virtualization approach to exchange the page usage and residency information between guest OS and hypervisor, thereby hypervisor makes use of this information to improve paging performance.

It can be observed that all the above proposed techniques adapt guest OS modification which is not viable to closed-source OSs due to the limitation of source code or copy-right issues.

2.2 Memory Scanning Based Deduplication Approaches

VMware ESX-Server [4] periodically scans the memory in a random way regardless of memory origin. During scanning, the hash value of each page is recorded in a hash table and a lookup is done to locate the sharing candidates. Upon a successful match, these identical pages can be shared using CoW fashion. KSM [8]is a memory deduplication daemon in the Linux kernel. It uses red-black trees as a data structure to discover potential sharing candidates where page content itself is used as a node key. However, these memory deduplication scanners(VMWare, KSM) are restricted by its two configurable parameters such as page scan rate and sleep time between scans which in turn used to retain a trade-off between saved memory and associated CPU overhead.

Difference Engine [9] enhances the sharing efficiency by leveraging sub-page sharing through page patching, and in-core memory compression techniques. XLH [10]substantially optimizes the performance of KSM by affording a hinting mechanism that prioritizes the memory scanners to process hint pages followed by other pages. Chiang *et al.* [11] proposed the bootstrapping VM Introspection (VMI) technique to extract the free memory pool information of guest memory, which can be offered to the memory scanner as hints to improve the deduplication efficiency.

Stability is the crucial utility for page sharing mechanism. If sharing is short lived, it triggers CoW exception handler instantly which in turn increases the memory scanning overhead. While serveral techniques exist in the existing literature, none of them have dealt with this issue of instability pages. *The proposed approach excludes the instability pages from deduplication process through the temporal analysis of page sharing behaviors, without any guest OS modification.*

3 The Temporal Based Memory Deduplication

The proposed approach (temporal based memory deduplicaiton) is implemented using KSM-a memory deduplication daemon in the linux kernel. In KSM, a reverse map item (rmap_item) is created for every virtual page. This rmap_item is used to store the checksum of the virtual page content and page sharing details. In the proposed approach, the two fields viz., time-stamp and VMs id fields are added to the KSM′s existing rmap_item to capture the temporal behaviour(lifetime of shared pages)of each shared page.The information about page sharing duration is profiled in the form of tuples ⟨ *rmap_item, physical page address, virtual page address, sharing duration* ⟩ and low sharing duration pages are

Algorithm 1. Temporal based Memory Scanning

Input: $Scan\_Pages = 100$ //Number of pages to scan on every wake up
M_p //Current page under memory deduplication process
$P \leftarrow \{P_1, P_2, P_3, ..., P_m\}$ // Registered pages list with page size(4 KB)
$I \leftarrow \{I_1, I_2, I_3, ..., I_n\}$ // Instability pages list
$R \leftarrow \{R_1, R_2, R_3, ..., R_m\}$// Rmap_item list

while $(Scan\_Pages - - < 0)$ **do**
  $R_p \leftarrow scan\_get\_next\_rmap\_item(M_p)$; //gets the rmap_item for M_p
  if $R_p \in I$ **then**
    continue;

    //The next condition checks whether M_p is already merged page or not.
    //If not, call *cmp_and_merge_page*() to merge M_p with tree pages.

  else $\{!PageKsm(M_p) || !in\_stable\_tree(R_p))\}$
    $cmp\_and\_merge\_page(M_p, R_p)$;
  end if
end while

filtered out based on the threshold and stored in instability list (I). By empirically, this threshold value is fixed as 15s based on deduplication performance against various threshold values.The detailed algorithm is described as follows.

4 Evaluation

In this section, the effectiveness and efficiency of the proposed approach is evaluated under different workload scenarios. The different benchmark workloads are chosen to cover wide range of memory usage pattern (Table 1). The key performance indicators such as amount of page sharing achieved, and CPU overhead are chosen to quantify the efficiency and effectiveness of the following configurations:

1. **Baseline:** The vanilla KSM configuration .
2. **Temporal:** The temporal based deduplication is enabled in vanilla KSM.

All experiments were conducted on the Intel Xeon E5507 processor (4 cores) machines with 8 GB of physical memory. Ubuntu 12.04 Linux distribution and Linux kernel v3.2.52 can be used in both host and guest. The KVM hypervisor was used to create virtual machines.

The KSM′s parameters such as Pages_to_Scan = 100, Sleep_Time= 20(ms) and Page_Size = 4(KB) have been used as a default parameters for evaluation.

Table 1. Various workloads for evaluation

Workload	Description
Linux Idle	An idle linux installation without hosting any application on VMs. This workload was primarily used as a baseline for comparison.
Kernel Compile	The Linux kernel v 3.2.52 is compiled with default configuration.
RUBiS	An auction site prototype modeled after eBay.com which consists of web and database servers. The Apache/PHP implementation of RUBiS v1.4.3 with MySQL database has been used in experiments. The workload is varied by varying the number of clients in the RUBiS workload generator.

4.1 Memory Deduplication Effectiveness

The sharing effectiveness(amount of page sharing achieved over the period of time)of various configurations (Baseline, Temporal)under both homogeneous and heterogeneous workload scenarios are evaluated as follows.

The evaluation begins with an analysis of homogeneous workloads where all VMs running the same workloads. In this scenario, significant memory savings are possible due to workload homogeneity. Initially, the idle VMs are instantiated and measured the sharing potentials. From Fig. 1, it can be observed that both configurations performed equally well due to the workload stability.

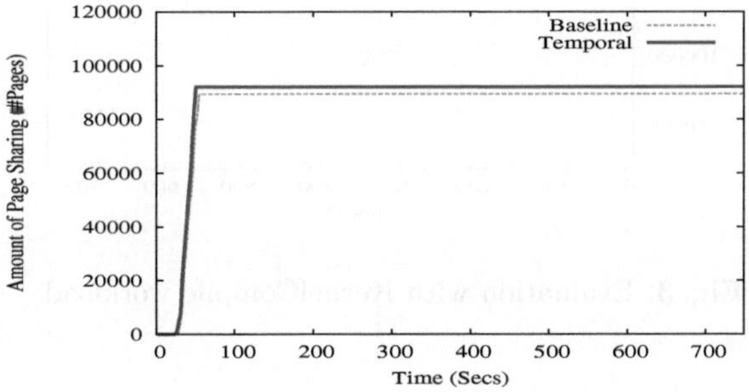

Fig. 1. Evaluation with idle workload

The same experiment has been repeated with another set of VMs where each VM hosting RUBiS workload [12]. The client workload generator in each VM instance simulates 1200 user sessions per run. As part of evaluation, the RUBiS workload is configured with mixture of read and writes requests. In this workload, most of requests are configured as read type which results stablility in sharing opportunities (Fig. 2). Similarly, the experiments have been conducted with kernel compile [13] workloads that runs two parallel gcc threads for about 40 minutes. This workload fetches the file from disk, compiles them and writes the resultant object files back to the disk. In this workload, the working set of

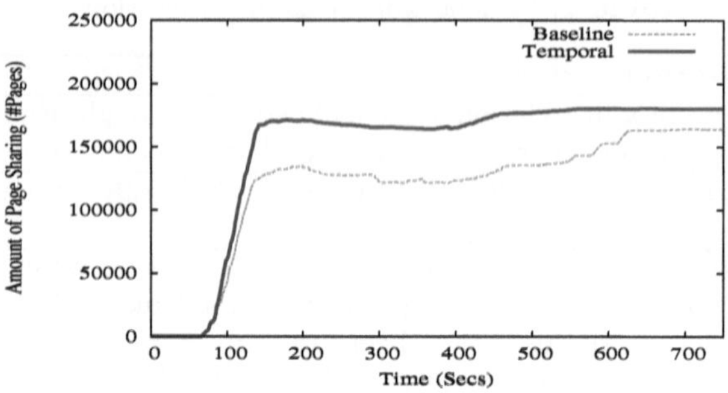

Fig. 2. Evaluation with RUBiS workload

VMs changes more rapidly results in fluctuation in page sharing opportunities (Fig. 3). As more number of short-lived pages occur in this workload, the proposed approach eliminates those short-lived pages and outperforms the base-line approach.

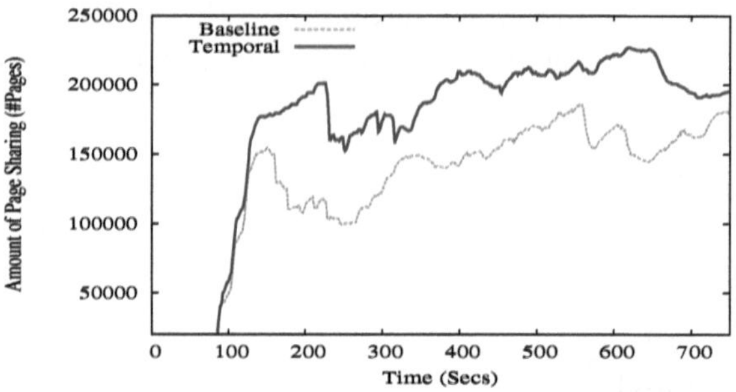

Fig. 3. Evaluation with KernelCompile workload

For heterogeneous (mixed) wokloads, an experiment has been conducted where each VM hosts different workload, i.e., VM1 hosts idle workload, VM2 hosts RUBiS workload and VM3 hosts Kernel Compile workload. In this mixed workload, the probability of identical pages is very low resuting in less sharing opportunities compared to all other workloads and also the kernel compile workload variability causes the fluctuation in resultant page sharing opportunities(Fig. 4).

The results of above experiments are shown in Fig. 1-4. It can be observed that, the proposed approach achieves significant sharing effectiveness in all workload scenarios. The reason is that there is a significant amount of instability pages occur in each workload. The proposed approach takes advantage of this instability memory pages and achieves a notable memroy savings in all workload scenarios whereas vanilla KSM fails to focus this issue leads to lower memory savings.

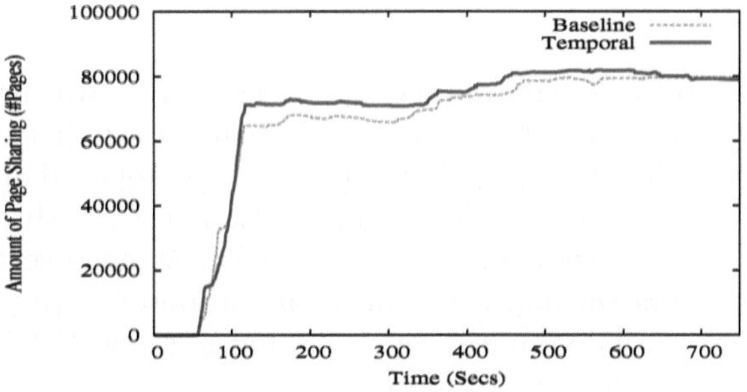

Fig. 4. Evaluation with mixed workload

4.2 Deduplication Efficiency

The efficiency in memory deduplication means detecting high page sharing potentials without compromising the computational cycles. The experiment has been conducted towards this perspective for evaluating the overhead involved in each configuration under different workload scenarios. It can be observed from the Fig. 5 that the proposed approach consumes less CPU cycles in all scenarios compared to Baseline approach. The reason is that the proposed approach obviates the instability page′s overhead.

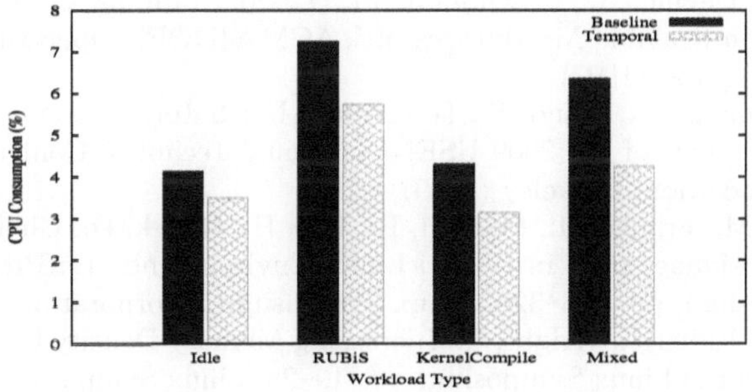

Fig. 5. CPU overhead of various workloads

4.3 Comparison with Existing Works

The prior work XLH [10] also used the hinting mechanism for enhancing memory deduplication efficiency. The XLH generates the page cache pages as hints and offers the hints to KSM scanner to find sharing opportunities more quickly. The hint pages which are short-lived introduce additional overhead through hint processing. The proposed approach eliminates all in-stability pages and further improves the above exising hinting mechanism.

5 Conclusion

Memory deduplication is a widely used approach to improve memory savings in virtualized systems. However, this approach incurs a significant overhead because of the short-lived(instability) shared pages. The proposed approach evades these instability pages form memroy deduplication process through temporal analysis of page sharing behaviors. The evaluation using several benchmarks demonstrates the significant improvement of the proposed approach in terms of memory savings, memory deduplication speed and minimal CPU overhead over vanilla KSM memory deduplication process.

References

1. Armbrust, M., Fox, A., Griffith, R., Joseph, A.D., Katz, R.H., Konwinski, A., Lee, G., Patterson, D.A., Rabkin, A., Stoica, I., Zaharia, M.: Above the Clouds: A Berkeley View of Cloud Computing. Technical Report, UCB/EECS-2009-28 (2009)
2. Barker, S., Wood, T., Shenoy, P., Sitaraman, R.: An Empirical Study of Memory Sharing in Virtual Machines. In: Proceedings of the 2012 USENIX Annual Technical Conference, pp. 273–284. USENIX Association, Berkeley (2012)
3. Chang, C., Wu, J., Liu, P.: An Empirical Study on Memory Sharing of Virtual Machines for Server Consolidation. In: Proceedings of the Ninth IEEE International Symposium on Parallel and Distributed Processing with Applications Workshops, pp. 244–249 (2011)
4. Waldspurger, C.: Memory Resource Management in VMWare ESX Server. ACM SIGOPS Operating Systems Review 36, 181–194 (2002)
5. Bugnion, E., Devine, S., Rosenblum, M.: Disco: Running Commodity Operating Systems on Scalable Multiprocessors. ACM SIGOPS Operating Systems Review 31(5), 143–156 (1997)
6. Milos, G., Murray, D., Hand, S., Fetterman, M.: Satori: Enlightened Page Sharing. In: Proceedings of the 2009 USENIX Annual Technical Conference, pp. 1–10. USENIX Association, Berkeley (2009)
7. Schwidefsky, M., Franke, H., Mansell, R., Raj, H., Osisek, D., Choi, J.: Collaborative Memory Management in Hosted Linux Environments. In: Proceedings of the Linux Symposium, pp. 313–328. Linux Symposium Incorporation, Ottawa (2006)
8. Arcangeli, A., Eidus, I., Wright, C.: Increasing Memory Density by Using KSM. In: Proceedings of the Linux Symposium, pp. 19–28. Linux Symposium Incorporation, Ottawa (2009)
9. Gupta, D., Lee, S., Vrable, M., Savage, S., Snoeren, A., Varghese, G., Voelker, G., Vahdat, A.: Difference Engine: Harnessing Memory Redundancy in Virtual Machines. Communications of the ACM 53(10), 85–93 (2010)
10. Konrad, M., Fabian, F., Rittinghaus, M., Hillenbrand, M., Bellosa, F.: XLH: More Effective Memory Deduplication Scanners through Cross-Layer Hints. In: Proceedings of the 2013 USENIX Annual Technical Conference, pp. 279–290. USENIX Association, Berkeley (2013)
11. Chiang, J., Li, H., Chiueh, T.: Introspection Based Memory De-duplication and Migration. In: Proceedings of the Ninth ACM SIGPLAN/SIGOPS International Conference on Virtual Execution Environments, pp. 51–61 (2013)
12. An auction site prototype modeled after ebay, `http://rubis.ow2.org`
13. Linux kernel archives, `http://www.kernel.org`

Performance Augmentation of a FAT Filesystem by a Hybrid Storage System

Wasim Ahmad Bhat and S.M.K. Quadri

Department of Computer Sciences
University of Kashmir
wab.cs@uok.edu.in, quadrismk@kashmiruniversity.ac.in

Abstract. In this paper, we propose segregation and dispersal of *hot-zone* & *cold-zone* of a FAT filesystem over a hybrid-storage system for performance gains. Specifically, we propose *hFAT*, a high performance FAT32 filesystem design, that stores the most frequently accessed metadata of files on a solid-state storage drive while as actual contents on the magnetic drive. The idea is to eliminate the head positioning latency incurred by FAT filesystem operations while accessing metadata & userdata disk areas. After exercising the *hFAT* filesystem using *Sprite LFS small-file* benchmark, we found that *hFAT* design can reduce the latency incurred by FAT32 filesystem operations by a minimum of 25%, 10% and 90% during writing, reading and deleting a large number of small files respectively, if a solid-state storage device having latency lesser or equal to 10% of that of magnetic disk is used in addition.

Keywords: FAT filesystem, hybrid storage, performance.

1 Introduction

In a FAT filesystem, the metadata necessary to locate the contents of every file and directory are placed at the beginning of a volume. As a consequence, the seek distance between this metadata and actual contents of files and directories is large. Moreover, no effort is made to place the contents of a file or directory at rotationally optimal positions. Furthermore, the directory entries are arranged as an unordered linear list. Therefore, finding a particular file within a directory requires a linear search with an algorithm complexity of `O(n)`. Worst, the metadata necessary to locate all the clusters belonging to a file (or directory) is scattered in a long FAT Table. As a consequence, this table needs to traversed from head to tail until sufficient entries are read to locate the contents of that file (or directory). Though being *update-in-place* filesystem, FAT filesystem does not achieve *logical locality*. However, depending upon the state of the filesystem, *temporal locality* may be possible for userdata but can't be exploited for performance.

Since, the inception of FAT filesystem in 1981, this performance problem has been there in its design. Although, old Unix filesystem (UFS) received many performance patches to address such problems, however no such effort was made for

M.K. Kundu et al. (eds.), *Advanced Computing, Networking and Informatics - Volume 2*, Smart Innovation, Systems and Technologies 28,
DOI: 10.1007/978-3-319-07350-7_54, © Springer International Publishing Switzerland 2014

FAT filesystem. The concepts of Fast filesystem (FFS) and other related proven techniques, which enhanced the performance of UFS, can be applied to FAT filesystem; however this will create a high performance but incompatible version of FAT filesystem which is not desirable after achieving the spot of highly compatible filesystem. Nevertheless, other techniques which do not demand modification of filesystem design or need little affordable source modification are feasible.

This paper describes the design, simulation and evaluation of a new FAT filesystem design, namely *hFAT*, which stores the most frequently accessed metadata of files in a FAT filesystem on a flash drive while as userdata on the magnetic drive. Specifically, the idea is to store the contents of directories and small amount of most frequently accessed FAT Table of FAT filesystems on a solid-state storage device to eliminate the head positioning latency incurred by FAT filesystem operations.

2 Related Work

Based on the idea of hybrid storage system many proposals have been made. Baker et al. [1] proposed using a small amount of battery-backed RAM to act as a small write buffer to reduce disk accesses. The motive was to prevent losing recent updates to file caches without having to continuously write data back to the disks as soon as updates occur. Similarly, Miller et al. [2] designed a system called *HeRMES* which uses a form of non volatile RAM called Magnetic RAM (MRAM) to act as a persistent cache for a magnetic disk drive. They used MRAM to cache the filesystem metadata and also buffer writes to the magnetic disk drive. Furthermore, Wang et al. [3] proposed *Conquest* which uses a simple partitioning approach in which they place all small files and metadata (e.g. directories and file attributes) in NVRAM (Non-Volatile Random Access Memory) while the remaining large files and their associated metadata are assigned to the magnetic disk drive. Though these approaches significantly improve the performance of a filesystem, however these demand some filesystem source modification in addition to physical overhaul of the machine to support new pieces of hardware.

Moreover, Microsoft Windows 7 utilises hybrid drives with an option known as *ready boost* [4]. The idea is to use flash memory as a persistent buffer or cache to absorb all read and write requests. This was primarily developed to improve the power usage of laptop systems by allowing the magnetic disk drive to spin down during low workload times. Similarly, Soundararajan et al. [5] proposed a hybrid solid-state device and magnetic disk drive system that accumulates a log of changes on the magnetic disk drive before writing them in bulk onto the solid-state device at a later time. Finally, Fisher et al. [6] proposed to optimise the I/O performance of a system by using a large magnetic disk drive and limited size solid-state drive in tandem to store data. They proposed a drive assignment algorithm which determines which device to place data on in order to take advantage of their desirable characteristics while trying to overcome some of their undesirable characteristics. Though these proposals can be added to any filesystem on the fly without any hardware overhaul, however they only use the other device as an auxiliary log until the updates are committed to the actual device. Hence, this procedure demands house-keeping to be done at sometime later.

3 *hFAT* Filesystem Design

As an analogy, FAT32 filesystem keeps FAT Table at the beginning of the volume like UFS keeps the *inode* store. Moreover, in both filesystems directory entries hold filenames. However, *inodes* only maintain allocation information related to a particular file or directory while as the allocation/deallocation list is maintained individually by UFS within *Superblock*. In contrast, the FAT Table records both free and allocated clusters within the volume along with the chain-of-clusters associated with a particular file or directory, while as directory entries (in addition to filename) contains other information such as attributes, size, first cluster in chain and so on [7]. So logically, the design of FAT32 filesystem is bit better than UFS. In-spite of this, FAT32 filesystem suffers from performance problem mainly because of the large seek distance between FAT Table and clusters. Furthermore, the percentage of storage space occupied by FAT Table is very small as compared to clusters (`32 bits/cluster`). As such, as the size of volume increases, the seek distance increases significantly. Furthermore, this seek distance is covered twice as FAT Table is located at the beginning of the volume. The Figure 1 shows the impact of the volume size on maximum seek distance between FAT Table and clusters.

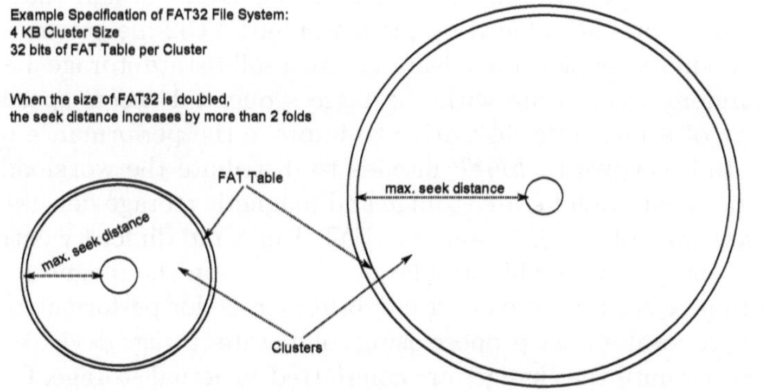

Fig. 1. Impact of volume size on seek distance between FAT Table & Clusters

Also, because directory entries hold actual key for locating the clusters allocated to a file (or directory), by containing the first cluster in the chain, access to this metadata is mandatory. As clusters allocated to a directory can also be scattered, latency is inevitable. Furthermore, in a workload wherein large number of small files, whose size is less than cluster size of FAT32 volume, are accessed, there are more frequent accesses to directory entries than FAT Table. In fact, all the requests can be satisfied from directory entries. However, deletion operations require access to the FAT Table. In contrast, a workload wherein large files are accessed, there are more frequent requests to FAT Table than directory entries. Nevertheless, in both cases, the latency incurred during access to FAT Table and directory entries deteriorate the performance of a FAT32 filesystem.

One possible solution to this would be to modify the FAT32 filesystem the same way as UFS was modified. In other words, re-engineering FAT32 filesystem using the concepts of FFS and C-FFS can yield significant performance gains. However, this is much complex as FAT Table can't be put near every file and directory. Furthermore, this will rigorously modify the FAT32 filesystem design and the modified FAT filesystem will lose its identity. Moreover, FFS overcomes latency problems by making certain assumptions about magnetic disks which no more hold true.

Another possible solution to reduce these latencies would be to *pre-cache* and *delay-write* the whole FAT Table. This is beneficial because this FAT Table is a central metadata store that is needed by every individual file and directory residing on the volume. However, the size of FAT Table can impose serious limitations. As an example, for a `2 TB` FAT32 volume having `8 KB` cluster size, cache required for FAT Table is `1 GB`. Furthermore, FAT32 filesystem is not reliable and thus, any unanticipated crash can result in data loss and possible filesystem corruption. Finally, directory clusters can be placed near the contents of files and directories they list by modifying the source of FAT32 filesystem. However, the algorithmic complexity involved and assumptions to be made make it less appealing.

hFAT is a high performance hybrid FAT filesystem design that overcomes the performance problems faced by plain vanilla FAT32 filesystems. The *hFAT* filesystem intends to exploit the advantages of a solid-state storage device's small and flat random access time with the large sequential access speed and storage capacity of a magnetic disk drive to improve the performance of a FAT32 filesystem. In other words, *hFAT* intends to distribute the workload of FAT32 filesystem between a solid-state storage and magnetic storage devices by placing the small *hot zone* of the filesystem i.e. FAT Table and directory entries, on the solid-state storage device while the large *cold zone* on the magnetic disk drive. Although there exist proposals to use hybrid storage for performance gains, but all of these proposals either propose using solid-state storage device as a cache or an auxiliary log until the changes are committed to actual storage. Furthermore, they demand design and/or source modification. However, *hFAT* proposes using solid-state storage device as a persistent store for the most frequently used FAT Table and directory entries of a FAT32 filesystem. The motive of this design is to satisfy all the accesses to FAT Table and directory entries from a small and low latency solid-state storage device without modifying the design or source. As such, it totally eliminates the head positioning latencies incurred during access

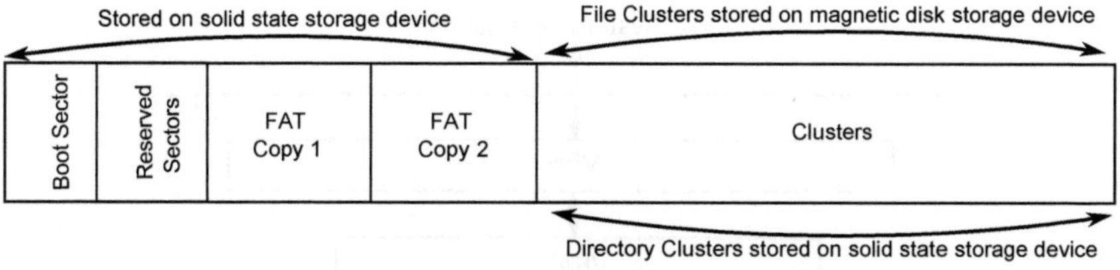

Fig. 2. Logical on-disk layout of *hFAT* design

to FAT Table and directory entries. The Figure 2 shows the logical layout of *hFAT* design.

4 *hFAT* Filesystem Implementation

In order to accomplish this, *hFAT* has two choices; 1) modify the source of FAT32 filesystem, or 2) modify the block device driver. As a matter of fact, modifying the source of either FAT32 filesystem or block device driver have well known limitations and thus is not feasible. However, adding a layer of abstraction just below the FAT32 filesystem and above the block device driver can be exploited. The technique called *driver stacking* enables one driver to be stacked on top of another driver just like filesystem stacking, in which one filesystem module can be stacked on top of another filesystem module. filesystem stacking is not feasible here as the requests made by FAT32 filesystem should be processed rather than processing the requests made to FAT32 filesystem. The stacked block device driver acts like fan-out filesystem by overlaying over two block device drivers; one responsible for doing I/O with solid-state storage drive and another responsible for doing I/O with magnetic disk drive. The driver so stacked can forward the request to appropriate block device driver depending upon the request made by FAT32 filesystem and return the results to upper layers.

The Figure 3 shows the logical design of *hFAT* with a driver stacked on top of two block device drivers.

One may argue that this may add performance overhead to the working of FAT32 filesystem. However, because most of the processing is simple, the performance overhead is expected to be low and will be amortised by faster hardware. Furthermore, it is clear that the space actually meant for FAT Table and directory clusters on magnetic disk is lost. However, in general scenario, the space lost is less than 1% of total space of magnetic disk and thus, is affordable. Moreover, it is mandatory for *hFAT* that the FAT32 filesystem be created newly.

5 Simulation of *hFAT* Stackable Device Driver

The behaviour of *hFAT* stackable device driver can be simulated because of the minimal functionality embedded in it. *hFAT* stackable device driver has to perform only one task; forward the request of upper layer (i.e. FAT32 filesystem)

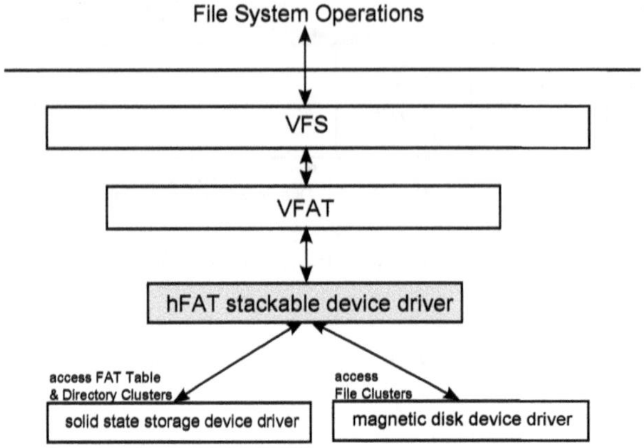

Fig. 3. Design of *hFAT* filesystem using driver stacking

to one of the two below existing device drivers depending upon the type of request. This behaviour can be simulated if FAT32 filesystem block trace of some workload or benchmark is fed to the simulator along with the information necessary to qualify the request as metadata or userdata.

As the motive of this simulation is to identify the number of requests satisfied from solid-state storage device and magnetic disk drive, we provide both block trace and the necessary information. Also, as *hFAT* makes no changes to FAT32 filesystem design and works below a level at which FAT32 filesystem operates, the efficiency will be achieved if *hFAT* sends more requests to solid-state storage device than magnetic disk drive. Indeed, this information is reported by our simulator. However, we need to further evaluate the efficiency of *hFAT* by assigning each block access to solid-state device and magnetic disk drive some latency values to identify the reduction in latency of operations.

6 Experiment

In order to illuminate specific operations which are enhanced by *hFAT*, the FAT32 filesystem was exercised using *Sprite LFS small-file* microbenchmark to get block trace. However, instead of writing, reading and deleting `1 KB` files, the benchmark wrote, read and deleted `4 KB` files. This is in accordance to repeatedly reported observation that filesystem workloads are dominated by accesses to small files, typically `4 KB` or less in size [8]. Furthermore, *Sprite LFS small-file* microbenchmark was implemented as a simple shell script using `bash-version 4.1.7(1)`. Moreover, the caches were flushed after every phase of *Sprite LFS small-file* benchmark. The block trace was captured via `/proc/sys/vm/block_dump` interface of Linux kernel to log only those blocks whose request was not satisfied by the buffer cache. Furthermore, because of the simplicity of the simulation, the simulator was implemented in-house as a shell script using `bash-version 4.1.7(1)`.

The experiment was conducted on an Intel based PC with `Intel Core i3 CPU 550 @ 3.20 GHz` processor with total `4 MB cache` and `2 GB DDR3 1333 MHz SDRAM`. The hard drive used is a `320 GB SATA` drive with on board cache of `8 MB` and `15.3 ms` of reported average access time. The drive is partitioned into 20 `GB` primary partition to hose the Fedora Core 14 operating system kernel version `2.6.35.14-95.fc.14.i686` and another `5 GB` partition to mount FAT32 filesystem. The partition is large enough to hold all the files and small enough to fit in one zone of the disk. Moreover, the experiment was run on a newly created filesystem. Furthermore, during the execution of *Sprite LFS small-file* microbenchmark, the Linux was set to run at `run-level 1` to reduce the random effect of other applications and demons. Also, the experiment was repeated `5 times` and all the results were averaged; the standard deviation was less than `3 %` of the average in all cases.

7 Results and Discussion

The result of the simulation is shown in Table 1 and 2. Table 1 shows the number of block reads satisfied by both non-HDD device and by HDD device against each operation of the *Sprite LFS small-file* microbenchmark. Similarly, Table 2 reports the same statistics but for blocks written. Moreover, in each category, the percentage of block accesses satisfied by non-HDD device is calculated. There are many things worth noticing here.

Table 1. Simulation report showing distribution of blocks read

Operation	Blocks Read (non-HDD)	Blocks Read (HDD)	Access %age of non-HDD
Create 10,000 4 KB Files	1380	0	100 %
Read 10,000 4 KB Files	643	10,000	6.04 %
Delete 10,000 4 KB Files	721	0	100 %

First, in case of blocks being read, in two phases `100 %` of block reads are satisfied from non-HDD device while in one phase the percentage is as low as `6.04 %`. This means that `66 %` of the operations of this workload can effectively exploit the `100 %` benefits of low and flat latency of non-HDD device for reading.

Second, in case of blocks being written, in two phases `100 %` of block writes are satisfied from non-HDD device while in one phase the percentage is as low as `17.10 %`. Again, this means that `66 %` of the operations of this workload can effectively exploit the `100 %` benefits of low and flat latency of non-HDD device for writing.

Table 2. Simulation report showing distribution of blocks written

Operation	Blocks Written (non-HDD)	Blocks Written (HDD)	Access %age of non-HDD
Create 10,000 4 KB Files	2063	10,000	17.10 %
Read 10,000 4 KB Files	626	0	100 %
Delete 10,000 4 KB Files	785	0	100 %

Third, the phases in which block reads or writes are not 100 % satisfied from non-HDD device correspond to reading and writing 10,000 files each of size 4 KB. This is expected as in these two phases the ratio of metadata-access to userdata-access is very low. However, these phases are mutually exclusive and such an operation within each phase is benefited from other operation of the phase that exploits 100 % benefits of non-HDD device. As an example, the first phase wherein 10,000 files are created, the percentage of blocks read from non-HDD device is 100 % while as the percentage of blocks written to non-HDD device is 17.10 % (expected as ratio of metadata to userdata written is low). However, the 100 % block reads from non-HDD device augments the 17.10 % blocks writes in two ways; 1) the reads in operations experience low and flat latency of non-HDD device, and 2) the repositioning of the read/write head of HDD device during such reads is eliminated which otherwise would have created latency for blocks to be written. Similar benefits are exploited in second phase.

Finally, in third phase 100 % block reads and writes are satisfied from non-HDD device. This is expected as this phase only deals with metadata.

Furthermore, we evaluated the performance of *hFAT* by assigning various access latencies to non-HDD device. To evaluate over a range of access latencies, we assigned weights in terms of percentage of access latency of HDD device with the granularity of 1 %. In other words, we assigned 100 possible latency values to non-HDD device where each value corresponds to some percent of latency of HDD. As an example, if HDD device has access latency of 200 ms, then we assigned latencies to non-HDD device ranging from 2 ms to 200 ms with the step size of 2.

Moreover, the read/write latency is not symmetric in neither solid-state storage device nor magnetic disk drive. Although, less expensive solid-state storage devices typically have write speeds significantly lower than their read speeds, but higher performing ones have similar read and write speeds. Furthermore, HDDs generally have slightly lower write speeds than their read speeds. Nevertheless, we assigned same latency values to both types of operations in each type of device. Furthermore, we assigned average access latency of HDD to each accessed block of HDD. Unfortunately, this way the benefits gained by removing the repositioning of the head which reduces the inter-userdata block latency can't be calculated. This means that the evaluation will yield the upper bound on the latency incurred by operations and can be expected to go down.

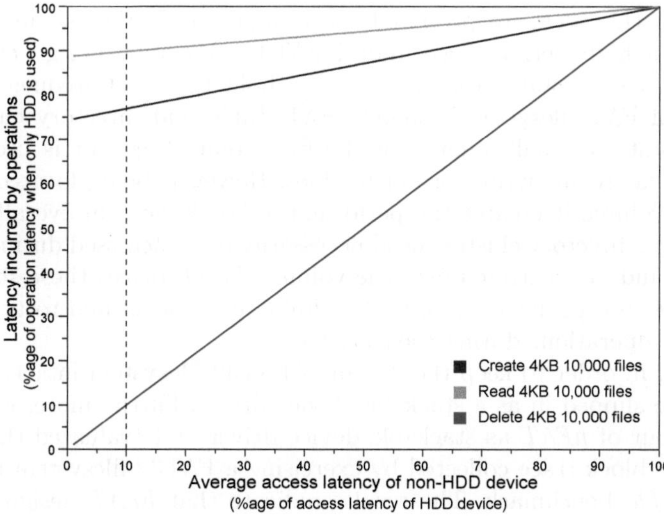

Fig. 4. Affect of various latencies on total latency of *hFAT* operations

Figure 4 shows the graph plotted for three phases of the benchmark against the range of access latencies across x-axis and the latency incurred by operations in terms of percentage of actual latency when only HDD is used. This graph indicates that those phases which have steep slope are highly affected by the latency of HDD device while as others are less affected. In the figure, the phase corresponding to deletion of files is highly affected by the latency of HDD device. Next comes the phase that creates the files followed by the phase that reads these files. This graph also shows that using *hFAT* with a non-HDD device having latency 10 % of that of HDD can reduce the latency of write operations by as minimum as 10 %, of read operations by a minimum of 25 % and of delete operations by 90 %. It is to be noted that this reduction is the lower bound as the average access time of HDD is taken into consideration.

Furthermore, the typical access latency of non-HDD is about 0.1 ms while as for HDDs the average access time ranges between 5 to 10 ms. The ratio of 0.1 ms to 5 ms corresponds to 2 % while as the ratio of 0.1 ms to 10 ms corresponds to 2 %, both being lesser than 10 %. This is even lesser than what is required as per our observation.

8 Conclusion

We can conclude that using hybrid storage system for enhancing the performance of disk filesystems is valuable. The fact is that the most frequently accessed structure of a filesystem includes a small amount of filesystem wide metadata. This metadata if stored on solid-state storage device can not only eliminate the seeks and rotational delays incurred during its access but will significantly reduce the total latency incurred by operations.

Based on this idea, we proposed the design and further simulated the behaviour of a high performance hybrid FAT32 filesystem design called *hFAT*. *hFAT* uses a solid-state storage device to hold the most frequently accessed metadata of FAT filesystems, namely FAT Table and directory clusters. FAT Table accounts to small amount of the filesystem space but is used by every operation that reads, writes or deletes files. However, being located at the beginning of volume it creates the performance bottle neck in overall filesystem. Furthermore, directory clusters need necessarily to be accessed during filesystem operations and are scattered over the volume. *hFAT* places these structures on a solid-state storage device in order to eliminate the seeks and rotational delays incurred by operations during their access.

However, in order to keep the design of FAT32 filesystem intact, we propose *hFAT* to be slipped in as a stackable device driver. Furthermore, we simulated the behaviour of *hFAT* as stackable device driver and evaluated the design by running the block trace collected by exercising a FAT32 filesystem using *Sprite LFS small-file* benchmark. The results indicate that *hFAT* design can reduce the latency incurred by FAT32 filesystem operations by a minimum of 25 %, 10 % and 90 % during writing, reading and deleting large number of small files respectively, if a solid-state storage device having latency lesser or equal to 10 % of that of magnetic disk is used as an addition.

References

1. Baker, M., Asami, S., Deprit, E., Ouseterhout, J., Seltzer, M.: Non-volatile memory for fast, reliable file systems. In: Proceedings of the Fifth International Conference on Architectural Support for Programming Languages and Operating Systems, ASPLOS-V, pp. 10–22. ACM, New York (1992)
2. Miller, E.L., Brandt, S.A., Long, D.D.E.: Hermes: High-performance reliable mram-enabled storage. In: Proceedings of the Eighth Workshop on Hot Topics in Operating Systems, pp. 95–99. IEEE Computer Society, Washington, DC (2001)
3. Wang, A.-I.A., Kuenning, G., Reiher, P., Popek, G.: The conquest file system: Better performance through a disk/persistent-ram hybrid design. Transactions on Storage 2, 309–348 (2006)
4. Matthews, J., Trika, S., Hensgen, D., Coulson, R., Grimsrud, K.: Intel turbo memory: Nonvolatile disk caches in the storage hierarchy of mainstream computer systems. Transactions on Storage 4, 4:1–4:24 (2008)
5. Soundararajan, G., Prabhakaran, V., Balakrishnan, M., Wobber, T.: Extending ssd lifetimes with disk-based write caches. In: Proceedings of the 8th USENIX Conference on File and Storage Technologies, FAST 2010, p. 8. USENIX Association, Berkeley (2010)
6. Fisher, N., He, Z., McCarthy, M.: A hybrid filesystem for hard disk drives in tandem with flash memory. Computing 94, 21–68 (2012)
7. Heybruck, W.F.: An introduction to fat 16/fat 32 file systems, http://www.hitachigst.com/ (accessed August 2010)
8. Agrawal, N., Bolosky, W.J., Douceur, J.R., Lorch, J.R.: A five-year study of file-system metadata. Transactions on Storage 3(3), 338–346 (2007)

Efficient Parallel Heuristic Graph Matching Approach for Solving Task Assignment Problem in Distributed Processor System

Pradeep Varma and Swaathi Ramesh

Department of Computer Science and Engineering,
National Institute of Technology, Tiruchirapalli, India
pradeep6varma@gmail.com, swaathiramesh@outlook.com

Abstract. Solving the Task Assignment Problem is important for many real time and computational scenarios where many small tasks need to be solved by multiple processors simultaneously. In this paper, a Heuristic and Parallel Algorithm for Task Assignment Problem is proposed. Results obtained for certain cases are presented and compared with the optimal solutions obtained by already available algorithms. It is observed that the proposed algorithm works much faster and efficient than the existing algorithms. The paper also demonstrates how the proposed algorithm could be extended to multiple distributed processors.

1 Introduction

The Task Assignment Problem plays an important role in recent computational systems which involve processing multiple tasks in a multiprocessor environment. The Task Assignment Problem has been proved to be a NP-Hard problem. Several Algorithms and methodologies [4–8] have been proposed to solve the Task Assignment Problem. Most Algorithms use Graph Partitioning and Graph Matching Techniques. Significant research has been carried out in solving the Task Assignment Problem in a parallel environment. This paper discusses the Shen Tsai's Algorithm [1] on Task Assignment. A detailed description about the parallel algorithm suggested in the paper [3], HGM Algorithm to solve the Task Assignment Problem in a distributed environment is given. The observations made in the above algorithms are analyzed and scope for improvisations has been identified and a new Parallel Heuristic Graph Matching Algorithm is proposed to improvise the current techniques. Further, this paper contains a detailed analysis, both qualitative and quantitative, about the algorithm. This signifies how efficient the proposed algorithm is, compared to the existing ones.

The proposed algorithm tries to solve the basic Task Assignment Problem of mapping k distinct tasks to p different processors using n processor distributed system. An analysis on Shen Tsai's Algorithm for Task Assignment [1] based on A* Algorithm [2] describes how the sequence of assigning various tasks to the processors in the solution tree formed in the algorithm contributes to the total computation time of deciding the most optimal assignment. Hence, the

M.K. Kundu et al. (eds.), *Advanced Computing, Networking and Informatics - Volume 2*, Smart Innovation, Systems and Technologies 28,
DOI: 10.1007/978-3-319-07350-7_55,

proposed algorithm tries to identify the optimal sequence of Tasks for assignment to the processors in the solution tree so that the Total Computational Time is optimized. Further, a qualitative analysis on [3] shows how the HGM algorithm deals with the Task Assignment Problem of mapping k distinct tasks to p different processors using n processor distributed system where n> p only. But our proposed algorithm is designed to work for all general Task Assignment Problem cases where n>p, n<p or n=p. This paper proposes a new algorithm which identifies the optimal sequence of tasks in a sequential manner and solves the task assignment problem in a distributed environment for any number of processors in the distributed environment.

2 Related Work

2.1 Shen Tsai's Article on Task Assignment Based on A* Algorithm

Consider the Task Assignment Problem of mapping k distinct tasks to p different processors. Let the tasks and available processors be represented by following Task and Processor sets T and P.

$$\text{Task Set (T)} = \{t_1, t_2, t_3, ..., t_k\}$$

$$\text{Processor Set (P)} = \{p_1, p_2, p_3, ..., p_p\}$$

According to Shen and Tsai, all tasks can be represented in a single task graph. Each node in the task graph corresponds to a particular task while an edge between them corresponds to the communication cost between 2 tasks when processed on 2 different processors Fig. 1. Similarly all the processors can be represented by a processor graph where the graph represents various interconnections between processors in the actual distributed system Fig. 1.

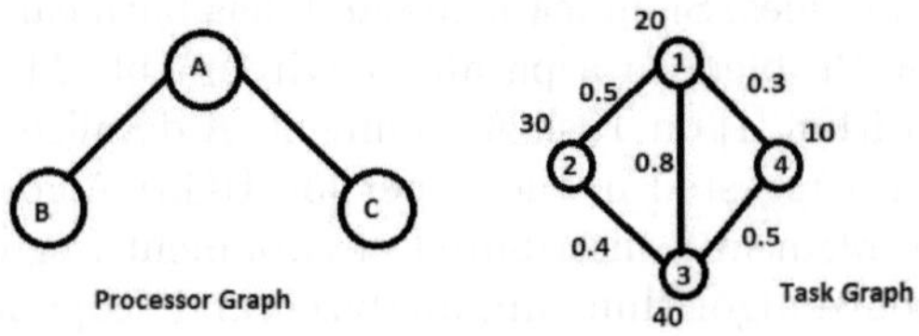

Fig. 1. Representation of task and processor graphs

Shen and Tsai propose that the task assignment of T to P is nothing but a homomorphic mapping of the task graph onto the processor graph. So the objective is to find an ideal homomorphic mapping between the task graph and the processor graph such that the total completion time of all the tasks is minimum and most optimized.

2.2 Mathematical Formulation

Let us consider any general mapping M, M: T to P, where T and P correspond to the Task Set and Processor Set respectively.

Let the completion time of a particular mapping M is denoted by Time (M) function.

1. To calculate Time(M), we need to calculate Time(M, Pk), the time of completion for a processor Pk in a given mapping M, for all Pk belongs to P. We need to calculate the time of completion individually for each processor because in any parallel environment the total time for completion of all tasks is the maximum of the times taken by each processor to solve all the tasks assigned to the processor. Consider a mapping of Tasks 1, 2, 3, 4, 5 to processors p1, p2 as $1_{p1}, 2_{p2}, 3_{p1}, 4_{p1}, 5_{p2}$.

Time (M) = Maximum (Time (M, P_1), Time (M, P_2))

where, Time (M, P_k) $= \sum_{t_i}(T_{i,k} + \sum_{t_l}$ communication $\text{cost}_{i,l})$ and $T_{i,k}$ is the computation time for the ith task on the kth processor

such that t_i belongs to T, t_i is allocated to P_k, t_l belongs to T and t_l is not allocated to P_k.

2. Now, we know that Time (M) = Maximum (Time (M, P_k)) for all P_k belongs to P.

Hence, we calculate the total time involved for computation for any given mapping. An optimal mapping (M) corresponds to minium Time(M).

Solution for Optimal Mapping (M):

The solution is represented by a tree, called the *Solution Tree* built according to the following rule:

Let $\{t_1, t_2, t_3, ..., t_k\}$ be a permutation of tasks in task set T. At any level i of the tree only task t_i is assigned to all the processors. Each node in the solution tree has (tasks, processors) as its attributes. So in building the solution tree we start with a dummy root node (level zero) and proceed with level 1 by assigning Task t_1 to all processors and get the initial nodes. From here we calculate f value for each node in level 1 and expand the node with least f value.

After expanding a node, we then scan through the entire tree for minimum f value and go expanding it. This is continued until the goal state where all tasks are assigned to a processor.

Calculation of f value is done in a heuristic fashion:

$$f_n = g_n + h_n$$

where,

g_n is the computation time involved for reaching a particular node n from start node,

h_n is the Heuristic approximate for present node to reach final goal state.

Calculation of the heuristic part (h) of the f value is an interesting topic of research and many techniques have been proposed to find the heuristic. It is assumed that the heuristic is obtained from one of the existing algorithms [3].

Observations
The efficiency of Shen Tsai's Algorithm is determined by the number of nodes generated in the solution tree. Our analysis proves that the number of nodes generated depends on the permutation of T, i.e., the sequence in which we assign the tasks to the processors in the solution tree reflects the number of nodes generated in the solution tree.

i.e., $T_1 = \{t_1, t_2, t_3....t_n\}$
$T_n = \{t_n, t_{n-1}, t_{n-2}....t_1\}$ etc.

All of the above permutations generate a different number of nodes in the solution tree thereby resulting in different computational times to get the optimal assignment. **Hence, finding the optimal permutation, i.e., the permutation which results in the most optimal solution is an interesting point of research.**

2.3 HGM Algorithm for Task Assignment Proposed by R. Mohan

This paper discusses about parallelizing the Shen Tsai's Algorithm to solve the Task Assignment Problem of mapping k distinct tasks to p different processors in an n processor distributed system. The algorithm works as follows:

- Assumptions: The number of processors available in the distributed system to solve the task assignment problem is greater than or equal to the number of processors involved in the task assignment problem.
- Algorithm:
  1. Initially consider the Task Assignment Problem of mapping k tasks to n different processors to be solved on an n processor distributed system.
  2. Each processor in the Task Assignment Problem is assigned to a single processor in the distributed system. If processor A of the problem is assigned to processor 1 in the distributed system, then Processor 1 can assign tasks only to Processor A.
  3. The processors in the distributed system start assigning tasks to the processors in the problem until a fixed interval (3 or 4 tasks) after which the f value is calculated for the assignment in each of the processor in the distributed system.
  4. All the processors in the distributed system communicate and interchange the f values to decide the optimal assignment. Once the optimal assignment is decided all the processors proceed with assigning the tasks about this optimal assignment.
  5. Steps (3) and (4) are repeated until all the tasks are assigned and an optimal mapping is achieved.

Example: Consider 2 processors and 5 tasks TASK ASSIGNMENT problem. We a have 2 distributed processors to solve this problem. Let us assign processor A to processor 1 and processor B to processor 2 (Fig. 2).

So 1A 2A 3A is much better. Both the processors start expanding about 1A 2A 3A (Fig. 3).

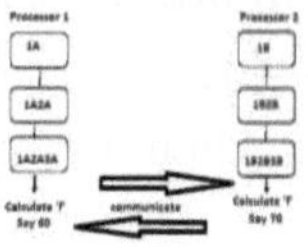

Fig. 2. Task assignment -1

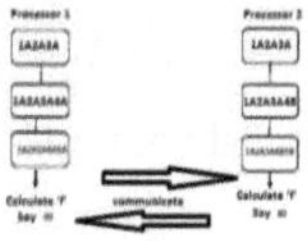

Fig. 3. Task assignment-2

Hence the optimal assignment is 1A2A3A4B5B.

2.4 Observations

We observe that the above parallel algorithm can be extended to the following cases only:

1. n tasks - p processors Task Assignment Problem and you have p processors to solve the Task assignment problem, or,

2. n tasks- p processors Task Assignment Problem and you have greater than p distributed processors to solve the task assignment problem

Hence, finding the parallel algorithm which works for solving all general Task Assignment Problems of mapping k distinct tasks to p different processors in n processor distributed system where $n>p$, $n<p$ or $n=p$ is an interesting point of research.

3 Problem Statement

To develop a modified parallel heuristic task assignment algorithm which:

- Identifies the ideal permutation of tasks that results in the least computational time to find the most optimal assignment, and
- Extends the parallel HGM Algorithm to run for any number of Generic Processors in the Distributed System.

4 Solution

1. Identify the ideal permutation :
   Consider Task Set (T) $=\{t_1, t_2, t_3, ..., t_k\}$,Processor Set (P) $=\{p_1, p_2, p_3....p_p\}$
   It is needed to map k Tasks to p processors. Now to obtain the solution tree, we should decide on the optimal permutation,

$\pi = \{Y_1, Y_2, Y_3, ..., Y_k\}$,Such that π is a parameter of $\{t_1, t_2, t_3....t_k\}$.
The sequence of the tasks is significant (as discussed) because in the solution tree only the kth task in permutation is mapped to all processors in the kth level of the solution tree.Now the various ways of choosing the permutation are:

(a) Based On Computation Time: Choose π such that
Mean computation time (Y$_j$) > Mean computation time (Y$_{j+1}$)
i.e, $\left(\frac{T_{j,1}+T_{j,2}+Tj,p}{p}\right) > \left(\frac{T_{j+1,1}+T_{j+1,2}+Tj+1,p}{p}\right)$,
where, T$_{j,i}$ refers to the cost of processing when Y$_j$ is executed on Processor i.
(b) Based On Communication Cost: Choose π such that
Total Communication time (Y$_j$) > Total Communication time (Y$_{j+1}$)
i.e, CC [j] > CC [j+1]
Where CC [j] corresponds to the summation of all the communication costs of Task Y$_j$ with all other tasks which do not run in the same processor as that of Y$_j$.The first way of choosing the permutation is useful if the computation costs of tasks are more significant than the inter task communication costs. The second way is better when the inter task communication cost is greater than the computation cost of tasks. This paper proposes an alternate way which takes care of all average cases where both computation cost and communication cost are equally significant.

At this stage, a quantity, α is defined for each task which proves helpful in deciding the most optimal permutation. α is defined for any particular task, t$_j$ as:
$\alpha(Y_j)=\left(\frac{T_{j,1}+T_{j,2}+Tj,p}{p}\right) + \left(\frac{CC[j]}{k}\right)$.
π is chosen such that: $\alpha(Y_j) > \alpha(Y_{j+1})$
Hence, the ideal permutation which gives an optimal assignment in the most optimal time is found.

2. Modified HGM Algorithm :
Consider the Task Assignment Problem of mapping k distinct tasks to p different processors in an n processor distributed system.
Algorithm:
(a) Distribute the p processors of the problem into n processors.
(b) Any processor in the distributed system can assign the tasks only to those which are assigned to it. i.e., if processors A, B, C are assigned to Processor1 in the distributed system, Processor1 can assign tasks only to A, B and C.
(c) Now in each of the processor 1, 2, 3, ..., N of the distributed system a solution tree is built individually, containing only the assignments to the processors which are assigned to this particular processor up to 3 levels, break and find the best node space from each solution tree based on the f value.
(d) Each of the distributed processors communicates to decide on the ideal state space and start expanding about it.

(e) Step(c) and Step(d) are repeated until all the tasks are assigned to the processors and an optimal assignment is obtained.
Example:
Consider,Task Set(T) = {1, 2, 3, 4, 5, 6, 7, 8, 9, 10},
Processor Set(P) = {A, B, C, D, E, F}
Let p1, p2, p3 be the Processors in the distributed System to solve the Task Assignment problem.
Then, Fig. 4 represents the state spaces after 3 allocations based on above algorithm in the 3 processors of the distributed system,Say at the end of 3rd level fp1, fp2 and fp3 correspond to the ideal f values from Processors P1, P2 and P3 respectively, amongst which fp2 has the least f value.
Let us say fp2 corresponds to the node 1D2D3C in P2's solution tree. Now, all 3 processors p1, p2, and p3 start expanding about 1D 2C 3D state space. This process is repeated until a goal state where all the tasks are assigned to processors is achieved.

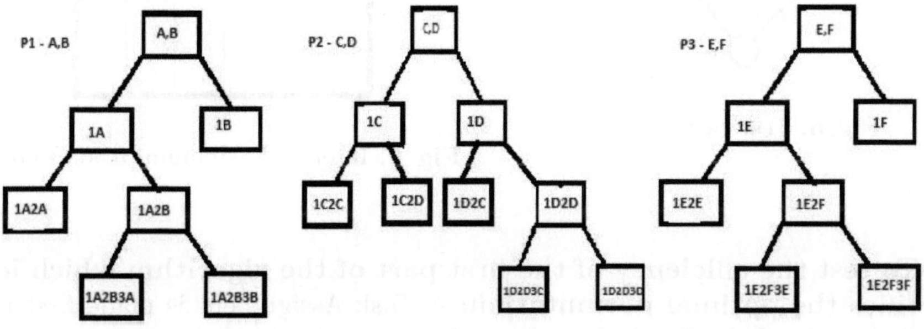

Fig. 4. Representation of task and processor graphs

5 Analysis

1. The efficiency of the first part of the algorithm to find the most optimal task permutation is very much obvious and evident. It is well supported by the experimental data represented in the next section.
2. The parallel part of the algorithm is very efficient because,in a general problem of k tasks- p processors Task problem to be solved using n processors, the proposed algorithm will work for all:
   (a) p>n
   (b) p=n
   (c) p<n
   (Though it is highly efficient in case of p>n.)

This algorithm uses Shen Tsai's algorithm to the maximum by implementing it in finding intermediate ideal state spaces in each of the distributed processors unlike the HGM algorithm which uses a brute force approach and simple heuristic based on A Algorithm.*

6 Experimentation

The test case in Fig. 5 represents a task graph with vertices pointing to tasks and the edges pointing to inter task communication cost when processed on different processors.The number of nodes in the task graph is 6, which means that there are 6 tasks defined by T = 0, 1, 2, 3, ..., 5 which need to be mapped. The computation time associated with these tasks is defined by the set TP = 10.0, 15.0, 5.0, 20.0, 15.0, 10.0. The inter task communication is defined by the matrix C (Refer Fig. 6). Let us assume the tasks should be mapped on to 2 processors.

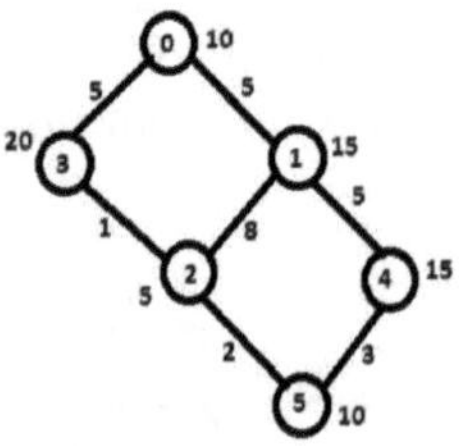

Fig. 5. Test case

0	5	0	5	0	0
5	0	8	0	2	0
0	8	0	1	0	2
5	0	1	0	0	0
0	2	0	0	0	3
0	0	2	0	3	0

Fig. 6. Inter-task communication matrix

1. **To test the efficiency of the first part of the algorithm which identifies the optimal permutation:** A Task Assignment is made first using the regular Shen Tsai's Algorithm. Later task assignment is done using the most optimal assignment.An index called the optimality index denoted by σ is introduced as,
$$\sigma = \frac{\text{Optimal Turnaround Time in the proposed Algorithm}}{\text{Optimal Turnaround Time in Shen Tsai's Algorithm}}$$
Similarly an index, θ is defined as,
$$\theta = \frac{\text{Number of Nodes Generated in proposed Algorithm}}{\text{Number of Nodes Generated in Shen Tsai's Algorithm}}$$
The results are expressed in the following Table 1, where,
ω-Turnaround Time In Shen Tsai's Algorithm
λ-Turnaround Time In Proposed Algorithm
n_s-No. Of Nodes Generated In Shen Tsai's Algorithm
n_p-No. Of Nodes Generated In Proposed Algorithm
The indices θ and σ represent time comparison factor and space comparison factor of both the algorithms. We observe that both θ and σ fall below 1 signifying the superiority of proposed algorithm over the Shen Tsai's Algorithm. It is observed that though the change in time complexity is less significant, the change in space complexity is very significant and appreciable.
2. **To test the efficiency of the parallel part of the proposed algorithm:** Consider the following two scenarios:

Table 1. Results

Optimal Mapping	ω	λ	n_s	n_p	σ	θ
0A1B2B3A4B5A	42.09	41.8	32	12	0.993	0.375

Table 2. Results-Parallel algorithm

Optimal Mapping	ω	λ	n_s	n_p	σ	θ
0A1B2B3A4B5A	21.2	16.38	24	19	0.77	0.79

Scenario 1: The above task problem is solved in a parallel way with HGM Algorithm using 4 processors.
Scenario 2: We solve the above task problem in a parallel way using the proposed parallel algorithm with 2 processors.
Let σ and θ be redefined to this context as follows,

$$\theta = \frac{\text{Effective Turnaround Time in the proposed Algorithm}}{\text{Effective Turnaround Time in HGM Algorithm}}$$

where, Effective Turnaround time is defined as Optimal Turnaround Time per processor used in the distributed system

$$\sigma = \frac{\text{Number of Nodes Generated in proposed Algorithm}}{\text{Number of Nodes Generated in HGM Algorithm}}$$

The results are expressed in Table 2.
Both σ and θ fall below 1 signifying the superiority of proposed algorithm over the HGM algorithm proposed in the research paper [3]. A very significant change in time and space complexities is observed experimentally thus establishing the efficiency of the proposed algorithm. It should also be noted that only half the number of processors were used in the distributed system to solve the task assignment problem. The algorithm is still more efficient than traditional HGM algorithm.

7 Conclusion and Future Work

This paper establishes a methodology to parallelize the Heuristic Graph Matching Algorithm proposed by Shen Tsai, which can be solved with the help of any generic distributed system containing any number of processors. Further the paper provides an approach to minimize the number of nodes generated in the solution tree developed to obtain the Optimal Task Assignment Mapping in the Shen Tsai's algorithm. Due to parallelizing and proceeding with an optimum Permutation of tasks the number of state spaces generated is reduced significantly and hence the complexity reduces. The proposed Parallel Algorithm follows a divide and conquer approach to solve the discussed Task-Assignment Problem.

Some of the further research areas are listed below:

1. To investigate the algorithm for larger test cases
2. To identify the ideal permutation based on other properties specific to the Task Assignment Problem
3. To devise an algorithm for heterogeneous processor systems.

References

1. Shen, C.-C., Tsai, W.-H.: A Graph Matching Approach to Optimal task assignment in Distributed computing systems using a Minimax Criterion. IEEE Transactions on Computers 34(3) (1985)
2. Cormen, L., Rivest, S.: A star Algorithm. Introduction To Algorithms (2001)
3. Mohan, R.: A parallel Heuristic Graph Matching Method for Task Assignment, pp. 575–579. IEEE
4. Chen, W.-H., Lin, C.-S.: A hybrid heuristic to solve a task allocation problem. Comput. Oper. Res. 27(3), 287–303 (2000)
5. Efe, K.: Heuristic models of task assignment scheduling in distributed systems. IEEE Comput. 15(6), 50–56 (1982)
6. El-Rewini, H., Lewis, T.G., Ali, H.H.: Task Scheduling in Parallel and Distributed Systems. Prentice-Hall (1994)
7. Giersch, A., Robert, Y., Vivien, F.: Scheduling tasks sharing les on heterogeneous master-slave platforms. In: 12th Euromicro Workshop on Parallel Distributed and Network-Based Processing (2004)
8. Hamam, Y., Hindi, K.S.: Assignment of program modules to Processors: A simulated annealing approach. European J. Oper. Res. 122(2), 509–513 (2000)

An Approach for Compiler Optimization to Exploit Instruction Level Parallelism

Rajendra Kumar[1] and P.K. Singh[2]

[1] Uttar Pradesh Technical University, India
[2] Madan Mohan Malviya University of Technology, Uttar Pradesh, India
rajendra04@gmail.com, topksingh@gmail.com

Abstract. Instruction Level Parallelism (ILP) is not the new idea. Unfortunately ILP architecture not well suited to for all conventional high level language compilers and compiles optimization technique. Instruction Level Parallelism is the technique that allows a sequence of instructions derived from a sequential program (without rewriting) to be parallelized for its execution on multiple pipelining functional units. As a result, the performance is increased while working with current softwares. At implicit level it initiates by modifying the compiler and at explicit level it is done by exploiting the parallelism available with the hardware. To achieve high degree of instruction level parallelism, it is necessary to analyze and evaluate the technique of speculative execution control dependence analysis and to follow multiple flows of control. The researchers are continuously discovering the ways to increase parallelism by an order of magnitude beyond the current approaches. In this paper we present impact of control flow support on highly parallel architecture with 2-core and 4-core. We also investigated the scope of parallelism explicitly and implicitly. For our experiments we used trimaran simulator. The benchmarks are tested on abstract machine models created through trimaran simulator.

Keywords: Control flow Graph (CFG), Edition Based Redefinition (EBR), Intermediate Representation (IR), Very Large Instruction Word (VLIW).

1 Introduction

Instruction Level Parallelism [1] represents the typical example that redefines the traditional field of compilation. It raises the issues and challenges that are not addressed in traditional compilers. To scale up the amount of parallelism at hardware level, the compiler takes on increasingly complex responsibilities to ensure the efficient utilization of hardware resources [3].New strategies may result in long compilation time to speed-up the compilation, for this two things are necessary to be considered:

1. Careful partitioning of application
2. Selection of better algorithm for the purpose of branch prediction analysis and optimization.

M.K. Kundu et al. (eds.), *Advanced Computing, Networking and Informatics - Volume 2*,
Smart Innovation, Systems and Technologies 28,
DOI: 10.1007/978-3-319-07350-7_56,

The major outcome of ILP compilers is to enhance the performance by elimination the complex processing needed to be parallelized the program. ILP compilers accelerate the non-looping codes widespread in most of the applications.For analysis purposes, we need statistical information (extracted through the trimaran simulator). The statistical compilation [4] improves the program optimization and scheduling. The improvement in performance of frequently taken path is also supported by statistical compilation. The conventional compilers and optimizers do not produce optimal code for ILP processor. Therefore the designers of processors and compiler would have to find useful methods for ILP compiler optimization which produce maximally efficient ILP processor code for processing references to subscripted array variable.To achieve high performance of ILP, the compiler must jointly schedule multiple basicblock [5]. The compiler optimization includes:

1. Basicblock formation and their optimization
2. Superblock optimization
3. Hyperblock optimization

A superblock [6] is a structure in the form of control flow with a single entry and multiple exits and it has no side entrances. A hyperblock [10] is predicted region of code that contains straight-line sequence of instruction with a single entry point and possibly multiple exit points. The formation of hyperblock is due to modification in if-conversion.

The hyperblock optimization adds the if-conversion to superblock optimization. The if-conversion is the process of replacing the branch statements with compare operations and associated operations with predicate defined by the comparisons.The exploitation of ILP is increased as early as the branches are predicted. The Control Flow Graph (CFG) and Predicated Hyperblockinitiate this process. Fig.1 shows a Control Flow Graph.

The predicated hyperblock of Fig 1 is as follows:

$v = rand()$	if true
$v = q$	if c_1 (if c_1 is true, $v = q$ else nullify)
$v = v + 3$	if c_2 (if c_2 is true, $w = v + 3$ else nullify)
$x = v * 3$	if true

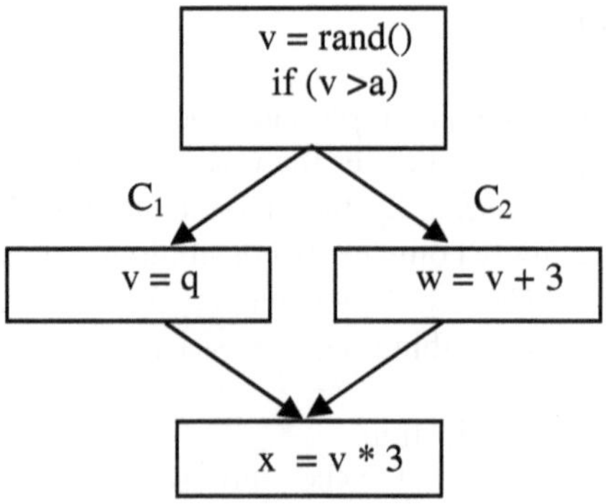

Fig. 1. Control Flow Graph

2 Related Work

Instruction-level parallel processing has established itself as the only viable approach for achieving the goal of providing continuously increasing performance without having to fundamentally re-write the application. The code generation for parallel register share architecture involves some issues that are not present in sequential code compilation and is inherently complex. To resolve such issues, a consistency contract between the code and the machine can be defined and a compiler is required to preserve the contract during the transformation of code. [7] has proposed a Parallel Register Sharing Architecture for Code Compilation. The navigation bandwidth of prediction mechanism depends upon the degree of ILP. It can be increased by increasing control flow prediction [2] at compile time. In [8], the author has presented the Role of multiblocks in Control Flow Prediction using Parallel Register Sharing Architecture.

There are two major questions regarding if-conversion: (i) when to if-convert, and (ii) what to if-convert. [11] indicates that performing if-conversion early in the process of compilation has benefit of enabling classical optimization of predicated instructions. As the control flow prediction is increases, the size of initiation is increased that permit the overlapped execution of multiple independent flow of control. [9] presented Control Flow Prediction through Multiblock Formation in Parallel Register Sharing Architecture.

The impact of ILP processors on the performance of shared memory multiprocessors with and without latency hiding optimizing software prefetching is represented in [12].

3 Our Approach

Our work aims to explore the parallelism at compiler (software) and hardware (architecture) level.

3.1 At Software Level

For our purpose we have modified the compiler that uses an Intermediate Representation (IR). There are three basic steps that our compiler includes:

1. record information about the Control Flow Graph so that the step 2 can work with Control Flow Graph and compute the Denominator Tree (DT).
2. introduce function into CFG to modify the compiler.
3. map the function to nodes of basic blocks that are handled by the code generator.

Construction of CFG and DT

The constriction of Control Flow Graph and Denominator Tree are put in separate phases of the compiler. This task is performed after the semantic analysis. This includes the sequences of steps that this phase generation is supported to:

1. Build a CFG for each function.
   The graph will be stored in new data structure separate from the abstract syntax tree (AST). The package from the trimaran simulator in the framework contains the classes for representing the control flow graphs. The framework includes the code to determine the set of variables that are modified or used for each basic block.
2. Construction of Denominator Tree.
3. Computation of dominance frontier.

Formation of SSA **(*Static Single Assignment*).** Following steps have been applied for each function added:

1. For each simple source language variable we determine the set of nodes where compiler-modified functions are inserted.
2. The compiler ensures the allocation of space for each newly inserted variable. To keep track of variable version, a stack data structure is used.

The above steps convert the intermediate representation in to SSA form. SSA form is optimal and has no unnecessary terms. As a next task we exploit the SSA form to implement code optimization phase.

Modify the Compiler's Backend. For production of executable code, it is necessary to modify the backend of the compiler. The easiest way we use to get a working backend is to use the code generation phase already provided. As next step we used to convert each modified function into copy statement. Prior to code generation we translate each modified function into a sequence of assignment statements. These assignment statements are place at the end of predecessor block.

3.2 At Hardware Level (ILP Processor)

The processor supporting ILP [13] is known as ILP processor. Its performance can be enhanced through compiler optimization. In an ILP processor, the basic unit of computation is a processor instruction, having operations like add, multiply, load or store. Non-interdependent instructions lead to load and execute in parallel. With the help of ILP processor the instruction scheduling [3] needs not to be done during program execution rather it can be done during compilation process. One possibility to optimize ILP processor's operations is to create a compiler which generates effective code on the assumption that no run-time decision are possible; it is only the responsibility of compiler to take all the scheduling and synchronization decision. This shows that the processor has very little task of reordering of code at run time. The multi core systems provide remarkable efficiency as compare to single core. For our experimental purpose, we compared speed-up performance of 2-core and 4-core systems with single core.

4 How the Optimized Compiler Helps to Exploit ILP

The optimized compiler created has a function of determining whether the result of a statement to be executed precedent to an if-condition is not affected by the execution result of traditional branch statement. If the branch statement is not affected by the execution result of the precedent statement, the branch statement is shifted in front of the precedent statement, to suppress the execution of unwanted statements [11]. This way the branch statements are shifted or copied by the optimizing compiler to minimize the execution time of the object code.

The setup we have from the Trimaran Simulator [14] is simulation of a computer system with parallel processor capable of executing two or more procedures in parallel. The assumed compiler (optimized), as shown in Fig. 2, is comprised of:

1. A syntax analysis unit (to interpret the statements of the source code and translate into the Intermediate Representation (IR).
2. An optimization unit (to optimize the use of parallel processor at the level of intermediate representation).
3. A unit for producing the object program.

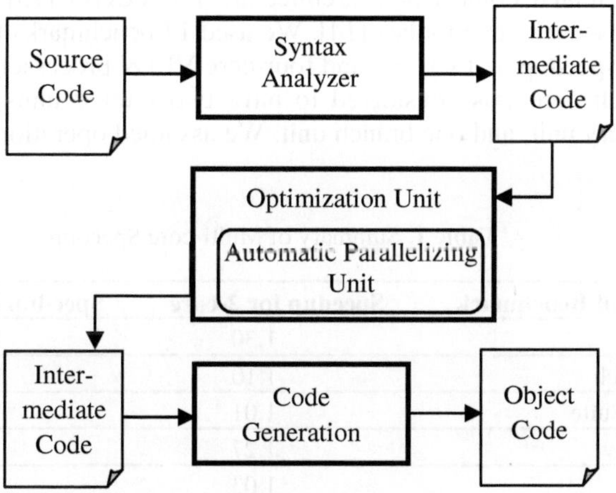

Fig. 2. Model of Optimizing Compiler

The automatic parallelization unit (as shown in Fig. 3) consist of:

1. A detection unit for detecting and recording the IR corresponding to source code.
2. A conversion unit for intermediate code conversion adding a different intermediate code resulting the similar result as of detection unit.

In Fig.3, the broad arrow represents the control flow while normal arrows represent data flow.

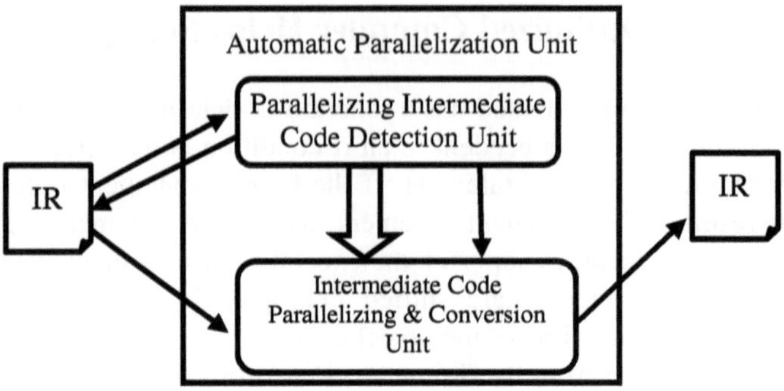

Fig. 3. Automatic Parallelization Unit

5 Experiments

For our experimental purpose we modified our compiler with EBR (Edition based Redefinition) operation. It allows toupgrade the database component (symbol table and library) of the compiler while it is in use. For evaluation of ILP exploitation on DVLIW (Distributed control path architecture for VLIW) [13] with modified ILP compiler, we used trimaran toolset [14]. We used 17 benchmarks for our experiments. We measured speedup on two core and four-core VLIW processors against a one-core processor. Each core was considered to have two integer units, one floating point unit,one memory unit, and one branch unit. We assumed operation latencies similar to

Table 1. Summary of Multi-core Speedup

Name of Benchmark	Speedup for 2-core	Speedup for 4-core
SPEC	1.30	1.45
JetBench	1.10	1.05
CloudSuite	1.01	1.01
Bitarray	1.27	1.27
Bitcnt	1.03	1.03
Cjpeg	1.50	1.50
Jcapistd	1.05	1.05
Rdbmp	1.06	1.06
Rdgif	1.07	1.07
Wrbmp	1.15	1.15
Wrppm	1.07	1.07
Correct	1.32	1.32
Dump	1.35	1.35
Hash	1.15	1.15
Gsmdecode	1.66	2.13
Gsmencode	1.58	2.07
Xgets	1.33	1.39

Table 2. Impact of Hardware and Software on Parallelism

Name of Benchmark	Hardware style model	Software style model
SPEC	7.0	7.0
JetBench	5.0	5.0
CloudSuite	6.0	6.0
Bitarray	5.0	4.5
Bitcnt	5.5	5.0
Cjpeg	6.5	6.0
Jcapistd	6.0	6.5
Rdbmp	5.0	4.0
Rdgif	5.5	5.0
Wrbmp	5.0	4.5
Wrppm	11.0	13.5
Correct	5.0	5.0
Dump	6.0	5.0
Fpppp	19.0	27.0
Gsmdecode	7.0	8.0
Gsmencode	5.0	5.0
Tomcat	21.0	44.0

those of the Intel Itanium. We compare DVLIW processor with two or four cores to multi cluster VLIW machine with centralized control path. The compiler employed the hyperblock region formation. Table 1 shows the speedup the ILP execution on two and four-core systems.

The average speed-up measure for 2-core system was 1.24 and for 4-core was1.30. The speedup achieve in our experiments is found closely related to the amount of ILP in the benchmarks that can be exploited by the ILP compiler. The benchmark like gsmdecode and gsmencode exposing high ILP achieve high speedup while benchmarks CloudSuite and bitcnt show low ILP. In order to achieve ILP, we must not have the dependencies among the instructions executing in parallel.

By taking selected 17 benchmarks, we compared parallelism achieved for hardware and software oriented models. For hardware model we considered zero-conflict branch and jump predictions; and for software style model we considered static branch and jump predictions. The Table 2 shows the summary of experiments.

The average result comparison shows the better results for software style model. This shows more scope of ILP exploitation at software level. The average speed up for hardware style system was measured 7.68 and 9.47 for software style system.

6 Conclusions

In our experiments we noticed that some benchmarks suffer slight slow-down to expose ILP. This is due to the EBR operation inserted in ILP compiler which maintains the correct control flow.

It increases pressure on the I-Cache and caused more I-Cache misses. The ILP compiler is not aware of this phenomenon and could possibly slow down the execution. The results show that VLIW architecture provides the mechanism for multi-core system to enforce existing ILP compiler to exploit ILP in the applications. We applied bottom up Greedy (BUG) algorithm for partitioning the operations to multiple cores. The ILP compiler ensured the control flow in the multiple cores for synchronization and operation insertion. The experiments conducted for hardware and software style models proved that much scope of ILP exploitation is at compiler level.

References

1. Carr, S.: Combining Optimization for Cache and Instruction Level Parallelism. In: Proceedings of the 1996 Conference on Parallel Architectures and Compilation Techniques (1996)
2. Pnevmatikatos, D.N., Franklin, M., Sohi, G.S.: Control flow prediction for dynamic ILP processors. In: Proceedings of the 26th Annual International Symposium on Microarchitecture, pp. 153–163 (1993)
3. Lo, J., Eggers, S.: Improving Balanced Scheduling with Compiler Optimizations that Increase Instruction-Level Parallelism. In: Proceedings of the Conference on Programming Language Design and Implementation (1995)
4. Zhong, H., Lieberman, S.A., Mahlke, S.A.: Extending Multicore Architectures to Exploit Hybrid Parallelismin Single-thread Applications. In: IEEE 13th International Symposium on High Performance Computer Architecture, pp. 25–36 (2007)
5. Posti, M.A., Greene, D.A., Tyson, G.S., Mudge, T.N.: The Limits of Instruction Level Parallelism in SPEC95 Applications. Advanced Computer Architecture Lab (2000)
6. Hwu, W.-M.W., Mahlke, S.A., Chen, W.Y., Chang, P.P.: The Superblock: An Effective Technique for VLIW and Superscalar Compilation. The Journal of Supercomputing 7, 227–248 (1993)
7. Kumar, R., Singh, P.K.: A Modern Parallel Register Sharing Architecture for Code Compilation. International Journal of Computer Applications 1(16) (2010)
8. Kumar, R., Singh, P.K.: Role of multiblocks in Control Flow Prediction using Parallel Register Sharing Architecture. International Journal of Computer Applications 4(4), 28–31 (2010)
9. Kumar, R., Singh, P.K.: Control Flow Prediction through Multiblock Formation in Parallel Register Sharing Architecture. Journal on Computer Science and Engineering 2(4), 1179–1183 (2010)
10. Kumar, R., Saxena, A., Singh, P.K.: A Novel Heuristic for Selection of Hyperblock in If-Conversion. In: 2011 3rd International Conference on Electronics Computer Technology, pp. 232–235 (2011)
11. August, D.I., Hwu, W.-M.W., Mahlke, S.A.: A Framework for Balancing Control Flow and Predication. In: Proceedings of 30th Annual IEEE/ACM International Symposium on Microarchitecture, pp. 92–103 (1997)
12. Pai, V.S., Ranganathan, P., Abdel-Shafi, H., Adve, S.: The Impact of Exploiting Instruction-Level Parallelism on Shared-Memory Multiprocessors. IEEE Transactions on Computers 48(2), 218–226 (1999)
13. Zhong, H.: Architectural and Compiler Mechanisms for Accelerating Single Thread Applications on Multicore Processors. PhD thesis, The University of Michigan (2008)
14. `http://www.trimaran.org`

Architecting Fabricated Implant in Interconnect for Multi-core Processor

Ram Prasad Mohanty, Ashok Kumar Turuk, and Bibhudatta Sahoo

Department of Computer Science and Engineering,
National Institute of Technology,
Rourkela, Odisha, India
ramprasadmohanty@gmail.com, {akturuk,bdsahu}@nitrkl.com

Abstract. In this paper, we analyze the effect of interconnect on multi-core processors and have proposed a novel highly scalable on-chip interconnection mechanism for multi-core processors. As the number of cores increases, traditional on-chip interconnect like bus and crossbar proves to be low in efficiency as well as suffer from poor scalability. In order to get rid of the scalability and efficiency issues in these traditional interconnects, ring based design has been proposed. But with the steady growth in number of cores have rendered the ring interconnect too infeasible. Thus, novel interconnect designs are proposed for the future multi-core processors for enhancement in the scalability. In this paper, we analyze and compare the interconnect of two existing multi-core processors named Multi-core Processor with Internal Network(MPIN) and Mult-core processor with Ring Network(MPRN). We have also proposed a highly scalable and efficient interconnect named as *fabricated Implant in Interconnect* for multi-core processors. The placement of cores and cache in a network is proved to be highly crucial for system performance. The benchmark results are presented by using a full system simulator. Results show that, by using the proposed on-chip interconnect, compared with the MPIN and MPRN, the execution time are significantly reduced for three applications.

Keywords: Multi-core Processor, Performance analysis, Interconnect, Cache Dependency.

1 Introduction

Multi-core processors with greater number of cores and complex on-chip interconnect are recent trend since the past few years. The constraints with respect to power consumption, chip clock frequency and heat dissipation have made the chip designers to evolve from improvement in the single-core processors to integration of multiple cores on a single chip. A recent trend of enhancement in performance is to enhance the number of cores per chip. [1] This enhancement in the number of cores lead to the proposal of concept of network-on-chip (NoC). Before this concept was proposed, system-on-chips (SoCs) took the aid of complex traditional interconnects like bus structures for connection between the

M.K. Kundu et al. (eds.), *Advanced Computing, Networking and Informatics - Volume 2*, Smart Innovation, Systems and Technologies 28,
DOI: 10.1007/978-3-319-07350-7_57,

cores to memory and I/O. The traditional bus structures were improved to be used as interconnect in the Multi-core processors. But with enhancement in the number of cores these bus designs were not able to sustain the interconnect scaling as well as complexity. Eventually NoC was used as a solution to the scalability issues [2] [3]. Multi-threading/multi-core technology increases performance, but doing so requires more power than single threading/core computers. Power was not an issue at the beginning of computer era. However, power has become a critical design issue in computer systems [4]. Multi-threaded and multi-core systems also requires more space (area) than a single threaded or single core system [5]. Cores in multi-core system have hardware resources for themselves and use them each for processing [6]. In this paper, we analyze the effect of interconnect on the multi-core processors and have proposed a novel highly scalable on-chip interconnection mechanism for multi-core processors.

The paper has been organized as follows: Section 2 gives a brief description on the existing architectures and related works done in Multi-core Processor Technology, Section 3 describes the proposed work, Section 4 provides a detailed description of the simulation results and Section 5 gives a concluding remark and the future direction of our work.

2 Architecture and Background

Various work in current literature has explored the multi-core architecture utilizing various performance metrics and application domain. D.M. and Ranganathan [7] have analyzed a single-ISA heterogeneous multi-core architecture for multi-threaded workload performance. The objective was to analyze the performance of multi-core architectures for multi-threaded workloads. This section details the benefits of variation in the interconnection network in the multi-core architecture with multi-threaded workloads.

Various works has analyzed the performance in both single core and multi-core architectures. Julian *et al.* [8] determined the relationship between performance and memory system in single core as well as multi-core architecture. They utilized multiple performance parameters like cache size, core complexity. The author have discussed the effect of variation in cache size and core complexity across the single core and multi-core architecture.

Multi-core architecture with multiple types of on-chip interconnect network [9] are the recent trend of multi-core architecture. This type of architecture have different type of interconnect networks with different core complexity and cache configuration.

3 Proposed Work

In this paper, we propose an interconnect architecture for multi-core processors. We name the proposed architecture as Multi-core Processor with Fabricated Implant in Interconnect (MPFII). A block diagram of the proposed interconnect is shown in Fig. 1. It can be configured with different numbers of cores. The

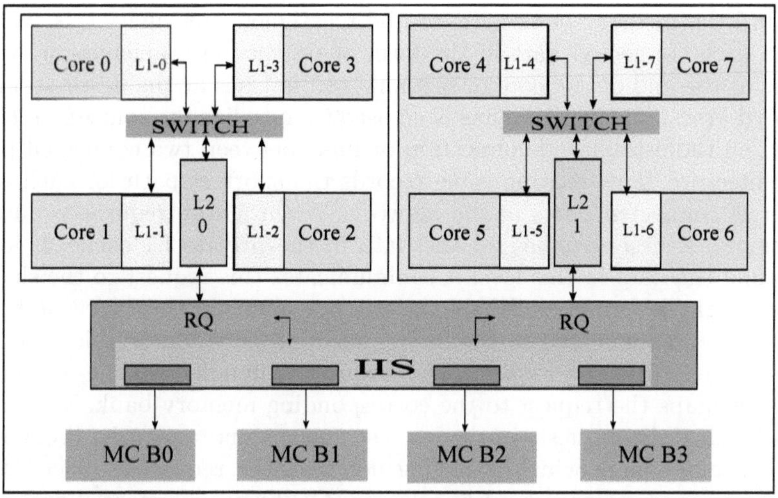

Fig. 1. Multi-core Processor with Fabricated Implant in Interconnect

figure shows an example with 8 cores. Each core is a out of order super-scalar SMT core capable of running more than one thread at once.

Each core has a private L1 cache and is shared between the multiple threads in that core. The L1 cache has the following configuration: Block size is 256 bytes for both data and instruction cache. Associativity: L1 caches are 2-way associative, so cache lines from the L2 cache can be mapped to any L1 cache line. Replacement Algorithm : LRU replacement. Ports: Number of ports for each L1 cache is 2. Each L2 cache shares the full address range thus isolating any coherent issues to a local region with which the L2 cache is associated with.

The MPFII uses a hybrid mode of connection. It uses a crossbar switch to connect every four L1 data and instruction cache to its corresponding L2 cache. L1 cache communicate with the L2 cache through the switch. No two L1 cache can communicate with each other. Each L2 cache has the following configuration: Block Size is 256 bytes. Associativity: This cache is 4-way set associative. Replacement Algorithm : LRU Latency: 20 cycles for a L2 cache miss. Ports: Number of ports for each L2 cache is 4. Each core communicates with the corresponding L2 cache through a non-blocking crossbar which can let simultaneous message passing as long as each message is headed for a unique output. Each switch is having input and output buffers that lets the message to be stored temporarily at times of contention. Main memory is divided into four memory banks. L2 cache communicate with the main memory through a On-Chip network called fabricated implant interconnect. This interconnect consists of request queue for individual L2 cache. These request queue (RQ) are connected to the intelligent interface section (IS) of the interconnect. This intelligent IS is capable of mapping a particular request with the corresponding memory bank. If the next request in the queue is requesting for a memory bank, which is being accessed by some other request then this request status is updated to waiting state until the memory bank is free to be reassigned. Information is exchanged between each connected part in the form of packets. A transmission of packet initiates at the end of one core or memory and finishes at the destination memory. Floyd-Warshall algorithm has been used to initialize the routing table which is based on the established connection or links between two connected parts of the architecture. This routing table records the shortest path for each and every pair of connected parts in the network. When a core requires a data from the memory, it first searches for the data in the private L1 cache. If it is not able to find the data in this level it communicates this request to the shared L2 cache through the crossbar. IF the data is not found in the L2 cache too then a request is communicated to the interconnection network. In this network the request queue stores the request. This request is handled by the interface section and it maps the request to the corresponding memory bank. The interface section handles the requests in the request queue simultaneously. If one request is for a memory bank which is having not free, the request is placed on hold until the required memory bank is released. Coherence is enforced with the aid of directory-based MOESI protocol at all caches connected to the upper link of the interconnect with respect to the single cache connected to the lower link [10]. Each block of the L2 cache has a directory entry which contains two fields. The first field is an identifier that specifies the single upper level cache which is the owner of the block that is , it holds the information whether an L1 cache has the block in exclusive, owned or modified state.The second one is a bitmap with as many bits as upper level caches, with those bits set to one that corresponds to the caches having a copy of the block.

3.1 Advantage

Once a memory address is presented each memory module returns with one word per cycle. It is possible to present different addresses to different memory modules in order to enable parallel access of multiple words simultaneously or in a pipelined fashion. Memory banking enhances the parallelism as well as effectively improves the effective memory bandwidth [11]. This parallelism is effectively implemented in the MPFII as shown in the Fig: 1. This leads to an improved performance over few existing architectures. The scalability of the interconnect is enhanced because of the usage of lower configurations of the crossbar. Multiple lower configured crossbar proves to be economical as well as more scalable as compared to high configurations of crossbar [12].

4 Simulation and Results

For the simulation we have used SPLASH2 benchmark suite [13–15] and multi2Sim 4.0.2 simulator [9] [16]. We have compared the proposed architecture with multi-core processor with internal network and multi-core architecture with ring network. The metrics considered for comparison are execution time and speedup. We varied the cache size to evaluate the above architectures [7].

4.1 Impact of Cache Size

To study the impact of cache on the performance of multi-core processors, the number of cores in each architecture was kept constant as 32 and the size of L1 and L2 cache was varied. L2 cache size was varied first keeping the L1 cache size constant. Then L1 cache size was varied keeping L2 cache size constant. The execution time for FFT, cholesky, and barnes benchmark program of the SPLASH2 benchmark suite was analyzed [17].

Fig. 2, Fig. 3, and Fig. 4 shows the CPU execution time for multi-core architecture with Proposed Interconnect on execution of FFT, cholesky and barnes program of the Splash2 benchmark suite. With the enhancement in the cache size the number of misses reduced thus resulting in the reduction in the total CPU execution time. But after certain size the impact reduced. Beyond the size of 512 KB for L1 cache the execution time almost remained constant. Similarly beyond L2 = 8 MB the execution time almost remained constant.

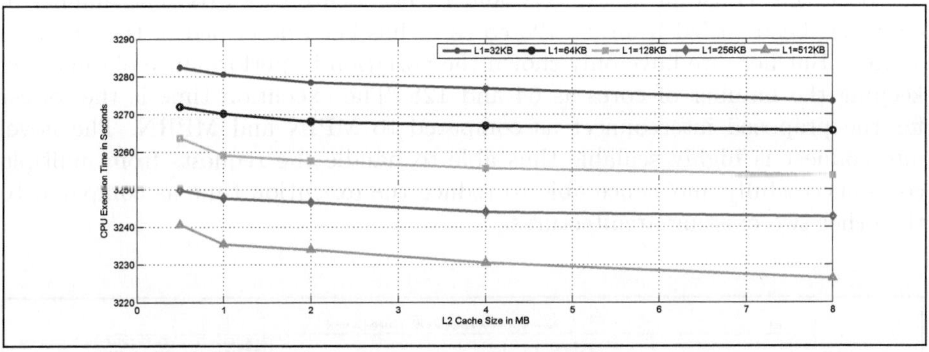

Fig. 2. Execution Time for MPFI

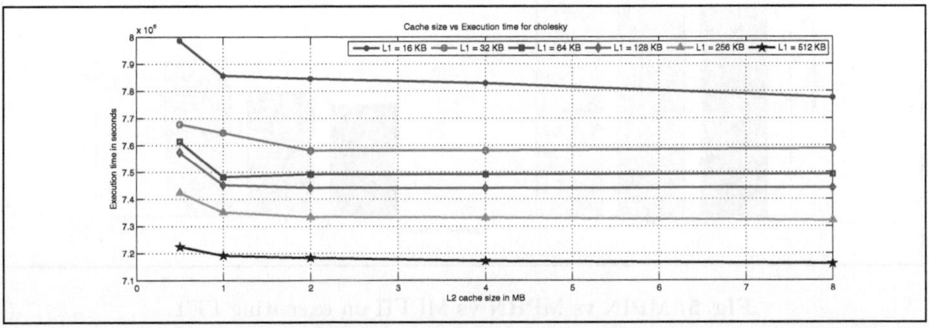

Fig. 3. Execution Time of MPFII for cholesky

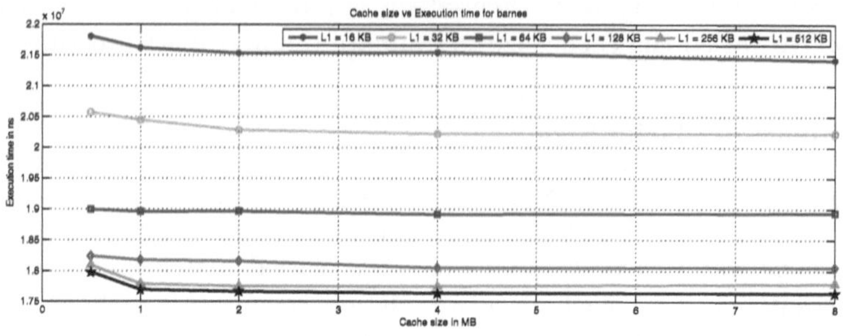

Fig. 4. Execution Time of MPFII for barnes

4.2 Performance Comparison of Proposed Architecture with Existing Architectures

The performance of MPFII has been successfully compared with few of the existing architectures. Fig. 5 shows the execution time for MPIN, MPRN, and MPFII on execution of the FFT benchmark program. Here the L1 cache size has been kept constant as 512 KB, and L2 cache size as 8 MB. The number of cores has been varied from 2 - 128 cores as has been described in the previous section. But here we have only shown the variation in performance obtained by keeping the number of cores as 64 and 128. The execution time is the lowest for the proposed interconnect as compared to MPIN and MPRN. The novel interconnect is highly scalable thus able to handle the requests from multiple cores successfully and hence able to reduce the execution time as compared to the other two existing architectures.

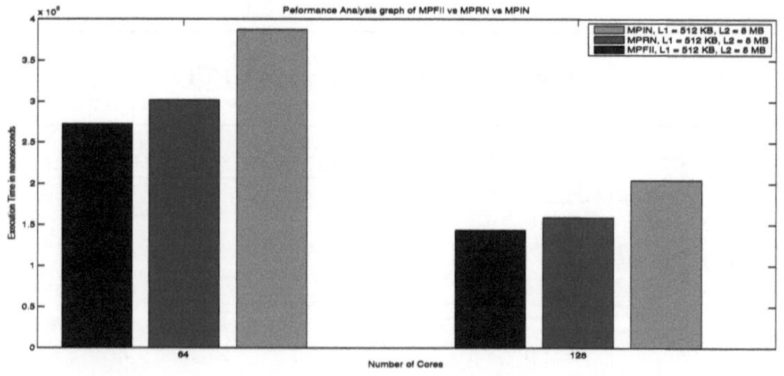

Fig. 5. MPIN vs MPRN vs MPFII on executing FFT

5 Conclusions

The problem of performance evaluation of Multi-core Organization being most challenging, is interesting too. Keeping a view of the literature the Various Multi-core Organization systems are modeled and few of them have been analyzed. The interconnect network as well as the memory system requires much more enhancement to match with the current trend of increasing number of cores. In the current work, the primary objective has been to reduce the delay in the core to memory or memory to memory communication. The second objective has been to analyze the performance of multi-core architecture with internal network, ring network and proposed interconnect. The objective could be achieved successfully by being able to enhance the performance with the proposed architecture. And secondly by analyzing the performance for varying size of L1, L2 cache, and number of cores. The performance of the processor is dependent on the cache size, but only by increasing the cache size the performance of the processor is not enhanced. This can be concluded by the simulation results obtained. By varying the interconnect network we are able to get a better performance for the proposed architecture as compared to the existing architectures.

References

1. Xu, T.C., Liljeberg, P., Tenhunen, H.: Explorations of optimal core and cache placements for chip multiprocessor. In: NORCHIP, pp. 1–6. IEEE (2011)
2. Buckler, M., Burleson, W., Sadowski, G.: Low-power networks-on-chip: Progress and remaining challenges. In: IEEE International Symposium on Low Power Electronics and Design, pp. 132–134 (2013)
3. Huang, T., Zhu, Y., Qiu, M., Yin, X., Wang, X.: Extending Amdahl's law and Gustafson's law by evaluating interconnections on multi-core processors. The Journal of Supercomputing, 1–15 (2013)
4. Pham, D., Asano, S., Bolliger, M., Day, M., Hofstee, H., Johns, C., Kahle, J., Kameyama, A., Keaty, J., Masubuchi, Y.: The design and implementation of a first-generation CELL processor. In: 2005 IEEE International Solid-State Circuits Conference, pp. 184–592 (2005)
5. Brey, B.B.: The Intel Microprocessors. Prentice Hall Press (2008)
6. Geer, D.: Chip makers turn to multicore processors. Computer 38(5), 11–13 (2005)
7. Kumar, R., Tullsen, D., Ranganathan, P., Jouppi, N., Farkas, K.: Single-ISA heterogeneous multi-core architectures for multithreaded workload performance. ACM SIGARCH Computer Architecture News 32 (2004)
8. Bui, J., Xu, C., Gurumurthi, S.: Understanding Performance Issues on both Single Core and Multi-core Architecture. Technical report, University of Virginia, Department of Computer Science, Charlottesville (2007)
9. Ubal, R., Sahuquillo, J., Petit, S., López, P., Chen, Z., Kaeli, D.: The Multi2Sim Simulation Framework
10. Ubal, R., Jang, B., Mistry, P., Schaa, D., Kaeli, D.: Multi2Sim: a simulation framework for CPU-GPU computing. In: Proceedings of the 21st International Conference on Parallel Architectures and Compilation Techniques, pp. 335–344 (2012)
11. Hwang, K.: Advanced computer architecture. Tata McGraw-Hill Education (2003)

12. Dally, W.J., Towles, B.P.: Principles and practices of interconnection networks. Elsevier (2004)
13. Woo, S.C., Ohara, M., Torrie, E., Singh, J.P., Gupta, A.: The SPLASH-2 programs: Characterization and methodological considerations. ACM SIGARCH Computer Architecture News 23, 24–36 (1995)
14. Akhter, S., Roberts, S.: Multi-core programming, vol. 33. Intel Press (2006)
15. Bienia, C., Kumar, S., Li, K.: PARSEC vs. SPLASH-2: A quantitative comparison of two multithreaded benchmark suites on chip-multiprocessors. In: IEEE International Symposium on Workload Characterization, pp. 47–56 (2008)
16. Ubal, R., Sahuquillo, J., Petit, S., López, P.: Multi2Sim: A Simulation Framework to Evaluate Multicore-Multithread Processors. In: IEEE 19th International Symposium on Computer Architecture and High Performance Computing, pp. 62–68 (2007)
17. Mohanty, R.P., Turuk, A.K., Sahoo, B.: Performance evaluation of multi-core processors with varied interconnect networks. In: 2nd International Conference on Advanced Computing, Networking and Security, pp. 7–11 (2013)

Data Concentration on Torus Embedded Hypercube Network

Uday Kumar Sinha[1], Sudhanshu Kumar Jha[2], and Hrishikesh Mandal[1]

[1] Department of Computer Science and Engineering,
Bengal College of Engineering and Technology, Durgapur –713212, West Bengal, India
[2] Department of Computer Applications,
National Institute of Technology, Jamshedpur – 831014, Jharkhand, India
{udaydhn, sudhanshukumarjha}@gmail.com, hrishiman@yhaoo.co.in

Abstract. Data concentration is an important tool for various scientific and engineering applications. Recently, torus embedded hypercube have attracted much attention among researchers due to its inherent architectural property of two different interconnection networks. In this paper we present an algorithm to perform data concentration on torus embedded hypercube network. Our proposed algorithm takes $d(5.5n + 3 \log n)$ time to perform data concentration of d ($d < N$) datum on torus embedded hypercube network having N ($= n \times n \times n$) processing elements. Our proposed algorithm can be compared with other data concentration algorithm designed for various other interconnection networks.

Keywords: Data concentration, parallel prefix, torus embedded hypercube, interconnection network, routing in interconnection network.

1 Introduction

A concentration or packing problem is one that involves how to route k packets that is arbitrarily distributed over any k (source) processors in an interconnection network to some fixed k consecutive destination processors on the same network without the ability to distinguish among those destinations [1]. This is an important task that is used frequently to reduce the communication delay among the processors for future processing of the data [2]. In a simple way, for a given d data elements distributed arbitrarily over a network of p processors ($d \leq p$), the problem of data concentration is to move the data in the first d locations of the network [3] in linear fashion starting from first processor [4], an example, data concentration of 7 data elements (a – g) randomly stored in a 5×5 grid is shown in Fig. 1.

It is easier to notice that the data concentration reduce the communication delay among the processors, thus the energy consumption can also be reduced drastically [5]. Since the effectiveness of this technique depends upon the workload and data popularity and it is also common that such statistics can be dynamically changed during run time, data re-organization may need to applied periodically or on-demand to the system. As a result, the performance overhead and penalty of data placement techniques could become a concern for more dynamic systems or workloads.

M.K. Kundu et al. (eds.), *Advanced Computing, Networking and Informatics - Volume 2*, Smart Innovation, Systems and Technologies 28,
DOI: 10.1007/978-3-319-07350-7_58, © Springer International Publishing Switzerland 2014

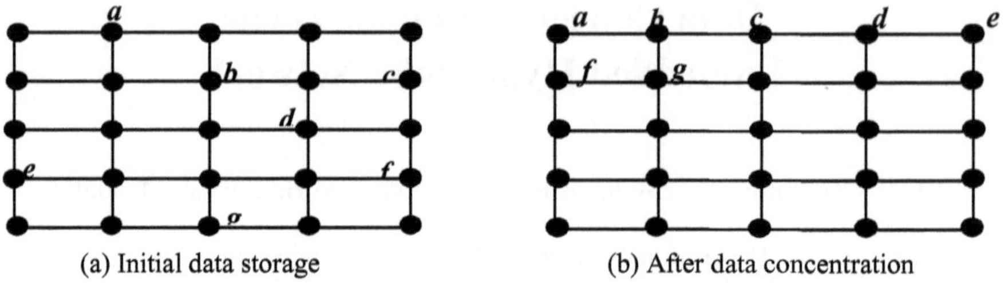

Fig. 1. Data Conentration 5 × 5 grid network

Therefore, it remains an interesting research problem to explore the possibility of minimizing the overheads and penalties for this class of techniques [1]. A great deal of research has been devoted to this subject. Sahni *et al.* [2] has shown that the data concentration of d data elements scattered on p processors linear array can be performed straightforwardly in $2p$ time. Jha and Jana [6] have shown that the data concentration of d ($\leq n^2$) data elements can be done in $O(d \log n)$ time on an $n \times n$ mesh of trees network in their first algorithm. In the same paper, they have also implemented the same problem on an $n \times n$ OTIS-mesh of trees network that requires $O(d \log n)$ electronic moves and 2 OTIS moves for the same number of elements. In [7] Jha and Reddy have presented a parallel algorithm for data concentration on hierarchical ring network in $O(d) + O(n)$ time, where d ($< N$) is the size of data for n^2 processors. Jan, Lin and Lin [1] shown that data concentration problem can be solved in $O(n)$ time on n processors hypercube network.

Torus embedded hypercube network [8] is an efficient interconnection network, which has been benefited by both torus network and hypercube network. Mesh is an interconnection network with constant node degree in its internal nodes where as torus network has constant node degree with all its nodes [9-10]. The advantages of these networks can be combined by embedding torus with hypercube to give rise to embedded architecture known as Torus embedded hypercube scalable interconnection network [8]. Parallel prefix is an effective tool that has been used to perform data concentration on any network [2]. Recently, Sinha *et. al.* [11] have presented an algorithm to perform prefix computation on torus embedded hypercube network that requires $3(n + \log n) + 2$ time to compute the prefix for N data elements on N processors torus embedded hypercube network.

In this paper we will present an algorithm to perform data concentration on torus embedded hypercube network [8]. Our proposed algorithm will use prefix computation technique [11] as a basic tool to find the rank of the data elements that can be further use to know about the address of destination processor.

The rest of the paper is organized as follows. In Section 2, we will present the topological structure of torus embedded hypercube network. Section 3 discusses about the proposed algorithm followed by the conclusion in Section 4.

2 Topology of Torus Embedded Hypercube

In this section, we describe the topology of torus embedded hypercube network. As we know that a torus embedded hypercube network is a mixture of torus and hypercube interconnection network. Let us assume here, that the size of concurrent torus network is $l \times m$ and the number of nodes connected in the hypercube network is K. Then the size of torus embedded hypercube network will be (l, m, K). Nodes with identical positions in the torus network will form a hypercube group of K nodes. Such node can be addressed with three components i.e. row number i, column number j of torus network and address of node k ($k \in K$) of hypercube using its binary notation. Hence (l, m, K) torus embedded hypercube network will have $l \times m \times K$ number of nodes. The node or processing elements (PE) can be addressed as (i, j, k) where $0 \leq i < l$, $0 \leq j < m$ and $0 \leq k < K$. As an example (2, 2, 8) torus embedded hypercube is shown in Fig 2.

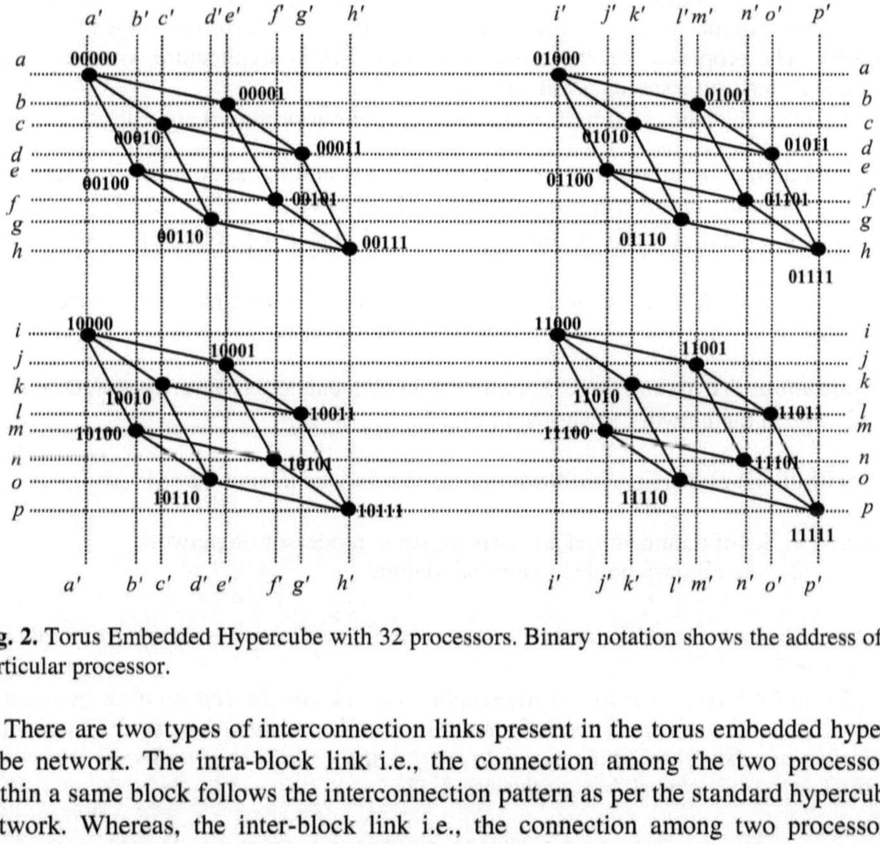

Fig. 2. Torus Embedded Hypercube with 32 processors. Binary notation shows the address of a particular processor.

There are two types of interconnection links present in the torus embedded hypercube network. The intra-block link i.e., the connection among the two processors within a same block follows the interconnection pattern as per the standard hypercube network. Whereas, the inter-block link i.e., the connection among two processors between two different blocks follows the interconnection pattern as per standard 2D torus network which can be further classified as horizontal wraparound link and vertical wraparound link. The horizontal link (a, b, c...) is shown as wraparound and the vertical link (a', b', c'...) also shown to wraparound i.e., the link a is connected to link a (in horizontal) and a' is connected to a' (in vertical).

3 Proposed Algorithm

In general, data concentration on any interconnection network can be achieved by first performing the prefix sum to determine the rank of destination processor of each data elements, next the routing of the data elements can be done using the appropriate shortest path routing algorithm for the network [12-13]. In the similar way, to perform the data concentration on torus embedded hypercube network we will first perform a prefix computation as described in [11] to determine the rank of destination processor of each data element and then we apply routing of the packets using the appropriate routing algorithm.

For the simplicity and reader's motivation, here we will present the figurative illustration of the proposed algorithm on (2, 2, 8) torus embedded hypercube, however, our proposed algorithm can be mapped on arbitrary size of torus embedded hypercube network. The proposed algorithm is based on the SIMD architecture where all active processors perform a same task at a time.

The following steps summarize the basic idea of our proposed algorithm.

1. We initialize the processors with the tag bits '1' and '0' as follows. We set a processor with tag bit '1' if it holds a datum, '0' otherwise.
2. Perform a parallel prefix on the tag bits stored in step 1 to obtain the rank of each datum.
3. Perform a shortest path routing to send the data as per their individual rank derived in step 2.

Initialization: We assume here that a few data elements are arbitrarily stored on some processors in the network as shown in Fig. 2.

Algorithm – I: *Data_Concentration_Torus_Embedded_Hypercube* ()

Input: 1. Set of datum stored arbitarly on some processor on network
2. A temproary boolean variable 'datum'

Output: Datum placed to target processor in linear fashion starting from the first processor

/ The address of processor in hypercube network can be represented by a set of tuples, if the size of hypercube is 2^n, then we need* log *n number of tuple. Thus in our algorithm we use three tuples i, j, k to represent the address of indivudial processor in the hypercube network having the size 2^3. */*

Step 1: *for all processor* $p(x, y, i, j, k)\forall 1 \le x \le l, 1 \le y \le m, 0 \le i, j, k \le 1$ *do in parallel*

if $p[x, y, i, j, k]$ *holds a datum*
set $p[x, y, i, j, k] = 1$
else
set $p[x, y, i, j, k] = 0$

Note: Situation after this step is shown in Fig. 3.

Step 2: *for all processor* $p(x, y, i, j, k)\forall 1 \le x \le l, 1 \le y \le m, 0 \le i, j, k \le 1$ *do in parallel*

Perform parallel prefix computation on torus embedded hypercube network as presented in [11] and store the partial result in its temporary register A. (This partial result value is now treated as Rank of the datum)

Note: Situation after this step is shown in Fig. 4 where, number denotes the *Rank*.

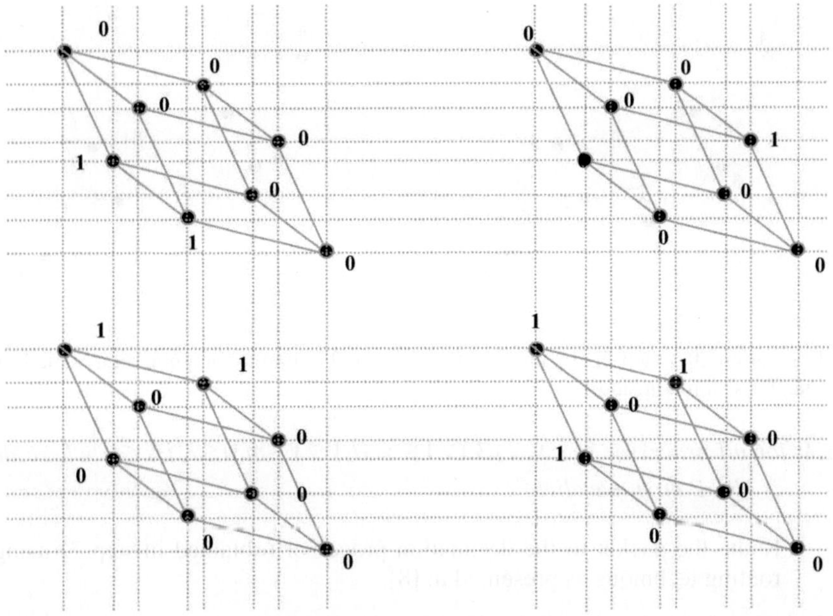

Fig. 3. Initializations of tag bit's

Step 3: *for all processor* $p(x, y, i, j, k)\forall 1 \le x \le l, 1 \le y \le m, 0 \le i, j, k \le 1$ *do in parallel*

if datum == TRUE //i.e., where, datum is 1

Find the address of destination processor using *Find_Destination_ Address*(*Rank*) procedure as described later, and store the result in register *B*.

else

Do nothing // *i.e.*, where, datum is 0

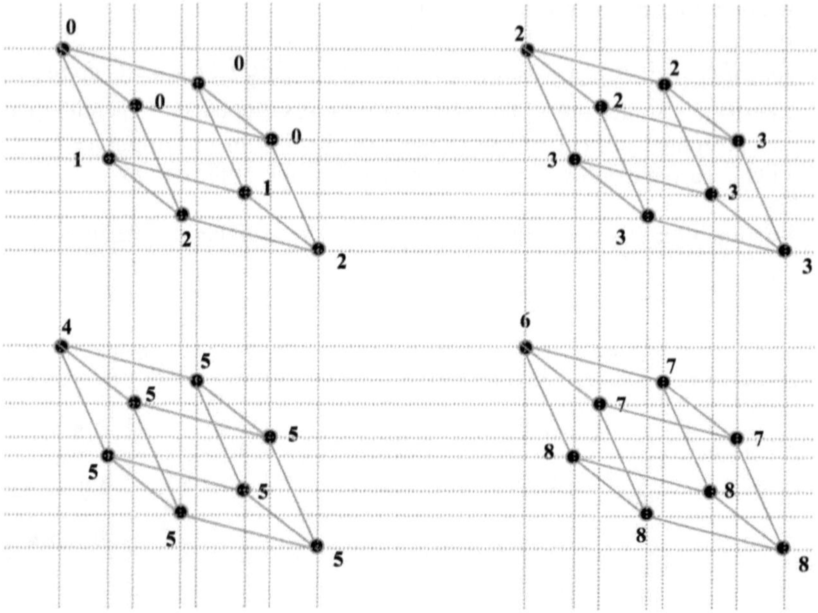

Fig. 4. Situation after step 2, i.e., after performing parallel prefix computation as stated in [11] on the tag bit's stored in step 1

Step 4: *for all processor* $p(x, y, i, j, k) \forall 1 \le x \le l, 1 \le y \le m, 0 \le i, j, k \le 1$ *where datum is TRUE do in parallel*

Route the packet to the destination processor (obtained in step 3) using the routing technique as presented in [8].

Step 5: Stop

Algorithm – II: *Find_Destination_ Address(Rank)*

/* Basic purpose of this algorithm to find the address of the destination processor. This algorithm takes Rank as an input and returns the address of processor in binary notation. Algorithm runs in two phases. Phase – I finds the row, column address of the block and the address of processor within a block in decimal format while phase – II engage with concatenating of the equivalent binary notation of the row address, column address and the address of the processor within a block. */

Input : *rank* : rank (obtaied in step 2 in Algorithm – I)
l : number of rows in network
m : number of columns in network
k : number of nodes in each hyperube network
Output : Return the address of the processor in binary format

Phase - I

$r = Rank - 1$
$if\,(r \bullet k)$
$row = r\,/\,(m \times k)$
$column = (r\,/\,k) \bmod m$
$hypercube\_node\_address = r \bmod k$
else
$row = 0$
$column = 0$
$hypercube\_node\_address = r$

Phase - II

Destination_address = concatenate (binary_of (row, column, hypercube_node_address))

return (Destination_address)

End_of_Algorithm – II

Time Complexity

For the generalization of time complexity, let us assume the size of concurrent torus network is $n \times n$ and the number of processors in individual hypercube is also n. Thus the total number of processing elements in torus embedded hypercube is N ($= n \times n \times n$). In other words, the number of processor in a hypercube is equal to the number of rows and columns in concurrent torus (for example 8, 8, 8 torus embedded hypercube). It is clear that Algorithm – II will find the destination processor address in constant time while in Algorithm – I, step 1 and 3 can perform initialization and address finding also in constant time. Step 2 require $3(N + \log N) + 2$ to perform parallel prefix computation as discussed in [11] to find the rank. Step 5 needs $2.5N - 2$ times [8, 12] to route the one packet from source to destination processor. Thus overall time complexity to perform data concentration on torus embedded hypercube network is $d(5.5n + 3 \log n)$ time to move d datum to its appropriate destination in worst case.

4 Conclusion

Torus embedded hypercube network is an efficient interconnection network which inherit the properties of two independent (torus and hypercube) interconnection network. In this paper we have proposed an algorithm to perform data concentration on torus embedded hypercube network. Our proposed algorithm require $d(5.5n + 3 \log n)$ time to move d data item to its destination processor in worst case.

References

1. Jan, G., Lin, F.Y.S., Lin, M.B., Liang, D.: Concentrations, load balancing, multicasting and partial permutation routing on hypercube parallel computers. Journal of Information Science and Engineering 18, 693–712 (2012)

2. Horowitz, E., Sahni, S., Rajasekaran, S.: Fundamental of Computer Algorithms. Galgotia Publication (2008)
3. Lee, C.-Y., Oruc, A.Y.: Design of efficient and easily routable generalized connectors. IEEE Transactions on Communications 43(234), 646–650 (1995)
4. Horowitz, E., Sahni, S., Rajasekaran, S.: Computer Algorithms. Silicon Press (2008)
5. Ahmad, I., Ranka, S.: Handbook of Energy-Aware and Green Computing. Champman and HallComputer and Information Science Series, vol. 1. CRC Press, Taylor and Francis Group (2012)
6. Jha, S.K., Jana, P.K.: Fast Data Concentration on OTIS-Mesh of Trees. In: Frontiers in Electronics, Communications, Instrumentation and Information Technology, pp. 232–237 (2009)
7. Jha, S.K., Reddy, M.R.: Parallel Prefix on Hierarchical Ring with its Application to Data Concentration. In: Recent Advances in Information Technology, pp. 293–300 (2009)
8. Kini, N.G., Kumar, M.S., Mruthyunjaya, H.S.: A Torus Embedded Hypercube Scalable Interconnection Network for Parallel Architecture. In: IEEE International Advance Computing Conference, pp. 858–861 (2009)
9. Hwang, K.: Advanced Computer Architecture: Parallelism, Scalability, Programmability. McGraw-Hill (1993)
10. Jong-Seok, K., Hyeong-OK, L., Yeong-Nam, H.: Embedding among HCN(*n*, *n*), HFN(*n*, *n*) and hypercube. In: Eighth IEEE International Conference on Parallel and Distributed System, pp. 533–540 (2001)
11. Sinha, U.K., Jha, S.K., Mondal, H.: Parallel Prefix Computation on Torus Embedded Hypercube. In: International Conference on Recent Trends in Engineering and Technology (2014)
12. Akl, S.G.: The Design and Analysis of Parallel Algorithms. Prentice - Hall (1989)
13. Jha, S.K., Jana, P.K.: Study and Design of Parallel Algorithms for Interconnection Networks. Lambert Academic Publishing (2011)

SAT Based Scheduling in High Level Synthesis

Sudeshna Kundu, Khushbu Chandrakar, and Suchismita Roy

Department of Computer Science and Engineering,
National Institute of Technology Durgapur, India
{sudeshna.buie,1kh.chandrakar}@gmail.com, suchismita27@yahoo.com

Abstract. High level synthesis is the process of generating the register transfer level (RTL) design from the behavioural description. Time-constrained scheduling minimizes the requirement of functional units under a given time constraint and Resource-constrained scheduling minimizes the number of control steps under given resource constraint. A PB-SAT based approach which concentrates on operation scheduling, and also optimizes the number of resources and control steps is proposed here. Time-constrained and Resource-constrained based scheduling is formulated as a Pseudo-boolean satisfiability (PB-SAT) based problem and a SAT solver is used for finding the optimum schedule and minimum number of functional unit and control steps satisfying all constraints.

Keywords: High-level synthesis, Time-constrained scheduling, Resource-constrained scheduling, PB-SAT.

1 Introduction

High level synthesis is the process of generating the register transfer level (RTL) design from the behavioural description. The synthesis process consists of several interdependent phases: Preprocessing, Scheduling, Register Allocation and Binding of variables, Control Path and Data Path generation, and generation of Synthesizable code (RTL). Among the above steps, operation scheduling and hardware allocation are the two major task. These two subtasks are interdependent. In order to have an optimal design, a system should perform both subtask simultaneously. Operation scheduling determines the cost-speed tradeoff of the design. If the design emphasizes on speed constraints, the scheduling algorithm will attempt to parallelize the operations to meet the timing constraint. Conversely, if there is limit in the cost(area or resources), the algorithm will serialize operations to meet the resource constraint. Once the operations are scheduled,the number and types of functional units, the lifetime of variables, and the timing constraint are fixed. Thus a good scheduler is very necessary in high level synthesis.

Two basic scheduling problems with different requirements is addressed in this paper.

(1)Time-Constrained Scheduling: Given constraints on the maximum number of control steps, find the cost-efficient schedule which satisfies the constraints.

M.K. Kundu et al. (eds.), *Advanced Computing, Networking and Informatics - Volume 2*,
Smart Innovation, Systems and Technologies 28,
DOI: 10.1007/978-3-319-07350-7_59, © Springer International Publishing Switzerland 2014

(2)Resource-Constrained Scheduling: Given constraints on the resources, find the fastest scheduling which satisfies the constraints.

The proposed approach introduces a Pseudo-Boolean (PB) Sat based formulation for solving time-constrained and resource-constrained based scheduling problem. Solving this problem using PB-SAT based approach gives the flavours of both optimization problem and decision problems.The approach is complete and hence examines the entire search space defined by the problem to prove that either (i) the problem has no solution, i.e., the problem is unsatisfiable, or (ii) that a solution does exist, i.e., the problem is satisfiable. If the problem is satisfiable, the proposed approach will search all possible solutions to find the optimal solution.

2 Related Work

In the past decade there has been a lot of experiments on different aspects of high-level synthesis. Several scheduling algorithms exits to solve the scheduling problem [1],[2]. ILP-based high-level synthesis has been exploited for several years. Hwang, Lee, Hsu [3], Chaudhuri [4] gave a formal analysis of the constraints of ILP-base scheduling, and presented a well-structured ILP formulation of the scheduling problem to reduce the computation time. A SAT-based approach to the scheduling problem in high-level synthesis is presented by Memik [5] considering resource-constrained scheduling as a satisfiability (SAT) problem. Recent advances and ongoing research on advanced Boolean Satisfiability (SAT) solvers have been extended to solve 0-1 ILP problems [6] enabled the successful deployment of SAT technology in a wide range of applications domains, and particularly in electronic design automation (EDA). Many complex engineering problems have been successfully solved using SAT. Such problems include routing [7] scheduling problem, power optimization [8],[9], [10] verification [11], and graph colouring [12], optimization [13] etc. Today, several powerful SAT solvers exist and are capable of handling problems consisting of thousands of variables and millions of constraints [14], [15], [16]. They can also compete with the best available generic ILP solvers.

This paper shows a novel application to solve the scheduling problems in high-level synthesis and introduces a PB-SAT based formulation for solving the scheduling problem. The main motive of this paper is using PB-SAT optimizing solver to estimate (1) the minimum number of functional unit used and optimize the cost satisfying the latency constraint in time-constrained scheduling (2) the minimum number of control steps required to schedule the DFG satisfying the resource constraint in resource-constraint scheduling. PB-SAT based solvers can handle this type of problem very effectively. The paper is organised as follows. PB-SAT based formulation for time-constrained and resource-constrained scheduling is formulated in Section 3. Experimental results are presented in Section 4. Finally conclusions are drawn in Section 5.

3 Problem Formulation

The proposed methodology produces a PB-SAT based formulation for time-constrained scheduling and resource-constrained scheduling. Few definitions and notations are provided that required to this study.

3.1 Definitions

Definition 1.(Pseudo Boolean optimization). Besides solving decision problems, handling PB constraints expands the ability of SAT solvers to solve Boolean optimization problems, which call for the minimization or maximization of a linear objective as follows:

$$\sum_{i=1}^{n} a_i x_i \tag{1}$$

where, a_i is an integer coefficient x_i is a boolean variable and there are n terms in the objective function, subject to a set of m linear constraints.

$$Ax \leq b \tag{2}$$

where $b \in Z^n, A \in Z^m \times Z^n$, and $x \in \{0,1\}^n$.

Definition 2.(Data flow graph). A data flow graph DFG is a directed acyclic graph $G = (V, E)$, where V is a set of nodes, and E is a set of edges between nodes.Here each node represents an operation, and a directed edge from node v_i to node v_j represented as $v_i \rightarrow v_j$ means execution of v_i must precede that of v_j.

Definition 3.(ASAP schedule time). ASAP schedule time of any node is the soonest time at which it can be scheduled.

Definition 4.(ALAP schedule time). ALAP schedule time of any node is the latest time at which it can be scheduled.

Definition 5.(Mobility). The mobility of a node is the difference between its ALAP schedule time and ASAP schedule time.

3.2 Notation

The variables used in the formulations are defined as follows:
n = number of operations in the data flow graph.
o_i = operation $i, 1 \leq i \leq n$

$o_i \rightarrow o_l, o_i$ = an immediate predecessor of o_l
FU_k = functional unit of type k
c_k = cost of FU_k
$o_i \in FU_k$ if o_i can be executed by FU_k
s = number of scheduling steps
M_k = integer variables that denote the maximum number of functional units of type k required in all steps
$x_{i,j}$ = Boolean variables associated with o_i. $x_{i,j} = 1$ if o_i is scheduled into step j and implemented using a functional unit; otherwise, $x_{i,j} = 0$.
S_i = starting step of o_i
E_i = ending step of o_i
C_{total} = total number of control steps
$mrange(o_i) = \{s_j | E_i \leq j \leq L_k\}$, mobility range of operation o_i
r = range to represent boolean variables

3.3 Time-Constrained Scheduling

The time-constrained scheduling algorithm can be defined as follows: *Given a data flow graph and the maximum number of control steps, find a minimal cost schedule that satisfies the given time constraint.* Here the cost of a data-path may be the cost of functional units, interconnections and registers. But only the hardware cost is considered here. It is obvious that the hardware cost is minimized if all the functional units are fully utilized in the design. In other words, operations of the same type should be evenly distributed among all control steps.

In order to trim the solution space of the SAT formulation, we first perform ASAP and ALAP calculations for all operations to determine the earliest and latest possible scheduling steps of all operations. After computing the ASAP and ALAP of all operations, we can formulate the problem as follows:

(a) Objective Function : The objective function minimizes the total cost of functional units and can be written as follows.

$$minimize \sum_{k=1}^{m} \left(c_k * \left(\sum_{b=0}^{r-1} 2^b * M_k \right) \right) \tag{3}$$

(b) Uniqueness Constraints : These constraints ensure that every operation o_i is scheduled to one unique control step within the mobility range (S_i, E_i).

$$\sum_{j=S_i}^{L_i} x_{i,j} = 1, \forall 1 \leq i \leq N \tag{4}$$

(c) Precedence Constraints : These constraints guarantee that for an operation o_i, all its predecessors are scheduled in an earlier control step and its successors are scheduled in an later control step.

$$\sum_{j=S_i}^{L_i}(j * x_{i,j}) - \sum_{j=S_k}^{L_k}(j * x_{k,j}) \leq -1 \tag{5}$$

(d) Resource Constraints : These constraints make sure that no control step contains more than FU_k operations of type k,

$$\sum_{j=o_i \in FU_k} x_{i,j} - \left(\sum_{b=0}^{r-1} 2^b * M_k\right) \leq 0, \forall 1 \leq j \leq s \tag{6}$$

3.4 Resource-Constrained Scheduling

The resource-constrained scheduling problem is defined as follows. *given a data flow graph and a set of functional units, find a minimal cost schedule that satisfies the given resource constraint.* The cost includes the required control steps. ASAP and ALAP calculations is also performed here. The formulation is as follows.
(a) Objective Function : The objective function minimizes the total number of control steps satisfying all resource constraints.

$$minimize \sum_{b=0}^{r-1} 2^b * C_{step,b} \tag{7}$$

(b) Uniqueness Constraints : These constraints ensure that every operation o_i is scheduled to one unique control step within the mobility range (S_i, E_i).

$$\sum_{j=S_i}^{L_i} x_{i,j} = 1, \forall 1 \leq i \leq N \tag{8}$$

(c) Precedence Constraints : These constraints guarantee that for an operation o_i, all its predecessors are scheduled in an earlier control step and its successors are scheduled in an later control step.

$$\sum_{j=S_i}^{L_i}(j * x_{i,j}) - \sum_{j=S_k}^{L_k}(j * x_{k,j}) \leq -1 \tag{9}$$

(d) Resource Constraints : These constraints make sure that no control step contains more than FU_k operations of type k, here M_k is constant.

$$\sum_{j=o_i \in FU_k} x_{i,j} - M_k \leq 0, \forall 1 \leq j \leq s \tag{10}$$

(e) Latency Constraints : Ensures that no operations should be scheduled after C_{step}.

$$\sum_{j=S_i}^{L_i} x_{i,j} - \sum_{b=0}^{r-1} 2^b * C_{step,b} \leq 0, \forall o_i without successor \tag{11}$$

The steps followed to obtain the scheduling are as follows:

Step 1: Find ASAP and ALAP schedule of the UDFG.
Step 2: Determine the mobility graph of each node.
Step 3: Construct the PB-SAT formulations for the DFG.
Step 4: Solve the PB-SAT formulations using backend SAT solver.
Step 5: Find the scheduled DFG.
Step 6: Find minimum number of resources used and optimize cost in time-constrained scheduling.
Step 7: Find minimum number of control steps used in resource-constrained scheduling.

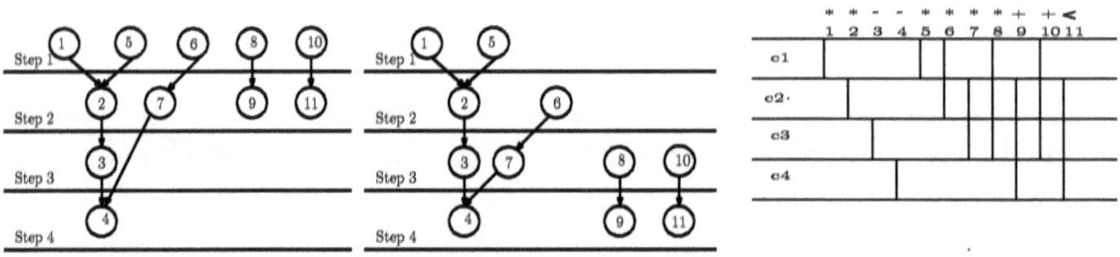

Fig. 1. (a) ASAP schedule, (b) ALAP schedule, (c) Mobility graph

Example: Consider a DFG consisting of 11 nodes. Fig. 1(a) represents ASAP schedule and Fig. 1(b) represents ALAP schedule of the DFG. From ASAP and ALAP schedule mobility graph is obtained as depicted in Fig. 1(c). PB-constraints are formulated for time-constrained scheduling using these mobility ranges. To obtain the total cost of each schedule, we consider the cost of multiplier as 2, and for adder, subtracter, and comparator as 1. The constraint generated are as shown below:

/* Objective Function
$min: +8\ Nm2 + 4\ Nm1 + 2\ Nm0 + 4\ Ns2 + 2\ Ns1 + 1\ Ns0 + 4\ Na2 + 2\ Na1 + 1\ Na0 + 4\ Nc2 + 2\ Nc1 + 1\ Nc0;$
/* Uniqeness Constraint
$+1\ x11 = 1;$
$+1\ x22 = 1;$
$+1\ x33 = 1;$
$+1\ x44 = 1;$
$+1\ x51 = 1;$
$+1\ x61 + 1\ x62 = 1;$
$+1\ x72 + 1\ x73 = 1;$
$+1\ x81 + 1\ x82 + 1\ x83 = 1;$
$+1\ x92 + 1\ x93 + 1\ x94 = 1;$
$+1\ x101 + 1\ x102 + 1\ x103 = 1;$
$+1\ x112 + 1\ x113 + 1\ x114 = 1;$
/* Precedence Constraint
$+1\ x11 - 2\ x22 \leqslant -1;$
$+2\ x22 - 3\ x33 \leqslant -1;$
$+3\ x33 - 4\ x44 \leqslant -1;$
$+1\ x51 - 2\ x22 \leqslant -1;$
$+1\ x61 + 2\ x62 - 2\ x72 - 3\ x73 \leqslant -1;$
$+2\ x72 + 3\ x73 - 4\ x44 \leqslant -1;$
$+1\ x81 + 2\ x82 + 3x83 - 2\ x92 - 3\ x93 - 4\ x94 \leqslant -1;$
$+1\ x101 + 2\ x102 + 3\ x103 - 2\ x112 - 3\ x113 - 4\ x114 \leqslant -1;$
/* Resource Constraint
$-1\ x11 - 1\ x51 - 1\ x61 - 1\ x81 + 4\ Nm2 + 2\ Nm1 + 1\ Nm0 \geqslant 0;$
$-1\ x101 + 4\ Na2 + 2\ Na1 + 1\ Na0 \geqslant 0;$
$-1\ x22 - 1\ x62 - 1\ x72 - 1\ x82 + 4\ Nm2 + 2\ Nm1 + 1\ Nm0 \geqslant 0;$
$-1\ x92 - 1\ x102 + 4\ Na2 + 2\ Na1 + 1\ Na0 \geqslant 0;$
$-1\ x112 + 4\ Nc2 + 2\ Nc1 + 1\ Nc0 \geqslant 0;$
$-1\ x73 - 1\ x83 + 4\ Nm2 + 2\ Nm1 + 1\ Nm0 \geqslant 0;$
$-1\ x33 + 4\ Ns2 + 2\ Ns1 + 1\ Ns0 \geqslant 0;$
$-1\ x93 - 1\ x103 + 4\ Na2 + 2\ Na1 + 1\ Na0 \geqslant 0;$
$-1\ x113 + 4\ Nc2 + 2\ Nc1 + 1\ Nc0 \geqslant 0;$
$-1\ x44 + 4\ Ns2 + 2\ Ns1 + 1\ Ns0 \geqslant 0;$
$-1\ x94 + 4\ Na2 + 2\ Na1 + 1\ Na0 \geqslant 0;$
$-1\ x114 + 4\ Nc2 + 2\ Nc1 + 1\ Nc0 \geqslant 0;$

On solving the PB-SAT instances. The result obtained as follows:

```
c Optimal solution: 7 (2*2+1*1+1*1+1*1)
s OPTIMUM FOUND
v -Nm2 Nm1 -Nm0 -Ns2 -Ns1 Ns0 -Na2 -Na1 Na0 -Nc2 -Nc1 Nc0 x11 x22
x33 x44
x51 -x61 x62 -x72 x73 -x81 -x82 x83 -x92 -x93 x94 x101 -x102 -x103 -x112
x113 -x114
```

Here 7 represents the optimum cost. It is obtained by multiplying cost of each type of FU with minimum number of resources used to schedule the DFG. The assignment of the variable gives the minimum number of resources of each type required to schedule the DFG and it also gives the optimized schedule. The optimized schedule for time-constrained scheduling is presented in Fig. 2(a). The PB-SAT constraints for resource-constrained scheduling are created similarly. The optimized schedule for resource-constrained scheduling is shown in Fig. 2(b).

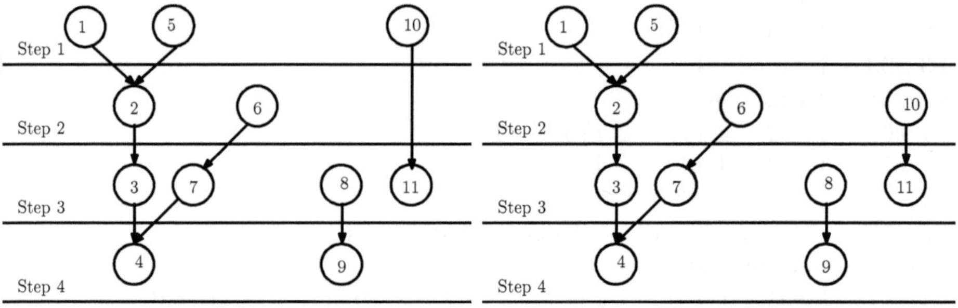

Fig. 2. (a) Time-constrained Scheduling (b) Resource-constrained scheduling

4 Experimental Results

The use of PB constraints in scheduling problem is evaluated here. The proposed PB-SAT based approach has been applied to different benchmark circuits such as the HAL differential equation solver, Infinite Impulse Response (IIR) Filter, Finite Impulse Response (FIR) Filter, Auto Regression Filter(ARF). The SAT instances were solved using the Minisat+ [17] as the backend solver. The constraints are generated using C++ on a 32 bit Intel Pentium Dual CPU T3400 @2.16 GHz 2.17 GHz. The experimental results for time-constrained scheduling applied on various benchmark circuits are reported in Table 1. Where first column represents the benchmark DFG, second column signifies the control steps of all the given DFG, third column represents optimum resources and fourth defines the optimum cost calculated as explained in previous section and the fifth column represents average CPU time for solving the PB instances. Table 2 present the results for resource-constrained scheduling by using the PB-SAT solver optimum resources and cost is obtained, satisfying the time constraints. The solver is able to give the optimal solution if the resource constraints are met. Otherwise, solver gives an unsatisfiable instances.

Table 1. Time-constrained Scheduling

Bench mark	Control Step	Optimized Resources	Optimum Cost	CPU time(s)
HAL	4	2*,1+,1-,1<	7	0.004
IIR	4	3*,2+	8	0.005
FIR	9	2*,2+	6	0.012
ARF	8	4*,2+	10	0.020

Table 2. Resource-constrained scheduling

Bench mark	Resource Constraints	SAT Solution	Control Step	CPU time(s)
	1*,1+,1-,1<	U	-	0
HAL	2*,1+,1-,1<	OPT. FOUND	4	0.001
	3*,1+,1-,1<	OPT. FOUND	4	0.002
	3*,2+	U	-	0.004
ARF	4*,2+	OPT. FOUND	8	0.006
	5*,2+	OPT. FOUND	8	0.008
	2*,2+	U	-	0.0
IIR	3*,2+	OPT. FOUND	4	0.004
	4*,2+	OPT. FOUND	4	0.004
	1*,1+	U	-	0.004
FIR	2*,2+	OPT. FOUND	9	0.0160
	3*,3+	OPT. FOUND	9	0.0160

5 Conclusion and Future Work

This paper explores a new approach for solving scheduling problem using PB-SAT solver. Implementing PB constraints, gives the optimum solution to the NP-complete time-constraint and resource-constraint scheduling problem. Experimental results shows that the size of this formulation is quite acceptable for practical synthesis and always finds the optimal solutions for all the benchmarks.

The scheduling algorithms can be extended to more realistic design models, such as functional units with varying execution times, multi-functional units. Which may enable the scheduler to generate schedule for more realistic designs.

References

1. Gajski, D., Dutt, N., Wu, A.H., Lin, S.L.: High-level Synthesis: Introduction to Chip and System Design. Kluwer Academic Publishers (1992)
2. Baruch, Z.: Scheduling algorithms for high-level synthesis. ACAM Scientific Journal 5(1-2), 48–57 (1996)
3. Hwang, C.T., Lee, J.H., Hsu, Y.C.: A formal approach to the scheduling problem in high level synthesis. IEEE Transactions on Computer-Aided Design of Integrated Circuits and Systems 10(4), 464–475 (1991)
4. Chaudhuri, S., Walker, R.: Ilp-based scheduling with time and resource constraints in high level synthesis. In: VLSI Design, pp. 17–20 (1994)
5. Memik, S., Fallah, F.: Accelerated sat-based scheduling of control/data flow graphs. In: IEEE International Conference on Computer Design: VLSI in Computers and Processors, pp. 395–400 (2002)
6. Aloul, F., Ramani, A., Markov, I., Sakallah, K.: Generic ilp versus specialized 0-1 ilp. In: IEEE/ACM International Conference on Computer Aided Design, ICCAD, pp. 450–457 (2002)
7. Nam, G.J., Aloul, F., Sakallah, K., Rutenbar, R.: A comparative study of two boolean formulations of fpga detailed routing constraints. IEEE Transactions on Computers 53(6), 688–696 (2004)
8. Sagahyroon, A., Aloul, F.: Using sat-based techniques in power estimation. J. Microelectron. 38(6-7), 706–715 (2007)
9. Aloul, F., Hassoun, S., Sakallah, K., Blaauw, D.: Robust sat-based search algorithm for leakage power reduction. In: Hochet, B., Acosta, A.J., Bellido, M.J. (eds.) PATMOS 2002. LNCS, vol. 2451, pp. 167–177. Springer, Heidelberg (2002)
10. Mohanty, S., Ranganathan, N., Chappidi, S.: An ilp-based scheduling scheme for energy efficient high performance datapath synthesis. In: ISCAS, vol. (5), pp. 313–316 (2003)
11. Biere, A., Cimatti, A., Clarke, E., Fujita, M., Zhu, Y.: Symbolic model checking using sat procedures instead of bdds, pp. 317–320 (1999)
12. Ramani, A., Markov, I., Sakallah, K., Aloul, F.: Breaking instance-independent symmetries in exact graph coloring. J. Artif. Intell. Res (JAIR) 26, 289–322 (2006)
13. Aloul, F.: On solving optimization problems using boolean satisfiability (2005)
14. Moskewicz, M., Madigan, C., Zhao, Y., Zhang, L., Malik, S.: Chaff: Engineering an efficient sat solver. In: Proceedings of the 38th Annual Design Automation Conference, DAC, pp. 530–535. ACM (2001)
15. Sheini, H., Sakallah, K.: Pueblo: A modern pseudo-boolean sat solver. In: Proceedings of the Design, Automation and Test in Europe DATE, pp. 684–685 (2005)
16. Chai, D., Kuehlmann, A.: A fast pseudo-boolean constraint solver. IEEE Trans. on CAD of Integrated Circuits and Systems 24(3), 305–317 (2005)
17. `http://minisat.se/Main.html`

Hybrid Function Based Microprocessor Realisation of Closed Loop Analysis and Synthesis

Anusna Chakraborty[1] and Anindita Ganguly[2]

[1] Cognizant Technology Solutions, India
anusna90@gmail.com
[2] St. Thomas' College of Engineering and Technology, Kolkata, India
aninditaganguly80@gmail.com

Abstract. The present work proposes application of a set of orthogonal hybrid functions (HF) which evolved from the synthesis of orthogonal sample-and-hold functions (SHF) and orthogonal triangular functions (TF). This HF set is employed for determining the result of closed loop convolution and the result has been used for solving linear control system analysis and synthesis problems. The theory is supported by an example and the results are compared with the exact solution. It has been observed that for closed loop system identification, oscillation occurs due to numerical instability.

Keywords: Sample-and-hold functions, Triangular functions, Hybrid functions, Function approximation, Analysis, Synthesis.

1 Introduction

For more than four decades different piecewise constant basis functions (PCBF) [1] have been employed to solve problems in different fields of engineering including control theory. Of this class, the block pulse function (BPF) [2], [3] set proved to be the most efficient because of its simplicity and versatility in analysis as well as synthesis of control systems.

In 1998, an orthogonal set of sample-and-hold functions [4] were introduced by Deb et al and the same was applied to solve problems related to discrete time systems with zero order hold. The set of sample-and-hold functions approximate any square integrable function of Lebesgue measure [2] in a piecewise constant manner and was proved to be more convenient for solving problems related to sample-and-hold systems.In 2003, orthogonal triangular functions [5] were introduced by Deb et al and the same were applied to control system related problems including analysis and system identification. The set of triangular functions approximate any square integrable function in a piecewise linear manner.In control theory, we essentially need to design convolution algorithms for the analysis of control system problems. Application of a set of orthogonal hybrid functions (HF) [6], which is a combination of sample-and-hold function and triangular function is presented in this paper.

M.K. Kundu et al. (eds.), *Advanced Computing, Networking and Informatics - Volume 2*, Smart Innovation, Systems and Technologies 28,
DOI: 10.1007/978-3-319-07350-7_60, © Springer International Publishing Switzerland 2014

A convolution algorithm, which is computationally attractive, is proposed for the analysis of linear closed loop control problems.

With this innovative combination, Deb *et al.* studied control theory problems including function approximation, integration, differentiation and differential equations [5] as well. Using the HF set, transfer function of a system was identified [6] from plant impulse response data, state equation for a homogeneous system was solved recursively [5] and higher order differential equations were also solved [7] using HF domain one-shot integration operational matrices.

In this paper, the set of orthogonal hybrid functions (HF) is utilized for:

(i) Closed loop system analysis with the aid of convolution of time functions in HF using operational matrices,

(ii) Closed loop system identification via 'deconvolution' of time functions in HF domain using operational matrices

2 Hybrid Functions (HF) A Combination of SHF and TF

We can use a set of sample-and-hold functions and a set of triangular functions to form an orthogonal function set, which we call a 'Hybrid Function' set. To define this set, we express the (*i*+1)-th member *Hi(t)* of the m-set hybrid function H(m) (t) as

$$Hi\,(t) = a_i S_i\,(t) + b_i T_2 i\,(t) \qquad (1)$$

int=*[0,T)* , where, $i = 0, 1, 2, \ldots, (m-1)$,

a_i and b_i are scaling constants.

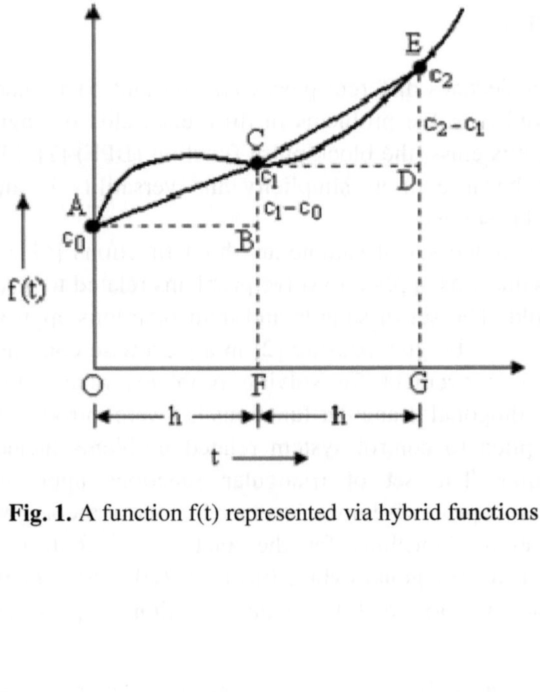

Fig. 1. A function f(t) represented via hybrid functions

We consider a square integrable function *f(t)* [2], [3] ofLebesgue measure and express it via orthogonal hybrid functions. The function *f(t)* in Fig. 1 is sampled at three equidistant points (sampling interval h) A, C and E and the sample values are c_0, c_1 and c_2. Now, *f(t)* can be expressed in a piecewise linear form by two straight lines AC and CE, which are the sides of two adjacent trapeziums shown in Fig. 1. That is

$$\begin{aligned} f(t) &= H_0(t) + H_1(t) \\ &= c_0 S_0(t) + c_1 S_1(t) + (c_1 - c_0) T_0(t) + (c_2 - c_1) T_1(t) \\ &\triangleq C^T S_{(2)}(t) + D^T T_{(2)}(t) \end{aligned} \quad (2)$$

where, $[\, c_0 \; c_1] = C^T$ and $[(\, c_1 - c_0\,)\; (\, c_2 - c_1)\,]^T = D^T$

and [···]T denotes transpose. $S_{(2)}(t)$ is a sample-and-hold function vector of dimension 2, $T_{(2)}(t)$ is a triangular function vector of dimension 2. Equation (2) represents *f(t)* in hybrid function domain.

3 Analysis of Closed Loop System

Consider a single-input-single-output (SISO) time-invariant system [8]. An input *r(t)* is applied at $t = 0$. The block diagram of the system using time variables is shown in Fig. 2. Application of *r(t)* to the system *g(t)* with feedback *h(t)* produces the corresponding output *y(t)* for $t = 0$.

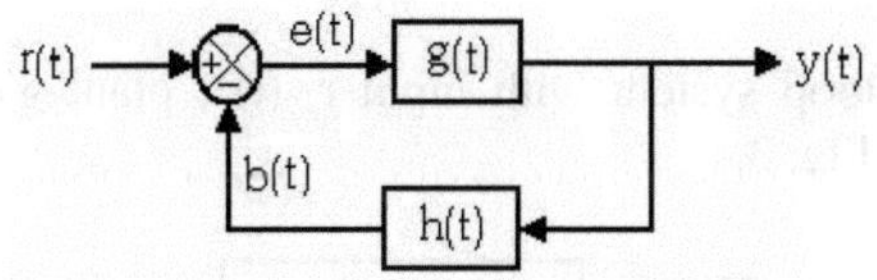

Fig. 2. Block diagram of a closed loop system

Considering *r(t), g(t), y(t)* and *h(t)* to be bounded (i.e. the system is BIBO stable) and absolutely integrable over $t \in [0,T)$, all these functions may be expanded via HF series. For m = 4, we can write

$$r(t) \triangleq \left[R_S^T S_{(4)} + R_T^T T_{(4)} \right], \quad g(t) \triangleq \left[G_S^T S_{(4)} + G_T^T T_{(4)} \right],$$
$$y(t) \triangleq \left[Y_S^T S_{(4)} + Y_T^T T_{(4)} \right] \text{ and } h(t) \triangleq \left[H_S^T S_{(4)} + H_T^T T_{(4)} \right]$$

Output of the feedback system is denoted by *b(t)*

$$\text{where, } b(t) = y(t) * h(t) \quad (3)$$

$$
=\frac{h}{6}\begin{bmatrix} y_0 & y_1 & y_2 & y_3 \end{bmatrix}\begin{bmatrix} 0 & H_0 & H_1 & H_2 \\ 0 & H_4 & H_5 & H_6 \\ 0 & 0 & H_4 & H_5 \\ 0 & 0 & 0 & H_4 \end{bmatrix} S_{(4)} + \frac{h}{6}\left\{ \begin{bmatrix} y_0 & y_1 & y_2 & y_3 \end{bmatrix}\begin{bmatrix} H_0 & (H_1-H_0) & (H_2-H_1) & (H_3-H_2) \\ 0 & H_0 & (H_1-H_0) & (H_2-H_1) \\ 0 & 0 & H_0 & (H_1-H_0) \\ 0 & 0 & 0 & H_0 \end{bmatrix} \right.
$$

$$
\left. +\begin{bmatrix} y_1 & y_2 & y_3 & y_4 \end{bmatrix}\begin{bmatrix} H_4 & H_8 & H_9 & H_{10} \\ 0 & H_4 & H_8 & H_9 \\ 0 & 0 & H_4 & H_8 \\ 0 & 0 & 0 & H_4 \end{bmatrix} \right\} T_{(4)} \tag{4}
$$

So we may write,

$$
b(t)=\frac{h}{6}\begin{bmatrix} 0 & H_0y_0+H_4y_1 & H_1y_0+H_5y_1+H_4y_2 & H_2y_0+H_6y_1+H_5y_2+H_4y_3 \end{bmatrix} S_{(4)}
$$

$$
+\frac{h}{6}[H_0y_0+H_4y_1 \quad (H_1-H_0)y_0+(H_0+H_8)y_1+H_4y_2 \quad (H_2-H_1)y_0+(H_1-H_0+H_9)y_1+(H_0+H_8)y_2+H_4y_3
$$

$$
(H_3-H_2)y_0+(H_2-H_1+H_{10})y_1+(H_1-H_0+H_9)y_2+(H_0+H_8)y_3+H_4y_4]T_{(4)} \tag{5}
$$

The generalized form of the output equation in Hybrid domain may be expressed as:

$$
y_i=\frac{\frac{h}{6}\left[r_0G_{(i-1)}+\sum_{p=1}^{i} r_pG_{(m+i-p)}-\frac{h}{6}\sum_{p=1}^{i} H_{(p-1)}G_{(m+i-p)}-\frac{h}{6}\sum_{p=1}^{i} K_{(i-p+2)}y_{(p-1)} \right]}{1+\frac{h^2}{36}(H_mG_m)} \tag{6}
$$

Consider a closed loop system with input r_c (t) , plant g_c(t) , feedback h(t) and output y_c (t) ,shown in Fig. 3.

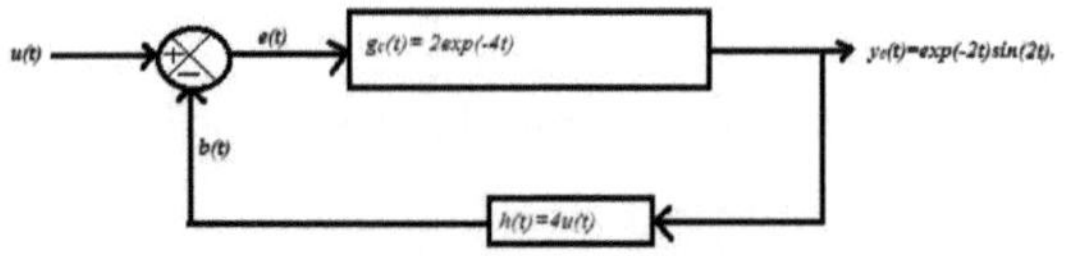

Fig. 3. A closed loop system with step input.

Table 1. Samples obtained via direct expansion and convolution of hybrid function method

t **(sec)**	***y*** **[Samples of $y_c(t)$] using exact method**	***y*** **[Samples of $y_c(t)$obtained via HF domain analysis**
0	0	0
1/4	0.2907	0.3112
2/4	0.3095	0.3390
3/4	0.2225	0.2446
4/4	0.1230	0.1307

This system is analyzed in HF domain with $r(t) = u(t)$, $g_c(t) = 2exp(-4t)$ and $h(t) = 4u(t)$ form=4with T=1 s. The exact output of the system is

$$y_c(t) = exp(-2t)sin(2t).$$

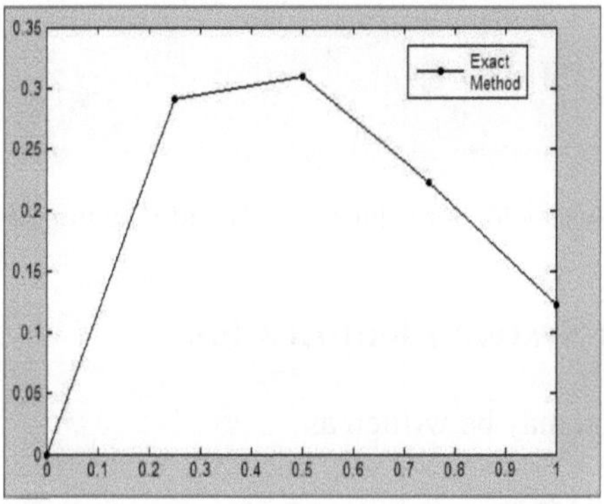

Fig. 4. Closed Loop Analysis with direct expansion of HF via MATLAB

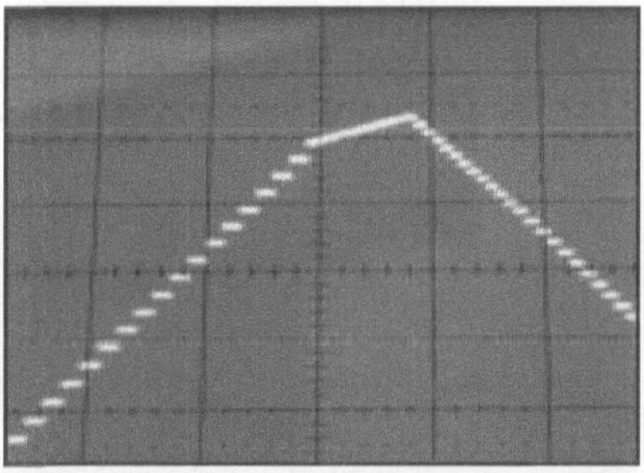

Fig. 5. Closed Loop Analysis with direct expansion of HF via Microprocessor

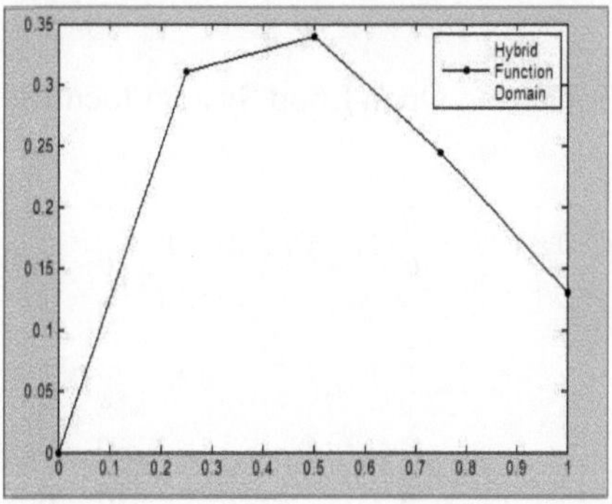

Fig. 6. Closed Loop Analysis with convolution of Hybrid Function Method via MATLAB

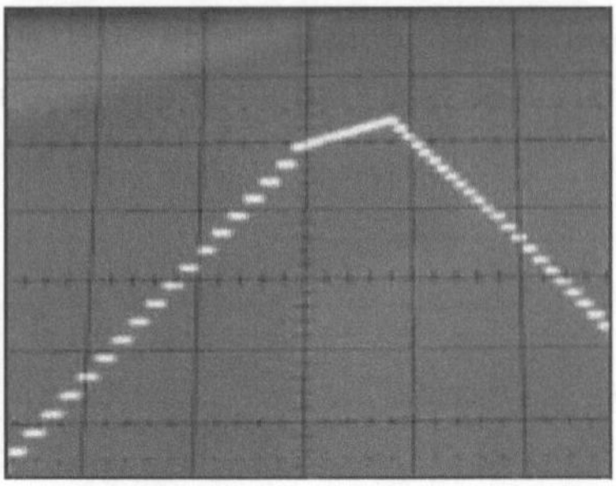

Fig. 7. Closed Loop Analysis with convolution of Hybrid Function Method via Microprocessor

4 Closed Loop System Identification

Closed loop convolution may be written as

$$y(t)=\frac{h}{6}[g_0 \quad g_1 \quad g_2 \quad g_3]\begin{bmatrix} 0 & 2e_1+e_0 & 2e_2+e_1 & 2e_3+e_2 \\ 0 & e_1+2e_0 & e_2+4e_1+e_0 & e_3+4e_2+e_1 \\ 0 & 0 & e_1+2e_0 & e_2+4e_1+e_0 \\ 0 & 0 & 0 & e_1+2e_0 \end{bmatrix} S_{(4)}$$

$$+\frac{h}{6}[g_0(2e_1+e_0)+g_1(e_1+2e_0) \quad g_0(2e_2-e_1-e_0)+g_1(e_2+3e_1-e_0)+g_2(e_1+2e_0) \quad g_0(2e_3-e_2-e_1)+g_1(e_3+3e_2-3e_1-e_0)+g_2(e_2+3e_1-e_0)+g_3(e_1+2e_0)$$
$$g_0(2e_4-e_3-e_2)+g_1(e_4+3e_3-3e_2-e_1)+g_2(e_3+3e_2-3e_1-e_0)+g_3(e_2+3e_1-e_0)+g_4(e_1+2e_0)]T_{(4)} \tag{7}$$

Comparing from the Open loop identification SHF Equation we may write :

$$[y_0 \quad y_1 \quad y_2 \quad y_3]=\frac{h}{6}[g_0 \quad g_1 \quad g_2 \quad g_3]\begin{bmatrix} 0 & 2e_1+e_0 & 2e_2+e_1 & 2e_3+e_2 \\ 0 & e_1+2e_0 & e_2+4e_1+e_0 & e_3+4e_2+e_1 \\ 0 & 0 & e_1+2e_0 & e_2+4e_1+e_0 \\ 0 & 0 & 0 & e_1+2e_0 \end{bmatrix} \tag{8}$$

Comparing the equations from Open Loop System Identification we may write

$$[y_0 \quad y_1 \quad y_2 \quad y_3]=\frac{h}{6}[g_0 \quad g_1 \quad g_2 \quad g_3]\begin{bmatrix} \varepsilon & E_0 & E_1 & E_2 \\ 0 & E_4 & E_5 & E_6 \\ 0 & 0 & E_4 & E_5 \\ 0 & 0 & 0 & E_4 \end{bmatrix} \tag{9}$$

$$[g_0 \quad g_1 \quad g_2 \quad g_3]=\frac{6}{h}[y_0 \quad y_1 \quad y_2 \quad y_3]\begin{bmatrix} \varepsilon & E_0 & E_1 & E_2 \\ 0 & E_4 & E_5 & E_6 \\ 0 & 0 & E_4 & E_5 \\ 0 & 0 & 0 & E_4 \end{bmatrix}^{-1} \tag{10}$$

Comparing between last elements of the second parts (TF) of equations (7) and (10), we get

$$y_4 - y_3 = \frac{h}{6}\left[g_0(2e_4 - e_3 - e_2) + g_1(e_4 + 3e_3 - 3e_2 - e_1) + g_2(e_3 + 3e_2 - 3e_1 - e_0) + g_3(e_2 + 3e_1 - e_0) + g_4(e_1 + 2e_0) \right] \quad (11)$$

So, final generalized output is as follows

$$g_m = \frac{[y_m - y_{(m-1)}] - \frac{h}{6}\left[g_0\left\{E_{(m-1)} - E_{(m-2)}\right\} + \sum_{i=1}^{m-1} g_i\left\{E_{(2m-i)} - E_{(2m-i-1)}\right\}\right]}{\frac{h}{6}E_m} \quad (12)$$

Illustrative Example 2

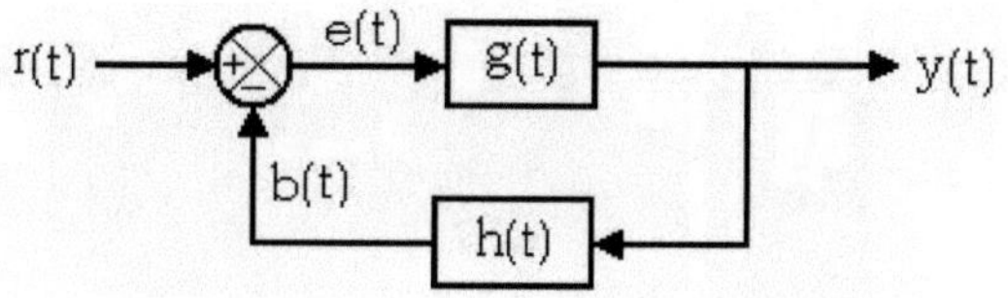

Fig. 8. Closed loop system with step input

Table 2. Samples obtained via direct expansion and de-convolution of hybrid function method

T (sec)	**ged [Samples of ge (t)]**	**gec [Samples of e g (t) obtained via HF domain analysis**
0	1.0000	24.0000
0.25	0.7788	-22.011
0.50	0.6065	23.5557
0.75	0.4723	-22.3578
1.00	0.3678	23.2867

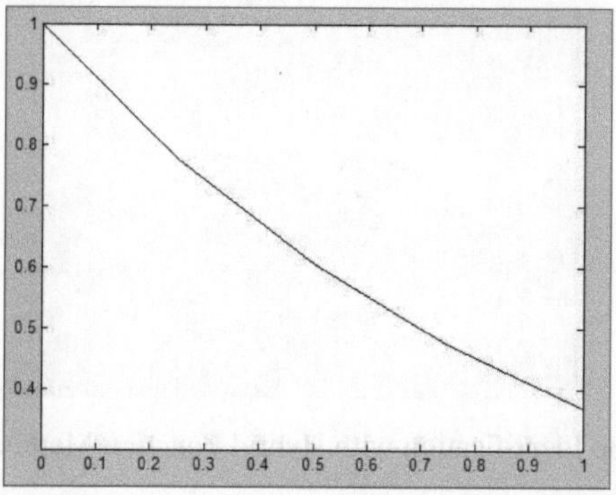

Fig. 9. Closed Loop Identification with Direct Expansion of HF via MTALAB

Consider a closed loop system of Fig. 8with input $r_e(t)= u(t)$, $h(t) = u(t)$ and output $y_e(t)= \frac{2}{\sqrt{3}} exp\left(-\frac{t}{2}\right) sin\left(\frac{\sqrt{3}}{2}\right)t$. The plant $g_e(t)= exp(-t)$ is calculated using HF domain via deconvolution and compared with the direct expansion of the plant. Let, time T = 1 s, m = 4 with $\varepsilon = 10^{-4}$.

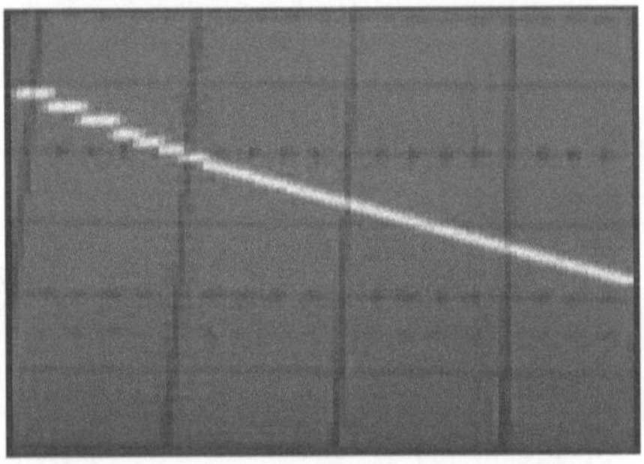

Fig. 10. Closed Loop Identification with direct expansion of HF via Microprocessor

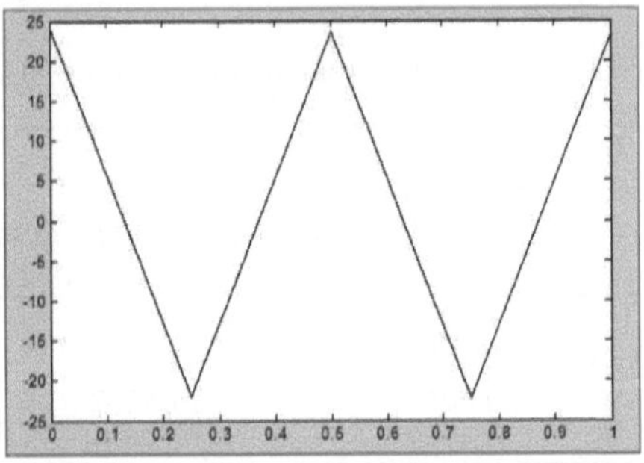

Fig. 11. Closed Loop Identification with Hybrid Function Method via MATLAB

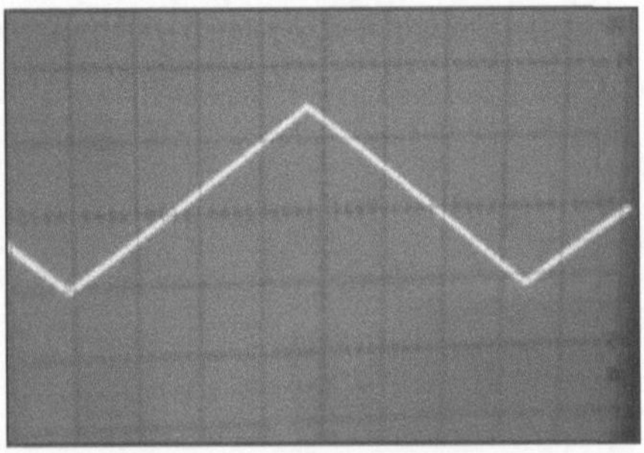

Fig. 12. Closed Loop Identification with Hybrid Function Method via Microprocessor

5 Conclusion

The presented work shows the application of a set of hybrid functions (HF) which evolved from the synthesis of sample-and-hold functions (SHF) and triangular functions (TF). This set is employed for determining the result of convolution as well as 'deconvolution' operation of two time functions and the same have been employed for solving control system analysis and synthesis problems. The theory is supported by examples and the results are compared with the exact solutions via microprocessor realization and MATLAB through Fig. 4, Fig. 5, Fig. 6, Fig. 7,Fig. 9, Fig. 10, Fig. 11, Fig. 12 and Table 1 and Table 2. The difference of values obtained via direct expansion and convolution of hybrid function method in Table 1explains that percentage errors of computation do exist when computed via the closed loop system analysis by HF with respect to the exact solution. Of course the error decreases on increasing the sample points. Table 2 shows closed loop system identification which is a pure case of numerical instability.

Acknowledgements. The authors are indebted to Professor Anish Deb, Department of Applied Physics, Calcutta University for using Hybrid function algorithm in this work.

References

1. Beauchamp, K.G.: Walsh and related functions and their applications. Academic Press (1984)
2. Jiang, J.H., Schaufelberger, W.: How to Multiply Matrices Faster. LNCS, vol. 179. Springer, Heidelberg (1992)
3. Deb, A., Sarkar, G., Sen, S.K.: Block pulse functions, the most fundamental of all piecewise constant basis functions. Int. J. Sys. Sci. 25(2), 351–363 (1994)
4. Deb, A., Sarkar, G., Bhattacharjee, M., Sen, S.K.: A new set of piecewise constant orthogonal functions for the analysis of linear SISO systems with sample-and-hold. J. Franklin Instt. 335B(2), 333–358 (1998)
5. Deb, A., Sarkar, G., Sengupta, A.: Triangular orthogonal functions for the analysis of continuous time systems. Anthem Press (2011)
6. Deb, A., Sarkar, G., Mandal, P., Biswas, A., Ganguly, A., Biswas, D.: Transfer function identification from impulse response via a new set of orthogonal hybrid functions (HF). Applied Mathematics and Computation 218(9), 4760–4787 (2012)
7. Brigham, O.E.: Fast Fourier Transforms and its applications. Prentice- Hall Inc. (1988)
8. Ogata, K.: Modern Control Engineering, 5th edn. Prentice Hall of India (2011)

Cognitive Radio: A Non-parametric Approach by Approximating Signal Plus Noise Distribution by Kaplansky Distributions in the Context of Spectrum Hole Search

Srijibendu Bagchi and Mahua Rakshit

RCC Institute of Information Technology
Kolkata, India
{srijibendu,mahuarakshit100}@gmail.com

Abstract. Cognitive Radio has been acknowledged to be the ultimate solution to meet the huge spectrum demand due to various state-of-the-art communication technologies. It exploits the underutilized frequency band of the legacy users for the unlicensed users opportunistically. This requires a sensible spectrum sensing technique, generally performed by binary hypotheses testing. Noise and signal plus noise distributions are important in this context. These are assumed to be Gaussian in the suboptimal energy detection technique whereas these assumptions may not be validated by practical data. In this paper, the signal plus noise distribution is approximated by four distributions, known as Kaplansky distributions that closely resemble with Gaussian distribution. Testing of hypothesis is performed by non-parametric Kolmogorov Smirnov test and power of the test is calculated for a specific false alarm probability. Numerical results are provided in support of our proposition.

Keywords: Cognitive Radio, Kaplansky distributions, non-parametric Kolmogorov Smirnov test, power of test, false alarm probability.

1 Introduction

Wireless technology is flourishing rapidly in recent times due to its diverse applications. This requires huge frequency band, whereas static spectrum allocation policy puts a barrier to cater a large number of users [1-6]. So, it is necessary to recycle the underutilized frequency bands from the legacy or primary users (PU) in an opportunistic as well as negotiating basis without affecting the other licensed users' transmission process, known as Dynamic spectrum access [7-9]. A careful searching methodology of vacant frequency band or spectrum hole is necessary to solve the spectral congestion problem. This is generally done by binary hypotheses testing [10-15]. Noise and signal plus noise distributions are significant in spectrum hole detection method. In energy detection technique, both are considered as Gaussian, whereas real time data may not support these underlying assumptions. In this paper, the noise distribution is granted as Gaussian (that may be justified by Central Limit

M.K. Kundu et al. (eds.), *Advanced Computing, Networking and Informatics - Volume 2*, Smart Innovation, Systems and Technologies 28,

DOI: 10.1007/978-3-319-07350-7_61,

Theorem), whereas signal plus noise distribution is approximated by a number of distributions, proposed by Kaplansky, that closely resemble the Gaussian distribution. Kaplansky graphs are some examples of distributions with various values of kurtosis discussed by Kaplansky in 1945 [16,17]. The binary hypothesis testing is performed by non-parametric statistical method [18,19]. The Kolmogorov-Smirnov (referred to as KS) test statistic is applied which belongs to the supremum class of EDF statistics and this class of statistic is based on the largest vertical difference between the hypothesized and empirical distribution [20]. Numerical results are presented to validate our proposition. Power of the test is then calculated for a specific false alarm probability. Numerical results are given to show the improvement of power of the test with enhancement of sampling points.

1.1 Earlier Works

Tandra and Sahai (2005) [14], and Tandra and Sahai (2008) [13] considered the detection of the presence/absence of signals in uncertain low SNR environments. They reported that noise uncertainty problem occurs if the noise distribution deviates significantly from its Gaussian nature. They also showed that the signal cannot be detected i.e. robust detection is impossible if SNR falls below a certain level. Recently Wang et.al (2009)[15] formulated the spectrum sensing problem in cognitive radio as a goodness of fit testing problem and proposed the Anderson – Darling (AD) sensing algorithm by applying the Anderson – Darling test to spectrum sensing. In their paper the observed data is compared with a specific distribution (Gaussian) of the received signal and decision is taken accordingly. They reported that AD sensing is an effective and sensitive method especially for small samples compared to Energy detection based spectrum sensing method.

1.2 Scope of the Present Work

In this paper, we propose a class of distributions like the Gaussian distribution as the distribution of the received signal under alternative hypothesis. This is logical because the signal plus noise distribution may deviate from the Gaussian nature but closely follow it. In Tandra and Sahai's paper [13] noise uncertainty was considered due to Gaussian nature. In our paper, noise uncertainty has not been considered. Wang et.al (2009) [15] proposed a non-parametric method but at the time of simulation H_1 was considered as Gaussian. They have showed that AD sensing is applicable especially for small samples. In our paper four types of Kaplansky distributions have proposed and also for large samples, our distributions closely resemble the Gaussian distribution. In this paper, random data are generated for different sample size to validate the distributions by plotting the respective histograms. Finally, it has demonstrated that the detection probabilities are improved for a specific false alarm probability with increasing sample sizes. The paper is organized as follows. Section 2 gives the detailed detection methodology. Numerical results are given in Section 3. Section 4 concludes the paper.

2 The Detection Methodology

2.1 Formation of Binary Hypotheses

The binary hypotheses regarding the existence of a spectrum hole can be expressed as follows:

H_0: $y(n) = w(n)$ decide frequency band is vacant

H_1: $y(n) = h(x(n)) + w(n)$ decide frequency band is occupied (1)

where $w(n)$ is the noise signal, $x(n)$ is the transmitted signal by the PU, $y(n)$ is the received signal by the energy detector, h being the channel response and it is considered that N number of sampling points are taken. Here, we consider $w(n) \sim N(0,1)$, a standard normal variate.

Table 1. List of distributions with μ_4 and maximum ordinate

Distribution	μ_4	Max ordinate
$f_1(y) = \frac{1}{3\sqrt{\pi}}(2.25 + y^4)e^{-y^2}$	2.75	0.423
$f_2(y) = \frac{3}{2\sqrt{2\pi}}e^{-\frac{y^2}{2}} - \frac{1}{6\sqrt{\pi}}(2.25 + y^4)e^{-y^2}$	3.125	0.387
$f_3(y) = \frac{1}{6\sqrt{\pi}}(e^{-\frac{y^2}{4}} + 4e^{-y^2})$	4.5	0.470
$f_4(y) = \frac{3\sqrt{3}}{16\sqrt{\pi}}(2 + y^2)e^{-\frac{3y^2}{4}}$	2.667	0.366

2.2 Kaplansky Distributions

All Kaplansky distributions are symmetric with mean 0 and variance 1 and have density functions, for variable x and c=√π. By default, densities are shown by Kaplansky for the range$0 \leq x \leq 44$.

The distribution of y (n) under H_1 can be approximated by any of the four distributions $f_K(y)$, K =1, 2, 3, 4 as given in Table 1 and omitting the index n. The Gaussian distribution is basically mesokurtic in nature with $\mu_4 = 3$, μ_4 being the fourth order central moment of a distribution. The maximum ordinate of a standard normal distribution is 0.399. Table 1 shows close resemblance of the four distributions with respect to μ_4 and maximum ordinate. Our assumption is logical because in energy detection technique, it is considered that x (n) also follows Gaussian distribution. If fading acts on the signal, it is very natural that the signal plus noise distribution may deviate from its Gaussian nature but closely follow it. The corresponding cumulative distribution functions ($F_K(y)$) can be expressed as,

$$F_1(y) = \tfrac{1}{2} - \tfrac{3}{8}\Gamma(y^2, \tfrac{1}{2}) - \tfrac{1}{8}\Gamma(y^2, \tfrac{5}{2}) \tag{2}$$

$$F_2(y) = \tfrac{1}{2} - \tfrac{3}{4}\Gamma(\tfrac{1}{2}y^2, \tfrac{1}{2}) + \tfrac{3}{16}\Gamma(y^2, \tfrac{1}{2}) + \tfrac{1}{16}\Gamma(y^2, \tfrac{5}{2}) \tag{3}$$

$$F_3(y) = \tfrac{1}{6\sqrt{2}} - \tfrac{1}{6\sqrt{2}}\Gamma(\tfrac{1}{2}y^2, \tfrac{1}{2}) + \tfrac{1}{3} - \tfrac{1}{3}\Gamma(y^2, \tfrac{1}{2}) \tag{4}$$

$$F_4(y) = \tfrac{1}{2} - \tfrac{3}{8}\Gamma(\tfrac{3}{4}y^2, \tfrac{1}{2}) - \tfrac{1}{8}\Gamma(\tfrac{3}{4}y^2, \tfrac{3}{2}) \tag{5}$$

Where $\Gamma(a,b) = \int_a^{\infty} t^{b-1}\exp(-t)dt$

2.3 Reframing the Hypotheses for Inference

The binary hypotheses can be reframed as

H_0: $F(y) = \Phi(y)$ decide frequency band is vacant

H_1: $F(y) = F_K(y)$ decide frequency band is occupied

Where $\Phi(y) = \frac{1}{\sqrt{2\pi}} \int_{-\infty}^{y} \exp\left(-\tfrac{1}{2}t^2\right)dt$ and $K = 1, 2, 3, 4$ (6)

Numerical data can be obtained for different N values and the empirical distribution can be obtained from the data. The test statistic can be formulated according to Kolmogorov-Smirnov test [20] as

$$D_N = \sup \left|F_N(y) - F(y)\right| \tag{7}$$

Where $F_N(y)$ is the cumulative distribution function of the empirical distribution obtained from the observed data. D_N is compared with a threshold γ and H_0 is rejected if $D_N > \gamma$.

The false alarm probability P_f is given by

$$P_f = P(D_N > \gamma \mid H_0) \tag{8}$$

Also, the probability of detection can be expressed as

$$P_d = 1 - P(D_N < \gamma \mid H_1) \tag{9}$$

P_f is considered constant (α). If $F(y)$ is assumed to be continuous, then for $\gamma >$ 0 and $N \rightarrow \infty$

$$P(D_N < \tfrac{y}{N}) = 1 - \exp(-2y^2) \tag{10}$$

The test statistic $T = 4ND_N^2$ follows chi-square distribution with two degrees of freedom.

For certain α value $P(T < \lambda)$ can be found from chi- square table. Consequently, D_N can be calculated as

$$D_N = \sqrt{\tfrac{1}{4N} P(T < \lambda)} \tag{11}$$

The probability of detection can be found by

$$P_d = 1 - F_k(D_N) \qquad K = 1, 2, 3, 4 \tag{12}$$

3 Numerical Results

In this section, we present simulation results to demonstrate the performance of detection probability with increasing sample-size discussed above. Random samples of different sizes (*N* up to 2000) are taken from Rayleigh faded channel.
We consider two cases: $\alpha = 0.05$ and $\alpha = 0.01$

Fig. 1 and Fig. 2 shows the Histogram for $\alpha = 0.05$ and $\alpha = 0.01$ respectively which is a graphical representation of the distribution of data. The height of a rectangle is equal to the frequency density of the interval and the total area of the histogram represents the number of data. These data help to calculate cumulative distribution function $F_N(y)$ by which we obtain Kolmogorov-Smirnov test statistic D_N values in equation (7).

When $\alpha = 0.05$, $P(T < \lambda) = 5.99$ and thus

$$D_{N,0.05} = 1.22/\sqrt{N} \tag{13}$$

Also, when $\alpha = 0.01$, $P(T < \lambda) = 9.21$ and

$$D_{N,0.01} = 1.517/\sqrt{N} \tag{14}$$

So, D_N can be calculated for different N values and consequently P_d can be calculated for a specific *K* value. Fig. 3 and Fig. 4 show the improvement of detection probability when the sample size is increased. So, it may be inferred that the choice of Kaplansky distribution under the alternative hypothesis is appropriate. One problem is that, in order to get high detection probability, the sample size should be high. This, in effect, may enhance the receiver complexity. Thus, a trade-off may occur between the detection probability and the receiver complexity in this detection technique.

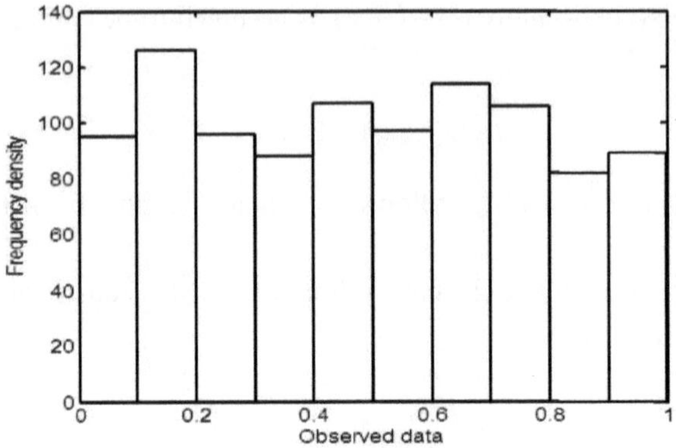

Fig. 1. Frequency density vs. observed data plot for α=0.05

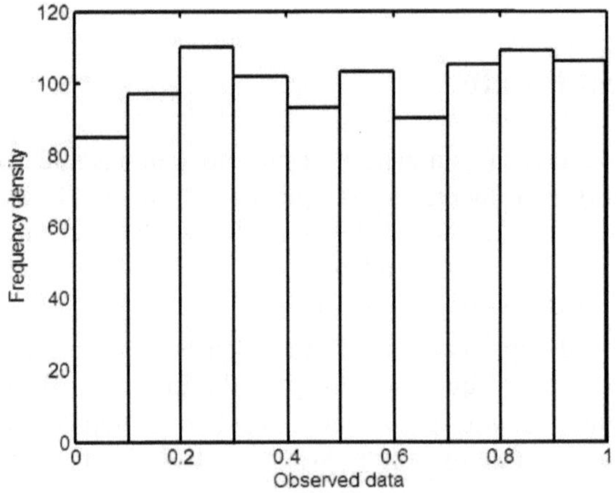

Fig. 2. Frequency density vs. observed data plot for α=0.01

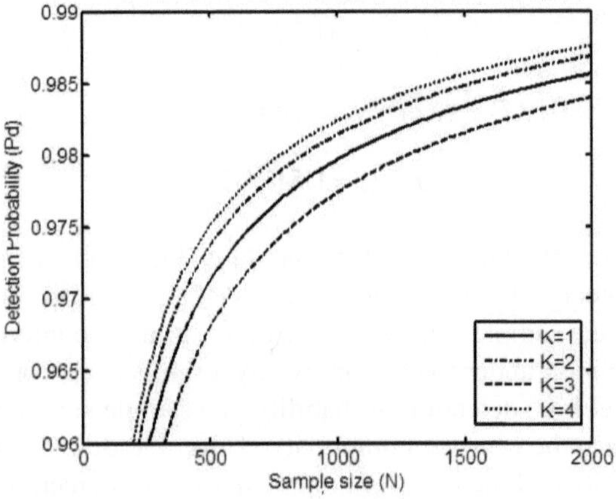

Fig. 3. Detection probabilities vs. sample size plot for α=0.05

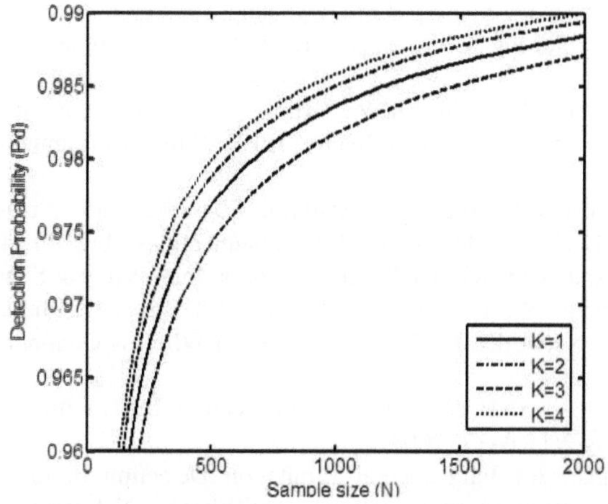

Fig. 4. Detection probabilities vs. sample size plot for α=0.01

4 Conclusion

It is a common phenomenon that signal plus noise distribution often deviates from its Gaussian nature. Kaplansky distributions closely follow the Gaussian distribution and thus may be treated as a suitable substitute under the alternative hypothesis. Numerical results also show that the detection probability also improves as the sample size increases. Therefore, the Kaplansky distributions can be exploited safely in the non-parametric case, that is, when the exact distribution of signal plus noise is unknown.

References

1. Federal Communications Commission. In: the Matter of Facilitating Opportunities for Flexible, Efficient and Reliable Spectrum Use Employing Cognitive Radio Technologies. ET Docket No.03-108 (2003)
2. Haykin, S.: Cognitive Radio: Brain-empowered wireless communications. IEEE J. Selected Areas in Communications 23(2), 201–220 (2005)
3. Arslan, H., Yücek, T.: Spectrum Sensing for Cognitive Radio Applications. Cognitive Radio, Software Defined Radio, and Adaptive Wireless Systems, 263–289 (2007)
4. Wang, J., Ghosh, M., Challapali, K.: Emerging Cognitive Radio Applications: A Survey. IEEE Communications Magazine 49(3), 74–81 (2011)
5. Akyildiz, I.F., Lee, W.Y., Vuran, M.C., Mohanty, S.: NeXt generation/dynamic spectrum access/cognitiveradio wireless networks: A survey. Computer Networks 50, 2127–2159 (2006)
6. Fitch, M., Nekovee, M., Kawade, S., Briggs, K., MacKenzie, R.: Wireless Service Provision in TV White Space with Cognitive Radio Technology: A Telecom Operator's Perspective and Experience. IEEE Communications Magazine 49(3), 64–73 (2011)
7. Tabakovic, Z., Grgicand, S., Grgic, M.: Dynamic Spectrum Access in Cognitive Radio. In: 51st International Symposium ELMAR, pp. 245–248 (2009)

8. Geirhofer, S., Tong, L., Sadler, B.M.: Dynamic Spectrum Access in theTime Domain: Modeling and Exploiting White Space. IEEE Communications Magazine 45(5), 66–72 (2007)
9. Shin, K.G., Kim, H., Min, A.W., Kumar, A.: Cognitive Radios for Dynamic SpectrumAccess: From Concept to Reality. IEEE Wireless Communications 17(6), 64–74 (2010)
10. Ghesami, A., Sousa, E.S.: Spectrum Sensing in Cognitive Radio Networks: Requirements, Challenges and Design Trade-offs. IEEE Communications Magazine 46(4), 32–39 (2008)
11. Chen, K.C., Prasad, R.: Cognitive Radio Networks. John Wiley & Sons. Ltd. (2009)
12. Liang, Y.C., Zeng, Y., Peh, E.C.Y., Hoang, A.T.: Sensing-Throughput Tradeoff for Cognitive Radio Networks. IEEE Transactions on Wireless Communications 7(4), 1326–1336 (2008)
13. Tandra, R., Sahai, A.: SNR Walls for Signal Detection. IEEE Journal of Selected Topics in Signal Processing 2(1), 4–17 (2008)
14. Tandra, R., Sahai, A.: Fundamental Limits on Detection in Low SNR under Noise Uncertainty. In: International Conference on Wireless Networks, Communication and Mobile Computing (2005)
15. Wang, H., Yang, E., Zhao, Z., Zhang, W.: Spectrum Sensing in Cognitive Radio Using Goodness of Fit Testing. IEEE Transactions on Wireless Communications 8(11), 5427–5430 (2009)
16. Kaplansky, I.: A common error concerning kurtosis. Journal of American Statistical Association 40 (1945)
17. Kaplansky, I.: The asymptotic distribution of runs of consecutive elements. Annals of Mathematical Statistics 16, 200–203 (1945)
18. Kendall, M.G., Stuart, A.: The Advanced Theory of Statistics. Charles Griffin & Company Limited (1961)
19. Gibbons, J.D.: Nonparametric Statistical Inference. McGraw-Hill (1971)
20. Lilliefors, H.W.: On the Kolmogorov-Smirnov test for Normality with Mean and Variance unknown. Journal of American Statistical Association 62(318), 399–402 (1967)

RFID for Indoor Position Determination

Piyali Das and Dharma P. Agrawal

University of Cincinnati, Center for Distributed and Mobile Computing, EECS
University of Cincinnati, Cincinnati, OH 45221-0030
daspi@mail.uc.edu,
agrawadp@mail.uc.edu

Abstract. As Global Positioning System (GPS) behaves irrationally due to poor satellite reception, we provide an overview of indoor position detection of objects to precisely find out positions of different objects. Here we come up with Radio Frequency Identification (RFID) as an important tool for detecting objects that are occluded from satellite visibility. We briefly describe underlying architecture of RFID technology. We show how many innovations RFID technique have made by combining with Ultrasonic sensors, Infrared sensors, Impulse-Radio Ultra wide band, and image sensors that lead to different models. We also discuss these approaches in this position determination scheme as well as provide an overview of the pros and cons of each system. We also compare many existing systems and based on the underlying drawbacks, we propose a novel system design, architecture and illustrate its usefulness.

1 Introduction

Global Positioning System (GPS) was primarily used in every sphere for location determination. GPS still serves its best when applied to outdoors applications but is often not possible in case of indoors. This is because the signals received from the satellite are very weak and uneven and is not adequate for the GPS enabled devices to accurately determine the location. Positioning or localization is the determination of location of a person or an object inside any closed domain of a building. Numerous methods and gadgets for location sensing have been proposed in the literature taking many factors into account. But, with technical advances we ought to increase the accuracy in locating the spatial positioning of objects. The traditional models used for actuation are radio signal based wireless networks. But, for these systems to work, the Receiver (Rx) obtained signals in the Line of Sight (LOS) with the Transmitter (Tx) and also in deflected and reflected paths, which posed the main challenge for distance and direction calculation. Though several methods exist for indoor position detection, but none has been standardized due to several other factors such as performance, cost, effectiveness, etc. We have chosen positioning system using RFID as the base due to its simplicity; ease of implementation, cost effectiveness as well as for improved performance. Simple RFID tag design has led to immensely low power consumption and thereby lowering the cost of the system.

M.K. Kundu et al. (eds.), *Advanced Computing, Networking and Informatics - Volume 2*,
Smart Innovation, Systems and Technologies 28,
DOI: 10.1007/978-3-319-07350-7_62, © Springer International Publishing Switzerland 2014

1.1 Parameters for RFID Localization

For determining location of RFID position based systems, some parameters are of particular interest. These are Time of Arrival (TOA), Angle of Arrival (AOA) and Received Signal Strength Indicator (RSSI). Though based on system architecture, researchers have developed several other ways of determining exact location of the objects. Time of Arrival, also called the Time of Flight (TOF), is the time required for a radio signal to travel from a single transmitter to a single receiver. But if multiple receivers are present, then the variation in TOA and time difference of the arrival of the signal are taken into consideration. Before we start measuring TOA, the receiver and the transmitter needs to be time synchronized. On the other hand, AOA measures the angle between the line joining the transmitter and receiver and reference line in a particular direction. This works best for rotating device or for a system having a number of receivers. While it is less accurate, RSSI is an important tool for measuring the power in the received RF signal. Its value varies inversely with the distance between the transmitter and receiver if path loss exponents are taken in to consideration.

1.2 Architecture of RFID System

The milestone of radio communication was laid long back in 19th century with the advent of electromagnetism. Modern day RFID technology is an extension of those principles with many new strata of improvement. Robert Watson-Watt obtained an important patent in 1935 for "Radio Detection and Ranging System" and could be said to be the pioneer of this system. A basic RFID system comprises of four fundamental components: RFID tags or transponders to carry permanent identification information of itself, hence "automatic identification" concept was introduced. RFID reader or transceiver can energize RFID tag and read tag's information. Antenna always remains attached to the reader and passes on energy to the RFID tags. Middleware or Reader Interface Layer collects the tag's value and maps its tag information and all the signals to a unique identification

RFID based system has been classified according to the presence or absence of a radio signal transceiver and an attached power supply with its tag. Active RFID based systems have an active RFID reader and an active RFID tag. The RFID tag has an internal power source within it to power the tag and for communication circuit to work in radio frequency. It is continuously powered on if it is in sensing proximity of the reader. In the idle mode, the tag uses less energy as compared to active state. A passive RFID system requires an active RFID reader and a passive RFID tag. A passive RFID tag does not need internal power source for it's working but uses the radio frequency energy of the reader to power up the tag. RFID reader broadcasts electromagnetic signal to the RF tag. The antenna of the tag stores charge into the capacitor on receiving the signal from the reader. When the capacitor attains sufficient energy, it releases it to the coil of the tag over time. This leads to an encoded radio wave signal containing information in the tag, which is then received by the reader and is demodulated. This technique is called "backscatter". The reader sends this data to middleware that then sends the data to the intended system for further interpretation. Generalized working principle of passive RFID and active RFID based system has been shown in Fig. 1.

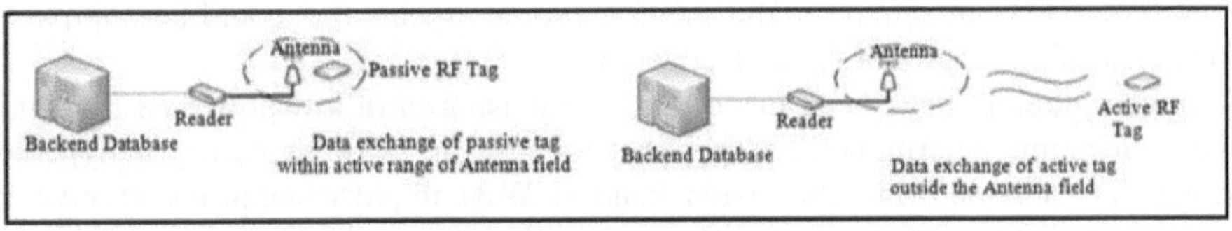

Fig. 1. Working of passive RFID and active RFID system

2 Related Work

There had been several schemes proposed for accurate indoor position determination and the most widely used technique had been the RFID. Olszewski *et al.*[3] have used the RFID to locate the area of interest along with indoor navigation with the help of a RFID positioning robot. They have also used the hybridization of ultrasonic and infrared (IR) sensors to intelligently traverse the pathway without any hassle and take diverse route to avoid any collision. The RFID antenna placed on readers according to triangulation pattern to be used by the TOA scheme and some technique of RSSI to locate an immediate position. The system integrates all the information about location, navigation and obstacle avoidance from the RFID. With the assumptions about the range of 2.5ft from the robot an accuracy of about 94% is achieved. But, there are certain limitations in the tag placement with the detection area and potential tag collision.

Similar attempt has been made [1], [2], [4] using a combination of RFID and an Ultrasonic Sensor to detect the location of objects and people indoor. The system is composed of caller, RF transceiver based sensors and transponders. In this active RFID based system, the transponder is used for emitting controlled ultrasonic waves that is used as transmitter, i.e., works similar to that of a tag in the RFID system. Caller broadcasts command to the transponder and sensor to set their timer to default setup. Ultrasonic signals are sent by the transponder to the fixed sensor, thereby distance is calculated using signals time difference. The sensor has two separate Rx/Tx modes for receiving and transmitting purposes with independent frequency channels. Information from a minimum of three sensors is needed in order to determine 3D location of the transponder using spatial coordinate values. The sensor and the transponder need to be within the transmission range of the caller and in LOS, which is an implied limitation to the system.

Zhou *et al.* [5] proposed another hybrid architecture where image sensors are utilized as a tool along with radio signal for adaptive indoor location detection. This system aims at location detection in order to offer smart and wide media access for digital home applications. The proposed system comprises of a RFID or WIFI transceiver and active RFID tags or any WIFI transmitters to send the radio signals across home. They used cameras having low complexity for interactive gesture recognition. With the variation of radio signal strength, any presence of human is detected, threshold for powering on the camera is calculated and threshold is updated periodically. Orientation of the users has been assumed for accuracy. By different probabilistic analysis of noise level between the range of 1dB and 6dB, success rate of about 95% has been achieved. But there are some aspects that could be improved in terms of sensing range and the area covered.

Another attempt have been made [6], [7] by using ultra low power RFID tags for precise location determination. They have come up with exploring a hyperbolic localization scheme with ultra wide band (UWB) impulse capturing receivers. It estimates the TOA and the time difference to find the tag location. The tag is active for a very short interval (0.72ms) for impulse radio UWB generator to feed in input signal and remains idle for rest of the time (1s). The cheap energy detection anchor receiver samples received signal and sends to a locator reader. After that, the received signal is sampled at high rate by integrating small time fragments. Proper measures are taken to reduce the total number of repetitive transmission, which is dependent on the pulse repetition interval (PRI). This also consumes current during frame transmission and retransmission and therefore needs to be optimized. Results obtained indicate prolonged system lifetime and geometric location analysis shows improved position accuracy in the range of 10cm.

Zhou *et al.*[8] have proposed a time and energy efficient system that aims towards 3-dimensional localization. Instead of all the available references they have considered interaction of reference tags with reference readers to find out precise location with higher efficiency. They used an efficient passive and active scheme for locating a target reader with fewer tag responses and reduced power computation. It selects the reference reader based on the minimum power level and thereon estimates the distance. Finally, the reading load is determined depending on the communication radius, reference tag density, power level and the distance between reader and tag. This system is quite promising in its effectiveness and accuracy of 95% with minimum incurred error. The time/energy cost is observed to be reasonably low thereby increasing the life span of the system.

The proposed localization method [9] aims at continuously changing the power levels of the RFID readers to determine distance and location of the target tag. They have used reference tags to minimize location uncertainty and enhance throughput in terms of cost, speed, accuracy and efficiency. Three localization algorithms have been proposed to reduce errors with power level as input but are flexible to support more algorithms in future. They have sorted the values of the power to converge to the minimum value and deploy schemes to minimize the location errors. This scheme is shown to have an average accuracy of 15cm with error of 0.31cm. But, the system fails to address the issue of tag collision and cannot minimize the tag detection time.

Holm *et al.* [10] have used hybrid methods of refined room-level ultrasound technique to go hand in hand with RFID system for motion tracking of any object or person in a room. They assumed every room to have a fixed ultrasound tag (transmitter) to communicate with multiple wearable receivers (without exceeding the permissible ultrasound tolerance) with a small battery. The change of direction property of ultrasound propagation minimizes power consumption of the battery for transmitter. They utilized ultrasound-RSSI and Doppler shift to enhance the accuracy. It employs RS-232 interface to obtain frequency shift keying instead of deriving the values of ID, RSSI and Doppler shift through RF-link. Results indicate successful detection of path tracing by velocity estimate graphs and RSSI plots.

A different concept of designing a fault tolerant RFID reader localization approach [11] has been proposed where the aim is to minimize the effect of errors in large region. In reader localization scheme, there are a number of tags placed and position of a portable reader is determined. The reader only provides information about the detected tags without using RSSI. Short fault can be eliminated by multiple reading methods and for long duration, faults are detected and corrected by incorporating geometric knowledge of all activated tags in the region. The activated region is determined by circular patterns of the signals (centroid method) and also by the key pair and quality index value of the two adjacent tags. As compared to the existing system, it is efficient in detecting regional faults using angle loss values.

3 Comparative Analysis of Existing Models

LANDMARC [13] and SpotON [12] are the two most famous models for location sensing using RFID technology. Table 1 shows comparison of qualitative and quantitative parameters of the two technologies. Numerous models have been designed till date but still there is a dearth of a complete system that efficiently manages indoor localization issue removing the potential challenges. From the analysis above it is found that issues are mainly with determination of localization parameters, anti-tag collision scheme, cost and power consumption. Hence a novel approach is desirable to mitigate all these issues.

Table 1. Table of comparison of two existing models

Qualitative	**SpotON**	**LANDMARK**
Localization technique	Triangulation	Using reference tags & RSSI
Battery life	10 hrs.	3.5yrs
Accuracy	Poor	Better
Reponse time	10 to 20 sec	7.5 sec
Cost	Costlier	Cheaper
Quantitative		
Read range	15ft	150 ft.
Operating frequency	916.5	308
Searching capability	3 dimensional	2 dimensional
Multiplexing techniques	Not used	Not used

4 Proposed Approach

RFID system face several technical challenges like tag collision and battery efficiency. Several tag anti-collision schemes are present and widely used is the time division multiple access (TDMA). Our proposed method uses a basic RFID system with RF tags and readers and RSSI value is used to detect location of an object. Also a combination of TDMA, code division multiple access (CDMA) and frequency-reuse technique is used to avoid multiple tag collision. To locate an object inside a large room, we divide the area into number of grids. Each grid is provided with a reader and the objects are having passive tags. The tags are capable of modulating signals carrying information about its identification and specific parameters. Readers broadcast radio signal to activate multiple tags tuned in a particular frequency called the downlink frequency. The time frame is broken into several slots based on the number of tags present. In the first slot, the activated tags send their identification information to the reader by using CDMA technique having similar pseudo code. By this technique several tags can send their information simultaneously over a single channel by generating orthogonal pseudo codes. In the second time slot, next set of activated tags transmits. By this technique tags are able to use only one frequency for receiving signal and one frequency for transmitting signal. After that the reader demodulates the received signal and can uniquely identify a tag and compute desired information.

Let the downlink frequency be 'F1' and uplink frequency be 'F2' and there be 'n' number of time slots 'TS' present for 'x' readers. We have chosen four orthogonal pseudo codes 'PC' and based on these codes the tags are assigned to each time slots. The schematic diagram has been depicted in Fig. 2. If the number of tags increases then the time slots needs to be reduced and more slots could be accommodated. The mentioned process is obtained in an optimized way based on the number of tags and readers present but number of frequency channel to be used remains constant.

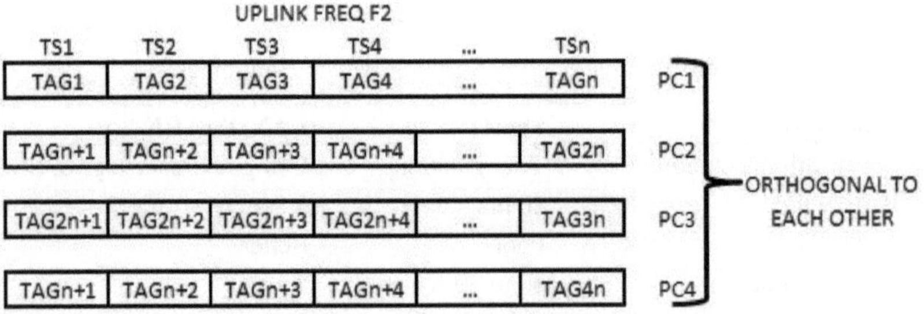

Fig. 2. Schematic representation of proposed uplink model

5 Conclusion

RFID system is an aspiring area for indoor positioning system and its application is increasing drastically. RFID finds its application in industrial spheres like health care, medical sector, animal rearing, transportation areas, parking garages and many more. Several methods have been proposed and each has their positive and negative aspects. Our novel approach will reduce the complexity of RFID systems considerably and will pave the way for many areas to be explored. One such field would be to apply this technique for communicating in highly congested and noisy areas like stock exchange and stock markets where cellular communication fails. The work could also be extended for facilitating communication with people having visual or hearing impairment.

Acknowledgment. We are thankful to our Center for Distributed and Mobile Computing (CDMC) lab members for their constant support and motivation. Their valuable insight and assistance were of immense help.

References

1. `http://www.veryfields.net/how-do-rfid-tags-work`
2. `http://en.wikipedia.org`
3. Olszewski, B., Fenton, S., Tworek, B., Liang, J., Yelamarthi, K.: RFID Positioning Robot: An Indoor Navigation System. In: IEEE International Conference on Electro-Information Technology, pp. 1–6 (2013)
4. Hepeng, D., Donglin, S.: Indoor Location System Using RFID and Ultrasonic Sensors. In: 8th International Symposium on Antennas, Propagation and EM Theory, pp. 1179–1181 (2008)
5. Zhou, W., Ma, X., Li, J.: An Adaptive Indoor Location Detection Method using Hybrid of Radio signal and Image Sensors. In: GreenCom, IEEE/ACM Int'l Conference on & Int'l Conference on Cyber, Physical and Social Computing (2010)
6. Mei-bin, Q., Rui, Z., Jian-guo, J., Xiang-tao, L.: Moving object localization with single camera based on height model in video surveillance. In: Proceedings of ICBBE 2007, pp. 1–5 (2007)
7. Zhou, Y., Look Law, C., Xia, J.: Ultra Low-Power UWB-RFID System for Precise Location-Aware Applications. In: IEEE WCNC Workshops, pp. 154–158 (2012)
8. Bu, K., Liu, J., Li, J., Xiao, B.: Less is More: Efficient RFID-based 3D Localization. In: IEEE 10th International Conference on Mobile Ad-Hoc and Sensor Systems (2013)
9. Chawla, K., Robins, G., Zhang, L.: Object Localization Using RFID. In: 5th ISWPC (2010)
10. Holm, S.: Hybrid Ultrasound–RFID Indoor Positioning: Combining the Best of Both Worlds. In: IEEE International Conference on RFID (2009)
11. Zhu, W., Cao, J., Xu, Y., Yang, L., Kong, J.: Fault-Tolerant RFID Reader Localization Based on Passive RFID Tags. In: IEEE INFOCOM (2012)
12. Hightower, J., Vakili, C., Borriello, G., Want, R.: Design and Calibration of the SpotON Ad-Hoc Location Sensing System (2001)
13. Lionel, M.N., Liu, Y., Lau, Y., Patil, A.: LANDMARC: Indoor Location Sensing using Active RFID. Kluwer Academic Publisher, Netherlands (2004)

STAGE Atmosphere: To Reduce Training Cost in EW Simulations

Lakshmana Phaneendra Maguluri[1], M. Vamsi Krishna[2], and G.V.S.N.R.V. Prasad[1]

Department of Computer Science and Engineering
Dept. of Information Technology,
Gudlavalleru Engineering College, India

Abstract. Now-a-days the increasing burden of maintaining both the topic-specific software and the framework that houses the basic components has become uncomfortably expensive. Most of the framework for legacy software has not been migrated to modern programming languages or new computational platforms. The flexibility of using this legacy infrastructure becomes more difficult. This situation is complicated by the use of government created software tools, which may also be out of date or generally are awkward to modify/manage. These conditions have driven an unstoppable movement to COTSS-based tools. In view of the above it is proposed to develop an Electronic Warfare simulator using COTS tools like STAGE for EW Scenario Generation and VAPS XT for building Dynamic displays.

Keywords: STAGE, Distributed Simulation, VAPSXT, Electronic Warfare.

1 Introduction

One way to potentially reduce training costs is to use EW simulator to provide a realistic mechanism for the EW unit in which to train all the facets of the EW process. Simulation, from an EW perspective, includes battle field dynamics and emitter environment. Achieving a high degree of competence in EW requires both the ability to read computer based presentations of threat activity cognitive and skills which the analyst must apply when time is short and environment is one of high stress. Training EW personal to properly interrupt The Electromagnetic Environment and make correct decisions in this difficult environment requires a simulator which is highly interactive and depicts the EW environment with a high degree of realism.

The primary purpose of EW simulation in the battle field scenario is, to model the effects of a tactical management between two teams in dense EW environment and to predict the likely outcomes. Simulation used by military commanders to prepare them there, staff and supporting command and communication systems without actual combat. The simulators can aid the EW commanders in threat perception, EW planning and deployment. Simulation is used to experiment with theories of warfare and refines the plans without the need for real resources.

Commercial-off-the-shelf (COTS) [11] software products are widely used now in software development, and their usage should increase quality of the product and

M.K. Kundu et al. (eds.), *Advanced Computing, Networking and Informatics - Volume 2*,
Smart Innovation, Systems and Technologies 28,
DOI: 10.1007/978-3-319-07350-7_63, © Springer International Publishing Switzerland 2014

reduce the time of its development. COTS have come to represent the solution when budgets and time scales are tight and engineering staff are becoming tight harder to come by. Remaining sections are as follows: Section 2 Presents Electronic Warfare Simulation, Section 3 Presents Interoperability between COTS Simulation Models, Section 4 Presents Scenario Toolkit and Generation Environment Section 5 Presents Conclusion.

2 Electronic Warfare Simulation

The term Simulation is the process of creation of an artificial situation or stimulus that causes an outcome to occur as through a corresponding real situation or stimulus was present. EW simulation is artificially created so that equipment can be tested and operators can be trained under realistic conditions. EW engagements are typically complex, with many threat emitters seen in constantly changing in [3]. Electronic Warfare is a military action that involves using electromagnetic spectrum to detect it, analyze and prevent the enemy's spectrum and also protecting our own spectrum. EW uses electromagnetic waves in battlefield. EW aims at reducing the enemy's electronic activity and simultaneously safeguarding own electronic systems from the enemy's EW activities in [3]. Electronic support includes signal intelligence (SIGNT), which consist communications intelligence (COMINT) and Electronic intelligence (ELINT). EW is divided into three measures (as shown in Fig. 1.).

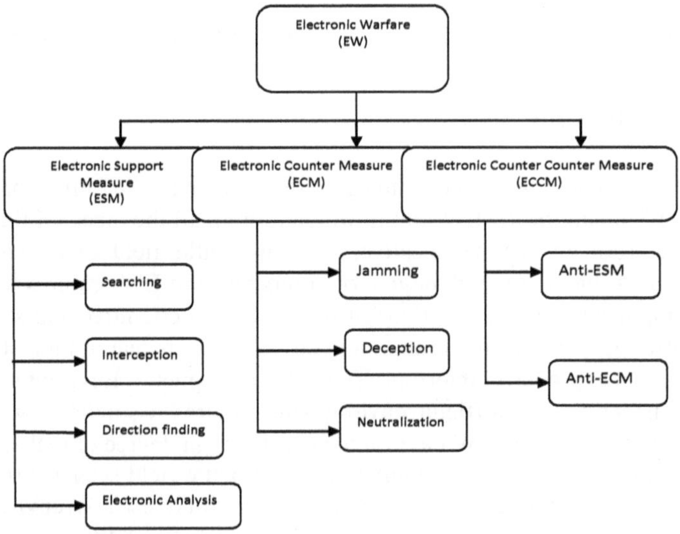

Fig. 1. Classification stages of EW Simulation

ESM supplies the necessary intelligence and threat reorganization to allow effective attack and protection. It allows us to search for, identify and locate sources of interaction and unintentional electromagnetic energy in [4]. ECM involves actions taken to prevent or reduce an adversary`s effective use of the electromagnetic spectrum through the use of electromagnetic energy. ECCM, which ranges from designing systems resistant to jamming, through hardening equipment to resist high power microwave attack, to the destruction of enemy jammers using anti-radiation missiles.

3 Interoperability between COTS Simulation Models

Today`s systems are mainly hybrid architectures in which part of the complete system is custom made and part is COTS. A new trend in software commerce is emerging: generic software components also called commercial-off-the-shelf components that contain fixed functionality. When a software component is integrated into system, it must support the style of integrations of the system architecture in order to work together with other components. If a COTS product has another style of interaction of the system architecture in order to work with other components. If a COTS product has another style of interactions, programmers must write integration software to allow this product to interact with other components of the systems. Since most COTS products cannot be changed by users because of absence of source code and other reasons. The simulation models designed and developed in COTS simulation packages cannot be achieve a direct interaction with other COTS simulation models. Therefore for those packages that are open or partially open a so called wrapper must be developed that take care of the interoperability with other model. The wrapper provides two way interactions:

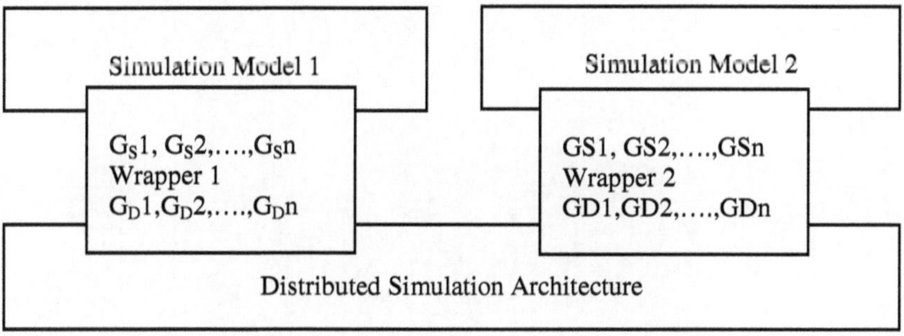

Fig. 2. Communication of Two COTS Simulation Models

1. Interaction toward COTS simulation models using the interface function for accessing the internal data.
2. Interaction with other models or wrapper of these models through a distributed simulation architecture using interoperability functions.

Most of the Cots Simulation Packages have possibilities for the modeler to define such a kind of wrapper around the simulation model. The Fig. 2 depicts an architecture where two models are connected through their wrappers to distributed simulation architecture. The interoperability between the simulation models is achieved by applying distributed simulation architecture, like HLA. The Distributed simulation [14] architecture the interoperability functions for simulation wrappers. As we stated before when two simulation models interact they might need to transfer the simulation entity from one model to another one, the receiver model must support the instantiation of the type of transferred entity.

4 STAGE (Scenario Toolkit and Generation Environment)

The stage, developed by PRESAGIS is another commercial simulation-oriented VE framework. STAGE is a software tool used to build and animate in real time synthetic environments containing both moving and stationary entities such as airplanes, ships, land, vehicles', missiles, radar sites, etc. that interact with one another as a function of pre-determined rule sets, or through operator intervention during execution of the simulation. STAGE scenario provides a GUI to enter entity parameters into XML Database and assembles them into dynamic, interactive, complex and tactical environment. It can be used as a totally stand-alone synthetic tactical environment generator, or as a fully integrated simulator for other applications. The simulated system design is modeled using UML diagrams. The Electronic Warfare service interactions and the actions performed are modeled using sequence diagrams.

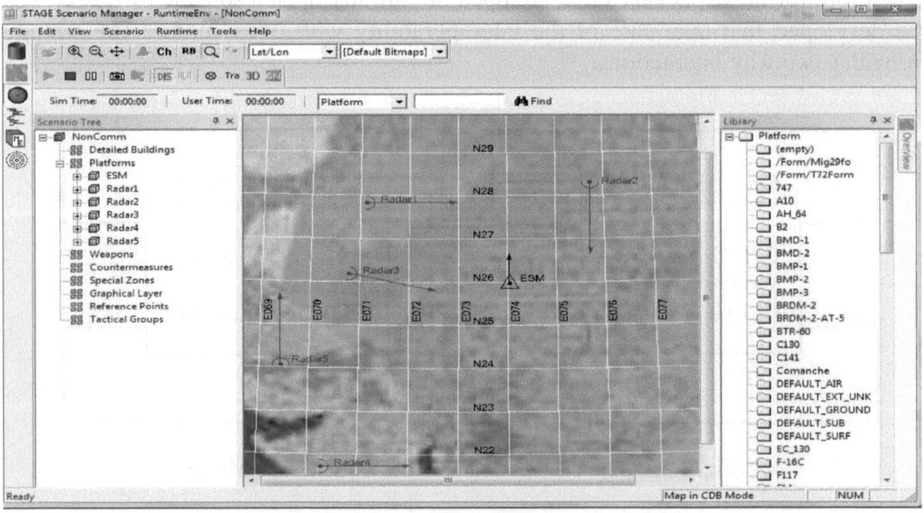

Fig. 3. Radar non communication developed using STAGE Scenario Editor

EW simulators which will Facilitates: Radar ES Simulator: This intercepts radar signals and displays the detected radars with bearing and their classification. The Scenario Generator application is developed using STAGE and VC++ using Microsoft Visual Studio 2008 IDE. Communication ES Simulator: Simulates an EW Receiver, which intercepts and analyzes hostel radio signals. This system consists of radio, Search receiver and monitors receiver simulations and associated GUI.

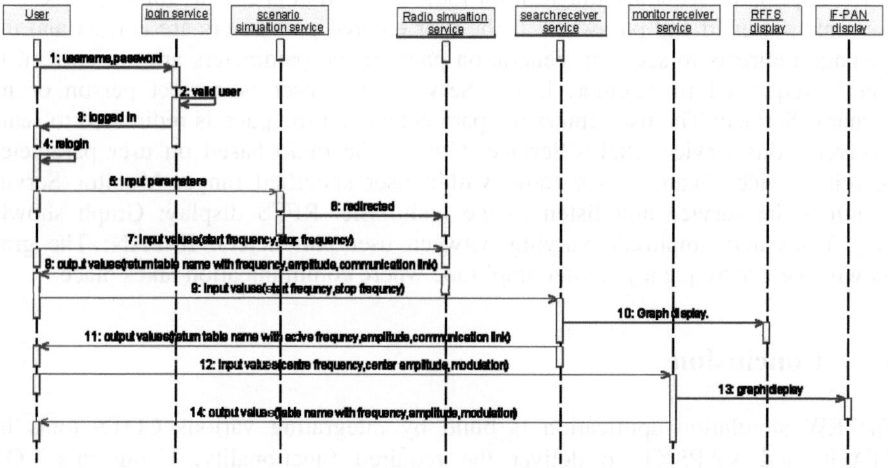

Fig. 4. Sequence diagram of radio simulation

Table 1. Test cases whether the radar developed using STAGE Scenario is pass or not

Test Criteria	Test input	Expected output	Actual output	Result
Change in bearing when platform is moved	Move the platform to perpendicular to the true north and ESM platform	Bearing has to be changed to 90	Bearing=90	pass
Calculation of range	Radar platform lat long and ESM platform lat long	Distance(lat1,long1,lat2,long2)	Distance (lat1,long1,lat2,long2)	pass
Effect in display when the sensor is inactive	Turn off the sensor of radar platform using hook window feature of stage	Radar 5 should be invisible	Radar 5 should be invisible	pass
Detection based on ESM sensitivity	ESM sensitivity=-40	Radar detected	Radar detected	pass
Detection based on ESM sensitivity	ESM sensitivity=-10	Radar detected	Radar detected	pass

Description of sequence diagram for radio simulation [12] is as follows: User: The user enters user id and password. If the user entered parameters are correct and then the page redirects to scenario generation page. If the parameters are not correct, the user is requested to re-enter. Login Service: The user is correct person or not. Scenario Service: The user enters the parameters and the page is redirected to search receiver radio service. Radio Service: Creates the radio based on user parameters. Search Service: Searches for radio within user specified range. Monitor Service: Monitors the service and listen to the audio file. RFFS display: Graph showing frequencies and amplitude varying between user parameters. IF-PAN: The graph showing center frequency, center amplitude where communication takes place.

5 Conclusion

The EW simulation application is build by integrating various COTS tools like STAGE and VAPSXT to deliver the required functionality. Using this COTS technology in implementing EW scenarios and Dynamic displays development become easier. In this way, COTS tools enable electronic warfare simulation functionalities to be provided and consumed linking libraries, which simulation developers make accessible as a plug in order to allow simulation Engine to combine and use them in the production of electronic simulation applications.

References

1. Schleher, D.C.: Introduction to Electronic Warfare, pp. 1–36. Artech House (1986)
2. Schleher, D.C.: Introduction to Electronic Warfare. Artech House (1999)
3. Adamy, D.L.: Introduction to Electronic ware fare modeling and simulation: British Literary Cataloguing Data (2003)
4. Department of defence. Electronic and Information warfare
5. Horton, I.: Begning Visual C++. Wiley publishing Inc. (2008)
6. Graham, A.: Communications Radar and Electronic Warfare. John Wiley and Sons Ltd. (2011)
7. EEE Std278.1-1995. IEEE Standard for Distributed Interactive Simulation-Application Protocals (1995)
8. IEEE Std1278.2-1995. IEEE Standard for Distributed Interactive Simulation Communication service and profiles (1995)
9. IEEE Std1516-2000: IEEE Standard for Medaling and Simulation (M&S) High Level Architecture (HLA) - Framework and Rules (2000)
10. IEEE Std1516.1-2000: IEEE Standard for Modeling and Simulation (M&S) High Level Architecture (HLA) - Federate Interface specification (2000)
11. IEEE Std1516-2000: IEEE Standard for Modeling and Simulation (M&S) High Level Architecture (HLA) - Object Model Template (OMT) Specification (2000)
12. Taylor, S.J.R., Sudra, R., Janahan, T., Tan, G., Ladbrook, J.: Towards COTS distributed simulation using GRIDS. In: Proceedings of the Winter Simulation Conference, pp. 1372–1379 (2001)
13. Paquette, D.L., Gouveia, L.M.: Virtual training for live fire naval gunfire support. In: OCEANS 2000 MTS/IEEE Conferenceand Exhibition, pp. 733–737 (2000)

14. Ying Miao, Y., Chen, C., Sun, Z.: A satellite system distributed simulation design and synchronous control. In: International Conference on Mechatronics and Automation, pp. 3889–3893 (2009)
15. Yen-Lung Chen, Y.-L., Wu, W.-R., Liu, C.-N.J., Li, J.C.-M.: Simultaneous Optimization of Analog Circuits with Reliability and Variability for Applications on Flexible Electronics. IEEE Transactions on Computer-Aided Design of Integrated Circuits and Systems 33(1), 24–35 (2014)

Analysis of Early Traffic Processing and Comparison of Machine Learning Algorithms for Real Time Internet Traffic Identification Using Statistical Approach

Jaiswal Rupesh Chandrakant[1] and D. Lokhande Shashikant[2]

[1] Pune Institute of Computer Technology, Pune, India
[2] Sinhgad College of Engineering, Pune, India
rcjaiswal@pict.edu, sdlokhande.scoe@sinhgad.edu

Abstract. In modern Internet, different protocols generate numerous traffic types with distinct service requirements. Therefore the Internet traffic identification plays an important role to improve the network performance as part of network measurement and network management task. Primarily well-known port based method was used. But latest services uses random and uncertain port numbers reduces the accuracy of identification. Consequently "payload based approach" also known as "deep packet inspection", used but still resulted less accuracy and required huge operational resources and are exposed to encrypted traffic flows. The recent techniques classify the application protocol based on statistical characteristics at packet level using network flow-based approach. Dealing with several datasets and millions of transaction of packets needs the use of Machine learning techniques for classification and identification of traffic. Our research shows the classification accuracy up to 99.7929%. In this paper we propose the statistical feature based approach for real-time network traffic classification. We compared the performance of three machine learning algorithms for the same. This mechanism of real time protocol identification confirms improved performance and reduced complexity.

1 Introduction

Packet loss and delay sensitive traffic like VoIP, Gaming and real time multimedia communication requires certain information about internet services types to decide their strategies and design of system parameters. Also automatic resource reservations-allocations, billing process are the major tasks of service providers as part of network and internet traffic management. Also, the Internet security provision for different services on the Internet is an upcoming issue handled by different intrusion detection and prevention system against several attacks and corresponding protocol traffic for service providers and for the common Internet users too. For these issues the internet traffic identification through classification process has been a challenging problem for the Internet community nowadays.

Sen [1] reported 30% accuracy for p2p traffic and Moore [2] reported 70% accuracy for port-based classification and hence Port based methods are not in much use

M.K. Kundu et al. (eds.), *Advanced Computing, Networking and Informatics - Volume 2*, Smart Innovation, Systems and Technologies 28,
DOI: 10.1007/978-3-319-07350-7_64, © Springer International Publishing Switzerland 2014

nowadays. Also because of heavy requirement of processing power, need of human intervention, large signature database handling requirement, inability to handle encrypted payloads and other privacy concerns, payload signature based method[3-6] is also rarely used. To overcome the constraints of the previous above techniques, newer methods with statistical attributes and machine learning algorithms are proposed to classify protocols depending upon statistical flow characteristics like unidirectional and bidirectional flow features. The idea behind this approach is that Internet traffic generated by different classes of protocols has unique statistical features for each source of application. The main inspiration of this investigation is the research papers of machine learning based internet traffic classification [7-15] and [16-23] for real time traffic classification of internet protocols.

Supervised classifiers like MultiboostAB, K-nearest neighbors and Naïve Bayes are selected for investigations. Machine learning, includes study of systems that can learn from given data. ML is trained on wanted protocol traffic to learn to distinguish between unwanted and wanted traffic. After training, it is used to identify new protocol traffic into unwanted and wanted traffic class.ML system deals with the processes like data representation of instances and data generalization. Learning from data instances is also known as training or Representation. Performance is examined or tested on unseen data instances, is known as generalization process. Data instances are also known as traffic features. Sometimes these features are also called as discriminators or traffic attributes. The rest of this paper is organized as follows: Section 2 shows literature survey of related research work. Section 3 gives brief explanation about ML classifiers. Section 4 tells about datasets used for training and testing. Research methodology and analysis of data is given in Section 5. Section 6 concludes the research work and says about future work.

2 Related Work

Previous investigations in the traffic identification field are discussed as follows:

Research papers [7-15] give the detailed idea about flow computation, features calculation, training and testing with supervised and unsupervised ML methods. They are accurate but not suitable for real time traffic recognition. But papers [16-23] indicate the classification of protocols for real time environment.

Bernaille [16] explained the experiment on an early TCP classification of flow based features on first few complete packets for prediction of the unknown traffic but gives poor performance of the ML classifier if early packets of flows are missed.

Statistical attributes are calculated by Nguyen [18] over multiple short sub-flows derived from full flows and showed the notable improvement in its performance and unaffected by missing of first few packets of flow. Supervised [18] and Un-Supervised [17] ML techniques performance has been proven consequently. Yu Wang [19] also evaluated the efficacy of ML techniques with traffic attributes derived from the first few packets of every flow and achieved high accuracy. System proposed by Xu TIAN [20] combines ML-based and behaviors-based method to give stable, accurate and efficient Real time traffic classification. Combination of payload based features and statistical attributes is done by Dehghani [21] and showed the efficient real time traffic identification system. Accurate real time traffic classification module for high speed links is developed by Alice Este [22] and offering promising results. BuyuQu [23] proposed the technique of real time classification when forged data packets are inserted before actual communication of protocol.

Thus, we have derived short sub flows from full flows and remarkable improvement with improved accuracy is obtained. The idea implemented is working well with prominent features and classifiers used in this research work. Thus there is scope of further improvement and advancements.

3 Machine Learning Classifiers

In this paper, MultiboostAB, K-nearest neighbors and Naïve Bayes ML techniques are explained in brief as follows:

3.1 MultiboostAB Classifier

MultiBoosting is an extension to AdaBoost classification algorithm by combining with wagging technique and base learning algorithm used is C4.5. In weka this is known as MultiboostAB. Significantly lower error and superior parallel execution than AdaBoost and Wagging is observed. This technique also boosts the performance of learning algorithm. Basically bagging and boosting used to create an ensemble of classifiers. Wagging assigns different weights to data instances whereas bagging assigns equal weights. Boosting is iterative approach whereas bagging builds the classifiers independently. Thus in multiboost algorithm random weights are assigned using continuous poisons probability distribution and gives more accuracy than bagging approach [24]. The algorithm is as follows:

Multiboosting Algorithm steps:

1. Start
2. Assign random weights (W_I) to data instances (D_I) using

Continuous Poisson distribution.

3. For (each of the 't' iterations)

{

Call C4.5 learning algorithm to form a model for D_I

Compute error 'E' for all D_I and Compute re-weighting of data instances D_I

Keep W_I unchanged for False Positives

```
Create set of D_IL for low weight and D_IH for high
weight.
Build classifier model C*(x) for re-weighted D_I
and classify D_IH correctly.
Combine C*(x) output using weighted vote to form a
prediction with-
      W_V = -log (E/1-E) formula.
Add max. Vote Classifiers weights by choosing-
class with greatest sum
If E is close to zero
{C*(x) receive a D_IH shows C*(x) performing well}
If E is close to 0.5
{C*(x) receive a D_IL shows C*(x) poor performance}
   Do the final Prediction of C*(x)
 }
4. End
```

3.2 K-Nearest Neighbors Classifier

The *k*-Nearest Neighbors algorithm (***k*-NN**) is simplest and a non-parametric, lazy learning or instance-based learning method used here for Internet traffic classification where input data instances consists of the *k* closest training examples of processed traffic. It is a non-parametric method and used when probability distributions are not known or difficult to calculate. Algorithm stores all available processed data instance cases and classifies new cases based on a similarity degree like distance functions (Euclidean, Minkowski Distance etc). Traffic Data instances are classified by a most vote of its neighbors, and given a required class. K-NN gives high flow accuracy in different domains of classification and hence it is taken as part of our investigations.

K-Nearest Neighbors classification algorithm is as follows:

Input: P, the set of k training instances,
and test instances Q = (a' , b'), Ψ = class label

```
1. Start
2. Process:
   {
      1. Compute d (a', a), the distance between Q
         and every Data instance, (a, b) ∈ P
```

2. Select PQ $\subseteq$ P, the set of k closest training data instances to Q

}

3. Output: $b' = \binom{argmax}{\Psi} \sum_{(a_i,\ b_i) \in P_Q} I(\Psi = b_i)$

4. End.

3.3 Naïve Bayes Classifier

The Naive Bayes algorithm uses Bayes' Theorem and is based on conditional probabilities with strong (naive) presumptions of independence and also known as independent model of features. In simple terms, in internet traffic classification we consider number of features or attributes of flows. It is assumed that the presence or absence of a particular attribute of one traffic class is not related to any other attributes. It considers all of these attributes and their independent probability distributions. This classifier can be trained effectively based on nature of probability model with fewer amounts of data for training to estimate the parameters like means and variances of features and not the entire covariance matrix. Thus it is fast to train and hence fast to classify the data instances and it is not sensitive to irrelevant traffic attributes. The algorithm steps are as follows:

1. Start
2. Consider a set of tuple : T
3. Set each tuple is an 'n' dimensional feature vetor
4. Y : (y1, y2, y3, y4, y5, ...yn)
5. Let there be 'k' classes : C1, C2, C3, C4, C5,….CK
6. Let NB classifier predicts Y belongs to class Ciiff
7. P (Ci/ Y) > P (Cj / Y) for 1 ≤ j ≤ k, j ≷ i
8. Max. Posteriori hypothesis with

   {

   P (Ci/ Y) = P (Y / Ci) (P (Ci) / P (Y)

   Maximize P (Y / Ci) (P (Ci) as P (Y) is constant

   }

9. Apply "independent feature model" assumption.
10. P (Y / .Ci) = $\prod_{k=1}^{n} P(Y_k/C_i)$

```
11. P (Y / Ci)  =  P (Y1 / Ci )  *  P (Y2 / Ci )*  P
    (Y3 / Ci)  * P (Y4 / Ci )  * P (Y5 / Ci )* ...... P (Yn
    / Ci)

12. End.
```

4 Data Preparation

We captured online internet traffic before router at junction point with mirror port strategy and stored for further investigations as proprietary dataset. Standard datasets for online games traces are used for Enemy Territory, Half Life and QUAKE-3 [26-28]. We have also used the dataset from MAWI Working Group [29]. Training and testing datasets details of MAWI traces are mentioned in Table 1.

Table 1. Traffic Protocols and Data Instances

Protocols	Training data instances	Testing data instances
Quake	11700	6027
Half life	7850	4043
DNS	14500	7469
HTTP	11000	5666
SSH	9579	4934
SMTP	13786	7101
POP3	10500	5409
FTP	11567	5958

The MAWI data set is a traffic trace from a trans-Pacific line (150Mbps link) in operation since 2006. This trace is publicly available in anonymized form, collected on December 07, 2013. Also we have used Manual-classification in our research and are referred from papers [25] and [26]. Full feature and Reduced feature dataset is computed as mentioned in work [25].

5 Implementation and Result Analysis

5.1 Methodology

Training and testing data instances from available datasets are computed and WEKA (Waikato Environment for Knowledge Analysis) tool is used for further research. Linux environment including GCC and Octave is also used for trace processing. We used MultiboostAB, K-nearest neighbors and Naïve Bayes classifiers and sixteen prominent statistical features like packet length (min, max, mean, sd), payload length (min, max, mean, sd), Inter-arrival time (min, max, mean, sd), flow duration (min, max, mean, sd) etc. 137089 samples are taken for training-testing purpose from MAWI and 87786 instances from proprietary and others mentioned sources. Different split options are used and ten folds cross validation is done. Cross validation is used to obtain estimates of classifier model parameters that are not known. Subset evaluator is applied to filtered data with Best first search method in attribute selection process. Parameters of evaluation like precision {TP/(TP+FP)}, recall {TP/(TP+FN)} and error of classifier is computed [15]. Further graphs are simulated for different evaluation parameters.

5.2 Result and Analysis for Full Flows

Fig. 1 indicates that Maximum accuracy of 99.7929% is obtained by K-NN classifier for full feature dataset and 98.8717% for reduced feature dataset. MultiboostAB classifier offers accuracy of 78.8923% and 74.837% for full and reduced feature dataset. Naïve bayes classifier provides 79.6147% and 82.6723% for full and reduced feature dataset. Better accuracy is offered for reduced feature dataset compared to full feature dataset by Naïve bayes classifier. Fig. 2 compares training time required. It is observed that least time is taken by Naïve bayes and maximum for K-NN. Training time of all ML classifiers shortens for reduced feature dataset in contrast with full feature dataset.

Table 2. Metrics of Performance Analysis

Parameters	M-AB	K-NN	NB
Accuracy (Full features)	78.8923%	99.7929%	79.6115%
Accuracy (Reduced features)	74.831%	98.8717%	82.6713%
Training Time (Full features)	16.45 Sec	70.07 Sec	8.25 Sec
Training Time (Reduced features)	12.97 Sec	46.28 Sec	4.44 Sec
MAE (Full features)	0.0488	0.0068	0.0461
MAE(Reduced features)	0.0478	0.0099	0.0804
Precision (Full features)			
Mean	0.732	0.9723	0.8118
SD	0.061	0.02251	0.06673
Precision (Reduced features)			
Mean	0.6763	0.9376	0.7831
SD	0.09471	0.03887	0.06077
Recall (Full features)			
Mean	0.630	0.9207	0.7137
SD	0.09615	0.04571	0.08364
Recall (Reduced features)			
Mean	0.5945	0.8801	0.682
SD	0.03526	0.03810	0.1053
Early packets to get max. Accuracy	5th Packet and 78%	4th Packet and 97.34%	8th Packet and 89.45%

Fig. 3 indicates that Mean Absolute Error (MAE) of K-NN is least and higher for MultiboostAB for full feature dataset. Also error is highest in naïve bayes for reduced feature dataset. Fig. 4, Fig. 5, and Fig. 6 gives Precision and Recall analysis for MultiboostAB, K-NN and Naïve Bayes classifier respectively. It is noted that minimum value of precision and recall is observed for ssh traffic for all classifiers.

5.3 Early Classification Process Results and Analysis

Here, we have evaluated required real-time traffic classification by detailing the early classification process. The idea is that the early network sub-flows attributes are taken from full-flows statistics. We have done several experiments for 1 to 50 packets of

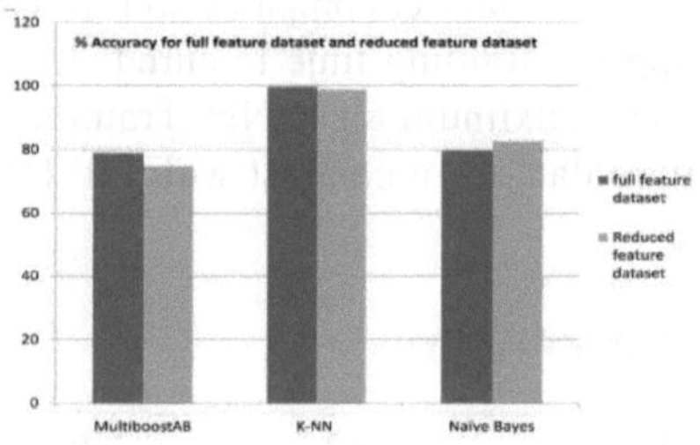

Fig. 1. Full Flows Accuracy Analysis

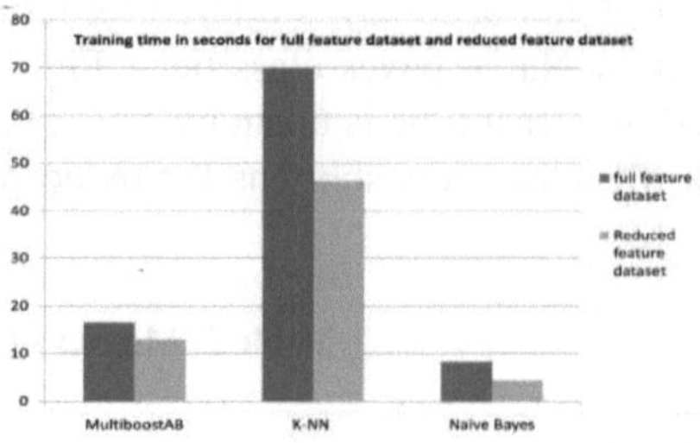

Fig. 2. Full Flows Training Time Analysis

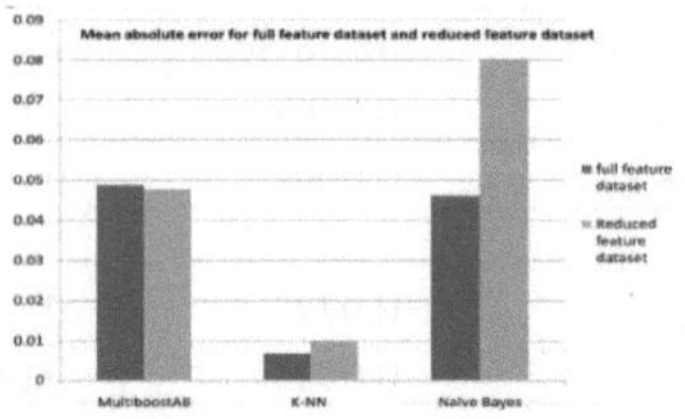

Fig. 3. Full flows MAE Analysis

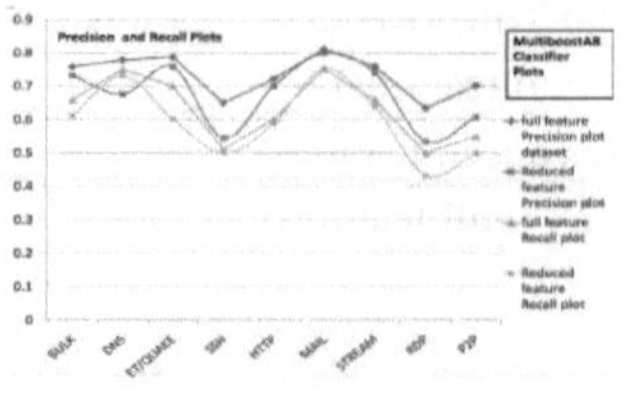

Fig. 4. P-R for MultiboostAB (Full Flows)

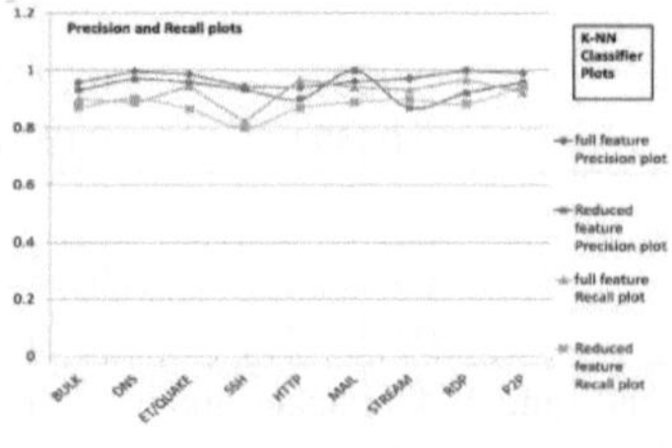

Fig. 5. P-R For K-NN (Full Flows)

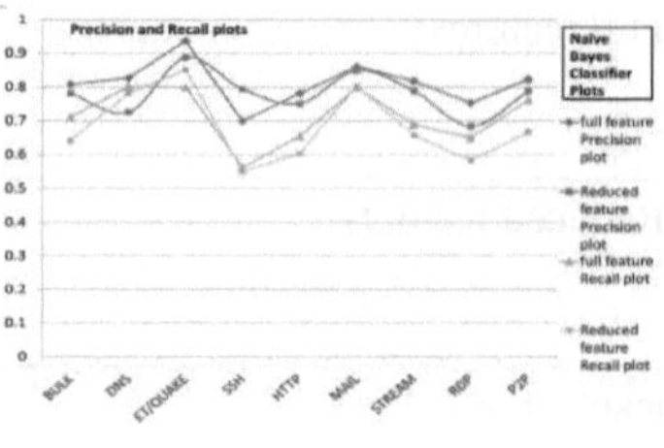

Fig. 6. P-R For Naïve Bayes (Full Flows)

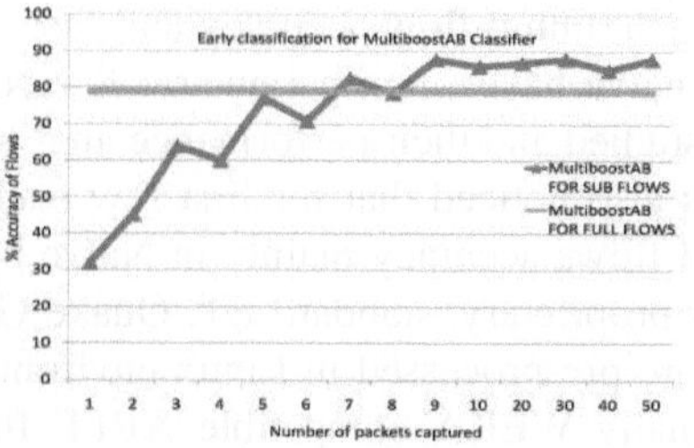

Fig. 7. Early Classification for MultiboostAB

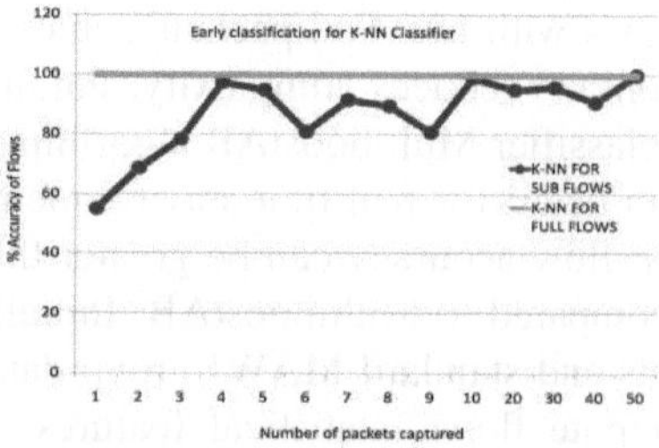

Fig. 8. Early Classification for K-NN

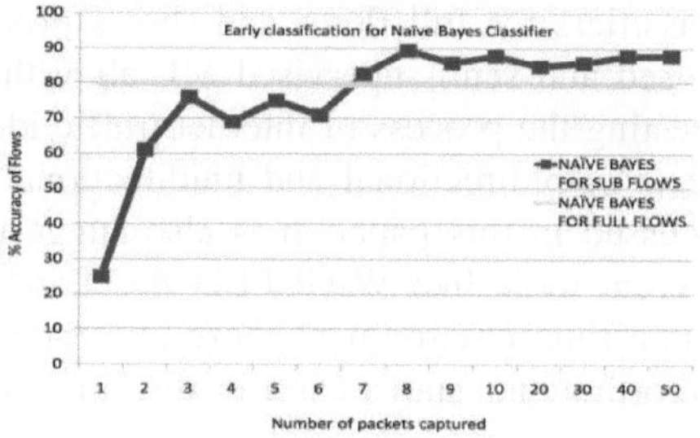

Fig. 9. Early Classification For Naïve Bayes

numerous traces as mentioned earlier until we get satisfactory and required performance. Surprisingly, we got promising results with precise observations that the attributes computed from the first 4 to 7 data packets are competent to train Multi boostAB, K-nearest neighbors and Naïve Bayes classifiers and offers improved performance and efficacy in real time environment.

In Fig. 7, it is observed that for MultiboostAB supervised learning classifier 5th packet is proficient to train MultiboostAB and reaches up to 78% of flow accuracy which is almost equal to full flow accuracy. In Fig. 8, it is evident that for K-NN classifier 4th packet is proficient to train K-NN classifier and provides 97.34% of flow accuracy which is almost equal to full flow accuracy. Also Fig. 9 shows the similar achievement as in above two cases. It is clearly noticed that the flow accuracy above 80% is accomplished by Naïve Bayes which is greater than full flow accuracy at 8thpacket it touches to 89.45% and which is remarkable achievement in our investigation. Also there after 8, 9,10th packet onwards the accuracy increases gradually and it is observed up to few packets thereafter.

6 Conclusion and Future Work

Thus in this research with rigorous experimentation, we compared the performance of MultiboostAB, K-nearest neighbors and Naïve Bayes classifiers for real-time traffic identification environment and noticed that promising results are achieved for sub-flows with few first packets achieves the maximum flow accuracy and fast computation and reduces complexity. Parametric naïve bayes, non parametric K-NN and Meta classifier MultiboostAB algorithms are studied and their performance analysis is done in detail for real time environment. It is also noticed that for first few packets, the sub-flow accuracy can be greater than full flows accuracy mainly in Naïve bayes case compared to MultiboostAB. Initially our proprietary, standard ET, Quake Game datasets and standard MAWI group datasets are pre-processed in Linux environment to compute flows, statistical features and finally WEKA compatible ARFF file is developed for further research. Performance evaluation parameters for full flows are computed and analyzed in detail like accuracy, training time, and MAE for all three ML classifiers and compared thereafter. Detail analysis of Precision-Recall parameters is done for ML classifiers for full flows and shows promising results. We also suggest using unsupervised and semi-supervised ML algorithms to identify the sub-flows for training and testing the process of internet traffic identification in real time environment. Also effects of bidirectional and unidirectional flows on performance parameters are not discussed in this paper. It is also suggested to see the effect of flooding traffic by different tools like WAR-FLOOD, Trinoo, TFN and metasploit scripts will make this real time environment more interesting using ML classifiers. Also other optimum and orthogonal statistical flow discriminators can be tried.

References

1. Sen, S., Spatscheck, O., Wang, D.: Accurate, scalable in network identification of P2P traffic using application signatures (2004)
2. Moore, A.W., Papagiannaki, K.: Toward the accurate identification of network applications. In: Dovrolis, C. (ed.) PAM 2005. LNCS, vol. 3431, pp. 41–54. Springer, Heidelberg (2005)
3. Bro intrusion detection system - Bro overview (2008), `http://broids.org`
4. Application specific bit strings, `http://www.cs.ucr.edu/tkarag/papers/strings.txt`
5. Ma, J., Levchenko, K., Kreibich, C., Savage, S., Voelker, G.M.: Unexpected means of protocol inference. In: 6th ACM SIGCOMM Internet Measurement Conference (IMC 2006), Rio de Janeiro, BR (2006)
6. Haffner, S.S., Spatscheck, O., Wang, D.: ACAS: automated construction of application signatures. In: MineNet 2005, Philadelphia, Pennsylvania, USA (2005)
7. Paxson, V.: Empirically derived analytic models of wide-area TCP connections. IEEE/ACM Trans. Networking 2(4), 316–336 (1994)
8. Zander, S., Nguyen, T., Armitage, G.: Automated traffic classification and application identification using machine learning. In: IEEE 30th Conference on Local Computer Networks (2005)
9. Nguyen, -T.T.T., Armitage, G.: A Survey of Techniques for Internet Traffic classification using Machine Learning. IEEE Communications Surveys & Tutorials 10(4) (2008)
10. Roughan, M., Sen, S., Spatscheck, O., Duffield, N.: Class-of-service mapping for QoS: A statistical signature-based approach to IP traffic classification. In: ProcessingInternet Measurement Conference (2004)

11. Singh, K., Agrawal, S.: Internet Traffic Classification using RBF Neural Network. In: International Conference on Communication and Computing Technologies, vol. 10, pp. 39–43 (2011)
12. Haffner, P., Sen, S., Spatscheck, O., Wang, D.: ACAS: automated construction of application signatures. In: MineNet 2005: Proceeding of the 2005 ACM SIGCOMM Workshop on Mining Network Data, pp. 197–202. ACM Press (2005)
13. Zander, S., Armitage, G.: A preliminary performance comparison of five machine learning algorithms for practical IP traffic flow classification. Special Interest Group on Data Communication (SIGCOMM) Computer Communication Review 36(5), 5–16 (2006)
14. Auld, T., Moore, A.W., Gull, S.F.: Bayesian neural networks for Internet traffic classification. IEEE Trans. Neural Networks 1, 223–239 (2007)
15. Chandrakant, J.R.: Lokhande Shashikant. D.: Machine Learning Based Internet Traffic Recognition with Statistical Approach. In: Annual IEEE India Conference (2013)
16. Bernaille, L., Teixeira, R., Akodkenou, I., Soule, A., Salamatian, K.: Traffic classification on the fly. ACM Special Interest Group on Data Communication (SIGCOMM) Computer Communication Review 36(2) (2006)
17. Nguyen,-T.T.T, Armitage, G.: Clustering to Assist Supervised Machine Learning for Real-Time IP Traffic Classification (2008)
18. Nguyen, T., Armitage, G.: Training on multiple sub-flows to optimize the use of Machine Learning classifiers in real-world IP networks. In: Proceeding IEEE 31st Conference on Local Computer Networks (2006)
19. Wang, Y., Yu, S.Z.: Machine Learned Real-time Traffic Classifiers. In: Second International Symposium on Intelligent Information Technology Application. IEEE (2008)
20. Tian, X., Sun, O., Huang, X., Ma, Y.: A Dynamic Online Traffic Classification Methodology based on Data Stream Mining. In: WRI World Congress on Computer Science and Information Engineering. IEEE (2009)
21. Dehghani, F., Movahhedinia, N., Khayyambashi, M.R., Kianian, S.: Real-time Traffic Classification Based on Statistical and Payload Content Features. In: 2nd International Workshop on Intelligent Systems and Applications. IEEE (2010)
22. Este, A., Gringoli, F., Salgarelli, L.: On-line SVM traffic classification. In: 7th International Wireless Communications and Mobile Computing Conference (IWCMC). IEEE (2011)
23. Qu, B., Zhang, Z., Guo, L., Meng, D.: On accuracy of early traffic classification. In: 7th International Conference on Networking, Architecture and Storage (NAS) (2012)
24. Witten, I.H., Frank, E., Hall, M.A.: Data Mining-Practical machine learning tools and techniques. Morgan Kaufmann Publishers, Elsevier Copyright (2012)
25. Moore, A.W., Zuev, D.: Discriminators for use in flow-based classification. Intel Research Technical Report (2005)
26. Auld, T., Moore, A.W., Gull, S.F.: Bayesian neural networks for Internet traffic classification. IEEE Trans. Neural Networks 1, 223–239 (2007)
27. Lang, T., Armitage, G., Branch, P., Choo, H.Y.: A synthetic traffic model for Half-life. In: Procceding Australian Telecommunications Networks and Applications Conference (2003)
28. Lang, T., Branch, P., Armitage, G.: A synthetic traffic model for Quake 3. In: Proc. ACM SIGCHI International Conference on Advances in Computer Entertainment Technology (ACE 2004), Singapore (2004)
29. MAWI Working Group Traffic Archive. Packet traces from wide backbone, http://mawi.wide.ad.jp/mawi/

Detection of Communities in Social Networks Using Spanning Tree

Partha Basuchowdhuri[1], Siddhartha Anand[1], Diksha Roy Srivastava[1], Khusbu Mishra[1], and Sanjoy Kumar Saha[2]

[1] Department of Computer Science and Engineering,
Heritage Institute of Technology
Kolkata - 700107, West Bengal, India
parthabasu.chowdhuri@heritageit.edu,
{siddharthalibra13,diksharsrivastava1992,meetkhusbumishra}@gmail.com
[2] Department of Computer Science and Engineering,
Jadavpur University
Kolkata - 700032, West Bengal, India
sks_ju@yahoo.co.in

Abstract. Communities are inherent substructures present in social networks. Yet finding communities from a social network can be a difficult task. Therefore, finding communities from a social network is an interesting problem. Also, due to its use in many practical applications, it is considered to be an important problem in social network analysis and is well-studied. In this paper, we propose a maximum spanning tree based method to detect communities from a social network. Experimental results show that this method can detect communities with high accuracy and with reasonably good efficiency compared to other existing community detection techniques.

Keywords: Social Networks, Community Detection, Maximum Spanning Tree.

1 Introduction

Social networks have become increasingly popular among users in recent years for its unique ability to bring geographically distant people closer to each other. A social network can be represented as a graph, with users as *nodes* and connections between them as *edges*. An important problem in the area of social network analysis is detection of communities. Social networks are graphs, which may be *directed* or *undirected* in nature depending upon the relationship between nodes. For example, in a telecom call graph, the caller and callee relation in an edge determines the directionality of that edge. The edge will be directed from caller to callee. In undirected graphs, the notion of directionality is ignored and edge represents presence of communication. For *unweighted* graph, frequency of communication may be used to assign weights to the edge. Users can be part of nodes having common interests, like supporters of a fan club, discussion forums,

Smart Innovation, Systems and Technologies 28,
DOI: 10.1007/978-3-319-07350-7_65,

movie addicts or celebrity fan following. In this paper, we try to partition social networks into dense regions or *communities*.

There is still a lot of ambiguity in scientific community regarding the definition of a community. The densely connected subgraphs, which share only a few edges between them, are termed as a *communities*. Such dense regions signify a group of related nodes and are inherent in social networks. Detection of communities in social networks may be useful in applications such as finding groups of individuals based on their culture, political affiliation, professional collaboration. Identifying the intrinsic community structure of a network leads to a better understanding of its underlying characteristics. As a result, a lot of effort has been put into devising efficient techniques for finding communities in social networks [1–5]. Not many of the detection techniques are known to be based on building spanning tree. In this paper, we propose a method to find communities using maximum spanning tree.

The remainder of the paper is structured as follows. First, we describe some prior works on community detection in social networks. Then, in the next section, we present our algorithm step by step for find communities using maximum spanning trees. In the next section, we present the experimental results on different benchmark data-sets and finish with concluding remarks.

2 Prior Works

Study of past works reveal the emergence of various approaches of community detection. In this section, we mention a few popular disjoint and overlapping community detection techniques which are popular among network science researchers and used in practice. One type of detection technique uses reachability based detection methods [1], [2] categorize densely connected nodes into one community using some modified traversal based methods. The main problem with these methods is that they expect the networks to have some strict structures like cliques which may not necessarily be the case throughout. The other type of community detection technique try to identify *brokers* or *bridges* and use them to find the communities. Categorizing links based on vertex and edge betweenness [3], [4] in practice take considerable amount of time thereby making these methods impractical for real time community detection. Another popular method is detecting communities by maximizing goodness of subgraphs or partitions of the networks. A community detection method devised by Clauset, Newman and Moore, popularly known as CNM method [5], is widely used for its low worst case runtime, i.e. $O(md \log n)$, where d is the number of divisions that led to the final cover. *Steiner tree* [6] has been used for community detection before but it is substantially different from our method.

3 Proposed Methodology

In this section, we explain the spanning tree based method we have used to find communities from social networks.

3.1 Problem Statement

Given a graph $G := (V, E)$, where $n = |V|$ and $m = |E|$, we intend to find a set of k subgraphs or a cover of k communities, represented by $G_s = \{G_{s_1}, G_{s_2}, G_{s_3}, ..., G_{s_k}\}$, $|G_s| = k$, $V = V_s = \cup_{i=1}^{k} V_{s_i}$ and $E = E_s = \cup_{i=1}^{k} E_{s_i}$, such that if the maximum value of the goodness function of a graph G over a cover of size k is given by $CM_{G_k}(G)$, then our aim is to find

$$\arg\max_{\forall 1 \leq k \leq n} CM_{G_k}(G) = \{CM_{G_i}(G) | \forall i, j, 1 \leq i, j \leq n, CM_{G_i}(G) \geq CM_{G_j}(G)\}$$

For any i, j, where, $1 \leq i, j \leq n$ and $G_{s_i}, G_{s_j} \in G_s$, $V_{s_i} \cap V_{s_j} = V_{s_{ij}}$ and $E_{s_i} \cap E_{s_j} = E_{s_{ij}}$. If, for all possible pairs of i, j, $V_{s_{ij}}$ and $E_{s_{ij}}$ are $\emptyset$, then cover G_s denotes a set of disjoint communities, otherwise G_s is considered to be a set of overlapping communities where $V_{s_{ij}}$ and $E_{s_{ij}}$ are the set of overlapping node set and edge set between G_{s_i} and G_{s_j}.

3.2 Spanning Tree Based Method

In this method, our assumption is that a social network consists of triangles. Scale-free property of a social network ensures that the distribution of number of k-sized cliques follows power law [7]. Therefore, our assumption is a natural one and it may not work only for special cases, where the graph is made by combining many star networks or some other networks devoid of cliques of size 3 or more.

In our method, we use well known minimum cost spanning tree algorithms to reduce the graph G into a maximum cost spanning tree T. Then we apply a hierarchical clustering algorithm to keep dividing the nodes into hierarchical groups by removal of *weak* edges.

Part 1: Transforming the Unweighted Graph into a Weighted Graph

In the first step of the process, we take an unweighted graph $G(V, E)$ and convert it into a weighted graph $G'(V, E, W)$, where W is the set of weights for all the edges and hence $|W|$ is same as m. Before starting the process, we remove all the pendant nodes as they do not contribute to the triangles. Here, the derived weight is representative of the number of common neighbors between the nodes on which the edge is incident. Hence, an weighted graph with apriori weights may not helpful as an input graph and the original weights may be ignored for using this algorithm.

In order to find the modified graph G' with weights, we take the graph G as input, and find out the value of the similarity metric $w(v_i, v_j)$ for every edge $e(v_i, v_j) \in E$, as described below,

$$w(v_i, v_j) = \frac{|Nghb(v_i) \cap Nghb(v_j)|}{|Nghb(v_i) \cup Nghb(v_j)|} \quad (1)$$

where $Nghb(v)$ is the set of neighboring nodes of v, i.e., the set of nodes $\in V$ that are connected to it using an edge $\in E$. Hence, W can be represented as, $W = \cup_{v_i, v_j \in V} w(v_i, v_j)$. This metric is analogous to *Jaccard* similarity measure [8] and measures how similar the neighborhoods of v_i and v_j are. Lower the overlap in the neighborhoods of v_i and v_j, lower the value of $w(v_i, v_j)$. The values of $w(v_i, v_j)$ could vary from 0 to 1.

Part 2: Building the Maximum Cost Spanning Tree

Now, we find maximum cost spanning tree T from the weighted graph G' and remove the edge with lowest $w(v_i, v_j)$ value to divide the nodes into multiple groups. So, we connect every node $v \in V$ with its neighbor which has shared most number of common neighbors with v. Say, using the same weights, we build another tree T', which is a minimum cost spanning tree of graph G'. If we define communities by a cover of sets of nodes that maximizes the number of triangles over all the induced subgraphs created from the node sets in the cover. Existence of triangles and hence existence of larger cliques, has been marked as an important feature to measure goodness of a community [9]. Hence, it is meaningful to try to understand why maximum cost spanning tree will be a more suitable structure than a minimum cost spanning tree in terms of building hierarchical structure.

Lemma 1. *For a weighted graph G', removal of the edge with minimum weight from its maximum cost spanning tree T would retain more or at least equal number of triangles in the induced subgraphs than that of the minimum cost spanning tree T'.*

Proof. In T, a node u is always connected to its neighbor v, which shares maximum common neighbors with it. On the other hand, in T', a node u is always connected to its neighbor v, which shares minimum common neighbors with it. During the process of building the trees T and T' from the weighted graph G, at the first instance we are choosing $e(s, v_i)$ and $e(s, v_j)$ respectively, where s is the starting node. Here, we can say that $w(s, v_i) \geq w(s, v_j)$. Similarly, for the pair of edges chosen for growing T and T' in subsequent steps also, we can draw the same conclusion. Hence, sum of all the edge weights in T will always be greater than or equal to sum of all the edge weights in T'. As edge weights are directly proportional to the number of triangles, we can say that for graph G, T will be able to retain either more or at least equal number of triangles as T', in their respective induced subgraphs.

3.3 Algorithm

The algorithm is mainly divided in three parts.

Pre-processing: In this part, first, the pendant nodes are removed and the unweighted graph is converted to a weighted graph based on the number of common neighbors shared by the nodes on which an edge is incident. In the next step, all edges with zero edge weight are removed.

Finding Maximum Spanning Tree: In this step, we use inverse of all the edge weights for applying one of the popular minimum cost spanning tree algorithms. As a result, our heuristic also inherits the worst case run-time of minimum spanning tree algorithms. In this way, we find the maximum spanning tree for the modified graph. From the maximum spanning tree, we keep removing the edge with lowest weight and check the weighted average goodness measures over all the clusters received in the cover. We keep dividing hierarchically, by removing the edge with least edge weight, until there is a fall in weighted average goodness measure from previous step. Note that clusters of size less than four are not allowed to be broken out of its existing group.

Post-processing: In this step, after the cover has been found, the induced subgraphs are finally created from the nodes in the clusters and the pendant nodes as well as the other excluded nodes are reattached to their neighbors and communities are updated accordingly. The nodes, which were excluded initially for only having edges with zero edge weight, are reattached to the community where most of its neighbors lie.

The community detection algorithm has been described in Algorithm 1. In lines 2-4, all the pendant nodes are removed from the graph. Pendant nodes will always be part of the community to which its only neighbor belongs and hence, it is not necessary for the community detection algorithm. In lines 5-8, the edge weight of every edge is assigned based on the number of common neighbors between the nodes on the ends of the edge. All edges with zero weight are removed from the edge set. For some nodes, all the edges connecting them to their neighbors may be assigned zero weight due to absence of common neighbors and hence the node may get disconnected from the network. Also, this step may lead to multiple components in the graph. In that case, the components are considered as communities and we proceed to the next steps accordingly to further divide the network. In line 9, using the inverse of the weight set W, we apply Prim's minimum cost spanning tree finding algorithm [10] for finding maximum cost spanning tree from the existing graph. A min-priority queue, Q, is built by using edge weights of the tree or forest T as keys and the edge with minimum edge weight is extracted from the priority queue in every iteration. Iterations are counted by s. After removal of the edge with minimum weight from T, we check whether any component has a size of less than four. If no new component is smaller than size four, then we increment s and store the new cover as G_s. Otherwise, we proceed to remove the edge with next lowest edge weight. This process is described in lines 9-23. The stopping criteria, as shown in line 13, is to check whether the goodness of the newly created cover is improving the goodness measure from the last updated cover. If goodness measure improves, we try to divide the network further to see if even better goodness measure is achievable

Algorithm 1. Finding communities using maximum spanning tree

Input : Undirected, unweighted network $G(V, E)$
Output: Cover of k communities, $G_s = \{G_{s_1}, G_{s_2}, ..., G_{s_k}\}$

```
1  begin
2      forall the v ∈ V do
3          if deg(v) equals to 1 then V ← V \ {v}
4      end
5      forall the e(v_i, v_j) ∈ E do
6          w(v_i, v_j) ← |Nghb(v_i) ∩ Nghb(v_j)| / |Nghb(v_i) ∪ Nghb(v_j)|
7          if w(v_i, v_j) equals to 0 then E ← E \ {e(v_i, v_j)}
8      end
9      T(V, E_T) ← MST-PRIM(G, W^{-1}, root)   ▷Any node v ∈ V is chosen as root
10     Q ← W_T              ▷Build a priority queue Q using edge weights of T (W_T)
11     s ← 0
12     GM(G_0) ← 0          ▷GM measures average weighted goodness of s-th cover
13     while s is FALSE OR GM(G_s) > GM(G_{s-1}) do
14         if Q is empty then return G_s
15         w(v_p, v_q) ← EXTRACT-MIN(Q)
16         if removing e(v_i, v_j) creates a new component of size < 4 then
17             continue
18         else
19             remove e(v_p, v_q) from E_T
20             s ← s + 1
21         end
22         store induced subgraphs from forest T as G_s
23     end
24     return G_{s-1}
25 end
```

by further partitioning. If goodness measure does not improve, we return G_{s-1} as the final cover.

The run-time of this algorithm will never exceed the worst case run-time of Prim's algorithm. The iterative statements used in lines 2-4 and 5-8 have linear worst case run-time. The while loop, in lines 13-23, has a worst case run-time of $O(d.\log|E|)$. Every time EXTRACT-MIN is called, it takes $O(\log|E|)$ time. If the loop is executed for d number of times, then the total time to execute the while loop becomes $O(d.\log|E|)$. Prim's algorithm has a worst case run-time of $O(|E|\log|V|)$ and by asymptotic notation we can say that our algorithm will also assume a worst case run-time of $O(|E|\log|V|)$, when implemented using binary heap.

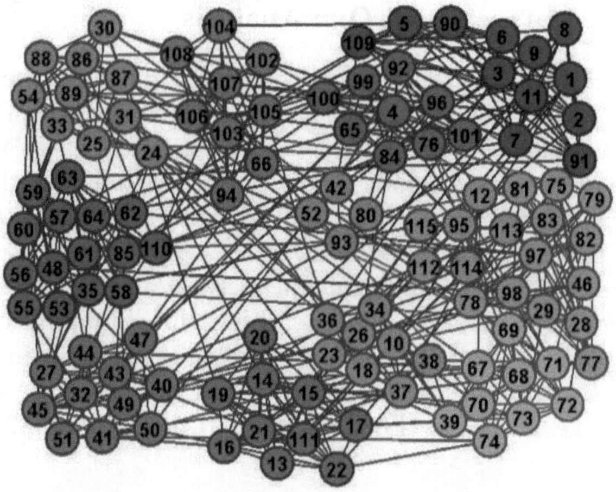

Fig. 1. American college football network, with benchmark communities shown in different colors, using Gephi [11]

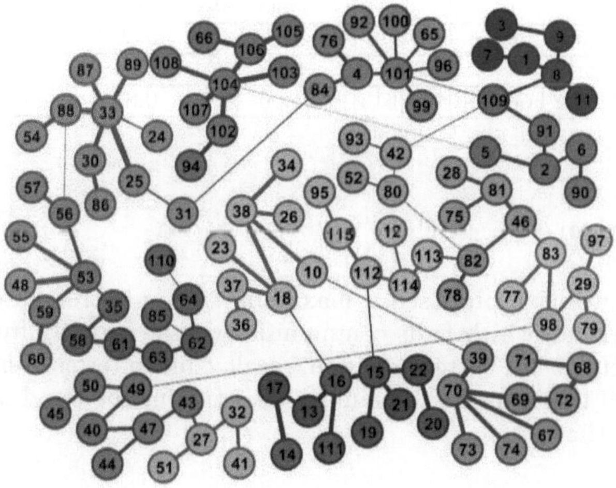

Fig. 2. Spanning tree for American college football network, with communities found by our algorithm marked in different colors

4 Results

Our analysis was performed on machine with Intel(R) Core(TM)2 Duo CPU 2.00GHz Processor and 3 GB RAM. Programs were written in Python 2.6 and C++, and Gephi [11] was used as the graph visualization and manipulation software. We applied our technique on three small benchmark datasets of different size and a summary has been provided in Table 1.

4.1 Benchmark Datasets and Our Analysis

Our method was tried and tested on three benchmark datasets, namely, Zachary's Karate Club [12], American College Football [13] and Dolphin social network [14]. We have used communities detected by Louvain method [4] as the benchmark and compared the communities detected by our algorithm with them. We found out that the precision and recall of the our community detection method is very high for all these three datasets and is listed in Table 1. In Figures 1 and 2, we can see the communities from the Louvain method and the communities from our method for football dataset. In some cases, the communities were further divided to increase the overall goodness value of the subgraphs induced from the final cover. All the communities generated by our method were together in the same communities generated by Louvain method, thereby making its precision to be 1.

Table 1. Precision and recall values for the maximum spanning based community detection technique

Name of Network Data	$\|V\|$	$\|E\|$	Precision Our method	Recall Our method	Precision CNM	Recall CNM
Zachary's Karate Club	34	78	0.835	0.89	0.72	0.88
Dolphin Network	62	159	0.85	0.81	0.66	0.73
American College Football	115	613	1.00	0.81	0.65	0.96

5 Conclusion

In this paper, we have proposed a maximum spanning tree based community detection algorithm, which finds communities very accurately from benchmark network datasets. It has better precision, recall value in terms of finding communities compared to a popular community detection method and has a log-linear worst case run-time.

References

1. Palla, G., Derenyi, I., Farkas, I., Vicsek, T.: Uncovering the overlapping community structure of complex networks in nature and society. Nature 435, 814–818 (2005)
2. Radicchi, F., Castellano, C., Cecconi, F., Loreto, V., Parisi, D.: Defining and identifying communities in networks. PNAS 101, 2658–2663 (2004)
3. Girvan, M., Newman, M.: Community structure in social and biological networks. PNAS 99, 7821–7826 (2002)
4. Blondel, V., Guillaume, J., Lambiotte, R., Mech, E.: Fast unfolding of communities in large networks. J. Stat. Mech. (2008)
5. Clauset, A., Newman, M., Moore, C.: Finding community structure in very large networks. Physical Review E 70 (2004)

6. Chiang, M., Lam, H., Liu, Z., Poor, V.: Why steiner-tree type algorithms work for community detection. In: 16th International Conference on Artificial Intelligence and Statistics (AISTATS) (2013)
7. Lancichinetti, A., Fortunato, S.: Community detection algorithms: a comparative analysis (2009)
8. Dunn, G., Everitt, B.: An Introduction to Mathematical Taxonomy. Cambridge University Press (1982)
9. Holland, P., Leinhardt, S.: Transitivity in structural models of small groups. Small Group Research 2, 107–124 (1971)
10. Prim, R.: Shortest connection networks and some generalizations. The Bell Systems Technical Journal 36, 1389–1401 (1957)
11. Bastian, M., Heymann, S., Jacomy, M.: Gephi: An open source software for exploring and manipulating networks. In: ICWSM. The AAAI Press (2009)
12. Zachary, W.: An information flow model for conflict and fission in small groups. Journal of Anthropological Research 33, 452–473 (1977)
13. Girvan, M., Newman, M.: Community structure in social and biological networks. PNAS 99, 7821–7826 (2002)
14. Lusseau, D., Newman, M.: Identifying the role that animals play in their social networks. Proceedings of the Royal Society of London. Series B: Biological Sciences 271, S477–S481 (2004)

Implementation of Compressed Brute-Force Pattern Search Algorithm Using VHDL

Lokesh Sharma[1], Bhawana Sharma[2], and Devi Prasad Sharma[1]

[1] Manipur University, Jaipur, India
[2] Amity University, Jaipur, India
{lokesh.sharma,deviprasad.sharma}@jaipur.manipal.edu,
bsharma@jpr.amity.edu

Abstract. High speed and always-on network access is becoming commonplace around the world, creating a demand for increased network security. Network Intrusion Detection Systems (NIDS) attempt to detect and prevent attacks from the network using pattern-matching rules. Data compression methods are used to reduce the data storage requirement. Searching a compressed pattern in the compressed text reduces the internal storage requirement and computation resources. In this paper we implemented search process to perform compressed pattern matching in binary Huffman encoded texts. Brute-Force Search algorithm is applied comparing a single bit per clock cycle and comparing an encoded character per clock cycle. Pattern matching processes are evaluated in terms of clock cycle.

Keywords: NIDS, Brute-Force Pattern Search, Huffman Coding.

1 Introduction

The intrusion detection system is based on the pattern matching technique. The compressed pattern matching problem [1] is to find the occurrence of compressed pattern C(P) in a compressed text C(T) without decompressing it. This work uses the pattern and text encoded by Huffman encoding technique [2], and search for encoded pattern using string search algorithm.

This paper presents hardware architecture of the signature matching process for Huffman encoded text. The designs are coded in VHDL, searched using Brute Force n pattern matching algorithm, and are evaluated in terms of the clock cycles.

The VLSI architecture is used to implement a memory-based compressed pattern matching chip. The VHDL coding based on the above architecture has been simulated using Active HDL 6.3 SE Tool [3-6].

2 Related Work

String matching is one of the well-studied classic problems in computer science, widely used in a range of applications, such as information retrieval, pattern recognition, and network security. In regular expression matching [12], the authors proposed

M.K. Kundu et al. (eds.), *Advanced Computing, Networking and Informatics - Volume 2*,
Smart Innovation, Systems and Technologies 28,
DOI: 10.1007/978-3-319-07350-7_66, © Springer International Publishing Switzerland 2014

to use Non-deterministic Finite Automaton in matching regular expressions and implemented it on a FPGA but this is complex. Disadvantage is whenever there are changes in the keyword set, the regular expressions have to be calculated again and the FPGA has to be reprogrammed. Another approach to string matching is the use of content addressable memories. Content addressable memories have long been used for fast string matching against multiple keywords. In [8], a KMP (a well-known software algorithm proposed in 1977, it was proposed by D.E. Knuth, J. Morris, and V.R. Pratt, shorted in KMP) algorithm based on FPGA implementation was proposed. In [13], compressed pattern is searched in the compressed text but a single bit is matched per clock cycle. In [14], used Half Byte Comparators in FPGA based pattern matching module. Because of the share of the comparing results pattern matching implementation can improve the area efficiency in FPGA significantly but they used two HBCs to match a character.

3 Compressed Pattern Matching Problem

The compressed pattern matching problem is defined as follows: The assumption is that, the compressed text is C (t) corresponding to text T [0...n-1] and compressed pattern C (p) corresponding to pattern P [0....m-1] where all symbols are taken from finite character set ∑. The problem is to search C (t) to find the occurrence of C (p). The patterns P may have don't care characters [7]. A don't care character, denoted by '$' is a character such that $ is not an element of ∑ and $ can match any character a belonging to ∑. So it says a= $ for all of which are elements of ∑.

The problem of searching a compressed text using Huffman [2] or run-length encoding seems superficial to be straight forward. The idea is to apply any well-known string search algorithm such as KMP algorithm on c(T) with respect to the compressed pattern c(P). A close examination of the algorithm reveals that such an approach is not very practical in many applications. Text T is encoded with Huffman encoding and applied KMP algorithm on C (t) to find C (p). But it raises the problem of false matches, i.e., finding an occurrence of C (p) in C (t) which does not correspond to an occurrence of pattern in text, due to crossing boundary. On using Huffman encoding, an implicit decompression process has to be performed to determine the character boundary (i.e. starting bit of each encoded symbol). One important practical consideration is to be able to determine the patterns with limited amount of main memory. The second major problem is handling don't care characters. The KMP algorithm with O (m +n) time complexity (n and m are text and pattern lengths, respectively) does not work for patterns with don't care characters.

For example the Huffman code {0-10-110-111} for character a, b, c, d respectively. The binary string 0-10-10-0 is encoding of the string *abba*. Suppose one is searching for binary string 0-10 encoding of pattern ab, he finds the encoded pattern at first bit and is a correct match and also finds at third bit and are a false match as shown in Fig.1. Modified KMP algorithm [8] was proposed to overcome false matches. To avoid processing any bit of the encoded text more than once, a pre-processed table is used to determine how far to back up when a mismatch is detected, and is defined so that we are always able to align the start of theencoded pattern with the start of a code word in the encoded text. But it cannot handle don't care characters.

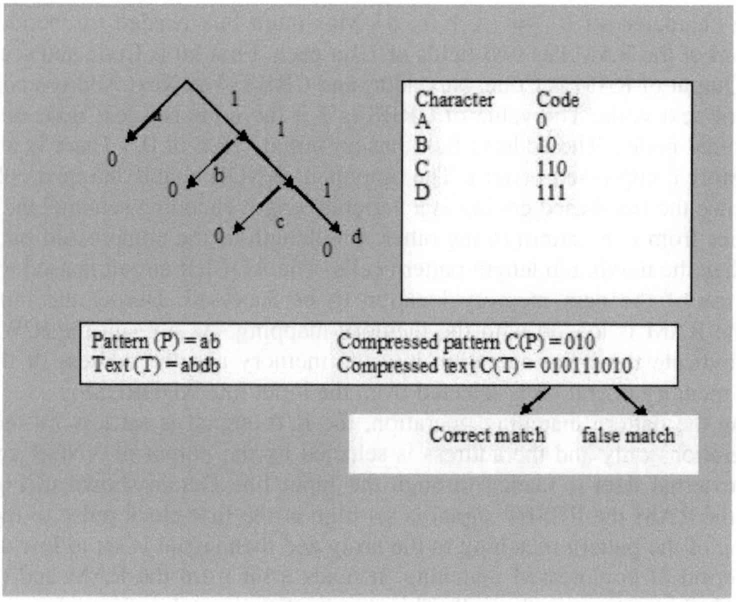

Fig. 1. Huffman coding

One needs to keep track of boundary of valid codes and it can be done by using binary signal called CRBS (character boundary signal). The algorithm uses an array of simple cells in which pattern is stored and the text to be searched is applied in serial fashion.

Each CAM cell is comparisons, storage element and generate signal which propagates across the array [7]. Comparator cell is a XNOR gate. Don't care characters are represented in the compressed pattern by only a single bit (either 0 or 1). As an example using Huffman code in fig 4.1 if pattern is p=a$ba then compressed pattern is C (p) = 0-0-10-0 or 0-1-10-0. The first bit correspond to a, second bit corresponds to don't care which can be represented as either 0 or 1. Now use the naïve algorithm to search for C (p) in C (t). In case of mismatch the encoded text is shifted one position to the right.

4 Brute Force Algorithm Implementation

The B. F. algorithms implementations consists various architectural modules which describe the overall operation of the architecture. The Huffman tree is mapped into the memory module (RAM).

4.1 Architecture of Matching 1-bit per Clock Cycle

Each character of text is encoded using Huffman encoding and then mapped into a fixed memory location in RAM.

Finite character set is ∑= {a, b, c, d} Maximum bits needed to encode is three. Each word of the RAM has two fields of 1-bit each. First bit is Code and second bit is CRBS. Output of RAM is Code, Nextaddr, and CRBS. The Next Address contains the address of next node. The value of CRBS is 1 if the node is a leaf node and is 0 for non-terminal nodes. The address field has an initial value of 0. There is an array of cells to store compressed pattern. The outputs of XNOR enable the next cell into the array. Since the tree-based coding is a variable-length encoding scheme, the length of c(P) varies from one pattern to the other. The length of the compressed pattern must be less than the maximum length pattern cells. The ADDER circuit is used to generate the address of the next memory location to be accessed. During the initialization phase, the RAM is loaded with the memory mapping. As a result the R/W signal is high to indicate the write operation into the memory and the address of the corresponding memory operation is selected from the input line ADDRESS.

During the pattern matching operation, the R/W signal is set low, to indicate the read operations only and the address is selected by the output of XNOR and CRBS. All the external data is loaded through the input line Datain shown in Fig.2. After loading the RAM the RESET signal is set high at the first clock pulse to indicate the beginning of the pattern matching in the array and then signal is set to low during rest of the period of compressed matching. It reads a bit from the RAM and then compared with the content in the cell. The output of the match information is stored in each clock pulse. If all cells match then MATCH signal is set high. Hence using one XNOR and matching 1 bit per clock cycle the compressed Pattern search operation is achieved in c(P)+1 clock pulses where, c(P) is the length of the compressed pattern.

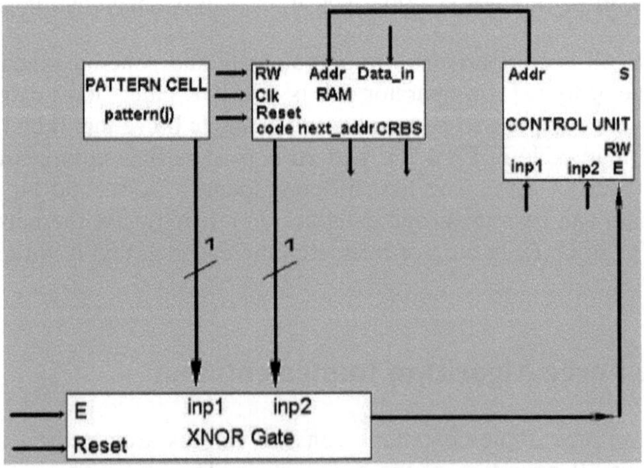

Fig. 2. Architecture of matching 1 bit per clock cycle

4.2 Architecture of Matching a Character per Clock Cycle

The compression scheme presented here is a variant of the word-based Huffman code [9, 10, 11]. The Huffman codeword assigned to each text word is a sequence of whole bytes, and the Huffman tree has degree either 128 (which we call "tagged Huffman code") or 256 (which we call "plain Huffman code"), instead of 2. In tagged Huffman coding each byte uses seven bits for the Huffman code and one bit to signal the beginning of a codeword. As we show later, using bytes instead of bits does not significantly degrade the amount of compression. In practice, byte processing is much faster than bit processing because bit shifts and masking operations are not necessary at compression, decompression, and search times.

Architecture is shown in Fig. 3. Finite character set is $\sum$= {a, b, c, d} maximum bits needed to encode is three. Each word of the RAM has two fields. First three bits are Code and last two bits are CRBS. Output of RAM is Code, Next addr, and CRBS. The Next Address contains the address of next character. The value of CRBS is the bit position where encoded character ends. The address field has an initial value of 0. There is an array of cells to store compressed pattern. Each cell store encoded character of pattern. The outputs of array of three XNOR gates enable the next cell into the array. The ADDER circuit is used to generate the address of the next memory location to be accessed. The ADDER can be implemented as incrementer. During the initialization phase, the RAM is loaded with the memory mapping. As a result the R/W signal is high to indicate the write operation into the memory and the address of the corresponding memory operation is selected from the input line ADDRESS. During the pattern matching operation, the R/W signal is set low, to indicate the read operations only and the address is selected by the output of XNOR array. All the external data is loaded through the input line data in shown in Fig. 3. After loading the RAM the RESET signal is set high at the first clock pulse to indicate the beginning of the pattern matching in the array and then signal is set to low during rest of the period of compressed matching. It reads a encoded character from the RAM and then compared with the content in the cell using XNOR array. The output of the match information is stored in each clock pulse. If all cells match then MATCH signal is set high. Hence using XNOR array and matching character per clock cycle the compressed Pattern search operation is achieved in m+1 clock pulses where, m is the length the pattern.

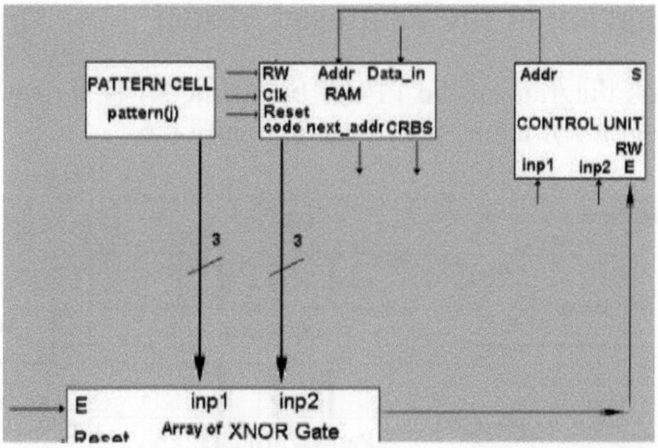

Fig. 3. Architecture of matching 1 character per clock cycle

For example if text is "abcdab" and pattern is "abcd" then compressed text C (t) is 0-10-110-111-0-10 and compressed pattern C (p) is 0-10-110-111 then by matching character per clock cycle it will take 4 clock cycles. Adder will be same as that used in matching 1 bit per clock cycle.

4.3 Comparison in the per Bit Matching and per Character Matching

Using bytes instead of bits does not significantly degrade the amount of compression. In practice, byte processing is much faster than bit processing as shown in Fig.4.The graph shown in the Fig. 4.is based on the pattern search and the output generated in the examples shown in the previous section.

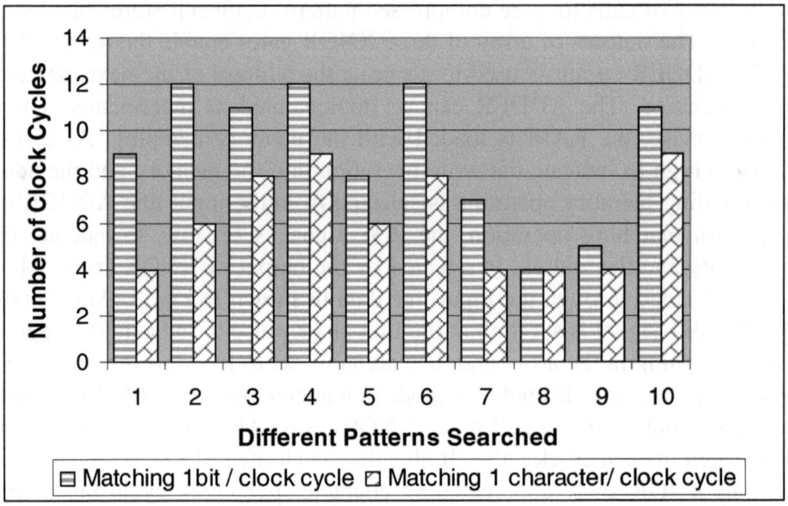

Fig. 4. Graph showing the comparsion between matching 1 bit and 1 character per clock cycle

5 Block Diagram of the Proposed Model

The proposed block diagram of the FPGA based model is shown in the Fig. 5. The proposed block diagram of main entity is shown in Fig. 6.

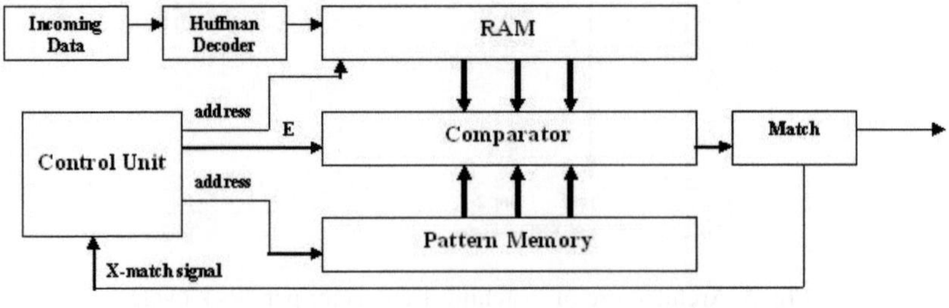

Fig. 5. Block Diagram of the FPGA based Hardware

6 Implementation Diagram of Main Entity of Proposed Model

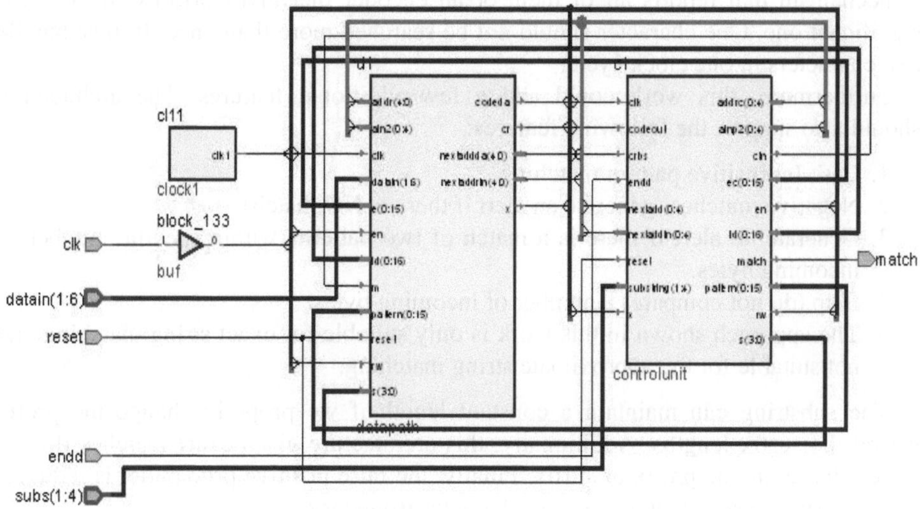

Fig. 6. FPGA based implementation diagram of the main entity of proposed model

7 Conclusion and Future Work

7.1 Conclusion

This work has studied exact string matching approaches from two perspectives, 1 bit per clock cycle and, 1 character per clock cycle using Brute Force Algorithm employed in designing a FPGA. Exact string matching approaches can be further divided into software-based and hardware-based algorithms approaches. Since software-based approaches are slower and less efficient, hardware-based approaches are highly preferred. Therefore these works used hardware-based exact string matching approach and discuss how this approach is used for pattern matching. It also proposed new string matching micro-architectures and computed the efficiency of the model in the perspective of 1 bit per clock cycle and 1 character per clock cycle. The 1 character per clock cycle provides the better performance than other.

This work presents hardware algorithms for compressed pattern matching using Huffman-type tree based codes. It also presents VLSI architecture to implement a memory-based compressed pattern matching chip. We have simulated the VHDL coding based on the above architecture using Active HDL 6.3 SE TOOL.

7.2 Future Work

Our proposed architecture achieves high-speed and low cost string matching. However the architecture presented in this work is not able to share entire substrings.

Just like other architectures, it must handle the problem of multiple matches in the same cycle. Since more than one pattern may match in one clock cycle, there must be a mechanism that reports all of them or an encoder that gives priority to the most significant one. One character should not be searched more than once. It must handle two characters in one clock cycle.

Furthermore, this work could add a few additional features. The architecture should also support the following features:

1. Case Insensitive pattern matching.
2. Negative matches (generate an alert if there is NO match).
3. Generate an alert if there is a match of two patterns within specific number of incoming bytes.
4. Skip (do not compare) a number of incoming bytes.
5. The approach shown in this work is only suitable for exact string matching and not suitable for the approximate string matching.

The substring can maintain a constant length if we properly change the prefix or/and the suffix lengths. Additionally, this architecture gives exact matches if each pattern has a unique prefix or suffix. Finally, the false positive probability is reduced, since every pattern is matched using Character Boundary.

References

1. Amir, A., Benson, G.: Two Dimensional Compressed Matching. In: Proceedings of Data Compression Conference (DCC), pp. 279–288 (1992)
2. Huffman, D.A.: A method for the construction of minimum-redundancy codes. In: Proceedings of the IRE, pp. 1098–1101 (1952)
3. `http://www.aldec.com/downloads`
4. Dubrawsky, I.: Firewall Evolution - Deep Packet Inspection (2003)
5. Mukherjee, B., Heberlein, L.T., Levitt, K.N.: Network Intrusion Detection. IEEE Network, 26–48 (1994)
6. Fisk, M., Varghese, G.: An analysis of fast string matching applied to content based forwarding and intrusion detection. In: Technical Report CS2001-0670, University of California (2002)
7. Buboltz, J., Kocak, T.: Front End Device for Content Networking. In: Proceedings of the Conference on Design, Automation and Test in Europe (2008)
8. Karp, R.M., Rabin, M.O.: Efficient randomized pattern-matching algorithms. IBM Journal of Research and Development 31(2), 249–260 (1987)
9. Knuth, D.E., Morris, J., Pratt, V.R.: Fast pattern matching in strings. SIAM Journal on Computing (1977)
10. Baker, Z.K., Prasanna, Z.K.,, V.K.: A methodology for synthesis of efficient intrusion detection systems on FPGAs. In: IEEE Symposium on Field-Programmable Custom Computing Machines (2004)
11. Sidhu, R., Mei, A., Prasanna, V.K.: String matching on multi-content FPGAs using self-reconfiguration. In: Proceedings of FPGA 1999 (1999)
12. Ashenden, P.J.: Modeling digital systems using VHDL. In: Proceedings of Potentials of IEE, vol. 17(2), pp. 27–30 (1998)

13. Sidhu, R., Prasanna, V.K.: Fast regular expression matching using FPGAs. In: IEEE Symposium on Field-Programmable Custom Computing Machines (2001)
14. Mukherjee, A., Acharya, T.: VlSI Algorithms for Compressed Pattern Search Using Tree Based Codes. In: Proceedings of the IEEE International Conference on Application Specific Array Processors (1995)
15. Clark, C.R., Schimmel, D.E.: Efficient reconfigurable logic circuit for matching complex network intrusion detection patterns. In: Cheung, P.Y.K., Constantinides, G.A. (eds.) FPL 2003. LNCS, vol. 2778, pp. 956–959. Springer, Heidelberg (2003)

A Modified Collaborative Filtering Approach for Collaborating Community

Pradnya Bhagat and Maruska Mascarenhas

Computer Engineering Department
Goa College of Engineering
Farmagudi, Ponda-Goa
pradnyabhagat91@gmail.com,
maruskha@gec.ac.in

Abstract. Web search is generally treated as a solitary service that operates in isolation servicing the requests of individual searchers. But in real world, searchers often collaborate to achieve their information need in a faster and efficient way. The paper attempts to harness the potential inherent in communities of like-minded searchers overcoming the limitations of conventional personalization methods. The community members can share their search experiences for the benefit of others while still maintaining their anonymity. The community based personalization is achieved by adding the benefits of reliability, efficiency and security to web search.

Keywords: community, personalization, collaborative filtering, collaborative web search, stemming, stopwords, lexical database.

1 Introduction

Web search is generally considered as an isolated activity. Modern search engines employ a strategy known as personalization to accommodate the differences between individuals. Several approaches have been adopted for personalization in the past; but all of these approaches have a serious limitation of treating web search as a solitary interaction between the searcher and the search engine.

In reality, web search has a distinctly collaborative flavor. Many tasks in professional and casual environment can benefit from the ability of jointly searching the web with others. Collaborative Filtering [12] is a methodology of filtering or evaluating items using the opinions of other people. The modified collaborative web search approach presented in this paper is inspired from collaborative filtering and is based on the approach followed in [3]. It tries to collaborate a community of like-minded searchers sharing similar interests to achieve personalization. The main areas focused include: the efficiency of the data structure, the reliability of the results and the security of the system from malicious uses.

M.K. Kundu et al. (eds.), *Advanced Computing, Networking and Informatics - Volume 2*, Smart Innovation, Systems and Technologies 28,
DOI: 10.1007/978-3-319-07350-7_67, © Springer International Publishing Switzerland 2014

2 Motivation

There are many scenarios where web search takes the form of a community oriented activity. For example students seeking for information on a weekly assignment, or the employees of a company working on a common project will have similar information needs during the project span. Similarly, searches originating from the search box of a themed website, or people with similar purchase history on an e-commerce web site show the potential for collaboration. A survey conducted by M. R. Morris [9] has revealed that a large proportion of users engage in searches that include collaborative activities. The results of the survey has shown that nearly 53% of the searchers involved in sharing either the process (search terms, sites etc) of the product (useful links, facts found within sites) of web search. The respondents even showed to adapt some form of strategy like brute force, divide conquer strategy, backseat driver approach [9] to achieve their required information need faster.

These scenarios clearly show that web search is astonishingly a collaborative task but yet it is not adequately supported by the existing search engines. Hence, the main motivation behind this paper is to allow to the searchers to collaborate irrespective of the time and place of searching provided they share similar interests.

3 Literature Survey

Personalization is proving to be a vital strategy in the success of any web search engine. Two approaches have been frequently adapted to implement personalization in the past: personalization based on content analysis [1] and personalization based on hyperlink structure [6] of the web. Both of these methods have proved successful to a great extent in delivering relevant results to the searchers, but they are limited by the constraint of treating web search as a solitary activity. They fail to identify the collaboration in which users naturally engage to further refine the quality of search results.

Personalization based on user groups is a methodology that incorporates the preferences of a group of users to accomplish personalized search. An approach that is based on this ideology is knows as Collaborative Filtering.

3.1 Collaborative Filtering

Collaborative Filtering is defined as the process of filtering or evaluating items based on the opinions of other people [12]. The fundamental assumption it holds is that if two people rate on n similar items similarly then and hence will rate or act on other future items similarly. But this approach has some drawbacks including: 1) One-to-one similarity calculation and 2) Privacy Violation.

The drawbacks pose serious limitations when it comes to the use of collaborative filtering in web search personalization where the user base is very large and also the uses prefer to stay anonymous.

To overcome with these problems, a modified collaborative web search approach is proposed in [3] called community based collaborative web search.Community based collaborative Web search is based on the principle of collaborative filtering, but instead of exploiting the graded mapping between users and items, it exploits a similar relationship between queries and result pages. It can work as a meta-search working on an underlying search engine and re-rank the results returned by the underlying search engine based on the learned preferences of the community of users. The approaches adopted in literature for collaborative information retrieval can be distinguished in terms of two dimensions: Time and place. Based on these dimensions, the search can be either co-located or remote, or synchronous or asynchronous. CoSearch [10] is an example of co-located, synchronous approach. SearchTogether [8] is an example of system supporting remote search collaboration (whether synchronous or asynchronous). I-Spy [5] is another search engine that is built on the community based collaborative web search.

4 Modified Collaborative Web Search Approach

The modified collaborative web search approach is based on [3] and it attempts to harness the asynchronous search experiences of a community of like-minded remote searchers to provide improved personalized results. It is based on case-based reasoning [2], an approach which uses previous search experiences of searchers to refine future searches. It is implemented as a meta-search engine working on a background search engine like Google to further refine the results returned by the underlying search engine.

The architecture of the Collaborative Web Search (CWS) is explained in Fig. 1. Whenever a searcher submits a query, the query is sent to Google and also to the collaborative web search meta-search engine. In collaborative web search meta-search engine, the query is first passed through the pre-processing block. The output query from pre-processing block and the results of the underlying Google search form the input to the Hit data structure which keeps a record of the number of hits a page has got for a particular query. The next processing block does all the computations and presents the promoted list of results R_P to the user.

At the same time, a list of normal results returned by Google is also collected. This forms the standard list R_S. Both promoted list and standard list and merged together and returned to the user as the final result R_{Final}. Normally the promoted results can be shown on top followed by normal Google search results. Otherwise, the promoted results can be shown in one column and standard results in another column.

4.1 Pre-processing

The query is first pre-processed to achieve efficiency in the search process. The pre-processing step consists of the following phases: 1) stopwords removal [4], 2) stemming [7] and 3) checking for synonyms [11] to avoid duplication of queries in the hit data structure.

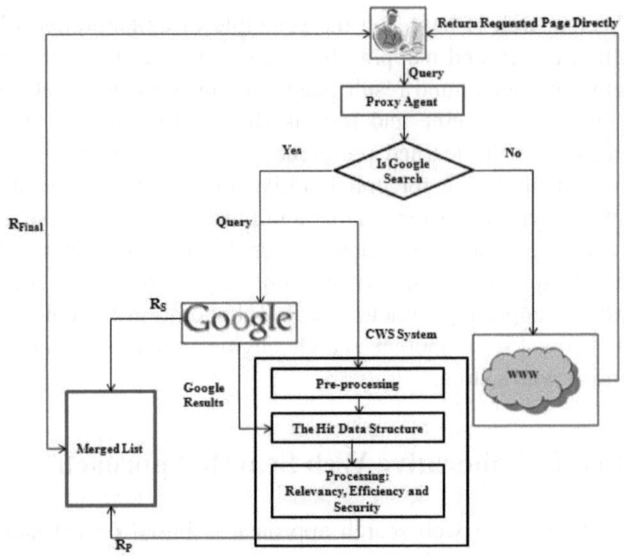

Fig. 1. Architecture of Modified Collaborative Web Search

For example, if "Pictures of Jaguar" is the target query (q_T) and "Jaguar photo" (q_i) is the one present in hit data structure then, without preprocessing, the similarity (*Sim*) computation using Jaccard correlation coefficient [3]between query q_T and q_i, equals 0.25 as given in Equation 1. The system fails to identify two exact similar queries.

$$Sim(q_T, q_i) = \frac{q_T \cap q_i}{q_T \cup q_i} = \frac{(Photoof Jaguar) \cap (JaguarPicture)}{(Photoof Jaguar) \cup (JaguarPicture)} = \frac{1}{4} = 0.25 \quad (1)$$

The pre-processing steps are as follows. The first step is to remove the stopwords in the tagert query q_T: So the "Pictures of jaguar" will get converted to "Pictures Jaguar". Next, using Porter Stemmer Algorithm [7] we can stem "Pictures Jaguar" to "Picture Jaguar". Finally using a lexical database we can convert "Picture" to "Photo" so that the two queries become similar. Now, using Jaccard correlation coefficient, the similarity (*Sim*) equals:

$$Sim(q_T, q_i) = \frac{q_T \cap q_i}{q_T \cup q_i} = \frac{(PhotoJaguar) \cap (JaguarPhoto)}{(PhotoJaguar) \cup (JaguarPhoto)} = \frac{2}{2} = 1$$

4.2 The Modified Data Structure

After pre-processing, the query is given as input to the hit data structure given in Fig. 2, along with the underlying search engine (Google) results. In this specially designed modified data structure, the pages are indexed on queries with the pointer from each query leading to a linked list of pages that are associated with that query. For example in the given figure, the node consisting of query q_1 consists of two pointers. One pointer points to the node containing the next query. The other pointer points to the

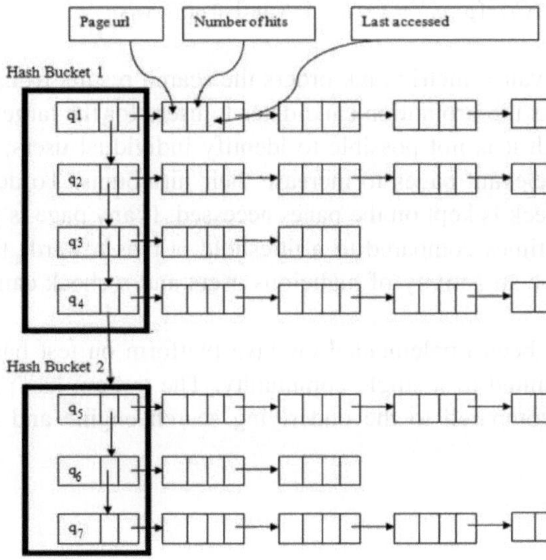

Fig. 2. Data Structure used in Modified Collaborative Web Search

corresponding linked list of pages associated with that query. The nodes in the linked list consist of the following four fields: 1) The URL of the Page, 2) The number of hits for the page, 3) Last Accessed Date and 4) Pointer to the Next Node.

Further, the queries are hashed into several buckets to increase the insert and retrieval efficiency. The pages are ordered in decreasing order whenever the load on the system is reduced, based on the number of hits so that pages having most number of hits are located in beginning and search time will be reduced to a significant extent.

4.3 Achieving Collaboration of Community

The relevance (*Rel*) of a page with some target query is calculated as given in Equation 2 where q_i refers to the target query which is already present in the data structure and p_j is the page whose relevance we are calculating:

$$Rel(p_j, q_i) = \frac{H_j \times \frac{1}{n_j}}{\sum_{\forall j} H_{ij} \times \frac{i}{n_{ij}}} \tag{2}$$

H_j refers to the number of hits that page has got for query and n_j refers to the number of days passed since its last access. This creates a bias towards never pages. Now, the Weighted Relevance (*WRel*) [3] of page p_j to some new target query is a combination of Rel(p_j, q_i) values for all cases $q_1, \ldots, q_n$ that are deemed to be similar to q_T and can be calculated as given in Equation 3:

$$WRel(p_j, q_1, \ldots, q_n) = \frac{\sum_{i=1 \ldots n} Rel(p_j, q_i) . Sim(q_T, q_i)}{\sum_{i=1 \ldots n} Sim(q_T, q_i)} \tag{3}$$

where, $Exists(p_j, q_i) = 1 \; if \, H_{ij} \neq 0$ and $0 \; otherwise$

The weighted relevance metric rank orders the search results from the community case base and presents the promotion candidates to users for the target query. Further, since in this approach it is not possible to identify individual users, malicious users may simply click irrelevant pages to increase their hit counts. To deal with this, instead of users, the check is kept on the pages accessed. If any page is getting accessed far more number of times compared to a threshold, a bias towards that page can be detected which can be an activity of malicious users and a check can be kept on that page.

The approach has been implemented on Java platform on test bases. The current implementation is limited to a single community. The system has proved to deliver better performance compared to the underlying search engine and the original approach [3].

5 Results

The dataset is selected from an online bookmarking service delicious.com [13]. Each of the bookmarks can be considered as the result selection and each tag as the query term. To find users having common information need, we tried to discover users having interest in a similar field. These users have the potential to show a significant overlap in searches giving evidence of collaboration.

The data set consisted of 30 users having interest in computer technologies from the delicious dataset. The total size of the dataset consists of nearly 4000 tagged keyword-url pairs with each user on an average having about 135 bookmarks. These 30 uses have proved to show surprising collaboration in their information need when identified properly. With the dataset consisting of about 30 users, if we take any random user, we found that at least 70 percent of the queries typed, are already searched by other community members, while only about 30 percent of the information need varies.

The graph in Fig. 3 shows the evidence of collaboration that can be harnessed to deliver better personalized results to users, hence saving their significant amount of search time. This huge amount of query overlap hints that there can be overlap in the solution need also. That is, given that the query typed is the same, the links selected also can be same. To study this behavior, we used our modified collaborative web search system. So, next we use these about 60 percent of the repeated queries only and try to find out how much percent of the solution overlap we can find. That is, given the query of the new user which query was already placed by the other users we need to find out how many percent of the times even the same link was presented for the same query.

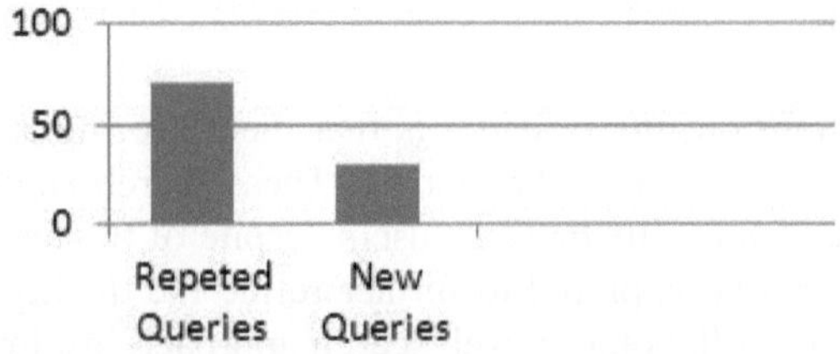

Fig. 3. Repetition in queries found

The CWS system rank orders the results in decreasing order of the weighted relevance. The top ten results are presented to the searcher as the promoted candidates. A search session is marked as successful if at least one result was selected by the users from the top 10 promoted results. The first case taken to find the results is, a user is taken as the test user and its queries are not included in the dataset consisting of result selections. So the promotions that user will be getting will be solely based on the promotions presented by other community member to which that user belongs. Finally, the list which is clicked by the test user for that query is checked if it is present in the top 10 promoted results. If it is present, the search session is considered to be a success.

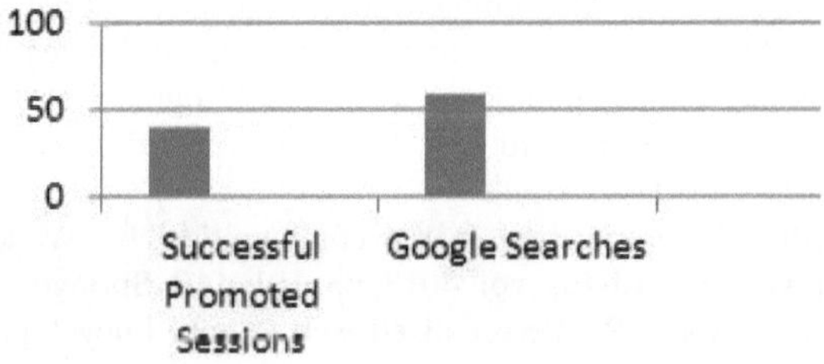

Fig. 4. Successful v/s Unsuccessful sessions when the queries of test user are excluded from the dataset

As can be seen from Fig. 4, about 40% percent of the search sessions were successful in promoting the clicked result of the test user in the top 10 results. Sometimes the promotions might come from the user himself. It refers to searching something which we have already searched before. If these are considered in the dataset, the success rate goes to noticeably higher showing that more than 50% of the search sessions were successful (Fig.5).

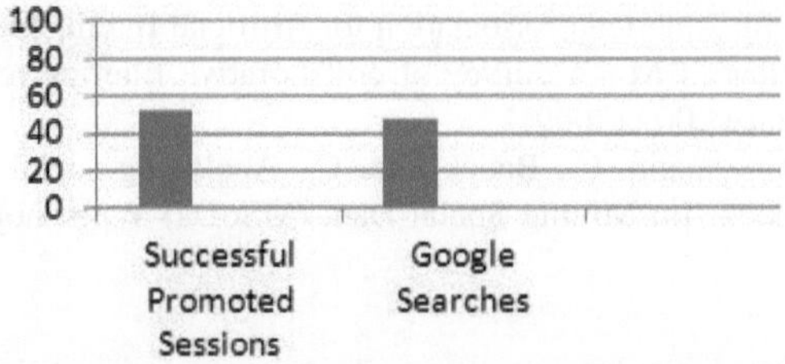

Fig. 5. Successful v/s Unsuccessful sessions when the queries of test user are included in the dataset

6 Conclusion

The motivating insight on this research is that there are important features missing from mainstream search engines like Google. These search engines offer no solution for sharing of the search results between users despite of the fact that there is tremendous potential that can be explored to further refine the quality of search results returned. The system of collaborative web search approach inspired from collaborative filtering allows members of a community of like-minded searchers to share their search experiences for the benefit of other community members. The members of the community can asynchronously collaborate irrespective of the distance between them to improve the search experience. The approach is proved to deliver better performance with respect to precision and recall in comparison to the other search engines.

References

1. Pretschner, A., Gauch, S.: Ontology based personalized search. In: Proceedings of 11th IEEE International Conferenceon Tools with Artificial Intelligence, pp. 391–398 (1999)
2. Aamodth, A., Plaza, E.: Case-based reasoning: foundational issues, methodological variations and system approaches. AI Communication 7(1), 39–59 (1994)
3. Smyth, B., Coyle, M., Briggs, P.: The altrustic seacher. In: Proceedings of 12th IEEE International Conference on Computational Science and Engineering (2009)
4. Buckley, C., Salton, G.: Stop Word List. SMARTInformation Retrieval System, Cornell University
5. Freyne, J., Smyth, B.: Cooperating search communities. In: Wade, V.P., Ashman, H., Smyth, B. (eds.) AH 2006. LNCS, vol. 4018, pp. 101–110. Springer, Heidelberg (2006)
6. Wen, J.-R., Dou, Z., Song, R.: Personalized web search. Encyclopedia of Database Systems, pp. 2099–2103 (2009)
7. Porter, M.F.: An algorithm for suffix stripping. Program 14(3), 130–137 (1980)
8. Morris, M.R., Horwitz, E.: Searchtogether: an interface for collaborative web search. In: Proceedings of the 20th Annual ACM Symposium on User Interface Software and Technology, UIST 2007 (2007)
9. Morris, M.R.: A survey of collaborative web search practices. In: Proceedings of the SIGCHI Conference on Human Factors in Computing Systems, pp. 1657–1660 (2008)
10. Amershi, S., Morris, M.R.: Cosearch: a system for co-located collaborative web search. In: Proceedings of the SIGCHI Conference on Human Factors in Computing Systems, pp. 1647–1656 (2008)
11. Peredson, T., Patwardhan, S., Michelizzi, J.: WordNet:Similarity - Measuring the Relatedness of Concepts. In: American Association for Artificial Intelligence, pp. 38–41 (2004)
12. Su, X., Khoshgoftaar, T.M.: A survey of collaborative filtering techniques. Advances in Artificial Intelligence 2009 (2009)
13. Wetzker, R., Zimmermann, C., Bauckhage, C.: Analyzing social bookmarking systems: A delicious cookbook. In: Mining Social Data (MSoDa) Workshop Proceedings, pp. 26–30 (2008)

Author Index

MIX
Papier aus verantwortungsvollen Quellen
Paper from responsible sources
FSC® C105338

If you have any concerns about our products,
you can contact us on
ProductSafety@springernature.com

In case Publisher is established outside the EU,
the EU authorized representative is:
Springer Nature Customer Service Center GmbH
Europaplatz 3, 69115 Heidelberg, Germany

Printed by Libri Plureos GmbH
in Hamburg, Germany